Food Engineering Innovations Across the Food Supply Chain

Food Engineering Innovations Across the Food Supply Chain

Edited by

Pablo Juliano

CSIRO Agriculture and Food, Australia

Kai Knoerzer

CSIRO Agriculture and Food, Australia

Jay Sellahewa

UNSW Sydney, Australia

Minh H. Nguyen

University of Western Sydney and University of Newcastle, Australia

Roman Buckow

University of Sydney, Australia

ACADEMIC PRESS

An imprint of Elsevier

elsevier.com/books-and-journals

Academic Press is an imprint of Elsevier
125 London Wall, London EC2Y 5AS, United Kingdom
525 B Street, Suite 1650, San Diego, CA 92101, United States
50 Hampshire Street, 5th Floor, Cambridge, MA 02139, United States
The Boulevard, Langford Lane, Kidlington, Oxford OX5 1GB, United Kingdom

Notices
Knowledge and best practice in this field are constantly changing. As new research and experience
broaden our understanding, changes in research methods, professional practices, or medical treatment
may become necessary.

Practitioners and researchers must always rely on their own experience and knowledge in evaluating and
using any information, methods, compounds, or experiments described herein. In using such information or methods they should be mindful of their own safety and the safety of others, including parties for
whom they have a professional responsibility.

To the fullest extent of the law, neither the Publisher nor the authors, contributors, or editors, assume any
liability for any injury and/or damage to persons or property as a matter of products liability, negligence
or otherwise, or from any use or operation of any methods, products, instructions, or ideas contained in
the material herein.

British Library Cataloguing-in-Publication Data
A catalogue record for this book is available from the British Library

Library of Congress Cataloging-in-Publication Data
A catalog record for this book is available from the Library of Congress

ISBN: 978-0-12-821292-9

For Information on all Academic Press publications visit our website at
https://www.elsevier.com/books-and-journals

Publisher: Charlotte Cockle
Acquisitions Editor: Nina Bandeira
Editorial Project Manager: Lindsay C Lawrence
Production Project Manager: R.Vijay Bharath
Cover Designer: Matthew Limbert

Typeset by Aptara, New Delhi, India

Contents

Contributors

Refat Al-Shannaq
Department of Chemical and Materials Engineering, The University of Auckland, Auckland, New Zealand

Simon Allen
CSIRO Agriculture and Food, Hobart, Tasmania, Australia

C. Anandharamakrishnan
Computational Modeling and Nanoscale Processing Unit, Indian Institute of Food Processing Technology (IIFPT), Ministry of Food Processing Industries, Govt. of India, Thanjavur, India

Vitus A. Apalangya
Department of Food Process Engineering, University of Ghana, Accra, Ghana

Ingrid Appelqvist
CSIRO Agriculture and Food, Werribee, VIC, Australia

Dimitrios Argyropoulos
School of Biosystems and Food Engineering, University College Dublin, Dublin, Ireland

Amar Auckaili
Department of Chemical and Materials Engineering, The University of Auckland, Auckland, New Zealand

Serafim Bakalis
Department of Food Science, University of Copenhagen, Rolighedsvej Frederiksberg, Denmark

Bhesh Bhandari
School of Agriculture and Food Sciences, University of Queensland, Brisbane, QLD, Australia

Katherine Blackshaw
School of Chemical and Biomolecular Engineering, The University of Sydney, Sydney, NSW, Australia; Centre for Advanced Food Engineering, The University of Sydney, Sydney, NSW, Australia

Paulomi (Polly) Burey
University of Southern Queensland, Toowoomba, QLD, Australia

C. Charette
Combat Capabilities Development Command Soldier Center (CCDC SC), Natick, MA, United States

Junlae Cho
School of Chemical and Biomolecular Engineering, The University of Sydney, Sydney, NSW, Australia; Centre for Advanced Food Engineering, The University of Sydney, Sydney, NSW, Australia

Ashim Datta
Department of Biological and Environmental Engineering, Cornell University, Ithaca, NY, United States

Hester De Wet
Aurecon, Energy, Resources and Water, Neutral Bay, VIC, Australia

Fariba Dehghani
School of Chemical and Biomolecular Engineering, The University of Sydney, Sydney, NSW, Australia; Centre for Advanced Food Engineering, The University of Sydney, Sydney, NSW, Australia

Fariba Dehghani
School of Chemical and Biomolecular Engineering, The University of Sydney, Sydney, NSW, Australia; Centre for Advanced Food Enginomics, The University of Sydney, Sydney, NSW, Australia

Christopher J. Doona
Massachusetts Institute of Technology, Cambridge, MA, United States; John A. Paulson School of Engineering and Applied Sciences, Harvard University, Cambridge, MA, United States; Combat Capabilities Development Command Soldier Center (CCDC SC), Natick, MA, United States

M. Azad Emin
Chair of Food Process Engineering, Karlsruhe Institute of Technology, Karlsruhe, Germany

Christos Emmanoulidis
School of Aerospace, Transport and Manufacturing, Cranfield University, Cranfield, United Kingdom

Ferruh Erdoğdu
Professor of Food Process Engineering, Ankara University, Turkey

Mohammed Farid
Department of Chemical and Materials Engineering, The University of Auckland, Auckland, New Zealand

F.E. Feeherry
Combat Capabilities Development Command Soldier Center (CCDC SC), Natick, MA, United States

E. Forster
Graduate School of Biomedical Sciences, Tufts University, Boston, MA, United States

Raj Gaire
CSIRO Data61, Canberra, ACT, Australia

R. García-Flores
CSIRO Data61, Docklands, VIC, Australia

Dimitrios Gerogiorgis
School of Engineering, University of Edinburgh, Edinburgh, United Kingdom

Jacopo E. Giaretta
School of Chemical and Biomolecular Engineering, The University of Sydney, Sydney, NSW, Australia; Centre for Advanced Food Enginomics, The University of Sydney, Sydney, NSW, Australia

Michel Havet
Oniris, Université de Nantes, CNRS, GEPEA, Nantes, France

Dennis R. Heldman
Department of Food Science and Technology, The Ohio State University, Columbus, OH, United States

Andreas Helwig
University of Southern Queensland, Toowoomba, QLD, Australia

Filip Janakievski
CSIRO Agriculture and Food, Werribee, VIC, Australia

Piyush Kumar Jha
ONIRIS, GEPEA CNRS 6144, Nantes, France; Department of Human Nutrition, Food and Animal Sciences, University of Hawaii, Honolulu, HI, United States

J. Johnston
New Zealand Institute for Plant & Food Research, Hawke's Bay, New Zealand

P. Johnstone
New Zealand Institute for Plant & Food Research, Hawke's Bay, New Zealand

Pablo Juliano
CSIRO Agriculture and Food, Australia

Soojin Jun
ONIRIS, GEPEA CNRS 6144, Nantes, France; Department of Human Nutrition, Food and Animal Sciences, University of Hawaii, Honolulu, HI, United States

Fanbin Kong
Department of Food Science and Technology, The University of Georgia, Athens, GA, United States

Nooshin Koolaji
School of Chemical and Biomolecular Engineering, The University of Sydney, Sydney, NSW, Australia; Centre for Advanced Food Engineering, The University of Sydney, Sydney, NSW, Australia

K. Kustin
Department of Chemistry, Brandeis University, Waltham, MA, United States

Alain Le-Bail
ONIRIS, GEPEA CNRS 6144, Nantes, France; Department of Human Nutrition, Food and Animal Sciences, University of Hawaii, Honolulu, HI, United States

M. Maria Leena
Computational Modeling and Nanoscale Processing Unit, Indian Institute of Food Processing Technology (IIFPT), Ministry of Food Processing Industries, Govt. of India, Thanjavur, India

Lilly Lim-Camacho
CSIRO Agriculture and Food, Brisbane, QLD, Australia

Yvan Llave
Food Thermal Engineering Laboratory, Department of Food Science and Technology, Tokyo University of Marine Science and Technology, Japan

Myriam Loeffler
KU Leuven, Department of Microbial and Molecular Systems, MTSP, Ghent Technology Campus, Ghent, Belgium

Amy Logan
CSIRO Agriculture and Food, Werribee, VIC, Australia

Francesco Marra
Dipartimento di Ingegneria Industriale, Università degli studi di Salerno, Fisciano, SA, Italy

Francesco Marra
Dipartimento di Ingegneria Industriale, Università degli studi di Salerno, Fisciano, SA, Italy

Slaven Marusic
Aurecon, Data and Analytics, Docklands, VIC, Australia

A.K.M. Masum
Spraying Systems Co., Fluid Air, Truganina VIC, Australia

J.A. Moses
Computational Modeling and Nanoscale Processing Unit, Indian Institute of Food Processing
Technology (IIFPT), Ministry of Food Processing Industries, Govt. of India, Thanjavur, India

Sina Naficy
School of Chemical and Biomolecular Engineering, The University of Sydney, Sydney, NSW, Australia;
Centre for Advanced Food Enginomics, The University of Sydney, Sydney, NSW, Australia

Long H. Nguyen
School of Chemical and Biomolecular Engineering, The University of Sydney, Sydney, NSW,
Australia
Centre for Advanced Food Enginomics, The University of Sydney, Sydney, NSW, Australia

Bart Nicolai
Division of Mechatronics, Biostatistics and Sensors (MeBioS), Biosystems Department,
KU Leuven, Leuven, Belgium

Keshavan Niranjan
Department of Food and Nutritional Sciences, University of Reading, Reading, United Kingdom

Farshad Oveissi
School of Chemical and Biomolecular Engineering, The University of Sydney, Sydney, NSW, Australia;
Centre for Advanced Food Enginomics, The University of Sydney, Sydney, NSW, Australia

Colm O'Donnell
School of Biosystems and Food Engineering, University College Dublin, Dublin, Ireland

Sunil K. Panchal
University of Southern Queensland, Toowoomba, QLD, Australia; Western Sydney University,
Richmond, NSW, Australia

Janet L. Paterson
The University of New South Wales, Sydney, Australia

Ronil J. Rath
School of Chemical and Biomolecular Engineering, The University of Sydney, Sydney, NSW, Australia;
Centre for Advanced Food Enginomics, The University of Sydney, Sydney, NSW, Australia

José I. Reyes-De-Corcuera
Department of Food Science and Technology, CAES, Athens, GA, United States

Olivier Rouaud
Oniris, Université de Nantes, CNRS, GEPEA, Nantes, France

Shyam S. Sablani
Department of Biological Systems Engineering, Washington State University, Pullman, WA,
United States

K.P. Sandeep
Department of Food, Bioprocessing and Nutrition Sciences, North Carolina State University, Raleigh, NC, United States

Periaswamy Sivagnanam Saravana
Department of Food Chemistry & Technology, Teagasc Food Research Centre, Ashtown, Dublin, Ireland

Fabrizio Sarghini
Department of Agricultural Sciences, University of Naples Federico II, Naples, Italy

Juhi Saxena
Spraying Systems Co., Fluid Air, Truganina VIC, Australia

Aaron Schindeler
School of Chemical and Biomolecular Engineering, The University of Sydney, Sydney, NSW, Australia; Centre for Advanced Food Engineering, The University of Sydney, Sydney, NSW, Australia; Bioengineering and Molecular Medicine Laboratory, The Children's Hospital at Westmead and the Westmead Institute for Medical Research, Westmead, NSW, Australia

D. Scotland
New Zealand Institute for Plant & Food Research, Hawke's Bay, New Zealand

Cordelia Selomulya
School of Chemical Engineering, UNSW, Kensington, NSW, Australia

Zahra Shahrbabaki
School of Chemical and Biomolecular Engineering, The University of Sydney, Sydney, NSW, Australia
Centre for Advanced Food Enginomics, The University of Sydney, Sydney, NSW, Australia

S. Shanthamma
Computational Modeling and Nanoscale Processing Unit, Indian Institute of Food Processing Technology (IIFPT), Ministry of Food Processing Industries, Govt. of India, Thanjavur, India

A. Shen
School of Medicine, Molecular Biology and Microbiology, Tufts University, Boston, MA, United States

Josip Simunovic
Department of Food, Bioprocessing and Nutrition Sciences, North Carolina State University, Raleigh, NC, United States

Chandrashekhar R. Sonar
Department of Biological Systems Engineering, Washington State University, Pullman, WA, United States

Juming Tang
Department of Biological Systems Engineering, Washington State University, Pullman, WA, United States

Netsanet Shiferaw Terefe
CSIRO Agriculture and Food, Werribee, VIC, Australia

Brijesh Tiwari
Department of Food Chemistry & Technology, Teagasc Food Research Centre, Ashtown, Dublin, Ireland

Viruja Ummat
Department of Food Chemistry & Technology, Teagasc Food Research Centre, Ashtown, Dublin, Ireland; School of Biosystems and Food Engineering, University College Dublin, Dublin, Ireland

Peter Valtchev
School of Chemical and Biomolecular Engineering, The University of Sydney, Sydney, NSW, Australia; Centre for Advanced Food Engineering, The University of Sydney, Sydney, NSW, Australia

Pieter Verboven
Division of Mechatronics, Biostatistics and Sensors (MeBioS), Biosystems Department, KU Leuven, Leuven, Belgium

Olivier Vitrac
UMR 0782 SayFood Paris-Saclay Food and Bioproduct Engineering Research Unit, INRAE, AgroParisTech, Université Paris-Saclay, Massy, France

Shaojin Wang
College of Mechanical and Electronic Engineering, Northwest A&F University, Yangling, Shaanxi, PR China; Department of Biological Systems Engineering, Washington State University, Pullman, WA, United States

Yong Wang
School of Chemical Engineering, UNSW, Kensington, NSW, Australia

Bo Wang
School of Behavioural and Health Sciences, Australian Catholic University, Kensington, NSW, Australia

Peter Watkins
CSIRO Agriculture and Food, Werribee, VIC, Australia

T. White
New Zealand Institute for Plant & Food Research, Ruakura, New Zealand

Jiadai Wu
School of Chemical and Biomolecular Engineering, The University of Sydney, Sydney, NSW, Australia; Centre for Advanced Food Engineering, The University of Sydney, Sydney, NSW, Australia

D. Yi
New Zealand Institute for Plant & Food Research, Auckland, New Zealand

Jimmy Yun
School of Chemical Engineering, The University of New South Wales, Sydney, NSW, Australia; Qingdao International Academician Park Research Institute, Qingdao, Shandong, PR China

Jimmy Yun
School of Chemical Engineering, The University of New South Wales, Sydney, NSW, Australia; Qingdao International Academician Park Research Institute, Qingdao, Shandong, PR China

Bogdan Zisu
Spraying Systems Co., Fluid Air, Truganina VIC, Australia

Maarten van der Kamp
European institution in Leuven, Leuven, Belgium

About the editors

Pablo Juliano

Dr. Pablo Juliano leads the Food Processing and Supply Chains Group at CSIRO, Australia, which aims at solving food industry challenges through science and innovation. He obtained his Ph.D. in Food Engineering at Washington State University in the United States and a Master of Business Administration at Deakin University in Australia. He also occupied management roles with Nestlé Uruguay and CONAPROLE, a major dairy exporter in Uruguay. With over 20 years of service to the food industry in seven countries, he develops and directs circular economy research programs through technology innovation. He has published over 100 peer-reviewed research in innovative food processing technologies and optimization of food supply chains. In particular, he led the development and commercialization of an unprecedented, patented technology for food waste recovery. He is the CSIRO representative on the Australian Food Waste Strategy together with industry peak bodies, the federal and state governments, and is working with food clusters toward the implementation of agricultural food processing hubs for regional development. He is the president of the Australian Food Engineering Association.

Kai Knoerzer

Dr. Kai Knoerzer is a Principal Research Scientist/Engineer in Food Engineering at CSIRO, Australia. He has a background in process engineering (B.Sc.), chemical engineering (M.Sc.), and food process engineering (Ph.D., summa cum laude), all awarded from the Karlsruhe Institute of Technology (Germany). In 2006, he commenced work with Food Science Australia (a joint venture of the CSIRO and the Victorian Government) as a postdoctoral fellow. He has since become a Principal Research Scientist in CSIRO Agriculture and Food. He has a proven track record in food process engineering research and development, particularly of innovative technologies. Currently, he is leading research and venture science activities across a number of innovative food processing technologies, including high pressure thermal processing and extrusion technology. His work has shown both science impact, with more than 100 peer-reviewed journal publications, conference proceedings and book chapters, seven patents/applications, four edited books, and over 100 oral and 50 poster presentations at national and international conferences, as well as commercial impact in the food industry. His work has also been recognized with various international awards for research excellence. He has been an active member of IFT's International Division in the leadership team for >10 years and is past chair of this division. He serves Elsevier's Food Science Reference Collections as the Food Process Engineering section editor.

Jayantha (Jay) Sellahewa

Mr. Jay Sellahewa is an Adjunct Senior Lecturer, UNSW Sydney, following a long career at CSIRO. He graduated in Chemical Engineering from Bath University followed by an M.Sc. in Food Engineering from Leeds University. He has a career spanning 45 years in research and industry in Australia and the United Kingdom, where he has undertaken assignments in Australia, Europe, USA, Africa, and Asia. His experience includes managing complex multidisciplinary projects, carrying out strategic planning and the successful commercialization of research in the agrifood industry. He has expertise in food process systems engineering, extrusion technology, nonthermal processing, sustainable food processing, and humanitarian food science & technology. He is a Fellow of the Institution of Chemical Engineers and the Australian Institute of Food Science and Technology.

Minh H. Nguyen

Prof. Nguyen is Adjunct Associate Professor at WSU and a Conjoint Associate Professor at the University of Newcastle. He has over 40 years of experience in food science and technology as lecturer and researcher at Hawkesbury Agricultural College/University of Western Sydney, National College of Food Technology, United Kingdom, Asian Institute of Technology, Thailand, Nestle Australia, Sandy Trout Food Preservation Research Laboratory Queensland. Held appointments as Member of Board of Directors, Hawkesbury Technologies Ltd (the university consulting company), Food Expert Consultant for the FAO of the United Nations and the Commonwealth Fund for Technical Co-operation and Food Biotechnology Expert for the International Development Program of Australian Universities. Elected Fellow of the International Academy of Food Science Food Science and Technology and Life Fellow of the Australian Institute of Food Science and Technology. Served as chair, Food Engineering Group of the Australian Institute of Food Science and Technology, president of the Australian Food Engineering Association, and the Australian representative, International Association of Engineering and Food. Authored and co-authored over 150 publications including refereed papers, refereed proceedings, patents, books, and book chapters.

Roman Buckow

Prof. Roman Buckow is a Professor of Practice in Food Engineering at the Centre for Advanced Food Engineering of the University of Sydney, Australia. He obtained his Ph.D. in Food Process Engineering and Biotechnology from the Berlin University of Technology in Germany. Before joining the University of Sydney, he held multiple science management positions at CSIRO to create future food processing platforms and technologies that deliver sustainable gains in productivity and efficiency across the food value-chain. He has a strong track record of successful commercial adaptation in alternative dairy, meat, sugar, grain, and legume processing and product innovations.

He has served as the president of the Australian Food Engineering Association and International Association of Engineering and Food. He has authored and co-authored over 100 publications including refereed papers, refereed proceedings, patents, book, and book chapters, and has been recognized with several CSIRO and international awards for his achievements.

Preface

Feeding the world's population with safe, nutritious, and affordable foods across the globe using finite resources is a challenge. The population of the world is increasing, and two distinct subpopulations can be considered: those who are more affluent and need to decrease their caloric intake and/or become healthier from the food they eat, and those who are malnourished and require more caloric and nutritional intake. The world suffers from food supply chain inefficiencies resulting from losses of valuable nutrients and waste generation as well as inadequate utilization of increasingly scarce water, land, and energy resources. Such inefficiencies are seen at various levels of the supply chain: during the raw material harvest, collection, and preprocessing, transportation, storage, or manufacturing, and at the consumer level. The extent to which this occurs depends on the level of development of countries and regions and on the specific food supply chains.

For sustainable growth, an increasingly integrated systems approach across the whole supply chain is required. The efficient use of resources and their transformation to manufacture foods that provide sustainable nutrition and health promoting foods require integrating the food engineering field with disciplines that encompass supply chain systems thinking, food safety, food chemistry and material science, sensory and consumer segment understanding, and human health research (e.g., understanding digestion and nutrient delivery in the human body). Supply chain systems research may integrate disciplines which at least include agriculture and agronomy, operations research, economic and environmental sustainability and life cycle assessment, industry 4.0 and data management, innovation business modeling, regional economics, social science, social licensing and law, cultural and political systems research, market research, among others.

Food engineers have traditionally focused on the understanding of food components and structures and in the development, modeling, and optimization of food processes, often operating within the food factory and equipment boundaries. Expansion of such boundaries is becoming essential to achieve cross food chain integration. Globalization, amalgamation of companies through mergers and acquisitions, and retailers holding the highest leverage in the food chain place considerable pressure on the food systems to lower costs. Greater integration is being achieved between on-farm and food factory operations, as well as factory and retail and/or consumer operations, by inclusion of Industry 4.0 sensing, data collection, automation, and blockchain technologies (including intelligent packaging). The Industry 4.0 implementation across the food chain allows more efficient utilization of labor resources while minimizing food wastage, as well as water and energy inputs, within intellectual property transaction frameworks. Other Industry 4.0 approaches that interphase with the food engineering field include machine learning and digital twin food factory design to tailor the factories of the future. The field of food engineering is also expanding by cross-linking with other disciplines to contribute to aspects of human health and nutrition, where food engineering approaches are applied in digestion and delivery models. Companies are becoming increasingly socially and environmentally responsible, and the interplay of technological implementation of food-driven solutions in cities and regions and the social license for acceptance, access, and benefit beyond the traditional enterprise profit model is providing a stronger need for such supply chain integration and diversification.

The present book aims to provide a broader view of *the state-of-the-art* of food engineering and their combined cross-disciplinary approaches to tackle the challenges that current supply chains pose for dynamic resilience and to meet the needs of today and tomorrow's consumers for adequate nutrition and health and food security. It discusses some of the latest technology advances and innovations into industrial applications to improve supply chain sustainability, food security, and human nutrition. The book touches on these various aspects of innovation across the food supply chain beyond the traditional food engineering discipline, while intersecting with other subject domains, as described earlier. The book also presents the latest resources and recommendations for educating others in this field while increasing public–private collaboration. It is therefore intended to suit and benefit a broad audience from industry, academia, and government.

The contributions to this book capture the highlights of the 13th International Congress of Engineering (ICEF13), which took place in Melbourne from 23 to 26 September 2019. This book also serves as a legacy document of this unique global congress which has, for the first-time, placed food engineering in the context of food supply chains and their relevance to food security and human health. ICEF13 was an outstanding success with more than 750 abstracts accepted, 280 posters presented, and 550 delegates attending from 40 countries and five continents. The delegates represented leading academics, practitioners from industry, equipment and software suppliers, and students. The depth and breadth of the latest food engineering topics covered in this congress and the networking with prominent global food engineers provided excellent opportunities for learning, growing, and connecting for the participating food engineers and related professionals and students.

The ICEF is known as the most premier event in the field of food engineering. The first ICEF was held in Boston, United States, in 1976, and the congress series has since been held 13 times in various locations identified and selected by the International Association for Engineering and Food (IAEF). IAEF is a global body of 35 delegates representing professional engineering societies including food engineering activities. The main objective of the IAEF is to identify and select a member to organize the next ICEF event. IAEF members are societies that promote the food engineering profession in the country, not individuals nor professional organizations.

ICEF13 was a result of the efforts of the Australian Food Engineering Association (AFEA), Engineers Australia (EA), and Waldron Smith Management, who supported all the conference logistics within and outside the Melbourne Convention Centre. This is a major milestone for AFEA since its incorporation in 2000 to represent Australia as an IAEF member, while also representing its forming professional societies, the Australian Institute of Food Science and Technology (AIFST), Engineers Australia and the Institution of Chemical Engineers (ICheME). The IAEF Executive supporting the organization of ICEF13 (Kim Staples, Pablo Juliano, Minh Nguyen, Michele Marcotte, Petros Taoukis, Mohamed Farid, Weibiao Zhou) was led by Prof. Roman Buckow as the IAEF president and he was also the convener of the ICEF13. IAEF member representatives also contributed to shaping the congress through several conference calls from 2016 and 2019 joining from more than 15 locations from around the globe. The ICEF13 Local Organizing Committee (Andrew Watkins, Benu Adhikari, Dennis Forte, Geoff Hurst, Gordon Young, Janet Paterson, Jay Sellahewa, Kim Staples, Kai Knoerzer, Meltem Bayrak, Minh Nguyen, Pablo Juliano, Roman Buckow) spent numerous efforts through their subcommittees including the Scientific Committee (Jay Sellahewa, Chair), Marketing Committee (Kim Staples, Chair), Publications Committee (Pablo Juliano, Chair), and Sponsorship Committee (Kai Knoerzer, Chair) in making this congress a great success.

The congress also delivered two special issues in the *Food Engineering Reviews Journal* and *Journal of Food Engineering*. The *Food Engineering Reviews* "Special issue based on the International Congress on Engineering and Food XIII" (https://link.springer.com/journal/12393/volumes-and-issues/13-1) with Gustavo V. Barbosa-Cánovas, Kezban Candöğan, Jorge Welti-Chanes, and Yrjö

Roos as guest editors, collected food engineering contributions from ICEF13 authors summarized in seven review articles and 13 research articles. The *Journal of Food Engineering* special issue encompasses 10 research articles from ICEF13 authors contributing in the field of "Food Engineering in Nutrition and Digestion" (https://www.sciencedirect.com/journal/journal-of-food-engineering/special-issue/10C07T2Z8H2), with Bhesh Bhandari, Pablo Juliano, Roman Buckow, Kai Knoerzer, and Minh Nguyen as guest editors. The concluding chapter will provide an overview of the content of the congress presentations and will connect the chapters contributed in this book by ICEF13 authors as well as the contributions to the said special issues.

The ICEF13 congress' theme was "Engineering Innovations for Food Supply Chains." Topics covered during ICEF13 were equally interesting and broad, and inspired the structure and content logic of this book:

- Food engineering across the supply chain
- Sustainable food systems
- Food security
- Advances in food process engineering
- Novel food processing technologies
- Food process systems engineering and modeling
- Engineering properties of food and packaging
- Food engineering for nutrition and health
- Food engineering education
- Innovations of food engineering in Australasia
- Industry 4.0
- Sensors

The chapters in this book are therefore structured under each of the above themes and the relevance of each chapter is described and highlighted below, with some of the chapters including substantial reviews on the most important technological advances and innovations applicable to industrial scale and with high sustainability standards for health impact. Moreover, this multidisciplinary book brings contributions from others outside the food engineering area with expertise in the operations research and Industry 4.0 domain to demonstrate the existing complex solutions across food systems.

Chapter 1 starts with an overview of the operations research within supply chains and highlight related challenges on a global and regional scale. The chapter reviews existing work and ideas relating to resilience and robustness from the perspective of fresh produce supply chains and explains a framework to understand how the structure of their supply networks may enhance their resilience when facing disruptions. It also includes a summary of the work that deterministic models can do in locating innovative food processing operations in supply chains using a case study on whey production from small cheesemakers.

Chapter 2 describes sustainable food systems, considerations for their implementation and how food engineering innovations may assist in solving issues within these systems. The chapter discusses supply chain management strategies to make food chains more efficient with flexible processing technologies. These types of modifications can potentially optimize food, energy, and water resource usage to minimize waste while also optimizing population health. Ideally, sustainable practices need to occur while maintaining economic viability of the food systems and their surrounding environment. This chapter discusses how dedicated food engineers, food technologists, and food scientists can contribute to attaining these lofty targets.

Chapter 3 goes into further detail about the sustainability of the food supply chain and its dependence on conservation of resources, primarily energy and water, and the reduction of food waste, as well as impacts on the environment. It describes the problems of food waste in developed economies and potential solutions. The chapter explains how basic food process design to ensure food safety or extend product shelf-life, can be combined with life cycle assessment methodologies to reduce energy, water, and waste in the food supply chain.

Chapter 4 addresses the challenge of food waste and provides a brief overview of the high value-compounds that have been successfully identified and isolated from food by-products. Further novel uses of food waste, that was previously overlooked due to limitations of the extraction techniques, are also summarized in this review. The overview includes recovery of antioxidant compounds, dietary fibers, plant-based proteins, and other bioactive compounds from plant-based by-products and bioactive peptides, protein, and fat recovery from animal by-products. Methods to recover antiviral compounds from by-products are also discussed in the context of pharmaceutical applications.

Chapter 5 introduces advances in food and process by-product fermentation, given the renewed interest in fermented foods for health and the application of synthetic biology for production of food as sustainable alternatives to traditional agriculture. The chapter includes a description of fermentation applications beyond kimchi, kombucha and sauerkraut, fermentation process, and bioreactor design in the context of traditional and industrial food fermentation. Developments in synthetic biology for the manufacture of ingredients analogous to animal derived and other food ingredients are introduced as a technology with untapped potential for a sustainable and food secure future.

Chapter 6 continues by addressing the engineering and other strategies to mitigate protein deficit in view of the increasing demand for animal-based proteins such as ruminant and porcine meat and dairy, which is not sustainable, both from an environmental and food security perspective. The chapter addresses alternative protein sources as a way of mitigating such demand, which in turn will reduce several factors such as greenhouse gas emissions, environmental impact and consumption of water and land, among others. Alternative and sustainable protein sources to meet the nutritional and sensory satisfaction requirements of consumers include plant-based sources such as vegetables, pulses and seeds, meat and fish by products, microbes, insects, algae, and in vitro meat. Innovative engineering solutions for alternative protein extraction and development of novel protein-based ingredients, concentrates and isolates and the key determinants for consumer acceptance and health considerations are also covered.

Chapters 7 through 15 cover the latest updates on advances in food process engineering and the application of key or innovative food unit operations such as extrusion, microwaves, radiofrequency waves, freezing, cooling, drying, encapsulation, and 3D printing.

Chapter 7 describes how extrusion has evolved from a simple forming/shaping process to a highly flexible and advanced process with many unit operations. It includes an update on its capability to the design of sustainable food systems such as meat analogues. The chapter brings a discussion on engineering tools to control this process and the corresponding products toward a less empirical and more mechanistic approach. The mechanistic analysis includes characterizing the decisive process parameters that are crucial for a complex food design and corresponding methods based on fractionation and analysis of the process through its dynamically interrelated sections.

Chapter 8 describes the advanced use of microwaves on continuous flow thermal processing of foods. It describes the evolution of the technology since its early benchtop developments in 1998 leading to the AseptiWave prototype including a positive displacement piston pump and the coupling with various types of aseptic pouch fillers. Applications include the commercial installations on thermally labile materials such as fruits and vegetables to make vegetable soups, fruit and vegetable purees, and fruit smoothies.

Chapter 9 describes the novel applications of radiofrequency for bulk size product processing at industrial scale, by making use of the larger penetration depth of the RF energy, compared to microwave energy. The chapter includes a review on the industrial systems with various electrode configurations. The applications discussed include disinfestation of agriculture products, pasteurization of food products, and tempering applications of frozen food commodities. The chapter finishes with a review of computational mathematical modeling approaches for process innovation, design, and optimization.

Chapter 10 describes the state-of-the-art of food freezing, technologies to reduce freeze damage mostly by controlling the size of ice crystals, and refrigeration technologies. The chapter includes descriptions of noninvasive innovative freezing methods beyond blast air and cryogenic freezing, including pressure-shift freezing and pressure-assisted freezing, opportunities to apply electric and magnetic fields, microwave, and radiofrequency-assisted freezing and ultrasound-assisted freezing. Substances regulating the freezing process and final product quality introduced or brought in contact with the sample to manipulate the freezing process and reduce the freeze damage are also described. The chapter also includes a section on chilling, super chilling, and supercooling alternatives.

Chapter 11 further expands the application of food engineering to dairy farms by discussing traditional and novel approaches for fast cooling milk during and after milking. The chapter describes the efficient and reliable cold energy storage systems required to meet New Zealand's new legislation for cooling down the milk (cool down to 6 °C or below within 6 h from the commencement of milking and within 2 h of the completion of milking). The chapter discusses options for further cooling with chilled water in heat exchangers for snap chilling and innovative approaches including using graphite/water composite sphere in a packed bed as phase change materials and ice slab encapsulated storage systems.

Chapters 12 and 13 address innovations in food drying by using electric and electromagnetic fields aiming either at reducing drying costs or developing drying of heat labile components such as micronutrients or probiotics. Chapter 12 includes recent experimental and numerical studies combining electric or electromagnetic fields with conventional air to reduce drying time and thereby increase the quality of dried products. Examples are provided on microwave-assisted drying, radiofrequency drying and electrohydrodynamic drying and the concept of multiphysical models is discussed for design and optimization of such innovative drying processes.

Chapter 13 describes the development and commercialization of electrostatic spray drying that enables water diffusion to the core of droplets at low temperatures. The technology allows agglomeration of dried particles and eliminates the need for a postdrying agglomeration step. Examples are provided for drying probiotic microorganisms, microorganisms associated with the human microbiota, heat sensitive proteins, microalgae, and other pharmaceutical products, and combinations with other components to achieve high oil load emulsions.

Chapter 14 takes the application of spray drying further into the microencapsulation applications. It presents an overview of atomization-based microencapsulation technology by using spray drying and further expands on other techniques such as spray chilling, and fluidized bed coating. The chapter brings examples of dairy ingredients used as encapsulants or wall materials (dairy proteins, lactose, milk fat, mixtures) and the entrapped bioactive compounds (lactoferrin, peptides).

Chapter 15 introduces the novel applications of 3D food printing as a method for computer-aided layer manufacturing, or additive manufacturing. The chapter describes the hardware used for printing applications including extrusion, inkjet printer, binder jets, and selective laser sintering and an overview of companies manufacturing 3D printers for food production. It describes the fabrication of food products or food product decorations with customized color, shape, flavor, texture, nutritional loading, and delivery systems. Some examples include the manufacture of meat constructs such as sausages, steak, recombined meats or patties for people with swallowing difficulties, or the novel use of tomato and cricket powdered blends to be included as part of the inks developed for personalized nutrition.

Chapter 16 describes the *state-of-the-art* in food process modeling to avoid trial and error experimentation in the food industry. The use of virtualization or computer simulations to perform "what-if" scenarios using mathematical models that define the food processes are described as enablers for faster time-to-market and reduced resource requirements in food product and process development. The chapter provides a summary of the strengths of physics based (mechanistic) modeling in virtualization, discusses the modeling frameworks and their extension to quality and safety outcomes, and introduces the multiphase transport in deformable porous media modeling framework. It also presents approaches to obtaining the properties necessary for such physics-based models, extension of the applications to multiphysics as well as multiscale. It concludes by discussing modern virtual or computer-aided food product and process optimization and packaging design.

Chapters 17 and 18 cover the area of food packaging innovations. Chapter 17 presents the use of active food packaging applications and other microbial decontamination applications using chlorine dioxide (ClO_2). The chapter describes the combination of novel chemical systems in conjunction with polymers for ensuring microbial safety, extending shelf-life, and reducing waste of perishable foods such as fresh fruits, vegetables, and berries. The chapter describes a system that produces ClO_2 in-packaging. A second application includes a superabsorbent hydrogel polymer that enables humidity-activated, time-released controlled and sustained production of low concentrations of gaseous ClO_2 in-container. Other spray applications of ClO_2 technologies include sanitizing food processing/handling surfaces, and cleaning-in-place methods for production lines.

Chapter 18 describes the advances in polymer packaging able to withstand in-pack thermal pasteurization technologies. The chapter describes flexible packaging material options that maintain visual integrity post thermal processing, and low and medium gas barrier properties to provide pasteurized products with a shelf life of 10 days to 12 weeks. Gas-barrier and migration properties resulting from both conventional and microwave-assisted pasteurization are discussed. Storage studies and the impact of oxygen migration on food components such as pigments, vitamins, and lipids are also discussed.

With ICEF13 held in Australia, Chapter 19 gives a historical perspective of Food Engineering innovations in Australia. It collects examples from ancient Aboriginal engineering works, developments during early colonial times, industrial food production development, and outlines how some of these pioneering technologies have had global impact. There have been other significant developments in Food Engineering in Australia during the last 50 years in various Australian food supply chains, which are briefly mentioned including separation technologies such as continues chromatographic separation, the industrialization of high pressure, the development of a canister for high pressure thermal processing, encapsulation of fish oils, and the use of megasonics for the enhanced extraction of palm and olive oils.

Chapter 20 introduces the concept of digitization of the agrifood ecosystem, potentially supporting a circular bioeconomy, and the application of Industry 4.0 including digital technologies to food manufacturing. Specific applications of IoT (internet of things) for sorting and inspection, food safety systems, machine learning, robotics, data visualization and analysis, and decision making with integrated IoT systems are included. The chapter also identifies the barriers to adoption by the agrifood industry and some of the potential benefits and impacts.

Chapter 21 delves further into the Industry 4.0 topic by mapping the key features of high performing food production chains and enabling Industry 4.0 technologies. The chapter aligns with Chapters 1, 2, 3, and 4, as supply chain resilience and food resource utilization become of key importance, particularly given the new challenges brought about by the ongoing COVID-19 pandemic. The Industry 4.0 technology enablers listed in the chapter are predicted to increase the performance of customer-oriented food chains and their specific food chain characteristics.

A feature of digitalization and data generation across the supply chain is the ability to implement adequate sensing technologies. Chapter 22 introduces micro- and nanosensors (optical, chemical, electrochemical, and biological) applicable for quality, safety and traceability of food from farm to fork. It presents the nanomaterials required and potential applications across the supply chain including crop cultivation, food processing, packaging, and traceability of food, including applications in detection of pesticides, pathogenic bacteria, food additives, dyes, sweeteners, and food packaging.

Chapter 23 complements Chapter 22, by describing the various sensors currently manufactured for assessing food quality and safety and their mechanisms of detection (i.e., colorimetric or electrochemical), beyond time–temperature indicators. Applications described include detection of gases and volatile organic compounds, toxic and antinutritional molecules, or unwanted microbial pathogens in-pack, many of which are commercialized, and others are at proof-of-concept stage.

Chapter 24 shifts into the topic of food engineering education and provides an overview of where food engineering education is now and how food engineering degrees should be re-engineered for graduates to be able to effectively tackle strategic innovation challenges required in the manufacture of food in the 21st century. Multinational companies, large retailers, or small and medium enterprises are finding it difficult to recruit food engineering graduates or understand how to utilize this resource in a cost-effective manner. The chapter identifies health, the environment and food security as the three key drivers of the discipline and proposes a new definition of food engineering, to encompass biophysical, biochemical, and health sciences in addition to engineering sciences and proposes creating a new subject in *food product realization engineering* for students to understand where they fit in the larger scheme. The chapter includes a critical assessment of the available resources, including online courses and content.

Chapter 25 further expands the scope of Chapter 24, by discussing how advanced food engineering students may access more experience-based learning by exposing them to projects with multidisciplinary teams able to provide strong academic and industrial mentorship. In particular, the chapter describes the European Institute for Innovation & Technology (EIT) Food Solutions programs for *experience-based learning*. The EIT promotes extracurricular project activities for students to receive industry exposure where they co-create innovative circular bioeconomy solutions while receiving mentoring from professionals in food production and retail. It also provides entrepreneurial and innovation training and promote the commercialization of ideas.

Chapter 26 is the concluding chapter, which provides an overview of ICEF13 outputs, presented as keynote, oral, and e-poster presentations. As mentioned earlier, this chapter attempts to summarize some of the most relevant contributions from the entire collection of abstracts submitted to ICEF13 from all over the world, and reflects on the main topics covered in this book and the special issue journal contributions, under the above-described ICEF13 topics.

The editors hope that this book covers the *state-of-the-art* of the application of food engineering across food supply chains with sufficient depth, while providing a number of current and future engineering-driven innovations, from farm to human digestion, to address a global need for highly resilient, digital and interconnected consumer-driven supply chains. While the book does not intend to be exhaustive in covering this broad topic, we expect that it illustrates the broader scope of the application of food engineering through the food supply chain across disciplines and the knowledge gaps in the various chapters covered.

The editors would like to take the opportunity to acknowledge and give special thanks to the ICEF13 organizers, management team, committee members, speakers, poster presenters, and delegates who have made this congress a great and unique food engineering event. We particularly would like to thank all the chapter contributors, who have made this book come true. Special thanks to the guest editors and editors of the *Food Engineering Reviews* and *Journal of Food Engineering* special issues for co-creating these legacy documents of the congress. We hope that this book will be a useful tool to educate and inspire the present and next generations of food engineers to design solutions for the challenges in the food supply chains of today and in the future.

Sincerely,
Pablo Juliano, Jay Sellahewa, Kai Knoerzer, Minh Nguyen, Roman Buckow

Understanding and building resilience in food supply chains

R. García-Flores[a], Simon Allen[b], Raj Gaire[c], P. Johnstone[d], D. Yi[e], J. Johnston[d], T. White[f], D. Scotland[d], Lilly Lim-Camacho[g]

[a]CSIRO Data61, Docklands, VIC, Australia
[b]CSIRO Agriculture and Food, Hobart, Tasmania, Australia
[c]CSIRO Data61, Canberra, ACT, Australia
[d]New Zealand Institute for Plant & Food Research, Hawke's Bay, New Zealand
[e]New Zealand Institute for Plant & Food Research, Auckland, New Zealand
[f]New Zealand Institute for Plant & Food Research, Ruakura, New Zealand
[g]CSIRO Agriculture and Food, Brisbane, QLD, Australia

1.1 Introduction

The economic and natural environments in which food supply chains operate are continuously changing, and more so with the challenges that climate change and a growing population impose. In order to adapt to these changes, there is a pressing need to quantify and understand the risks and identify the strategies that can minimize them. The International Standards Organization defines *risk* as the effect of uncertainty on objectives (ISO, 2009). *Risk assessment* is the systematic process of evaluating the potential risks that may be involved in a projected activity, and *risk management* is the identification of procedures to avoid or minimize the impact of identified risk's impact. Risk is everywhere and its sources are mostly probabilistic variation (aleatory uncertainty) and lack of knowledge (epistemic uncertainty).

Given the importance of the topic, several reviews on supply chain risk management (SCRM) are available. Ho et al. (2015) provided an exhaustive review with the aim of covering all aspects of SCRM, including supply chain risk types, risk factors, and risk management methods. Ho et al. (2015) also proposed a conceptual framework to classify the papers they reviewed, and which we will use as a guide to demonstrate our methodology to quantify resilience (Fig. 1.1). Bak (2018) reviewed the existing literature and identified knowledge gaps and research trends, noting that most existing publications are biased to case studies in the United States and the United Kingdom. Fan and Stevenson (2018) provided another review with the aspiration of providing a new definition of SCRM and contribute to its theory with an "object-process-outcome" framework. Another general review was completed by Zhu et al. (2017), who discussed integrated SCRM and point to potential research directions.

Reviews addressing specific aspects of SCRM are more numerous. For example, Negreiros de Oliveira et al. (2019) who, motivated by the many recent environmental scandals and accidents, analyzed environmental risk factors and provide a taxonomy and a framework that synthesized environmental risks, consequences, and strategies covered by academic research. The framework they developed is quite complete in that it explicitly considers that the consequences for businesses of environmental

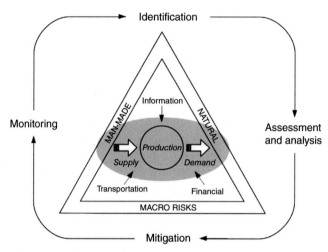

FIG. 1.1 The conceptual framework of supply chain risks, adapted from Ho et al. (2015).

negligence are not only financial in the shape of fines. Companies can also face severe reputational consequences, boycotts from customers, negative media exposure, and loss of credibility that can be irreversible.

Optimization methods are an important technology that is commonly used to assess and mitigate risk in supply chains. These can be *deterministic*, when it is assumed that all the information is known beforehand, or *stochastic*, which are methods that generate and use random variables that commonly represent scenarios. These methods used to inform strategic investment and operational decisions. Optimization can be used to analyze every stage or combination of stages in the supply chain, for example, transportation and logistics (Li et al., 2020; Alkaabneh et al., 2020), location of centralized collection, and processing facilities (Domingues-Zucchi et al., 2011; Garcia-Flores et al., 2014), reduction of losses prefarm and postfarm gate (Paam et al., 2019; Banasik et al., 2017) and others. These methods are becoming crucial given the increasing constraints imposed to production by scarcer resources and social needs. The most effective way to increase the amount of food produced is through efficiency gains, as clearing new land for agricultural activities stopped being sustainable a long time ago. For example, García-Flores (2015) carried out a study of the efficiency of investment to support small communities of cheese makers to add value to their whey by-product, reduce food losses and decrease environmental damage of disposal. The supply chain was designed as the optimal configuration of the whey supply chain and demonstrated that important savings and benefits to the community could be achieved by investing early on adequate processing facilities, which also increased the resilience of the system.

We will center the discussion in this chapter mostly on the concepts of *resilience* and *robustness* in the context of agricultural supply chains. Behzadi et al. (2018) reviewed quantitative models for agri-business SCRM with focus on these two terms, which they define as follows: robustness is "an ability to withstand disruption with an acceptable loss of performance" whereas resilience is "the potential to recover quickly from disruption." Robustness is a suitable capacity for managing business-as-usual risks (i.e., high probability, low impact risks to be *mitigated*), while resilience is suitable for disruption risks (i.e., low probability, high impact risks, or *contingencies*). Roy (2010) defines "robust" as the capacity for withstanding "vague approximations" and/or "zones of ignorance" in order to prevent undesirable impacts, notably the degradation of the properties of the system that the stakeholders want to be maintained. Ben-Haim (2012) (cited by Aven, 2016) affirms that in decisions under uncertainty, what should be optimized is robustness rather than performance, or in other words, the decision maker should prefer continuity and satisficing rather than optimizing. Resilience and robustness together ensure that a system is *reliable*, that is, it is free from failure and can perform consistently well. A useful framework to unify some of these concepts is the *risk matrix* (Fig. 1.2).

The level of resilience for a system or organization is linked to the ability to sustain or restore its basic functionality following a stressor. A resilient system can (Hollnagel et al., 2006, cited by Aven, 2016):

- respond to regular and irregular threats in a robust yet flexible (adaptive) manner,
- monitor what is going on, including its own performance,
- anticipate risk events and opportunities,
- learn from experience.

Although resilience is a generic term, it is most used in the safety domain, whereas robustness is most commonly referred to in business and operational research contexts.

A related term is *vulnerability*, which is defined in as "the propensity of risk sources and risk drivers to outweigh risk mitigation strategies, thus causing losses and adverse supply chain consequences" (Jüttner et al., 2003).

Resilience stands out as a promising research topic to address change, although it is inherently complex because it involves not only a response to disruption, but also touches on the cultural aspects of continuous improvement; these challenges are briefly discussed in the next section. Hamel and Valakingas (2003) include resilience as one of the three common forms of innovation:

1. *Revolution*, or creative destruction,
2. *Renewal*, or creative reconstruction, and
3. *Resilience*, or the capacity of continuous reconstruction.

In making the case for continuous reconstruction, Hamel and Valakingas (2003) note that, contrary to a very common business misconception, the idea of novelty is not fundamentally attached to risk, but

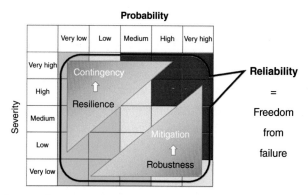

FIG. 1.2 The risk space framework.

that risk is a function of uncertainty multiplied by the size of the decision maker's financial exposure; novelty defies convention, and confusing newness with risk misleads companies to continue investing in old, decaying strategies based in past successes. According to the extensive literature review of Behzadi et al. (2018), strategies to enhance resilience have been studied far less than strategies to give robustness to supply chains, a similar conclusion as Barbosa-Póvoa et al.'s (2018), who state that work exploring sustainable supply chains' resilience is practically inexistent.

Regarding reports on case studies for making agricultural supply chains more resilient, there are abundant examples and case studies of supply chain adaptations in response to a wide range of disruptions. Fujisawa et al. (2015) studied the strategies for adaptation to climate change of apple farmers in three regions of Japan and South Africa, and found that their responses changed depending on the way their business is organized, with cooperatives being more likely to adopt a top-down approach to decision making, whereas farmers who have established their own sales channels tend to act on decisions that are born from the bottom up. Leguizamon et al. (2016) describes a program from a major retailer to buy directly from small farm suppliers, implementing a support strategy that commits to increase the resilience of the supply chain. Among Leguizamon et al.'s (2016) conclusions is that including a broad base of farmers in the supply chain is important for both suppliers' income and the supermarket's supply practices. Fujisawa et al. (2015) acknowledge that a combination of bottom-up and top-down leadership will facilitate more flexible and easily accepted policies for adaptation to a changing environment.

The purpose of the present chapter is, first, to review existing work and ideas relating to resilience and robustness from the perspective of fresh produce supply chains such as horticulture, grains, fish, and meat, and second, to propose a quantitative framework to measure resilience and help fresh food producers, packers or processors understand how the structure of their supply networks may enhance their resilience when facing disruptions based on the risk matrix summarized in Fig. 1.2. To that end, we discuss the *risk space* that supply chains are immersed in. We apply the framework to a case study of apple production and distribution in New Zealand.

1.2 The challenges for the supply chains of fresh produce

Disruptions to fresh produce supply chain operations may be produced by *economic, environmental*, or *social* factors. Identifying the most likely sources of disruption will depend on the type of supply chain under study, as each productive activity will be subject to its own specific challenges. To cite some examples, fruit and vegetable production is dominated by the effect of climate while being grown in the field; grains are very sensitive to the quality of transportation systems; the demand for beef has been increasing worldwide, but the cattle industry competes for resources such as land and water with society and other industries (García-Flores et al., 2015). To cope with disruptions and become more resilient, companies face the following internal challenges (Hamel and Valakingas, 2003):

1. The *strategic challenge*, which relates to medium- and long-term planning. Organizations must consider alternatives and create new options to update and replace strategies as they age and become less effective.
2. The *cognitive challenge*, which is a cultural concern. Companies must free themselves from denial, nostalgia, and arrogance.

3. *The political challenge*, which is related to resource allocation. Funding and skills must find their way to prototypes and experiments that may become tomorrow's products and services.
4. *The ideological challenge* is related to the search for efficiency. Optimization alone cannot make a business model more relevant when a changing environment dictates that the current operations are not effective anymore.

The social and cultural nature of these challenges, especially the last three, demonstrates that resilience is a complex concept that involves not only the operative response to disruption, but also the cultural aspects of continuous improvement and proactive anticipation, which makes it difficult to quantify. To complicate matters, these challenges are harder to attain for networks of companies than for individual players, as more coordination and a broader view of the system are necessary. The fact that cognitive, political, and ideological agreement is needed makes the design of a truly resilient supply chain a very difficult endeavor. Resilient supply chains must be proactive, as reactive strategies present many pitfalls; an enlightening example is presented in Rice and Gavin (2006). For the sake of the argument, we will consider in the remainder of this chapter that, regardless of the nature of the disruption, the ultimate effect will be the interruption of flow caused by the removal of one or more nodes or links from a network representation of the supply chain.

1.3 Quantifying resilience

Several metrics have been proposed to quantify resilience from a strategic perspective, mostly from the fields of ecology and telecommunications. The focus in ecology has been to understand the capability of populations to recover from threats. For example, Moore et al. (2016) analyzed 9 years of ostrich movement data to explore the resilience of the Western Cape ostrich industry, which was affected after an outbreak of avian influenza in 2011 and has gradually recovered. The network metrics used to assess resilience were density, network diameter, number of weak components, size of giant component and degree distribution. They found the ostrich movement network to be resilient. Despite a decrease in the number of farms after the outbreak, the system slowly returned to its former state. Plagányi and Essington (2014) developed an index to identify key species in an ecosystem in order to measure the repercussions of strengthening or weakening fish populations on the broader ecosystem structure as well as the overall economic value of fisheries.

There is also a lot of interest in defining resilience in telecommunications, especially for cybersecurity and infrastructure risk management. Alenazi and Sterbenz (2015) performed a comprehensive comparison of the most commonly used graph robustness metrics and found the most adequate predictors of resilience for different types of networks, including some generated artificially. In the context of Internet security, Salles and Merino (2012) introduced a resilience factor based in the concept of k-connectivity and propose strategies to improve resilience by altering the network topology. Rosenkrantz et al. (2009) developed a graph-theoretic model for service-oriented networks. In this type of network, when a user requests a service that is locally available at a node, the node provides the service directly. When the requested service is nonlocal, the node forwards the user's request to another node in the network where the service is available locally. Rosenkrantz et al. (2009) provided a resilience metric, as well as algorithms for designing resilient networks.

Regarding risk-related supply chain network metrics, Ortmann (2005) analyzed the South African fresh fruit export supply chain and modeled it as maximum flow and minimum cost flow problems, using optimization and graph theory methods. Barthélemy (2011) provided an extensive review of network metrics in the context of spatial networks, that is, networks where space is relevant and where topology alone does not contain all the information. Martin and Niemeyer (2019) discussed the implications of error measurements, which are unavoidable in networks representing real-world entities like supply chains, on the reliability of network metrics. More relevant to the methodology presented in this chapter is Lim-Camacho et al. (2015) and Lim-Camacho et al. (2017), who used a network-based simulation to estimate the resilience of supply chains and proposed four supply chain indices (SCIs) for evenness, resilience, continuity of supply, and climate resilience, to estimate the performance of fish, rice, and mineral supply chains. In particular, the resilience metric is based on the SCI introduced in Plagányi et al. (2014). This metric is described in detail in Section 4.1.

1.4 Methodology

In this Section we describe our methodology in detail, first by introducing the SCI of Plagányi et al. (2014) and then by explaining an approach to measure resilience empirically.

1.4.1 The supply chain index

The original SCI introduced in Plagányi et al. (2014) represents a measure of "connectance," with lower values indicating higher connectance. In a connected network G, let N and L be the set of nodes and links, respectively, and super-index S indicate the source nodes such that N^S is the set of all source nodes. Let $i, j \in N$ the source and destination of a product, respectively, $|L|$ the total number of links and $|N|$ the total number of vertices. Let f_{ij} the flow from i to j.

Input.

1. The elements of matrix S_{ij}, which represents the proportion of flows into each node respect to the total input flow,

$$s_{ij} = \frac{f_{ij}}{\sum_{i:(i,j) \in L} f_{ij}} \tag{1.1}$$

that is, S_{ij} is the flow from i to j divided by the total flow into node j from every supplier, such that $\sum_{i:(i,j) \in L} s_{ij} = 1$ for all nodes that are not sources in G.

2. The elements of matrix p_{ij}, which is the proportion of the total product in the supply chain that flows into receiver j,

$$p_{ij} = \frac{f_{ij}}{\sum_{(i,j) \in L: i \in N^S} f_{ij}} \tag{1.2}$$

that is, the flow from i to j divided by the total production of the sources, such that $\sum_{j:(i,j) \in L} p_{ij} = 0$ for all nodes that are sources in G. The element p_{ij} represents the proportion of the total product in the supply chain that flows into receiver j.

Output.
1. The SCI for each element j is

$$SCI_j = \sum_{(i,j) \in L} s_{ij} p_j^2 \tag{1.3}$$

The critical elements in G are identified as those with the highest SCI score.

2. The supply chain index total (SCIT) for the supply chain as a whole is obtained by summing over individual scores,

$$SCIT = \sum_{i \in N} SCI_j \tag{1.4}$$

3. The SCIT can be standardized by dividing it by the number of links |L|, thereby allowing comparisons across supply chains:

$$SCI^{std} = \sum_{i \in N} SCI_j \tag{1.5}$$

This index has been shown to provide insight as SCI based resilience through practical application. As supply chains are directional to defined end points it is crucial that the subnetwork used to supply each defined endpoint/sink be analyzed separately giving a distribution of resilience for an organization and enabling pressure points or critical nodes to be identified.

1.4.2 Testing resilience empirically

The methodology we use for our case study combines the well-known risk matrix in Fig. 1.2 and the strategy presented in Alenazi and Sterbenz (2015) consisting of pulling apart the supply network by removing the most critical node at a time and assessing the damage the removal caused. We extend the discrete risk matrix, which is normally used in risk assessment studies and relates the severity of a disruption to its probability of occurrence, into a continuous risk space. The methodology is as follows.

Let n be the number of nodes to remove, set initially to 1, and m the number of times to remove random n nodes, set to 100.

1. For each of the networks (Figs. 1.1–1.3 in Section 1.5):
   1.1 Select n nodes from the network to remove at random.
   1.2 Recalculate flows to make sure that the flow balances still hold.
   1.3 Calculate the severity of the disruption as the fraction of commodity that continues to flow from source (orchards) to sinks (customer) after the disruption. If there is no flow from at least one source to at least one sink, the network has been completely disrupted.
   1.4 Increase n. If n is equal to the number of nodes in the network or the network has been completely disrupted, stop.
   1.5 Go back to 1.1 and repeat m times.

2. Prepare output.

The methodology assumes, first, that the probabilities of disruptions of individual nodes are independent, and second, it assumes that the flows departing a node are split evenly between its descendants, none of which is necessarily true.

In the next section we apply the methodology to relate the SCI to the risk framework in Fig. 1.2.

1.4.3 Optimizing resilient supply chains

Once the supply chain has been designed, it is important to make the best use of the resources available. In Section 1.1 we mentioned study of the efficiency of investment to support small communities of cheese makers to add value to their whey by-product, reduce whey losses via value capturing through concentration and spray drying, and decrease environmental damage of disposal (García-Flores, 2015). In this problem, the objective is to select the sites to build processing facilities in from a set of candidate sites. Processing must transform sweet whey to a demineralized or nondemineralized concentrate, and then to dry whey powder. The *plants* are facilities that produce whey powder and can also produce concentrate, but *collection centers* can only produce concentrate via membranes. Thus, we have a globally inclusive service hierarchy. To illustrate the working principle of optimization techniques, consider the following model (Daskin, 2013). Let x_{jk} take the value 1 if a facility of type k is located at candidate site j and 0 otherwise, and y_{ijk} take the value 1 if demands at node i for type k services are satisfied by a facility at candidate site j, and 0 otherwise. Also let h_{jk} the demand for type k services at node i and P_k the number of type k facilities to locate. The problem is formulated as follows:

$$\text{Minimize} \quad \sum_{i \in I} \sum_{j \in J} \sum_{k \in K} h_{ik} d_{ij} Y_{ijk} \tag{4.1}$$

Subject to:

$$\sum_{j \in J} Y_{ijk} = 1 \ \forall \ i \in I; k \in K \tag{4.2}$$

$$\sum_{j \in J} X_{jk} = P_k \ \forall \ k \in K \tag{4.3}$$

$$Y_{ijk} \leq \sum_{g=k}^{m} X_{jg} \ \forall i \in I; j \in J; k \in K \tag{4.4}$$

$$X_{jk} \in \{0,1\} \ \forall j \in J; k \in K \tag{4.5}$$

$$Y_{ijk} \in \{0,1\} \ \forall i \in I; j \in J; k \in K \tag{4.6}$$

The objective function (4.1) minimizes the demand-weighted total distance. Constraint (4.2) ensures that all demand types in all locations must be assigned to some facility. Constraint (4.3) limits the total number of type k facilities located to P_k. Constraints (4.4) state that demands for service k originating at node i cannot be assigned to a facility at node j unless there is in j a service of type k or higher. Constraints (4.5) and 4.6) state the domain of the indicator variables. A problem like this can be solved using exact (e.g., simplex, branch, and bound) or approximate (heuristic) methods.

1.5 Case study and discussion

Figs. 1.3–1.5 show simplified schematics of actual operating models of apple supply chains, each of a different complexity. The impact of climate change on the apple industry in Hawkes Bay, New Zealand is less known when compared to other produce. It is known that rainfall trends have become more variable, that the mean temperatures have increased and that the diurnal temperature range is reducing (NIWA, 2019), and that this has an impact on fruit appearance, taste, and shelf life. Extreme climate events are also disrupting operations an impacting quality (e.g., cyclone Gita in Motueka 2018). Regional variability of the observed changes in weather patterns and in regional yield indicate that geographic risk spreading is a valid policy for NZ apples. (NIWA, 2019 and PFR data analyzed by CSIRO.)

　　Table 1.1 shows the type of nodes that compose the networks. These networks have flows assigned to every link, representing the flow of goods and materials in the supply chain. In this simplified representation of the network, flows are normalized, that is, it is assumed that all the commodities that leave

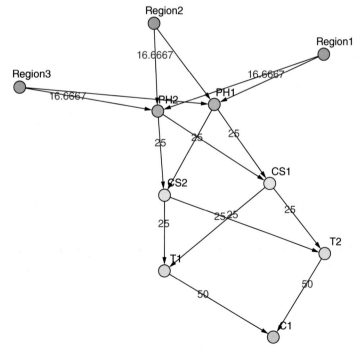

FIG. 1.3 Simplified schematic of the supply chain of apple variety *a1*.

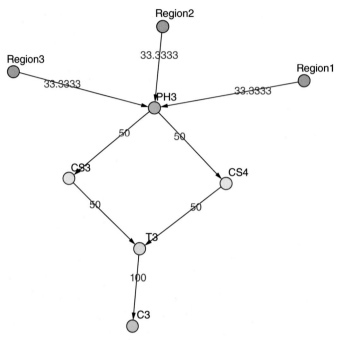

FIG. 1.4 Simplified schematic of the supply chain of apple variety *a2*.

the orchards add up to one hundred units. The value of the total index *SCIT* for these networks is 0.41, 1.19, and 4.00 for apple businesses *a1*, *a2*, and *a3*, respectively. These values indicate, as expected, that the supply network for variety *a1* is more connected than *a2*, and this in turn more connected than *a3*, which is a linear chain and therefore very easy to disrupt.

Fig. 1.5 shows the risk profiles for each variety's supply network. In these profiles, shown as box-plots, the *x* axis represents the probability of disruption: the right-most data point represents the effect of removing only one node at random, the second from left to right represents the effect of removing two nodes at random, and so on. The *y* axis represents the severity of the disruption as the initial flow of one hundred units minus the remaining flow from sources to sinks after the disruption divided by the total initial flow; a severity of one thus represents total disruption. The solid line represents the mean of the disruptions, so that a profile can be associated to each variety.

Variety *a1* is better connected, with a low *SCIT* value of 0.41, which reflects better connectance than for the other two apple varieties. Fig. 1.6A shows that variety *a1* can withstand disruptions better, as its risk profile shows that it can continue operating, on average, when 50% of its nodes have been removed (i.e., the severity of the disruption is less than 100% on average when four out of ten nodes have been removed). Variety *a2* is an intermediate case, with a *SCIT* of 1.19 and a profile indicating that 37.5% of its nodes can be removed on average and the network will continue operating. By contrast, variety *a3* has the highest *SCIT* (equal to 4) and is the most easily disrupted network. The removal of any single node will produce total disruption.

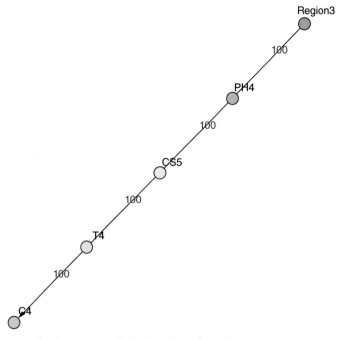

FIG. 1.5 Simplified schematic of the supply chain of apple variety *a3*.

Table 1.1 Types of nodes.

Node names	Type of facility
Region1, Region2, Region3	Orchards
PH1, PH2, PH3, PH4	Packing houses
CS1, CS2, CS3, CS4, CS5	Cold storage
T1, T2, T3, T4	Terminals
C1, C2, C3, C4	Customers

1.6 Concluding remarks

Climate change and the increasing demand for resources needed by a growing population bring many risks to food supply chains. In order to mitigate these risks, there is a pressing need to quantify resilience, robustness, and reliability. It has been reported that there is a gap in research about resilience in sustainable supply chains. In this chapter we have outlined an approach to fill a gap in resilience research by extending the concept of risk matrix to a risk space that is a continuous version of the commonly used risk matrix, and combining it with the SCI first reported by Plagányi et al. (2014). In this way, we can quantify the resilience and robustness of fresh produce supply chains (in our case, using

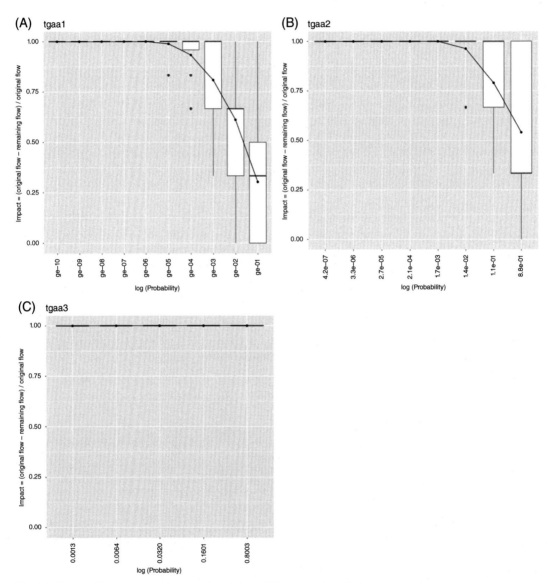

FIG. 1.6 Risk profiles of the supply networks of three different apple varieties.

Variety *a1* has risk profile (A) and is more reliable, as the severity of the disruption is lower for the same number of nodes removed when compared to *a2* and *a3*. Variety *a1* has the highest connectance with a *SCIT* = 0.41. The risk profile for variety *a2* is shown in (B), with intermediate resilience and a *SCIT* = 1.19, and finally the risk profile of variety *a3* (C) with a linear and very vulnerable supply chain. In this case, the removal of any one node disrupts the network completely. For this variety, *SCIT* = 4.

an exemplar sector looking to understand the impact of climate change) by associating a risk profile to network structure and flows. We would like to stress that the construction of this profile involves not simply an average metric (as, for example, *network centrality*) as commonly done in network analysis, but it considers the effect of connectedness on the supply network's flows. We have also discussed the cultural aspects associated with resilience and what represent opportunities for future research and, as recognized in Aven (2016), the scientific foundation of risk assessment and risk management is still somewhat weak and open to formalization on some issues, leaving the door open to more research in this field.

We finalize this chapter by noting that true resilience only comes from continuously asking the right questions proactively, building networks and partnerships and ensuring a variety of creative strategies as part of a broader entrepreneurial culture.

References

Alenazi, M.J.F., Sterbenz, J.P.G, 2015. Comprehensive comparison and accuracy of graph metrics in predicting network resilience, Proceedings of the 11th International Conference on the Design of Reliable Communication Networks (DRCN). Kansas City, KS, 157–164 March 24-27.

Alkaabneh, F., Diabat, A., Gao, H.O., 2020. Benders decomposition for the inventory vehicle routing problem with perishable products and environmental costs. Comput. Oper. Res. 113, 104751. UNSP. https://doi.org/10.1016/j.cor.2019.07.009.

Aven, T., 2016. Risk assessment and risk management: review of recent advances on their foundation. Eur. J. Oper. Res. 256, 1–13.

Bak, O., 2018. Supply chain risk management research agenda: from a literature review to a call for future research directions. Bus. Process Manage. J. 24 (2), 567–588. doi:10.1108/BPMJ-02-2017-0021.

Banasik, A., Kanellopoulos, A., Claasen, G.D.H., Bloemhof-Ruwaard, J.M., van der Vorst, J.G.A.J, 2017. Assessing alternative production options for eco-efficient food supply chains using multi-objective optimization. Ann. Oper. Res. 250 (2), 341–362. https://doi.org/10.1007/s10479-016-2199-z.

Barbosa-Póvoa, A.P., da Silva, C., Carvalho, A., 2018. Opportunities and challenges in sustainable supply chain: an operations research perspective. Eur. J. Oper. Res. 268, 399–431.

Barthélemy, M., 2011. Spatial networks. Phys. Rep. 499 (1-3), 1–101. doi:10.1016/j.physrep.2010.11.002.

Behzadi, G., O'Sullivan, M.J., Olsen, T.L., Zhang, A., 2018. Agribusiness supply chain risk management: a review of quantitative decision models. Omega (79), 21–42. http://dx.doi.org/10.1016/j.omega.2017.07.005.

Ben-Haim, Y., 2012. Doing our best: optimization and the management of risk. Risk Anal. 32 (8), 1326–1331.

Daskin, M.S., 2013. Network and Discrete Location: Models, Algorithms and Applications, second ed. Wiley, Hoboken, New Jersey.

Domingues-Zucchi, J., Zeng, A., Caixeta-Filho, J., 2011. Optimum location for export-oriented slaughterhouses in Mato Grosso, Brazil: a dynamic mathematical model. Int. J. Logistics: Res. Applic. 14, 135–148.

Fan, Y., Stevenson, M., 2018. A review of supply chain risk management: definition, theory, and research agenda. Int. J. Phys. Distrib. Logistics Manage. 48 (3), 205–230. doi:10.1108/IJPDLM-01-2017-0043.

Fujisawa, M., Kobayashi, K., Johston, P., New, M., 2015. What drives farmers to make top-down or bottom-up adaptation to climate change and fluctUATIOns? A cOmparative study on 3 cases of apple farming in Japan and South Africa. PLoS One 10 (3), e0120563. doi:10.1371/journal.pone.0120563.

Garc ía-Flores, R., Higgins, A., Prestwidge, D., S, McFallan, 2014. Optimal location of spelling yards for the northern Australian beef supply chain. Comput. Electron. Agric. 102, 134–145.

García-Flores, R., de Souza Filho, O.V., Martins, R.S., Martins, C.V.B., Juliano, P, 2015. Using Logistic models to optimize the food supply chain. In: Bakalis, S., Knoerzer, K., Fryer, P. (Eds.), Modelling Food Processing Operations. Woodhead Publishing, Sawston, Cambridge, England, pp. 307–330.

Hamel, G., Välikingas, L., 2003. The quest for resilience. Harv. Bus. Rev. 81 (9), 52–63.

Ho, W., Zheng, T., Yildiz, H., Talluri, S., 2015. Supply chain risk management: a review. Int. J. Prod. Res. 53 (16), 5031–5069. http://dx.doi.org/10.1080/00207543.2015.1030467.

Hollnagel, E., Woods, D., Leveson, N., 2006. Resilience Engineering: Concepts and Precepts. Ashgate, UK.

ISO (2009), 2009Risk Management—Vocabulary. Guide73. https://www.iso.org/standard/44651.html accessed on August 27, 2019.

Jüttner, U., Peck, H., Christopher, M., 2003. Supply chain risk management: outlining an agenda for future research. Int. J. Logistics Res. Applic. 6 (4), 197–210.

Leguizamon, F., Selva, F., Santos, M., 2016. Small farmer suppliers form local to global. J Bus Res 69, 4520–4525. http://dx.doi.org/10.1016/j.jbusres.2016.03.017.

Li, Y.T., Chu, F., Cote, J.F., Coelho, L.C., Chu, C.B., 2020. The multi-plant perishable food production routing with packaging consideration. Int. J. Prod. Econ. 107472, 221. UNSP. https://doi.org/10.1016/j.ijpe.2019.08.007.

Lim-Camacho, L., Hobday, A., Bustamante, R., Farmery, A., Fleming, A., Frusher, S., Green, B., Norman-López, A., Pecl, G., Plagányi, É., Schrobback, P., Thebaud, O., Thomas, L., van Putten, I., 2015. Facing the wave of change: stakeholder perspectives on climate adaptation for australian seafood supply chains. Reg. Environ. Change 15, 595–606.

Lim-Camacho, L., Plagányi, E., Crimp, S., Hodgkinson, J.H., Hobday, A.J., Howden, S.M., Loechel, B., 2017. Complex resource supply chains display higher resilience to simulated climate shocks. Global Environ. Change 46, 126–138.

Martin, C., Niemeyer, P., 2019. Influence of measurement errors on networks: estimating the robustness of centrality measures. Network Sci. 7 (2), 180–195. doi:10.1017/nws.2019.12.

Moore, C., Grewar, J., Cumming, G.S., 2016. Quantifying network resilience: comparison before and after major perturbation shows strengths and limitations of network metrics. J. Appl. Ecol. 53, 636–645.

Negreiros de Oliveira, F., Leiras, A., Ceryno, P., 2019. Environmental risk management in supply chains: a taxonomy, a framework and future research avenues. J. Cleaner Prod. 2032, 1257–1271. https://doi.org/10.1016/j.jclepro.2019.06.032.

NIWA 2019. https://data.niwa.co.nz/#/home/, accessed on May 24, 2019.

Ortmann, F.G., 2005. Modelling the South African Fresh Fruit Export Supply Chain. University of Stellenbosch, Stellenbosch Ph.D. thesis.

Paam, P., Berretta, R., Heydar, M., Garc ía-Flores, R., 2019. The impact of inventory management on economic and environmental sustainability in the apple industry. Comput. Electron. Agric. 163, 104848. https://doi.org/10.1016/j.compag.2019.06.003.

Plagányi, E.E., Essington, T.E., 2014. When the SURFs up, forage fish are key. Fish. Res. 159, 68–74.

Plagányi, É.E., van Putten, I., Thébaud, O., Hobday, A.J., Innes, J., Lim-Camacho, L., Norman-López, A., Bustamante, R.H., Farmery, A., Fleming, A., Frusher, S., Green, B., Hoshino, E., Jennings, S., Pecl, G., Pascoe, S., Schrobback, P., Thomas, L., 2014. A quantitative metric to identify critical elements within seafood supply networks. PLoS One 9, e91833.

Rice, J., Gavin, P., 2006. Alliance patterns during industry life cycle emergence: the case of ericsson and nokia. Technovation 26 (3), 384–395.

Rosenkrantz, D.J., Goel, S., Ravi, S.S., Gangolly, J., 2009. Resilience metrics for service-oriented networks: a service allocation approach. IEEE Trans. Serv. Comput. 2 (3), 183–196.

Roy, B., 2010. Robustness in operational research and decision aiding: a multi-faceted issue. Eur. J. Oper. Res. 200, 629–638.

Salles, R.M., Merino, D.A.M, 2012. Strategies and metric for resilience in computer networks. Comput. J. 55 (6), 728–739. doi:10.1093/comjnl/bxr110.

Zhu, Q., Krikke, H., Caniëls, M.C.J, 2017. Integrated supply chain risk management: a systematic review. Int. J. Logistics Manage. 28 (4), 1123–1141. doi:10.1108/IJLM-09-2016-0206.

Sustainable food systems

Paulomi (Polly) Burey[a], Sunil K. Panchal[a,b], Andreas Helwig[a]
[a]University of Southern Queensland, Toowoomba, QLD, Australia
[b]Western Sydney University, Richmond, NSW, Australia

2.1 Introduction

Food systems encompass all the activities, knowledge (and related cultural routines), relating to the production, processing, transport, consumption, and disposal of food (Carolan, 2018). Food systems also include supporting enabling mechanisms, such as economic and political systems, and the technologies to allow food to travel from producer to consumer (Block, 2015). As a result, they tend to be shaped by the societies in which they are formed, consisting of complex linkages between food supply chain points.

Historically food systems have been linear in nature, which can pose issues regarding their capability to provide enough food for future generations (Galanakis, 2018). In order to develop sustainable food systems (SFS) into the future, there needs to be a focus on social, economic, and environmental sustainability (SEES). To achieve this, supply chains must be modified through the development of new supply chain management strategies, and more efficient, flexible processing technologies. These types of modifications can potentially optimize food, energy, and water resource usage to minimize waste and optimize population health. Ideally this occurs while maintaining economic health of the food systems and their surrounding environment. These are very lofty targets, but with dedicated food engineers, food technologists, and food scientists, many of these goals can be attained. This chapter will focus considerations for SFS implementation, and how food engineering innovations may assist.

2.2 Sustainability of food systems

Historically, food systems have been linear in nature, consisting of resource inputs at the food production end, and outputs along the food supply chain, typically consisting of consumed food, food by-products such as peels, rinds, bones, and seeds, and excess food which may become food loss and waste (FLW) (Fig. 2.1).

2.2.1 Linear food system issues

In Australia and North America, linear food supply chains produce significant waste, up to 60% of food produced, which is a loss of nutrition as well as the resources used to produce that nutritional resource (Chen et al., 2020; FAO, 2011; Johnson et al., 2019). In Australia, the majority of FLW is generated

Food Engineering Innovations Across the Food Supply Chain. DOI: https://doi.org/10.1016/B978-0-12-821292-9.00015-7

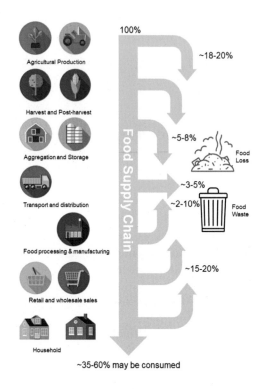

Agricultural Production

Harvest and Post-harvest

Aggregation and Storage

Transport and distribution

Food processing & manufacturing

Retail and wholesale sales

Household

100%

~18-20%

~5-8% Food Loss

~3-5%
~2-10% Food Waste

~15-20%

~35-60% may be consumed

Food Supply Chain

FIG. 2.1

Linear food systems with food loss and waste (FLW) figures representative of North America and Oceania (Australia for our illustrative purposes).

at the food production and food consumption points of the supply chain, where it is unused for human consumption, and often discarded (Arcadis, 2019). In life cycle analysis (LCA), this is known as a "Cradle-to-Grave" approach, or more colloquially as a "Take-Make-Dispose" approach, where resources have no use at the end of their supply chain and may end up in landfill (Babich and Smith, 2010; Tillman, 2010). There are many factors which cause such suboptimal linear food systems, including inefficient usage of water and energy, lack of pathways for excess food, and potential food safety issues (Béné et al., 2020).

A key inefficiency in food systems is where food is produced to meet retail standards but as some may naturally not meet requirements, oversupply is factored into production. The reasons for rejection at harvest include cosmetic physical characteristics (size, shape, and color) not meeting market requirements and considered "imperfect" produce. This problem is exacerbated if there is no alternative pathway for this food material, so it subsequently becomes waste. Animal feed is one option, but fresh produce must be consumed quickly, while processing into other products may provide a solution. Unexpected weather events may also affect food quality, safety, and harvest. If harvest timing does not match the original plans, labor availability may be affected and harvest may not occur (KPMG, 2020b).

Another influencing factor may be market variations in the costs of production, and the food's market value. These fluctuations may cause producers to decide not to harvest due to the lack of economic feasibility, thereby generating FLW (Johnson et al., 2018, 2019). This can also be influenced by FLW percentage, as there is a tipping point at which production is no longer profitable enough. For high volume Australian vegetable crops of potatoes, tomatoes, and carrots, this tipping point can range from 15% to 30%, where a financial loss commences (Fig. 2.2).

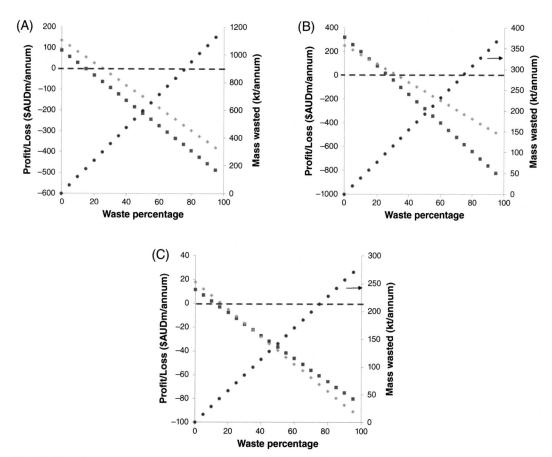

FIG. 2.2 The effects of food loss and waste (FLW) percentage on FLW mass generated and profit/loss for high volume Australian crops of (A) potatoes, (B) tomatoes, and (C) carrots.

● - waste kilo-tonnage, ■ – profit/loss assuming all farms are ≤40 hectares, ◆ - profit/loss in AUD million assuming all farms are >40 hectares. Production and cost data used to calculate these figures was from the Australian vegetable growing farms economic survey 2017–2018 and 2018–2019 plus waste levy charges. Costs of production were based on kilotons produced per annum and assumed as a "sunk" cost. An extra waste levy cost was added to account for kilotons that are assumed as FLW based on percentage of crop. Currency conversion 2015–2020 has typically been 1 AUD = 0.70-0.80 USD.

There are also many environmental and resource stressors on food systems causing sustainability issues. These include climate change, water resource depletion, soil degradation, biodiversity loss, deforestation, ecosystem stress, and a lack of nutrition security (KPMG, 2020b; Valentini et al., 2019). Lastly, there can also be acute shocks to food systems, which affect their resilience and sustainability, including biosecurity issues, natural disasters, drought, technology failures (particularly data related), and geopolitical influences including trade, regulation, and policy, internal conflict and as we have seen most recently, pandemics (KPMG, 2020b; Béné, 2020). An example of an acute shock is the needles in strawberry scare that occurred in Australia in 2018, which impacted an entire industry, not just the producers affected (Bavas, 2018).

With increasingly limited resources, unsustainable linear food systems cannot continue. Future food systems must meet the needs of growing populations and a move toward SFS would address many of the aforementioned issues. SFS encompass a "Cradle-to-Cradle" approach (Bjørn and Hauschild, 2018) for resource tracking that links into circular economy (CE) principles. This approach is where food excess and by-products (FE&Bs) are rescued before they become FLW. Instead, food chains mostly feedback rejects into the earlier stages, creating some resource recycle/repurpose loops (Fig. 2.3 as an example).

Although technical change can be developed by scientists, engineers, and technologists, regulatory and cultural change is also required to move toward more circular SFS.

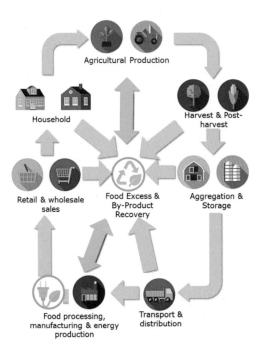

FIG. 2.3

Proposed circular food system loops with minimal to no food loss and waste (FLW). This example shows FLW circulating back to agricultural production, and to other parts of the supply chain, such as processing and manufacturing via the green arrows, where both energy and products could be produced from food excess and by-products (FE&Bs) instead of them being converted to FLW.

2.2.2 Definition of sustainable food systems

SFS are those where food is produced to provide global food and nutrition security, while also supporting sustainable economic, social, and environmental function (FAO, 2018; Morath, 2016; Anderson et al., 2019; Biel, 2016). The Food and Agriculture Organization of the United Nations (FAO) (FAO, 2018) elaborates on how sustainability needs to apply in these three areas:

a. There must be a broad range of benefits for society to support *social sustainability*.
b. Profits are sufficient to support *economic sustainability*; without this aspect the system may fail.
c. Impacts on the natural environment need to be positive or neutral to support *environmental sustainability*.

From a *social perspective*, food systems are considered sustainable when there is equitable distribution of value-add, while accounting for vulnerable populations (FAO, 2018). Healthy and sustainable diets may not be economically achievable for all levels of society (Barosh et al., 2014) and so there needs to be enabling mechanisms to ensure access for all, which may include policy and regulation (Boylan et al., 2019). Positive outcomes from ensuring social sustainability in an SFS include solid nutrition and health for all populations, including secure nutrition access (HLPE, 2020; Charlton, 2016; Smetana et al., 2019), sound animal welfare practices and fair labor conditions for all food system workers (Carino et al., 2020). This may require a change in global diets to ensure that these outcomes are achievable into the future (Caraher, 2013; Lawrence et al., 2019; Willett et al., 2019).

From an *economic perspective*, food systems are considered sustainable when the activities conducted by the stakeholders and actors are commercially and financially viable. Characteristic benefits of such a system should apply to all stakeholders and may include fair wages for workers, government taxes supporting social sustainability, profitable enterprise, and improvements to the food supply for consumers to support food and nutrition security (FAO, 2018). The profitability aspect is key—food charity has an important role to ensure food access for vulnerable populations, but some part of the food system will bear the cost of doing so. Government fiscal policy surrounding claimable donations of not only money, but also resources (donation of transport, labor, etc.) could help ease this burden.

From an *environmental perspective*, food systems are considered sustainable when their characteristics produce only positive or neutral impacts on their supporting environment. There needs to be a healthy people–industry–environment–symbiosis (PIES) to support sustainability which would include many aspects such as:

1. Consideration of maintenance of biodiversity
2. Sustainability of water resource usage and recycle
3. Maintenance of soil quality including crop nutrition
4. Promotion of plant and animal health in the SFS ecosystem
5. Reduction of carbon and water footprints and FLW
6. Minimal to no introduction of toxicity into the supporting environment.

Consequently, an SFS requires an optimal balance between the material and financial resources required for food production, the needs of the people who will consume the food, and the needs of the environment in which the food is produced. In essence, the analysis of SEES of food systems needs to be determined, taking into account demographic, geographical, and cultural factors. This can be achieved through proper accounting and appropriate tracking of all system components through a material flow analysis (MFA), technoeconomic analysis (TEA), and LCA (Halpern et al., 2019). To enable this tracking, a key component of successful sustainable food system design is the use of predictive models to determine how to provide food and nutrition security for future populations (Allen and Prosperi, 2016; Allievi et al., 2019). This modeling can be used to determine the optimal PIES via sensitivity analysis, and this may provide the framework for future SFS (Allen and Prosperi, 2016).

2.2.3 Technoeconomic analysis

TEA is used for comparative analysis of processing technologies based on their capabilities and advantages, in comparison to their economic impact on costs of production, target revenues, and profits (Barbiroli, 1997). TEA can analyze processes, products, or services and can assist with MFA, particularly for biorefineries where new technological approaches may be employed to tackle the problem of FLW and create new products to support SFS (Cristóbal et al., 2018). There are many parameters to be considered in such analyses, including crop production statistics, technologies employed to process food excess, and trade and sales data. To investigate economic feasibility of new technologies, operating and capital costs to set up a processing ecosystem need to be considered, as well as return on investment (ROI) (Fig. 2.4). More sophisticated TEA could allow analysis of the simultaneous production of multiple product streams via multiple technologies, to determine the best pathways for different food components, which could take into account economic flow-on effects through the use of regional processing hubs and microfactories (Kwan et al., 2015; Esposto et al., 2019; Augustin, 2019).

2.2.4 Life cycle analysis

Food systems can impact the environment and to measure this impact, LCA may be used. The current basic framework for performing an LCA looks at environmental impacts of production activities, no matter the material, and is outlined in the ISO standard ISO14040:2019 (ISO, ASISO 14040:2019, 2019). Due to the global nature of modern food systems, these impacts may be widely dispersed, so LCA can become quite complex (Cucurachi et al., 2019). To focus the LCA it may be necessary to

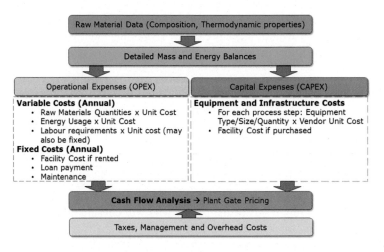

FIG. 2.4

Technoeconomic analysis cost considerations to support sustainable food systems.

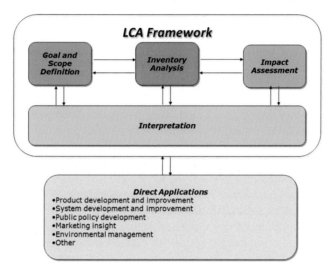

FIG. 2.5

Life cycle analysis framework. Modified from AS ISO 14040:2019 (ISO, ASISO 14040:2019 2019).

define the goals and scope well to avoid overcomplicating the analysis. There are three main components that contribute, which are goal and scope setting, a life cycle inventory, and a life cycle impact assessment. Together these are interpreted in order to evaluate what the environmental impact may be (Fig. 2.5) and some LCAs may even include societal impact in their scope.

Earlier food LCAs have focused on commodity products like bread, dairy, and meat, as well as associated packaging (Roy et al., 2009), where a wide range of factors were attributed to the major environmental impacts of producing these foods. Newer analysis approaches have combined TEA and LCA into a technoeconomic and life cycle analysis (TELCA) model to evaluate technoeconomics and environmental impact simultaneously, which allows greater sensitivity analysis of a wider range of influencing factors (Marx et al., 2019).

2.3 Features of a sustainable food system

There are several considerations for embedding sustainability in food systems, which include implementing CE principles; sustainable agricultural practices; localized and seasonal distribution to reduce "food miles"; production of healthy and sustainable foods; development of innovative increased shelf-life products to prevent food waste; and pathways for FE&Bs to support a "zero-waste" approach (Willett et al., 2019; Galanakis, 2018; Garcia-Gonzalez and Eakin, 2019; Jurgilevich et al., 2016; Mak et al., 2020; Pagotto and Halog, 2016; Friel et al., 2014). Some of these SFS features can improve the flexibility or agility of food systems in order to maintain SEES.

2.3.1 Circular economy principles

One of the main features of a CE is the sustainable generation, or regeneration, of resources. The goal is to design waste out of the value chain, and move toward "zero waste" using approaches which reduce material usage and encourage re-use or repurposing of material as depicted in Fig. 2.3 (Jurgilevich et al., 2016; Brears, 2015; Fassio and Tecco, 2019; Jagger, 2016). There are several strategies built on waste hierarchy principles (Teigiserova et al., 2020; Teigiserova et al., 2020; KPMG, 2020a) to support such a design, which includes prioritizing these strategies as well as identifying them (Fig. 2.6). The most preferred strategy is reduction of waste at the source, followed by food rescue for nutritional purposes for people. Although animal feed is touted as an option, this must involve preservation due to issues with diversion of fresh produce and food safety. The least preferred option is disposal or incineration. In the middle of the hierarchy, transformation into useful products or energy is also a useful strategy.

A key requirement for these strategies to work is the presence of appropriate infrastructure for collection, recovery, and processing of waste. An MFA can assist by looking at system inputs (water, energy, other resources), system outputs (greenhouse gas [GHG] emissions, waste) and also take into account logistics and economic indicators (production costs, total revenue and value-add) (Pagotto and Halog, 2016). One component not often included is food rescue which introduces another system flow based on social sustainability (Lindberg et al., 2014). The sustainable development goals (SDGs) provide some contributing areas of focus to support SEES (Fig. 2.7; Fassio and Tecco, 2019).

2.3.2 Sustainable agriculture

Future agricultural practices must provide nutritional and fiber resources to an increasing population and for this to occur, finite resources must be used in a sustainable manner. The EAT-Lancet report commissioned in 2019 (Willett et al., 2019) outlined the nutritional requirements for a healthy diet and explored the resource usage and associated impacts to produce such a diet. The analysis considered inputs like water and crop nutrient usage, and environmental impacts like GHG emissions, adverse biodiversity changes, and nitrogen and phosphorous pollution (Willett et al., 2019). Other studies have explored the environmental impacts of different food and beverage types, including meat, dairy, seafood,

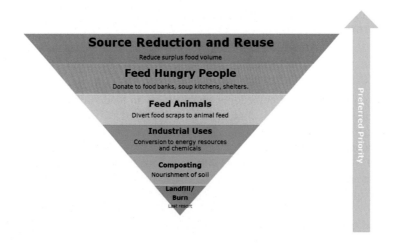

FIG. 2.6

Waste hierarchy depicting options for handling of food to prevent waste. Modified from (KPMG, 2020).

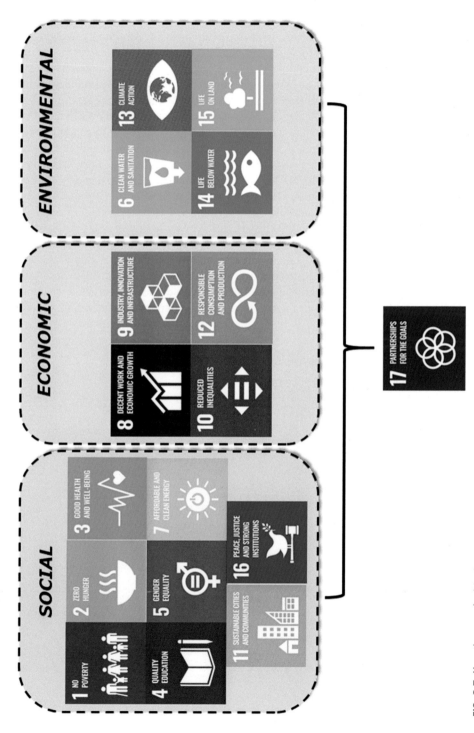

FIG. 2.7 How the sustainable development goals (SDGs) can support social, economic, and environmental stability for sustainable food systems.

Clustering is based on the "wedding cake" system from Fassio and Tecco (2019) and UN (2020). United Nations Sustainable Development Goals, https://www.un.org/sustainabledevelopment/. The content of this publication has not been approved by the United Nations and does not reflect the views of the United Nations or its officials or Member States.

starchy foods, oils, fruit, sugar, alcohol, chocolate, and coffee. The impacts compared across these food and beverage types included GHG emissions, land use requirements, water scarcity impacts, and acidifying and eutrophying emissions with a focus on phosphates (Poore and Nemecek, 2018). Overuse of crop nutrients leads to excessive eutrophication, possibly caused by runoff. These nutrients are also "wasted" as they do not go toward crop growth. Minimizing or reducing environmental emissions like this, and recovery of these components is imperative to prevent pollution and resource loss.

Another aspect to consider for sustainable agriculture is The water–energy–food (WEF) nexus described by the FAO (FAO, 2014), which emphasizes that these resources are vital to support human wellbeing. Their usage and related production activities intersect; therefore they should be analyzed simultaneously in a holistic manner, hence a nexus analysis. FAO also acknowledged that agricultural activity was responsible for 70% of global freshwater withdrawal and 30% of global energy usage at the time of reporting, which is unsustainable. A WEF nexus assessment attempts to optimize the balance of resource usage including energy, water, land, and soil, as well as consideration of socioeconomic factors to support usage decisions (Fig. 2.8) (Brears, 2015; FAO, 2014; Cai et al., 2018; Chang et al., 2016; OECD, 2017).

Modern strategies have focused on the digitalization of agriculture to support tracking of resource usage, leading to the enabling of WEF nexus footprint analysis for individual foods (Basso and Antle, 2020; Walter et al., 2017). This can potentially be supported by Industry 4.0 infrastructure to inform personal food choices and development of sustainable diets, which is important as an average plate of

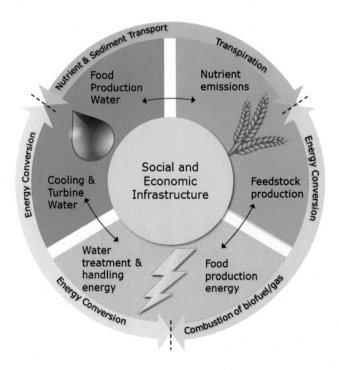

FIG. 2.8

Water–energy–food nexus and how the components interact. Modified from (Cai et al., 2018).

Resource	Amount
Nutritional Calories available	676 kcal/ 3 MJ
Energy used to produce meal	6 MJ
Water used to produce meal	1090 L

FIG. 2.9

Nexus footprint of an average meal according to FAO (FAO, 2014).

food can contain less nutritional energy than the energy used to produce it (Fig. 2.9) (Barosh et al., 2014; Lawrence et al., 2019; Willett et al., 2019; Holzeis et al., 2019). More complex interactions within food systems need to be elucidated which may require consideration of other resources beyond energy and water (Valentini et al., 2019; Namany et al., 2019), and enable development of technology, policies, and regulation to ensure sustainable resource and food usage into the future.

An important consideration for food production planning is ensuring that local populations have access to the food, and associated nutrition, that they require for health. From an Australian perspective, Australia produces more food than it consumes, and food export is one of the major economic industries. While food production levels in Australia are enough that >70% of all food production is being exported, indications show that enough food is still available for its citizens (ABARES, 2020). However, access may be affected particularly for those experiencing food insecurity (Foodbank, 2019). This may be due to the flexibility needed for food rescue and transport, which can easily handle longer shelf-life products, but fresh produce rescue requires significant storage resources and rapid response to ensure that the food is safe, and the quality is maintained.

Food production planning involves many activities to support decisions that ensure timely and sustainable production occur. Decision support tools supported by whole on-farm sensor connection to manufacturing can aid planning and scheduling of agricultural production (Namany et al., 2019; Kath and Pembleton, 2019), however, the mix of activities can be complex and may include many components (Yuana et al., 2017) such as impact of weather and genetics, soil and nutrient support in the soil, optimizing water holding capacity, energy recovery and energy demand management, crop planning and a new component of postharvest processing planning. Consideration of all these factors using a digital agriculture approach may support the first priority in the waste hierarchy which is to prevent FLW at the source and best match supply to demand.

The mix of activities varies for livestock and aquaculture production, but there are still many steps to complete for those resources. Once these components are balanced for optimum agricultural production and minimal environmental impact, then the production point of the food system may be considered sustainable. There is a complex relationship between all planning components which can be difficult to elucidate but could be enabled through use of data collection and analytics, possibly supported via Industry 4.0 platforms.

2.3.3 Localized food systems

Globalization of food systems has led to availability of foods "out of season" due to the ability to import and export. This means food may travel large distances; adding to their unsustainability; there is an argument for producing food close to where it is consumed to optimize resource use and reduce fuel usage and transport costs. In Australia, food may sometimes need to cover vast distances during distribution (>1500 km), which means that if the market value is insufficient, costs of distribution may outweigh potential return (Deloitte Access Economics, 2019). To remain economically sustainable this may mean the food is not transported for consumption, and subsequently becomes FLW. The percentage of food that is FLW at the point of production can range from 15% to 75% dependent on the crop (FAO, 2011), the weather, and any market forces in the year of harvest.

An alternative may be that fresh food is distributed for food rescue, which is the second priority in modern food waste hierarchies, but this must occur rapidly to maintain food safety and quality, and this is true for both human and animal nutrition purposes. This may not always be feasible due to tight timeframes, and availability of transport and labor. Because of these reasons it may be beneficial if more food can be produced close to where it is needed. When food is produced some distance from the consumer, significant volumes of reject produce may occur during harvest and postharvest, which means it can no longer be transported as fresh produce via this avenue. The food may still be edible and have a market; sometimes retail supply chains may allow for this produce to continue to the consumer; however, this is not yet standard practice. The remaining saleable food will continue in the food supply chain to consumers, either as fresh produce or processed food products. However, an increase in enabling pathways for excess "imperfect" produce to processors and consumers is imperative to prevent or reduce FLW.

In more recent times, the advent of increased protected agriculture has driven some food production to be closer to consumers. Protected agriculture, urban, and periurban agriculture are not only overcoming issues associated with distance to consumer, but also "imperfect" produce, and overcoming some of the issues associated with traditional crop production. These include coping with natural variations in rainfall, temperature, wind, wild weather events, as well as enabling a reduction in associated costs and environmental impact (Mok et al., 2014; Carey et al., 2011). The utilization of "imperfect" produce, and thus reduction of FLW requires not only technical solutions, but also changes to consumer behavior and expectations.

2.3.4 Innovative increased shelf-life products to prevent waste

One major cause of food waste for some commodities is their limited short shelf life, particularly for items like meat, dairy, and fresh horticultural produce. Shelf life indicators for consumers are via stamped "Use-by" and "Best Before" dates, but these are not specific, and are determined by average shelf life periods. If food is not sold in time, true or perceived quality and safety becomes compromised, and it becomes waste. To overcome this, innovations in "smart" packaging that directly indicate safety and quality of food could help prevent waste (Madhusudan et al., 2018; Müller and Schmid, 2019).

Food preservation techniques were established a few centuries ago to prolong the shelf-life of food, usually to be taken on exploration expeditions (Thorne, 1986). In the modern world, food preservation technologies are used to extend shelf-life, so that foods can cover vast distances within our food supply chains (Bentley, 2015), but in an SFS, food preservation technologies should also have a purpose in preventing FLW. This may also tie into SDGs focused on increasing access to food and nutrition (Torres-León et al., 2018). During food processing and preservation there may be excess food and by-products produced which are usually wasted but may have applications in other products.

Food excess and by-products (FE&Bs) produced along the supply chain are a rich source of water, biopolymers (starch, protein, celluloses, pectins, etc.), and bioactive compounds, which could be used to produce nutritional and food products (Galanakis, 2012, Mirabella et al., 2014; Schieber et al., 2001). Many examples of these applications have been outlined for different sources including fruit, vegetables, meat, and dairy in previous research (Mirabella et al., 2014). As well as nutritional products, they may also be used as a feedstock for herbicides and pesticides, antioxidants, fermentation substrates, enzymes, and sanitizing agents (Mirabella et al., 2014). FE&Bs may also have applications in other areas such as composite materials, packaging, chemicals, and even pharmaceutical products in some instances. Using a biorefinery approach, there is also the potential to produce bio-colorants, biopesticides, intermediate chemicals, bioplastics, biochemicals, biofuels, and enzymes (Maina et al., 2017).

To develop such innovative products from FE&Bs this may also include the recovery and use of items such as leaves and stems during primary production, which could be used as a source of nutritional components. There may be instances where the leaves and stems are not considered the primary human nutrition source, which instead may be a fruit, vegetable or tuber. Valorization of leaf biomass could occur for crops like berries, sweet potato, cauliflower, and many others (Olson et al., 2016; Biel and Jaroszewska, 2017; Suárez et al., 2020; Singh et al., 2019; Sagar et al., 2018; Oyeyinka and Oyeyinka, 2018; Belguith-Hadriche et al., 2017; Fasuyi, 2006). Some of this material is already used as a source of organic matter for soil conditioning, but often there is still excess biomass that could be used for human and animal nutritional purposes. Leaf biomass is a good source of carbohydrates and bioactive molecules, but some are even a good source of plant protein as dry matter (Sun et al., 2014).

During factory processing, skins, rind, seeds, and pomace may be removed, although they are a good source of compounds useful for nutritional, and other purposes. These could be extracted and dried as food powders with a much longer shelf-life than fresh produce and could potentially be used as nutritional supplements or ingredients in food products, or as materials for packaging with antimicrobial properties (Sagar et al., 2018; Çam et al., 2014; Bemfeito et al., 2020; Conesa et al., 2020; Huang et al., 2020; Nazeam et al., 2020; Zhang et al., 2020; Ajila et al., 2007; Chantaro et al., 2008; Yu and Ahmedna, 2013). The extent of FE&Bs from processing of fruit and vegetables can be 20–30% of the total mass, and up to >50% depending on the produce being processed (Sagar et al., 2018; Ayala-Zavala et al., 2010; Juliano et al., 2019).

While there has been extensive research done in the space of transforming food waste into products, one hurdle to industrial commercialization is the need for suitable scale-up technoeconomics. There is also a need for establishment of missing links in food supply chains, to provide enabling infrastructure for food waste transformation. This would involve the creation of entirely new industries to deal with this feedstock of FE&Bs.

2.3.5 Integrated valorization pathways for food excess and by-products

Using an integrated approach to valorize food components that may otherwise be disposed of, it is possible to encompass some key approaches as outlined in Fig. 2.10. However long transport pathways to traditional large factories may not be cost-effective in dealing with the variable level and timing of FE&Bs generation; distributed processing may be more applicable (Almena et al., 2019). This may be possible using a micro-factory approach which deploys mobile modular processing infrastructure with reduced energy usage, as well as regional processing hubs which centralize processing capability close to larger volumes of produce. This is now possible with some renewable energy strategies or a microgrid approach. A mobile processing platform approach was tested in Australia from 2017 through a mobile abattoir strategy, which was developed to support animal welfare and provenance as well as meet food regulation needs. Such an approach could be translated into processing of other raw food materials (Provenir, 2020).

The foundation components of an integrated food valorization system could include the capture and preservation of FE&Bs, which then allows more time to extract and refine components from these materials. Other pathways could include the use of algae as a sink for some FE&Bs, while remaining biomass could be used to nourish animals and soil.

Capture and preserve (CaP). One of the core issues of FLW generation is that there may be no processing or preservation infrastructure close to where it is generated (e.g., on-farm). To overcome this issue, a good approach is to localize capture and preservation of this food excess at the site of production to extend shelf life and prevent spoilage. This can be achieved in a few ways which could be suitable in an agricultural environment or as an annexe to a typical food processing facility.

Drying offers a suitable option as the water removed could be captured, treated, and recycled for use in irrigation on-farm, or it may find applications in symbiotic fermentation processes (e.g., SAFER in Fig. 2.10). The remaining dried food could then be transported at a much lighter weight than the original wet produce, lowering costs of transportation. However, it is important to remember that the costs of drying, particularly for energy, must be weighed against the transportation cost and any financial return from sale of the dried produce. Opportunities exist to supplement energy needs for preservation processes by producing energy resources from food waste through biogas/biodiesel production (Oh et al., 2018; Achinas and Euverink, 2019). Soon we can see how a food waste valorization "ecosystem" can support a myriad of integrated activities to prevent food waste. A good example of alternative reduced

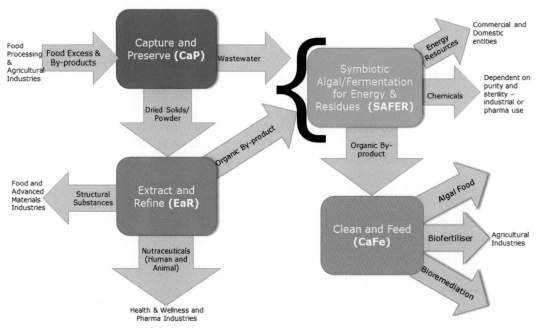

FIG. 2.10

Integrated diversion of food excess and by-products (FE&Bs) for food biomass valorization.

energy technologies is the integration of technologies for hybrid drying technologies, the use of solar, and also microwave-, infrared-, and ultrasonic-assisted drying technologies (Khaing Hnin et al., 2019).

Cold storage and frozen storage are two simple alternatives for food preservation. However, energy usage for these techniques may be significant, particularly for high volumes of produce. In order to combat this, there is scope to develop novel cooling technologies which use sustainable cooling mediums and one approach could also be to use the water removed from dried products. Consideration of energy sources and their associated capital and operating costs for drying must be taken into account for this to be cost-effective. Another approach to CaP FE&Bs could be to use preservatives that are derived from the FE&Bs. As an example, there are many flavonoids present in food, particularly fruit and vegetables. These flavonoids have antimicrobial and antioxidant functionality, and may be sourced from herbs, spices, leaves, and fruit (Gyawali and Ibrahim, 2014). In this way multiple FE&Bs streams could be combined to support FLW prevention.

Extract and refine (EaR). Once food has been captured and preserved, it is possible to build a stockpile of food material with an extended shelf life and process it progressively instead of a period of high intensity processing during the season. For this to be sustainable, the value of any extracted or refined items must be greater than the economic activity that went into its production (Fig. 2.11). Technoeconomics must be determined to decide how far to go with food material valorization.

Some products such as bioactive phytochemicals could be extracted from FE&Bs and include polyphenols, carotenoids, flavonoids, catechins, phenolic acids, and many more (Xiao and Bai, 2019). Some of these compounds have been shown to have potential positive health effects, including against cardiovascular disease, cancer, and metabolic syndrome (Kris-Etherton et al., 2002; Higdon and Frei, 2003; Hannum, 2004; Chattopadhyay et al., 2004; John et al., 2020), while other compounds have been known to have antioxidant or antimicrobial effects (Tongnuanchan and Benjakul, 2014; Shan et al., 2007; Negi, 2012; Falowo et al., 2014). While the focus here has mostly been on bioactive compounds from fruit and vegetables, bioactive peptides are also of interest and can be extracted from animal food sources,

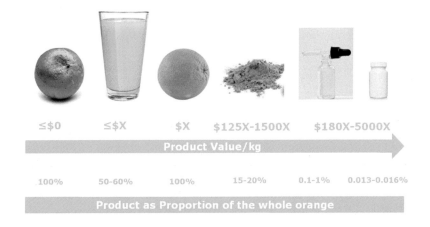

FIG. 2.11

Example of change in value when produce can be transformed into different products, using an orange as the basis. The orange decreases in value if it is rotten or is a juicing orange, while it increases in value if transformed into a higher value product such as powder, oils extracted from the orange, or nutritional supplements.

legumes, and grains (Yoshikawa et al., 2000; Udenigwe and Aluko, 2012; Moller et al., 2008; Li-Chan, 2015; Korhonen and Pihlanto, 2003; Kitts and Weiler, 2003; Gibbs et al., 2004; Erdmann et al., 2008).

There are many methods by which compounds could be extracted from FE&Bs, including solid–liquid extraction, supercritical fluid extraction, ultrasound-assisted extraction, pressurized liquid extraction, and microwave-assisted extraction (Sagar et al., 2018; Silva et al., 2017; Barba et al., 2016). While these processes have been undertaken from a research perspective, the cost to purify and refine the compounds may be significant. One option is to purify these compounds to a concentration <100%, to the point where the revenue obtained is still significantly more than the costs of production, and the process of EaR is not so onerous. Another approach could be to use some FE&Bs fermentation co-products like ethanol and (supercritical) CO_2 to aid extraction. Again, technoeconomics need to be fully determined before pursuing this avenue of transformation.

Symbiotic algal/fermentation for energy and residues (SAFER). As an example of this approach, algae is a sustainable "crop" and is a versatile feedstock for producing food, feed, energy, fertilizer, and chemicals, making it an attractive option for SFS into the future (Marx et al., 2019; Avagyan, 2008; Ronga et al., 2019). A key reason why microalgae is a beneficial choice is that it is a photosynthetic organism, that is also highly productive in producing useful biomass due to its fast growth and adaptability to the environment (Usher et al., 2014). Algae may require less production resources than agricultural crops and can also be "nourished" using FE&Bs, including items such as cheese whey wastewater or spent coffee grounds among others (Pereira et al., 2020; Pandey et al., 2020). Algae also has the potential to clean wastewater, making it an ideal material to produce in symbiosis with processing of FE&Bs. In parallel with algal production, digestion or fermentation transformation of food waste using bacteria, yeasts, fungi to produce a range of products could occur, where the fermentation by-products (e.g., organic matter and CO_2) could be used to nourish algal growth making it an ideal "sink" to process FLW.

Microalgae can convert atmospheric CO_2 into carbohydrates, lipids, and other bioactive metabolites (Khan et al., 2018). As it can act as a miniature "factory," this has led to a growing interest in the use of synthetic biology to improve algal production and its yields (Fabris et al., 2020). A biorefinery approach can help support this, where a range of products could be produced for chemical, pharmaceutical, energy, and material industries (Mirabella et al., 2014; Dahiya et al., 2018). While much of the research on algae has focused on production for biofuel feedstocks, there has been growing focus on algal nutrients as functional foods with some species being a strong source of bioactive molecules, lipids, and minerals (Wells et al., 2017; Plaza et al., 2008; Herrero et al., 2006). Algae has great potential to overcome many sustainability issues regarding the production of food and other substances, however there are potential issues, yet to be quantified or determined, that may be associated with large-scale algae production (Usher et al., 2014).

Clean and feed (CaFe). Integrating the diversion and transformation of FE&Bs as depicted in Fig. 2.10 can better allow development of a SFS that moves toward "zero-waste," where each food biomass component has a purpose which can be realized in different areas. Using this approach, a myriad of industries can be supported via transformation of food biomass that may otherwise be discarded. While the CaP, EaR, and SAFER approaches deal with many FE&Bs components, there may still be remaining biomass by-products that have further use. These materials could easily be used as standard compost, or aid in bioremediation, or may be used as nutrient-rich biofertilizers (Ronga et al., 2019), but they can also be used to clean up land and water, and feed animals, soil, insects, and plants. Potential applications to feed insects can be used for insect protein production (Halloran et al., 2018; Prabhu, 2019). In this way an integrated valorization CE could be established for food.

Clearly, there are no shortage of applications for FE&Bs from the food supply chain, however, full utilization of this resource still needs to improve. Landfill figures in Australia still show food organics wastage in the millions of tons annually, and they typically only quantify what happens at the "end" of the food supply chain (Arcadis, 2019). They do not necessarily take into account discarded food at other points of the food supply chain. Globally FLW accounts for approximately 8% of GHG emissions, approximately 25% of all agricultural water usage, and requires agricultural land the size of China (~9.6 million km^2); it also has an annual market value of almost $USD1 trillion (WRI, 2019b). More than 1 billion metric tons of food per year is never consumed—this loss of resource is highly significant, and the SDGs aim to halve per capita FLW by 2030.

In order to overcome these issues, it is necessary to develop some supply chain "loops" to ensure full use of FE&Bs and move toward "zero waste" underpinned by a CE approach. Some studies, including the authors' own (Fig. 2.10), have already started examining how to integrate a range of pathways for FE&Bs transformation (Kwan et al., 2015; Rana et al., 2020).

2.4 A "zero-waste" approach for sustainable food systems

As we have established, current food systems are in a state of flux, as they tend to produce FE&Bs, which may become FLW. By utilizing appropriate technologies and logistics to ensure that these FE&Bs are captured and preserved, they can be transformed into other products. Previous research has focused on different pathways and technologies for transforming FE&Bs to prevent them becoming waste. This has typically been with a single field focus, either environmental, or energy, or food. The future aim would be to develop several CE "loops" that are embedded in an SFS "ecosystem," so that this becomes the norm.

2.4.1 Transitioning to sustainable food systems

Fig. 2.10 illustrated four main areas of focus to prevent FLW, but it is beneficial to see how this might be put into practice to understand all the material flows not only at a single site, but potentially along the entire supply chain. A possible example of an integrated valorization supply chain for food is outlined in Fig. 2.12. At the top of the diagram, typical linear food systems are shown, with the terminal pathways (red dotted lines) reaching endpoints as FLW. Below the horizontal dashed line, a potential food supply "ecosystem" is outlined. Instead of being FLW, the material is an excess resource that could be transformed into several products, via a range of simultaneously utilized processes. Individual environments globally may drive the choice of combination of processes and would be based on the needs of the local food system.

The processes to be incorporated could include options such as solid–liquid separation, fermentation, extraction (ethanol or SCCO$_2$), digestion processes (anaerobic or aerobic) or black soldier fly production. There are a range of products 1–7 to be produced via these processes, which include the food consumed for nutrition, common to both supply chains, but also other products like food powders, fiber products, food-grade alcohol, animal feed, and nutrition, as well as nutritional supplement products, compost, and also recovered energy storage products (e.g., methane, ethane, hydrogen, acetylene, ethanol, etc.). As well as end-products there are co-products A–D, potentially to be used as reagents, which include collected water and CO$_2$, soil nutrients, and biomass to be recycled into other processes in the "ecosystem." Technoeconomics is needed to determine which is the best balance of material flows in such a system and is the focus of the authors' recent research.

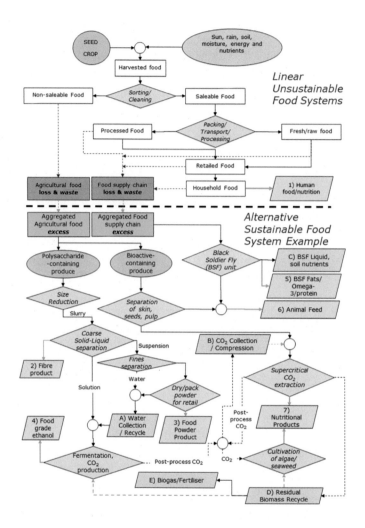

FIG. 2.12

Depiction of transition from an unsustainable food system to an alternative more sustainable food biomass valorization "ecosystem." Above the black dashed line represents linear food systems without valorization, while below the black dashed line depicts a potential "ecosystem" to support sustainable food systems (original work by authors).

2.4.2 System strategies for sustainability

Causative factors for FLW must be overcome with strategies that address the need for social and organizational change to reduce FLW. There are several beneficial interventions outlined by the World Resources Institute (Fig. 2.13; Mylan et al., 2016; WRI, 2019a; Yetkin Özbük and Coşkun, 2020). Many of them require strong leadership in this area, and collaborative partnerships across industry, society, and government. The World Resources Institute report divides the strategies into whole of supply chain approaches, hotspot-specific approaches, and enabling approaches.

Whole of supply chain strategies include the need for development of national strategies and public-private partnerships to reduce FLW, which requires collaboration across the supply chain to ensure success. The "10 × 20 × 30" initiative focuses on collaboration via a voluntary private sector campaign where *10* key players in the food supply chain, commit to setting FLW reduction targets, measuring FLW over time, and acting to reach those targets. Along with potentially leading by example, the drive is for these key players to encourage *20* of their major suppliers to also undertake a similar Target-Measure-Act approach to FLW and aim to halve FLW by 20*30* (WRI, 2019a).

Hotspot-specific strategies focus on collaborative efforts and strengthening of value chains to enable smallholders to prevent FLW generation. This may include development of climate-smart food storage solutions, strong commitment to GHG emissions reduction and a shifting of social norms to ensure that

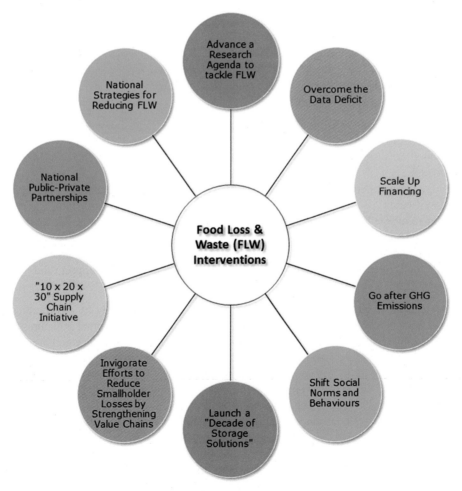

FIG. 2.13

Ten intervention strategies to tackle food loss and waste as outlined by the World Resources Institute categorized into which area needs to be heavily involved in each intervention (WRI, 2019).

from a consumer and industrial perspective FLW and associated environmental impacts are reduced (WRI, 2019a, 2019b,). Finally, enabling approaches include scale-up of financing to reduce FLW and transition to SFS, a commitment to gathering more data outlining the levels of FLW to support measurement of reduction progress, and advancement of a research agenda focusing on the development of next generation strategies and associated technologies to deal with FLW. As part of technical strategies to tackle FLW, food handling, storage, transport, and processing innovations have a role to play.

2.4.3 Sustainable food processing

Food processing has an important role to play in SFS, to support food and nutrition security, population health, and social justice but also to mitigate the effects of climate change and resource depletion on the environment (Tiwari et al., 2014). As the demand and need for sustainable food products grows, the environmental impact of foods and associated labeling may be required to help consumers make informed decisions about the impact of their diet on the environment.

In Australia, as the significant cost of energy has been a key driver toward the use of more sustainable energy sources, from an economic sustainability perspective it has been necessary to move in this direction. As consumers are becoming more interested in the provenance, sustainability, and impact of the food that they eat as well as the associated packaging. Food processors and packaging suppliers must continue to come up with innovations and technologies to meet these demands. Teixeira (Teixeira, 2018) outlined some of the challenges in sustainable food processing which align with the focus of this chapter and include:

1. Development of efficient, environmentally sustainable processing technologies with process integration as a key feature.
2. Minimization of waste including recovery and incorporation of FE&Bs in food processing using CE principles.
3. Technology development for sufficient nutrient delivery from food materials to consumers to support well-being.
4. Development of intelligent, functional packaging which can be edible, biodegradable, or recyclable and supports reduction of FLW.
5. Development and deployment of rapid food safety evaluation techniques.
6. Incorporation of flexibility to rapidly respond to consumer demands and support food personalization.

There are still many emerging food technologies which could potentially be incorporated in SFS of the future (Priyadarshini et al., 2019) and these can be thermal, or nonthermal in nature. In the development of such technologies the focus has typically been on process efficiency and productivity, as well as food quality, safety, and stability. Incorporating these in more sustainable processes would also be beneficial. A growing interest in vacuum, ultraviolet, microwave, and reduced-energy technologies has been of particular focus in the last decade, with a focus on minimal processing to prevent deterioration of nutrients, such as bioactive compounds (Priyadarshini et al., 2019). As a strategy to achieve these goals, Galanakis described a universal recovery strategy for recovery and transformation of food waste and also outlined some of the conventional and emerging technologies to implement the strategy (Fig. 2.14; Galanakis, 2015b). Some of these process options are more sustainable regarding resource and energy usage and are worth exploring further. Use of food waste and by-product streams as heating/cooling mediums and as process reagents is also worth exploring further.

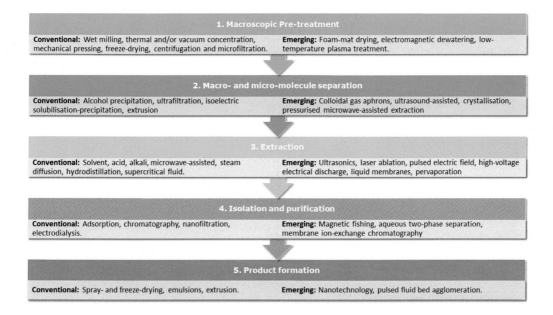

FIG. 2.14

The five-stage universal recovery process as outlined by Galanakis (Galanakis, 2012, 2015a) highlighting both contemporary and emerging technologies for processing food waste. Modified from Galanakis (2015a).

One of the hurdles for adoption of new technologies could be gaining consumer acceptance (Meijer et al., 2021) but others include the costs of implementation and existing industry capability to implement such technologies. Food laws and regulation may need modification, or may prevent implementation, and affect how food supply chains operate into the future (Priyadarshini et al., 2019). Different food commodities require a range of sustainable processing approaches; categories such as dairy, meat, seafood, horticultural produce, and grains are the most widely analyzed. They have some commonalities to be addressed including:

1. A required reduction in generated wastewater (or repurposing of wastewater)
2. A reduction in risk reduction related to emissions (GHG and waste) into the atmosphere and waterways
3. A reduction in solid and liquid pollutants into the environment
4. Increased usage of processing by-products and waste
5. The adoption of novel technologies to increase the recovery of residues from waste
6. Provision of viable labor opportunities, community development, and social welfare.

2.4.4 Sustainable food and beverage initiatives in Australia

There are several examples of initiatives in Australia fostering SFSs. These are outlined in Table 2.1 and encompass horticulture production, beverage supply chains and algae production.

Table 2.1 Example Australian Sustainable Food and Beverage Initiatives

Organisation	Initiative
Sundrop Farms Image used with permission from Sundrop Farms	Sundrop Farms produces tomatoes in a desert environment. This is achieved through a combination of hydroponic systems, desalinated seawater and a solar tower to produce evaporative cooling and steam to control the greenhouse climate. https://www.sundropfarms.com/ https://evokeag.com/evokeag/how-we-can-farm-australias-deserts/
Venus Shell Systems Image used with permission from Venus Shell Systems	Venus Shell Systems is an example of Integrated valorisation pathways as we depict in Figure 9. Wheat processing residue is used in tandem with kelp production with the kelp production component utilising this residue, CO2 and seawater. https://www.venusshellsystems.com.au/ https://www.abc.net.au/radionational/programs/science-show/new-seaweed-processing-plant-opens-in-southern-nsw/12512334
De Bortoli Wines Image used with permission from De Bortoli Wines	De Bortoli Wines has launched new wines in 2020 as part of its 17 Trees range which is underpinned by a range of sustainability initiatives throughout the supply chain including processing aid, water and energy recovery, composting and sustainable packaging and labelling. https://www.debortoli.com.au/17trees https://www.foodprocessing.com.au/content/sustainability/news/17-sustainability-intiative-stories-with-new-wine-range-749815082/
Knott Tara Farm, Barmedman, NSW Image used with permission from Knott Tara Farm. Credit: Benjamin Egge	Growers of Opuntia Ficus – Indica (Sweet prickly pear). It is a very high yielding crop with every part of the plant providing a marketable product similarly to the approach outlined in Figure 12. Cactus products include food, fruit, juice, faux leather, pharmaceutical products and fodder for livestock. *Source: Personal communication with Benjamin Egge*

2.5 The future of sustainable food systems

As the world faces an enormous problem in feeding a growing population with finite resources, the time is now to actively pursue social, economic, and environmental sustainability in food systems. It is important that there are continuing improvements in this area globally so that available resources could be utilized in an effective manner for the sustainability of future generations. This chapter has outlined how resources should be managed to produce, process, distribute, store, and consume food to satisfy nutritional and other consumer needs without compromising social, economic, and environmental sustainability. We have also addressed how adverse impacts on the environment by food supply chains could be reduced by optimizing the resource usage and minimizing the generation of food waste by the rescue of excess food and by-products. It is possible to add value to food waste and by-products through innovative processing technologies by producing a range of food and other products and ingredients with specific nutritional and other functionalities.

Digitalization of not only agriculture, but entire food supply chains through the "smart" use of sensors, automation, and machine learning approaches, can enable better tracking and monitoring of resource usage and movements, thus supporting better informed decision making for resource recovery and postharvest processing. Through the use of new and emerging technologies, distributed manufacturing, and mobile processing techniques, it will be possible to preserve and process food and handle food waste locally. These approaches are likely to improve sustainable food systems.

Future food manufacturing systems are likely to include 3D printing and other novel technologies using less energy from sustainable sources with minimal food waste. Sustainable diets that are nutrient-rich and low in resource usage are high on the agenda for the food industry and society, and the use of more sustainable resources in food systems are being closely examined. Through these initiatives and a collaborative approach on a global scale, we believe that it is possible to achieve a sustainable food system this century and Food Engineers will play a key role in realizing this vision.

Abbreviations

CE	circular economy
FAO	Food and Agriculture Organization of the United Nations
FE&Bs	food excess and by-products
FLW	food loss and waste
GHG	greenhouse gas(es)
LCA	life cycle analysis
MFA	material flow analysis
PIES	people–industry–environment–symbiosis
SCCO$_2$	supercritical carbon dioxide
SDGS	sustainable development goals
SEE	social, economic, and environmental
SEES	social, economic, and environmental sustainability
SFS	sustainable food systems

TEA	technoeconomic analysis
TELCA	technoeconomic and life cycle analysis
WEF	water–energy–food

References

ABARES, 2020. Analysis of Australian food security and the COVID-19 pandemicABARES Insights, Australian Bureau of Agricultural and Resource Economics and Sciences, Canberra, https://doi.org/10.25814/5e953830cb003. https://www.agriculture.gov.au/abares/products/insights/australian-food-security-and-COVID-19.

Achinas, S., Euverink, G.J.W., 2019. Elevated biogas production from the anaerobic co-digestion of farmhouse waste: insight into the process performance and kinetics. Waste Manage. Res. 37 (12), 1240–1249. doi:10.1177/0734242X19873383.

Ajila, C.M., Naidu, K.A., Bhat, S.G., Rao, U.P., 2007. Bioactive compounds and antioxidant potential of mango peel extract. Food Chem. 105 (3), 982–988. doi:10.1016/j.foodchem.2007.04.052.

Allen, T., Prosperi, P., 2016. Modeling sustainable food systems. Environ. Manage 57 (5), 956–975. doi:10.1007/s00267-016-0664-8.

Allievi, F., Antonelli, M., Dembska, K., Principato, L., et al., 2019. Understanding the global food system. In: Valentini, R. et al (Ed.), Achieving the Sustainable Development Goals Through Sustainable Food Systems. Springer International Publishing, Cham, pp. 3–23.

Almena, A., Fryer, P.J., Bakalis, S., Lopez-Quiroga, E., 2019. Centralized and distributed food manufacture: a modeling platform for technological, environmental and economic assessment at different production scales. Sustain. Prod. Consump. 19, 181–193. doi:10.1016/j.spc.2019.03.001.

Anderson, C.A.M., Thorndike, A.N., Lichtenstein, A.H., Van Horn, L., Kris-Etherton, P.M., Foraker, R., Spees, C., 2019. Innovation to create a healthy and sustainable food system: a science advisory from the American Heart Association. Circulation 139 (23), e1025–e1032. doi:10.1161/CIR.0000000000000686.

Arcadis, 2019. National Food Waste Baseline Final Assessment Report. 2019, Australia.

Augustin, M.A., 2019. Creating Value From Edible Vegetable Waste. CSIRO, North Sydney. https://www.horticulture.com.au/globalassets/laserfiche/assets/project-reports/vg15076/vg15076-final-report-complete.pdf.

Avagyan, A.B., 2008. A contribution to global sustainable development: inclusion of microalgae and their biomass in production and bio cycles. Clean Technol. Environ. Policy 10 (4), 313–317. doi:10.1007/s10098-008-0180-5.

Ayala-Zavala, J.F., Rosas-Domínguez, C., Vega-Vega, V., González-Aguilar, G.A., 2010. Antioxidant enrichment and antimicrobial protection of fresh-cut fruits using their own byproducts: looking for integral exploitation. J. Food Sci. 75 (8), R175–R181. doi:10.1111/j.1750-3841.2010.01792.x.

Babich, R., Smith, S., 2010. "Cradle to Grave": an analysis of sustainable food systems in a university setting. J Culinary Sci. Technol. 8 (4), 180–190. doi:10.1080/15428052.2010.535747.

Barba, F.J., Zhu, Z., Koubaa, M., Sant'Ana, A.S., Orlien, V., 2016. Green alternative methods for the extraction of antioxidant bioactive compounds from winery wastes and by-products: a review. Trends Food Sci. Technol. 49, 96–109. doi:10.1016/j.tifs.2016.01.006.

Barbiroli, G., 1997. The dynamics of technology: a methodological framework for techno-economic analyses. In: Julian, N.-R., Martin, R. (Eds.), Theory and Decision Library. Series A, Philosophy and Methodology of the Social Sciences 25. Springer, Netherlands.

Barosh, L., Friel, S., Engelhardt, K., Chan, L., 2014. The cost of a healthy and sustainable diet – who can afford it? Aust N Z J Public Health 38 (1), 7–12. doi:10.1111/1753-6405.12158.

Basso, B., Antle, J., 2020. Digital agriculture to design sustainable agricultural systems. Nature Sustain 3 (4), 254–256. doi:10.1038/s41893-020-0510-0.

Bavas, J., 2018. Strawberry needle tampering crisis: Queensland growers call for 'calm and common sense'. ABC News Australia.

Belguith-Hadriche, O., Ammar, S., Contreras, M.d.M., Fetoui, H., Segura-Carretero, A., El Feki, A., Bouaziz, M., 2017. HPLC-DAD-QTOF-MS profiling of phenolics from leaf extracts of two Tunisian fig cultivars: Potential as a functional food. Biomed. Pharmacother. 89, 185–193. doi:10.1016/j.biopha.2017.02.004.

Bemfeito, C.M., Carneiro, J.D.S., Carvalho, E.E.N., Coli, P.C., Pereira, R.C., Vilas Boas, E.V.B., 2020. Nutritional and functional potential of pumpkin (*Cucurbita moschata*) pulp and pequi (*Caryocar brasiliense Camb*.) peel flours. J. Food Sci. Technol. 57 (10), 3920–3925. doi:10.1007/s13197-020-04590-4. 32904012.

Béné, C., 2020. Resilience of local food systems and links to food security – a review of some important concepts in the context of COVID-19 and other shocks. Food Security 12 (4), 805–822. doi:10.1007/s12571-020-01076-1.

Béné, C., Fanzo, J., Prager, S.D., Achicanoy, H.A., Mapes, B.R., Alvarez Toro, P., Bonilla Cedrez, C., 2020. Global drivers of food system (un)sustainability: A multi-country correlation analysis. PLoS One 15 (4), e0231071. doi:10.1371/journal.pone.0231071.

Bentley, A., 2015. Food Systems: Central-Decentral Networks. A Cultural History of Food in the Modern Age. Bloomsbury Academic, London, United Kingdom, pp. 47–68.

Biel, R., 2016. Sustainable Food Systems. UCL Press, London.

Biel, W., Jaroszewska, A., 2017. The nutritional value of leaves of selected berry species. Sci. Agricola 74 (5), 405–410. doi:10.1590/1678-992X-2016-0314.

Bjørn, A., Hauschild, M.Z., 2018. Cradle to cradle and LCA. In: Hauschild, M.Z., Rosenbaum, R.K., Olsen, S.I. (Eds.), Life Cycle Assessment: Theory and Practice. Springer International Publishing, Cham, pp. 605–631.

Block, D., 2015. Food systems. In: Bentley, A. (Ed.), A Cultural History of Food in the Modern Age. Bloomsbury Academic, London, pp. 47–68.

Boylan, S., Sainsbury, E., Thow, A.M., Degeling, C., Craven, L., Stellmach, D., Gill, T.P., Zhang, Y., 2019. A healthy, sustainable and safe food system: examining the perceptions and role of the Australian policy actor using a Delphi survey. Public Health Nutr. 22 (16), 2921–2930. doi:10.1017/S136898001900185X.

Brears, R.C., 2015. The circular economy and the water-food nexus. Future Food J. Food Agric. Soc. 3 (2), 53–59.

Cai, X., Wallington, K., Shafiee-Jood, M., Marston, L., 2018. Understanding and managing the food-energy-water nexus – opportunities for water resources research. Adv. Water Res. 111, 259–273. doi:10.1016/j.advwatres.2017.11.014.

Çam, M., İçyer, N.C., Erdoğan, F., 2014. Pomegranate peel phenolics: Microencapsulation, storage stability and potential ingredient for functional food development. LWT - Food Sci. Technol. 55 (1), 117–123. doi:10.1016/j.lwt.2013.09.011.

Caraher, M., 2013. A global perspective: towards a healthy, fair and sustainable food system. J. Home Econ. Inst. Australia 20 (3), 9–12.

Carey, R., Krumholz, F., Duignan, K., McConell, K., Browne, J.L., Burns, C., Lawrence, M., 2011. Integrating agriculture and food policy to achieve sustainable peri-urban fruit and vegetable production in Victoria, Australia. J. Agric. Food Syst. Commun. Dev. 1 (3), 181–195. doi:10.5304/jafscd.2011.013.003.

Carino, S., McCartan, J., Barbour, L., 2020. The emerging landscape for sustainable food system education: mapping current higher education opportunities for Australia's future food and nutrition workforce. J. Hunger Environ. Nutr. 15 (2), 273–294. doi:10.1080/19320248.2019.1583621.

Carolan, M., 2018. Food systems. In: Castree, N., Hulme, M., Proctor, J.D. (Eds.), Companion to Environmental Studies. Routledge, London, pp. 192–198.

Chang, Y., Li, G., Yao, Y., Zhang, L., Yu, C., 2016. Quantifying the water-energy-food nexus: current status and trends. Energies 9 (2), 65. doi:10.3390/en9020065.

Chantaro, P., Devahastin, S., Chiewchan, N., 2008. Production of antioxidant high dietary fiber powder from carrot peels. LWT - Food Sci. Technol. 41 (10), 1987–1994. doi:10.1016/j.lwt.2007.11.013.

Charlton, K.E., 2016. Food security, food systems and food sovereignty in the 21st century: a new paradigm required to meet sustainable development goals. Nutr. Dietetics 73 (1), 3–12. doi:10.1111/1747-0080.12264.

Chattopadhyay, I., Biswas, K., Bandyopadhyay, U., Banerjee, R.K., 2004. Turmeric and curcumin: biological actions and medicinal applications. Curr. Sci. 87 (1), 44–53.

Chen, C., Chaudhary, A., Mathys, A., 2020. Nutritional and environmental losses embedded in global food waste. Resour. Conserv. Recycl. 160, 104912. doi:10.1016/j.resconrec.2020.104912.

Conesa, C., Laguarda-Miro, N., Fito, P., Segui, L., 2020. Evaluation of persimmon (*Diospyros kaki* Thunb. cv. Rojo Brillante) industrial residue as a source for value added products. Waste Biomass Valorization 11 (7), 3749–3760. doi:10.1007/s12649-019-00621-0.

Cristóbal, J., Caldeira, C., Corrado, S., Sala, S., 2018. Techno-economic and profitability analysis of food waste biorefineries at European level. Bioresour. Technol. 259, 244–252. doi:10.1016/j.biortech.2018.03.016.

Cucurachi, S., Scherer, L., Guinée, J., Tukker, A., 2019. Life cycle assessment of food systems. One Earth 1 (3), 292–297. doi:10.1016/j.oneear.2019.10.014.

Dahiya, S., Kumar, A.N., Shanthi Sravan, J., Chatterjee, S., Sarkar, O., Mohan, S.V., 2018. Food waste biorefinery: sustainable strategy for circular bioeconomy. Bioresour. Technol. 248 (Part A), 2–12. doi:10.1016/j.biortech.2017.07.176.

Deloitte Access Economics, The Impact of Freight Costs on Australian Farms. 2019, Agrifutures Australia, Australia. https://www.agrifutures.com.au/product/the-impact-of-freight-costs-on-australian-farms/.

Erdmann, K., Cheung, B.W.Y., Schroder, H., 2008. The possible roles of food-derived bioactive peptides in reducing the risk of cardiovascular disease. J. Nutr. Biochem. 19 (10), 643–654. doi:10.1016/j.jnutbio.2007.11.010.

Esposto, A.S., Abbott, M., Juliano, P., 2019. Growing regions through smart specialisation: a methodology for modelling the economic impact of a food processing hub in Australia. Econ. Pap. 38 (2), 114–130. doi:10.1111/1759-3441.12250.

Fabris, M., Abbriano, R.M., Pernice, M., Sutherland, D.L., Commault, A.S., Hall, C.C., Labeeuw, L., McCauley, J.I., Kuzhiuparambil, U., Ray, P., Kahlke, T., Ralph, P.J., 2020. Emerging technologies in algal biotechnology: toward the establishment of a sustainable, algae-based bioeconomy. Front. Plant Sci. 11, 279. doi:10.3389/fpls.2020.00279.

Falowo, A.B., Fayemi, P.O., Muchenje, V., 2014. Natural antioxidants against lipid-protein oxidative deterioration in meat and meat products: A review. Food Res. Int. 64, 171–181. doi:10.1016/j.foodres.2014.06.022.

FAO, 2011. Global Food Losses and Food Waste – Extent, Causes and Prevention. Food and Agriculture Organization of the United Nations, Rome, Italy.

FAO, 2014. Walking the Nexus Talk: Assessing the Water-Energy-Food Nexus in the Context of the Sustainable Energy For All Initiative. Food and Agriculture Organization of the United Nations, Rome, Italy.

FAO, 2018. Sustainable Food Systems: Concept and Framework. Food and Agriculture Organization of the United Nations, Rome, Italy.

Fassio, F., Tecco, N., 2019. Circular economy for food: a systemic interpretation of 40 case histories in the food system in their relationships with SDGs. Systems 7 (3), 43. doi:10.3390/systems7030043.

Fasuyi, A.O., 2006. Nutritional potentials of some tropical vegetable leaf meals: chemical characterization and functional properties. Afr. J. Biotechnol. 5 (1), 049–053.

Foodbank, 2019. Foodbank Hunger Report 2019. https://www.foodbank.org.au/wp-content/uploads/2019/10/Foodbank-Hunger-Report-2019.pdf.

Friel, S., Barosh, L.J., Lawrence, M., 2014. Towards healthy and sustainable food consumption: an Australian case study. Public Health Nutr. 17 (5), 1156–1166. doi:10.1017/S1368980013001523.

Galanakis, C.M., 2012. Recovery of high added-value components from food wastes: conventional, emerging technologies and commercialized applications. Trends Food Sci. Technol. 26 (2), 68–87. doi:10.1016/j.tifs.2012.03.003.

Galanakis, C.M., 2015a. Chapter 3 – the universal recovery strategy. In: Galanakis, C.M. (Ed.), Food Waste Recovery. Academic Press, San Diego, pp. 59–81.

Galanakis, C.M., 2015b. Food Waste Recovery, first ed. Academic Press, San Diego.

Galanakis, C.M., 2018. Sustainable Food Systems from Agriculture to Industry: Improving Production and Processing. Academic Press, doi:10.1016/C2016-0-02157-5.

Garcia-Gonzalez, J., Eakin, H., 2019. What can be: stakeholder perspectives for a sustainable food system. J. Agric. Food Syst. Commun. Dev. 8 (4), 61–82. doi:10.5304/jafscd.2019.084.010.

Gibbs, B.F., Zougman, A., Masse, R., Mulligan, C., 2004. Production and characterization of bioactive peptides from soy hydrolysate and soy-fermented food. Food Res. Int. 37 (2), 123–131. doi:10.1016/j.foodres.2003.09.010.

Gyawali, R., Ibrahim, S.A., 2014. Natural products as antimicrobial agents. Food Control 46, 412–429. doi:10.1016/j.foodcont.2014.05.047.

Halloran, A., Flore, R., Vantomme, P., Roos, N., 2018. Edible Insects in Sustainable Food Systems. Springer International Publishing, Cham, doi:10.1007/978-3-319-74011-9.

Halpern, B.S., Cottrell, R.S., Blanchard, J.L., Bouwman, L., Froehlich, H.E., Gephart, J.A., Jacobsen, N.S., Kuempel, C.D., McIntyre, P.B., Metian, M., Moran, D.D., Nash, K.L., Tobben, J., Williams, D.R., 2019. Opinion: Putting all foods on the same table: achieving sustainable food systems requires full accounting. Proc. Natl. Acad. Sci. U S A 116 (37), 18152–18156. doi:10.1073/pnas.1913308116.

Hannum, S.M., 2004. Potential impact of strawberries on human health: a review of the science. Crit. Rev. Food Sci. Nutr. 44 (1), 1–17. doi:10.1080/10408690490263756.

Herrero, M., Cifuentes, A., Ibanez, E., 2006. Sub- and supercritical fluid extraction of functional ingredients from different natural sources: Plants, food-by-products, algae and microalgae - a review. Food Chem. 98 (1), 136–148. doi:10.1016/j.foodchem.2005.05.058.

Higdon, J.V., Frei, B., 2003. Tea catechins and polyphenols: health effects, metabolism, and antioxidant functions. Crit. Rev. Food Sci. Nutr. 43 (1), 89–143. doi:10.1080/10408690390826464.

HLPE, 2020. Food Security and Nutrition: Building a Global Narrative Towards 2030. A Report by the High Level Panel of Experts on Food Security and Nutrition of the Committee on World Food Security. HLPE, Rome.

Holzeis, C.C., Fears, R., Moughan, P.J., Benton, T.G., Hendriks, S.L., Clegg, M., ter Meulen, V., von Braun, J., 2019. Food systems for delivering nutritious and sustainable diets: perspectives from the global network of science academies. Global Food Secur. 21 (1), 72–76. doi:10.1016/J.GFS.2019.05.002.

Huang, R., Zhang, Y., Shen, S., Zhi, Z., Cheng, H., Chen, S., Ye, X., 2020. Antioxidant and pancreatic lipase inhibitory effects of flavonoids from different citrus peel extracts: an *in vitro* study. Food Chem. 326, 126785. doi:10.1016/j.foodchem.2020.126785.

ISO, *ASISO 14040:2019*, 2019. Environmental Management - Life Cycle Assessment - Principles and Framework. ISO, Geneva, Switzerland.

Jagger, A., 2016. A circular economy: combined food and power projects. Biofuels Bioprod. Bioref. 10 (3), 202–203. doi:10.1002/bbb.1655.

John, O.D., du Preez, R., Panchal, S.K., Brown, L., 2020. Tropical foods as functional foods for metabolic syndrome. Food Funct. 11 (8), 6946–6960. doi:10.1039/d0fo01133a.

Johnson, L.K., Dara Bloom, J., Dunning, R.D., Gunter, C.C., Boyette, M.D., Creamer, N.G., 2019. Farmer harvest decisions and vegetable loss in primary production. Agric. Syst. 176, 102672. doi:10.1016/j.agsy.2019.102672.

Johnson, L.K., Dunning, R.D., Gunter, C.C., Dara Bloom, J., Boyette, M.D., Creamer, N.G., 2018. Field measurement in vegetable crops indicates need for reevaluation of on-farm food loss estimates in North America. Agric. Syst. 167, 136–142. doi:10.1016/j.agsy.2018.09.008.

Juliano, P., Sanguansri, P., Krause, D., Villaddara Gamage, M., Garcia-Flores, R., 2019. Mapping of Australian Fruit and Vegetable Losses Pre-Retail. CSIRO, Melbourne. https://doi.org/10.25919/5d28d8ba0fad5.

Jurgilevich, A., Birge, T., Kentala-Lehtonen, J., Korhonen-Kurki, K., Pietikainen, J., Saikku, L., Schosler, H., 2016. Transition towards Circular Economy in the Food System. Sustainability 8 (1), 69. doi:10.3390/su8010069.

Kath, J., Pembleton, K.G., 2019. A soil temperature decision support tool for agronomic research and management under climate variability: adapting to earlier and more variable planting conditions. Comput. Electron. Agric. 162, 783–792. doi:10.1016/j.compag.2019.05.030.

Khaing Hnin, K., Zhang, M., Mujumdar, A.S., Zhu, Y., 2019. Emerging food drying technologies with energy-saving characteristics: a review. Dry. Technol. 37 (12), 1465–1480. doi:10.1080/07373937.2018.1510417.

Khan, M.I., Shin, J.H., Kim, J.D., 2018. The promising future of microalgae: current status, challenges, and optimization of a sustainable and renewable industry for biofuels, feed, and other products. Microb. Cell Fact. 17 (1), 36. doi:10.1186/s12934-018-0879-x.

Kitts, D.D., Weiler, K., 2003. Bioactive proteins and peptides from food sources. Applications of bioprocesses used in isolation and recovery. Curr. Pharm. Des. 9 (16), 1309–1323. doi:10.2174/1381612033454883.

Korhonen, H., Pihlanto, A., 2003. Food-derived bioactive peptides - opportunities for designing future foods. Curr. Pharm. Des. 9 (16), 1297–1308. doi:10.2174/1381612033454892.

KPMG, 2020. Fighting Food Waste Using the Circular Economy. KPMG and FFW CRC. https://assets.kpmg/content/dam/kpmg/au/pdf/2019/fighting-food-waste-using-the-circular-economy-report.pdf.

KPMG, AgriFood Supply Chain Resilience. 2020, Australia.2020b.

Kris-Etherton, P.M., Hecker, K.D., Bonanome, A., Coval, S.M., Binkoski, A.E., Hilpert, K.F., Griel, A.E., Etherton, T.D., 2002. Bioactive compounds in foods: their role in the prevention of cardiovascular disease and cancer. Am. J. Med. 113 (Suppl 9B), 71S–88S. doi:10.1016/s0002-9343(01)00995-0.

Kwan, T.H., Pleissner, D., Lau, K.Y., Venus, J., Pommeret, A., Lin, C.S.K., 2015. Techno-economic analysis of a food waste valorization process via microalgae cultivation and co-production of plasticizer, lactic acid and animal feed from algal biomass and food waste. Bioresour. Technol. 198, 292–299. doi:10.1016/j.biortech.2015.09.003.

Lawrence, M.A., Baker, P.I., Pulker, C.E., Pollard, C.M., 2019. Sustainable, resilient food systems for healthy diets: the transformation agenda. Public Health Nutr. 22 (16), 2916–2920. doi:10.1017/S1368980019003112.

Li-Chan, E.C.Y., 2015. Bioactive peptides and protein hydrolysates: research trends and challenges for application as nutraceuticals and functional food ingredients. Curr. Opin. Food Sci. 1, 28–37. doi:10.1016/j.cofs.2014.09.005.

Lindberg, R., Lawrence, M., Gold Lisa, C., Friel, S., 2014. Food rescue – an Australian example. Br. Food J. 116 (9), 1478–1489. doi:10.1108/BFJ-01-2014-0053.

Madhusudan, P., Chellukuri, N., Shivakumar, N., 2018. Smart packaging of food for the 21st century – a review with futuristic trends, their feasibility and economics. Mater. Today: Proc. 5 (10, Part 1), 21018–21022. doi:10.1016/j.matpr.2018.06.494.

Maina, S., Kachrimanidou, V., Koutinas, A., 2017. A roadmap towards a circular and sustainable bioeconomy through waste valorization. Curr. Opin. Green Sustain. Chem. 8, 18–23. doi:10.1016/j.cogsc.2017.07.007.

Mak, T.M.W., Xiong, X., Tsang, D.C.W., Yu, I.K.M., Poon, C.S., 2020. Sustainable food waste management towards circular bioeconomy: policy review, limitations and opportunities. Bioresour. Technol. 297, 122497. doi:10.1016/j.biortech.2019.122497.

Marx, U.C., Roles, J., Hankamer, B., 2019. Analyzing the potential of algae farming to provide food, feed and energy from renewable sources. Agric. Res. Technol. Open Access J. 22 (2), 556194. doi:10.19080/ARTOAJ.2019.22.556194.

Meijer, G.W., Lähteenmäki, L., Stadler, R.H., Weiss, J., 2021. Issues surrounding consumer trust and acceptance of existing and emerging food processing technologies. Crit. Rev. Food Sci. Nutr. 61 (1), 97–115. doi:10.1080/10408398.2020.1718597. 32003225.

Mirabella, N., Castellani, V., Sala, S., 2014. Current options for the valorization of food manufacturing waste: a review. J. Cleaner Prod. 65, 28–41. doi:10.1016/j.jclepro.2013.10.051.

Mok, H.-F., Williamson, V.G., Grove, J.R., Burry, K., Barker, S.F., Hamilton, A.J., 2014. Strawberry fields forever? Urban agriculture in developed countries: a review. Agron. Sustainable Dev. 34 (1), 21–43. doi:10.1007/s13593-013-0156-7.

Moller, N.P., Scholz-Ahrens, K.E., Roos, N., Schrezenmeir, J., 2008. Bioactive peptides and proteins from foods: indication for health effects. Eur. J. Nutr. 47 (4), 171–182. doi:10.1007/s00394-008-0710-2.

Morath, S., 2016. IntroductionIn: Morath, S. (Ed.). Farm to Fork: Perspectives on Growing Sustainable Food Systems in the Twenty-First Century. University of Akron Press, Akron, Ohio, pp. 1–4.

Müller, P., Schmid, M., 2019. Intelligent packaging in the food sector: a brief overview. Foods 8 (1), 16. doi:10.3390/foods8010016.

Mylan, J., Holmes, H., Paddock, J., 2016. Re-introducing consumption to the 'circular economy': a sociotechnical analysis of domestic food provisioning. Sustainability 8 (8), 794. doi:10.3390/su8080794.

Namany, S., Al-Ansari, T., Govindan, R., 2019. Sustainable energy, water and food nexus systems: a focused review of decision-making tools for efficient resource management and governance. J. Cleaner Prod. 225, 610–626. doi:10.1016/j.jclepro.2019.03.304.

Nazeam, J.A., Al-Shareef, W.A., Helmy, M.W., El-Haddad, A.E., 2020. Bioassay-guided isolation of potential bioactive constituents from pomegranate agrifood by-product. Food Chem. 326, 126993. doi:10.1016/j.foodchem.2020.126993.

Negi, P.S., 2012. Plant extracts for the control of bacterial growth: efficacy, stability and safety issues for food application. Int. J. Food Microbiol. 156 (1), 7–17. doi:10.1016/j.ijfoodmicro.2012.03.006.

OECD, 2017. The nexus between land, water and energy. The Land-Water-Energy Nexus: Biophysical and Economic Consequences. OECD Publishing, Paris, France, pp. 17–34, doi:10.1787/9789264279360-en.

Oh, J.-I., Lee, J., Lin, K.-Y.A., Kwon, E.E., Tsang, Y.F., 2018. Biogas production from food waste via anaerobic digestion with wood chips. Energy Environ. 29 (8), 1365–1372. doi:10.1177/0958305X18777234.

Olson, M.E., Sankaran, R.P., Fahey, J.W., Grusak, M.A., Odee, D., Nouman, W., 2016. Leaf protein and mineral concentrations across the "miracle tree" genus *Moringa*. PLoS One 11 (7), e0159782. doi:10.1371/journal.pone.0159782.

Oyeyinka, A.T., Oyeyinka, S.A., 2018. *Moringa oleifera* as a food fortificant: recent trends and prospects. J. Saudi Soc. Agric. Sci. 17 (2), 127–136. doi:10.1016/j.jssas.2016.02.002.

Pagotto, M., Halog, A., 2016. Towards a circular economy in Australian agri-food industry an application of input-output oriented approaches for analyzing resource efficiency and competitiveness potential. J. Ind. Ecol. 20 (5), 1176–1186. doi:10.1111/jiec.12373.

Pandey, A., Srivastava, S., Kumar, S., 2020. Development and cost-benefit analysis of a novel process for biofuel production from microalgae using pre-treated high-strength fresh cheese whey wastewater. Environ. Sci. Pollut. Res. 27 (19), 23963–23980. doi:10.1007/s11356-020-08535-4.

Pereira, A.P., Dong, T., Knoshaug, E.P., Nagle, N., Spiller, R., Panczak, B., Chuck, C.J., Pienkos, P.T., 2020. An alternative biorefinery approach to address microalgal seasonality: blending with spent coffee grounds. Sustain. Energy Fuels 4 (7), 3400–3408. doi:10.1039/D0SE00164C.

Plaza, M., Cifuentes, A., Ibanez, E., 2008. In the search of new functional food ingredients from algae. Trends Food Sci. Technol. 19 (1), 31–39. doi:10.1016/j.tifs.2007.07.012.

Poore, J., Nemecek, T., 2018. Reducing food's environmental impacts through producers and consumers. Science 360 (6392), 987–992. doi:10.1126/science.aaq0216.

Prabhu, V.V., 2019. Book review: Edible insects in sustainable food systems. Future Food J. Food Agric. Soc. 7 (2), 4–5, https://www.thefutureoffoodjournal.com/index.php/FOFJ/article/view/282.

Priyadarshini, A., Rajauria, G., O'Donnell, C.P., Tiwari, B.K., 2019. Emerging food processing technologies and factors impacting their industrial adoption. Crit. Rev. Food Sci. Nutr. 59 (19), 3082–3101. doi:10.1080/10408398.2018.1483890.

Provenir. 2020 [cited 2020 27/07/2020]; Available from: https://provenir.com.au/.

Rana, M.S., Bhushan, S., Prajapati, S.K., Preethi, Kavitha, S., 2020. Chapter 15 - Techno-economic analysis and environmental aspects of food waste management. In: Banu, J.R., Kumar, G., Gunasekaran, M., Kavitha, S. (Eds.), Food Waste to Valuable Resources: Applications and Management. Academic Press, London, United Kingdom, pp. 325–342, doi:10.1016/B978-0-12-818353-3.00015-8.

Ronga, D., Biazzi, E., Parati, K., Carminati, D., Carminati, E., Tava, A., 2019. Microalgal biostimulants and biofertilisers in crop productions. Agronomy 9 (4), 192. doi:10.3390/agronomy9040192.

Roy, P., Nei, D., Orikasa, T., Xu, Q., Okadome, H., Nakamura, N., Shiina, T., 2009. A review of life cycle assessment (LCA) on some food products. J. Food Eng. 90 (1), 1–10. doi:10.1016/j.jfoodeng.2008.06.016.

Sagar, N.A., Pareek, S., Sharma, S., Yahia, E.M., Lobo, M.G., 2018. Fruit and vegetable waste: bioactive compounds, their extraction, and possible utilization. Comprehensive Rev. Food Sci. Food Saf. 17 (3), 512–531. doi:10.1111/1541-4337.12330.

Schieber, A., Stintzing, F.C., Carle, R., 2001. By-products of plant food processing as a source of functional compounds - recent developments. Trends Food Sci. Technol. 12 (11), 401–413. doi:10.1016/S0924-2244(02)00012-2.

Shan, B., Cai, Y.Z., Brooks, J.D., Corke, H., 2007. Antibacterial properties and major bioactive components of cinnamon stick (*Cinnamomum burmannii*): activity against foodborne pathogenic bacteria. J. Agric. Food Chem. 55 (14), 5484–5490. doi:10.1021/jf070424d.

Silva, S., Costa, E.M., Calhau, C., Morais, R.M., Pintado, M.E., 2017. Anthocyanin extraction from plant tissues: a review. Crit. Rev. Food Sci. Nutr. 57 (14), 3072–3083. doi:10.1080/10408398.2015.1087963.

Singh, S., Mishra, M.L., Singh, P., 2019. An intervention study on utilization of cauliflower leaf powder (CLP) on the nutritional status of selected rural school children of Kumarganj, Ayodhya, Uttar Pradesh, India. Biomed. Res. 30 (6), 875–881.

Smetana, S.M., Bornkessel, S., Heinz, V., 2019. A path from sustainable nutrition to nutritional sustainability of complex food systems. Front. Nutr. 6, 39. doi:10.3389/fnut.2019.00039.

Suárez, S., Mu, T., Sun, H., Añón, M.C., 2020. Antioxidant activity, nutritional, and phenolic composition of sweet potato leaves as affected by harvesting period. Int. J. Food Prop. 23 (1), 178–188. doi:10.1080/10942912.2020.1716796.

Sun, H., Mu, T., Xi, L., Zhang, M., Chen, J., 2014. Sweet potato (*Ipomoea batatas* L.) leaves as nutritional and functional foods. Food Chem. 156, 380–389. doi:10.1016/j.foodchem.2014.01.079.

Teigiserova, D.A., Hamelin, L., Thomsen, M., 2020. Towards transparent valorization of food surplus, waste and loss: clarifying definitions, food waste hierarchy, and role in the circular economy. Sci. Total Environ. 706, 136033. doi:10.1016/j.scitotenv.2019.136033.

Teixeira, J.A., 2018. Grand challenges in sustainable food processing Front. Sustain. Food Syst. 2 (19), 1–3. doi:10.3389/fsufs.2018.00019.

Thorne, S., 1986. The History of Food Preservation. Rowman & Littlefield Publishers, Maryland, USA.

Tillman, A.M., 2010. 4 - Methodology for life cycle assessment. In: Sonesson, U., Berlin, J., Ziegler, F. (Eds.), Environmental Assessment and Management in the Food Industry: Life Cycle Assessment and Related Approaches. Woodhead Publishing, Cambridge, UK, pp. 59–82. doi:10.1533/9780857090225.2.59.

Tiwari, B.K., Norton, T., Holden, N.M., 2014. Sustainable Food Processing. Wiley-Blackwell, Chichester, England.

Tongnuanchan, P., Benjakul, S., 2014. Essential oils: extraction, bioactivities, and their uses for food preservation. J. Food Sci. 79 (7), R1231–R1249. doi:10.1111/1750-3841.12492.

Torres-León, C., Ramírez-Guzman, N., Londoño-Hernandez, L., Martinez-Medina, G.A., Díaz-Herrera, R., Navarro-Macias, V., Alvarez-Pérez, O.B., Picazo, B., Villarreal-Vázquez, M., Ascacio-Valdes, J., Aguilar, C.N., 2018. Food waste and byproducts: an opportunity to minimize malnutrition and hunger in developing countries. Front. Sustain. Food Syst. 2 (52), 1–77. doi:10.3389/fsufs.2018.00052.

Udenigwe, C.C., Aluko, R.E., 2012. Food protein-derived bioactive peptides: production, processing, and potential health benefits. J. Food Sci. 77 (1), R11–R24. doi: 10.1111/j.1750-3841.2011.02455.x.

UN. *Sustainable Development Goals*. 2020 [cited 2020 31/07/2020] Available from: https://www.un.org/sustainabledevelopment/sustainable-development-goals/.

Usher, P.K., Ross, A.B., Camargo-Valero, M.A., Tomlin, A.S., Gale, W.F., 2014. An overview of the potential environmental impacts of large-scale microalgae cultivation. Biofuels 5 (3), 331–349. doi:10.1080/17597269.2014.913925.

Valentini, R., Sievenpiper, J.L., Antonelli, M., Dembska, K., 2019. Achieving the Sustainable Development Goals Through Sustainable Food Systems. Springer, Cham, doi:10.1007/978-3-030-23969-5.

Walter, A., Finger, R., Huber, R., Buchmann, N., 2017. Opinion: smart farming is key to developing sustainable agriculture. Proc. Natl. Acad. Sci. 114 (24), 6148–6150. doi:10.1073/pnas.1707462114.

Wells, M.L., Potin, P., Craigie, J.S., Raven, J.A., Merchant, S.S., Helliwell, K.E., Smith, A.G., Camire, M.E., Brawley, S.H., 2017. Algae as nutritional and functional food sources: revisiting our understanding. J. Appl. Phycol. 29 (2), 949–982. doi:10.1007/s10811-016-0974-5.

Willett, W., Rockström, J., Loken, B., Springmann, M., Lang, T., Vermeulen, S., Garnett, T., Tilman, D., DeClerck, F., Wood, A., Jonell, M., Clark, M., Gordon, L.J., Fanzo, J., Hawkes, C., Zurayk, R., Rivera, J.A., De Vries, W., Majele Sibanda, L., Afshin, A., Chaudhary, A., Herrero, M., Agustina, R., Branca, F., Lartey, A., Fan, S., Crona, B., Fox, E., Bignet, V., Troell, M., Lindahl, T., Singh, S., Cornell, S.E., Srinath Reddy, K., Narain, S., Nishtar, S., Murray, C.J.L., 2019. Food in the anthropocene: the Eat–Lancet Commission on Healthy Diets from sustainable food systems. Lancet 393 (10170), 447–492. doi:10.1016/S0140-6736(18)31788-4.

WRI, 2019a. Reducing Food Loss and Waste: Ten Interventions to Scale Impact. World Resources Institute, Washington DC, USA.

WRI, 2019b. Reducing Food Loss and Waste: Setting a Global Agenda. World Resources Institute, Washington DC, USA.

Xiao, J., Bai, W., 2019. Bioactive phytochemicals. Crit. Rev. Food Sci. Nutr. 59 (6), 827–829. doi:10.1080/10408398.2019.1601848.

Yetkin Özbük, R.M., Coşkun, A., 2020. Factors affecting food waste at the downstream entities of the supply chain: a critical review. J. Cleaner Prod. 244, 118628. doi:10.1016/j.jclepro.2019.118628.

Yoshikawa, M., Fujita, H., Matoba, N., Takenaka, Y., Yamamoto, T., Yamauchi, R., Tsuruki, H., Takahata, K., 2000. Bioactive peptides derived from food proteins preventing lifestyle-related diseases. Biofactors 12 (1-4), 143–146. doi:10.1002/biof.5520120122.

Yu, J.M., Ahmedna, M., 2013. Functional components of grape pomace: their composition, biological properties and potential applications. Int. J. Food Sci. Technol. 48 (2), 221–237. doi:10.1111/j.1365-2621.2012.03197.x.

Yuana, G., Tian, Y., Wang, S., 2017. A VaR-based optimization model for crop production planning under imprecise uncertainty. J. Intell. Fuzzy Syst. 33 (1), 1–14. doi:10.3233/JIFS-15982.

Zhang, W., Li, X., Jiang, W., 2020. Development of antioxidant chitosan film with banana peels extract and its application as coating in maintaining the storage quality of apple. Int. J. Biol. Macromol. 154, 1205–1214. doi:10.1016/j.ijbiomac.2019.10.275.

Sustainability of the food supply chain; energy, water and waste

3

Dennis R. Heldman

Department of Food Science and Technology, The Ohio State University, Columbus, OH, United States

3.1 Introduction

According to the most recent reports from the United Nations (2019), the world's population is projected to grow from 7.7 billion in 2019 to 8.5 billion in 2030, 9.7 billion by 2050 and 10.9 billion by 2100. This growth in population is occurring at different rates at different geographic locations, but the challenge of meeting the demands for sufficient quantities of high quality, nutritious, and safe food will impact all geographic regions. Sources (Institute for Integrated Economic Research, 2019) emphasize that future growth of world economies will be limited by the availability of energy, and only a continuing transition to renewable energy resources will assist in sustaining growth of all economies. Although the earth's surface displays 1.386×10^9 kilometers of water, only 2.53% or 35×10^6 kilometers is fresh water (Gleick and Palaniappan, 2010). The National Resources Defense Council (NRDC) Gunders (2012) estimated that nearly 40% of raw food produced in the United States and Canada is lost or not consumed. This represents approximately 133 billion pounds of food valued at $161 billion. Finally, Hall et al. (2009) have estimated that food waste contributes to the consumption of freshwater and fossil fuels, and the decomposition of food wastes contributes to the global decline in quality of the environment, as measured in terms of methane and CO_2 emissions. These same authors have estimated that the energy content of US per capita food waste has increased to over 6000 kJ per person per day or a total of 625 trillion kJ per year. Finally, food waste accounts for more than one quarter of all freshwater consumed, and the lost energy is equivalent to nearly 300 million barrels of oil per year.

All statistics and trends suggest that sustainability of the food supply chain depends as much on efficiency of the processes occurring within the food supply chain, as on the availability of raw food materials. The overall objective of this chapter is to develop a better understanding of sustainability as applied to the food supply chain, and the role of process analysis and design in the improvement of sustainability of the food supply. The specific objectives include: (a) a review the energy demand throughout the food supply chain on a process basis and development on strategies for reducing energy requirement for transforming raw materials into consumer products, (b) a discussion of fresh water used by individual processes throughout the food supply chain with consideration of opportunities for reclamation of water, (c) an analysis of food losses and wastes for each process throughout the food supply chain and the exploration of opportunities for reducing food waste, and with specific attention to waste after the products are acquired by the consumer, and (d) to explore applications of process analysis and design to the improvement of the overall efficiency of the food supply chain.

Sustainability has become a widely used term with many implications associated with energy conservation and the impacts on quality of the environment. The World Commission on Environment and Development (WCED, 1987) recommended the following definition of sustainability: "meeting present needs without compromising the ability of future generations to meet their needs." Later, Rosenbaum (1993) suggested that sustainability refers to using methods, systems, and materials that go not deplete resources or harm natural cycles. Finally, Bakshi and Fiskel (2003) recommended the use of the term "resilience" to describe the capacity of a system to tolerate disturbances while retaining structure.

3.2 Status of energy conservation

The production of raw food materials is directly dependent on availability of fossil resources and increases in production have been dependent on these resources since the 1960s (Woods et al., 2010). While the energy required for agricultural production in developed economies is significant, it represents a modest percentage of the total energy for the food supply chain. Heller and Keoleian (2000) estimated that the total energy requirements of the entire US food system (including household preparation) were 1.076×10^{16} kJ in 1995, and production represented just 20% of this total. Uhlin (1997) estimated that the entire chain of food production and consumption in Sweden was one-fifth of the total energy demand for that country. Canning et al. (2010) indicated that energy use in the US food chain increased six times more rapidly than total domestic energy use between 1997 and 2002. About 50% of this increase was attributed to the adoption of more energy-intensive technologies. Heller and Keoleian (2003) estimated the energy consumption for all sectors of the food supply system as illustrated by Fig. 3.1.

Energy requirements of the food system has been investigated and reported for over 50 years. Singh (1978) estimated the energy demand for the entire US food system as 16.5% of all energy consumption in the United States. Although food production and food manufacturing are energy-intensive sectors, it is obvious that food preparation is the most energy-intensive sector of the food supply chain. Food preparation in the home represents more than 32% of total energy demand, and the addition of food service indicates that food preparation is nearly 40% of total energy for the food supply chain in the United States. All statistics suggest many opportunities to improve the sustainability of the food supply chain by improving the energy-efficiency of each sector and the individual processes within each sector. Tassou (2014) indicated that energy efficiencies can be improved through design, optimization, and validation of new and modified processes, including process integration and intensification. The recommendations included the use of advanced sensors and on-line measurement and control of key process parameters. Given the significant contribution of food preparation, a more careful analysis of the processes used in this sector of the food supply chain should be given a priority.

3.3 Fresh water demand

As of June 2015, 70% of water available globally was used to produce raw agricultural products, including foods. Several parameters have evolved to express and describe the use of fresh water in the food supply chain. Schubert and Schuchmann (2011) used the term "water footprint" to define the

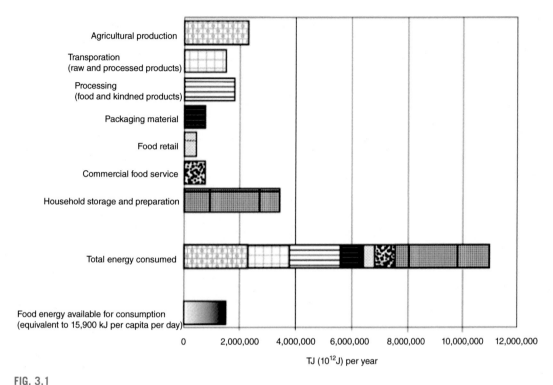

FIG. 3.1

Energy use by various sectors of the US food supply chain [from Heller and Koeleian (2003)].

volume of water required to produce and deliver the goods and services consumed by an individual or a community. Alternatively, "virtual water" has been defined as the volume of water used to produce and deliver a set quantity of a product. Current estimates are that between 4000 and 16,000 L of fresh water per capita per day are available for use by humans. Projections suggest the water consumption by humans will reach 3300 L per capita per day by 2025. These estimates and projections clearly demonstrate that fresh water will soon become a limiting resource, and more intense water conservation will be required for all sectors of society. The magnitudes of virtual water range from 150 to 15,000 L per kg for different food products. Although the significant range of virtual water may suggest the need for focus on selected products, the magnitudes of this parameter confirms that significant quantities of water are used throughout the food supply chain.

The approach to creation of water footprints has been demonstrated by Hoekstra et al. (2011). Although much of the focus has been on animal versus crop agriculture, the technics for development of comparable values is well documented. Although the outcomes indicate that significant portions of the total water footprint occur during production of raw materials, the analysis confirms that all additional sectors of the food supply chain, including food preparation, deserve analysis, and attention. This view becomes even more evident when all auxiliary inputs, such as packaging, are considered in the final water footprint.

3.4 Food waste

A detailed assessment of the flow of food materials through the food supply chain in the United States was completed and published by Heller and Keoleian (2000). The flow of materials illustrated in Fig. 3.2 is typical of the complexity of the food supply chain in a developed economy and provide insights into food losses and wastes.

Gunders (2012) has estimated that the food losses range from 20% for milk to 52% for fruits and vegetables in the United states, Canada, New Zealand, and Australia. As indicated in Fig. 3.3, the same author has illustrated that different magnitudes of food loss occur during different sectors of the food supply chain. The largest magnitudes of waste occur after the products are delivered to the consumer. These wastes range from 12% for meats to as high as 33% for seafood, and the magnitudes are nearly double the magnitude occurring in any other sector of the food supply system. The only exception to this relationship is fruits and vegetables where a similar percent occurs during production. The percentages of food waste occurring after the product is delivered to the consumer are consistent with the approximately 27% food waste suggested by the material flow analysis from Heller and Keoleian (2000)

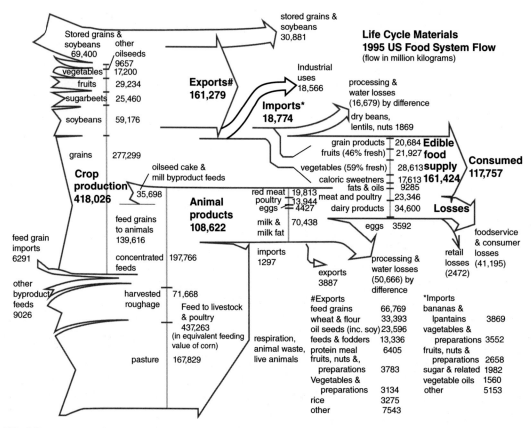

FIG. 3.2

Material flow in the US food supply system (Heller and Koeleian, 2000).

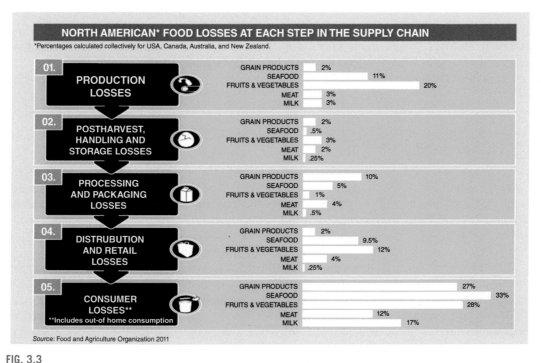

FIG. 3.3

Food losses during various sectors of the food supply chain Gunders (2012).

In many references, food losses are not distinguished from food wastes. According to the FAO (2019), a food loss is defined as "the decrease in edible food mass throughout the part of the supply chain that specifically leads to edible food for human consumption," during production, postharvest and processing. Alternatively, a food waste is food loss that occurs "at the end of the food chain," primarily during retail distribution or preparation for final consumption. These definitions provide an additional dimension to the challenge in reducing food losses and waste. It seems evident that a much greater impact on the total of losses and waste will occur by reducing food waste, and this distinction should be recognized when evaluating the impact of specific processes on food waste. The analysis provides detailed insights into the flow of foods and ingredients from the origin of the raw food material through the entire food chain to the preparation of the food for consumption. As indicated, this food waste is estimated at 27%. This magnitude of this food waste in the United States in 2000 was 4.37×10^{10} kg. Based on the estimates in Fig. 3.1, 1.076×10^{16} kJ of energy were used to produce, manufacture, deliver and prepare foods for consumption. Using the material flow from Fig. 3.2, an estimated 1.1757×10^{11} kg of foods were consumed in the United States. Using these estimates, the energy demand would be 9.152×10^4 kJ/kg of food product. These estimates should be similar to energy demand in other developed economies. It would follow that the energy efficiency of the food supply chain can be improved by reducing food waste. Using the same sources of information, the energy associated with food waste amounts to 4×10^{15} kJ/yr in the United States. This estimate is higher than the 625 trillion kJ/yr by Hall et al. (2009), but both emphasize the significant impact of food waste on efficient use of energy.

The impact of food waste on the efficient use of fresh water is similar. By assuming a modest water footprint of 400 L of water per kg of food delivered to the consumer, the amount of fresh water wasted annually in the United States due to food waste is 17.5 trillion L of water. These invisible wastes of energy and water associated with food waste can be reduced significantly by analysis of the processes used throughout the food supply chain. The proposed analysis should focus on all resources; energy, fresh water, and raw food materials.

3.5 Life cycle assessment

Life cycle assessment (LCA) is an analysis used to quantify the impact of products, processes or activities on resource consumption or environmental burdens. The quantification of these impacts is accomplished through mass and energy balances, and the translation of the outcomes into appropriate quantities. In many cases, the outcome from an LCA is expressed in terms of environmental impacts such as greenhouse gas emissions. An LCA involves the identification of (a) goal and scope, (b) life cycle inventory, (c) life cycle impact assessment, and (d) interpretation.

There have been numerous investigations to using LCA to evaluate environmental impacts of the food supply chain, including food production. Cucurachi et al. (2019) demonstrated that an LCA can capture the environmental impacts of foods, diets, and food production systems. Mogensen et al. (2012) provided a comprehensive review of the environmental impacts associated with food production, manufacturing, and consumption. McCarthy et al. (2015) reported that the assessment method did not influence outcomes, and that methods of food production have greater environmental impacts than the sectors of the food supply chain following production. Several of these investigations suggest that an LCA can provide insight into the relationships between vertically integrated supply chains and environmental performance. Applications of these tools to establish the specific impacts of livestock production systems or other agricultural products on environmental quality have been illustrated (Mogensen et al., 2012; McCarthy et al., 2015; Cucurachi et al., 2019).

An overall goal of LCA must be to assess the impact of the entire food supply system on the environment. In order to achieve this goal, the appropriate LCA tools must be used. Based on the contributions of Halberg et al. (2005) , these tools can focus on either area-based or product-based outcomes. An area-based indicator would include impacts such as nitrate leaching per unit area of land from an animal production facility, while a product-based indicator would estimate the global warming potential per unit of product from the animals in the production facility (Dalgaard, et al. 2007). The applications of these indicators depend on the focus of the environmental impact being explored. The area-based indicator provides insights into the local environment near the production facility. The product-based indicator provides information about the impact of a consumer food product on global environmental quality. These outcomes have different implications in terms of application; one being more useful to the improvement of environmental quality in local communities, while the product-based outcomes provide direction on the impacts on the entire food supply chain on the quality of the environment. The LCA case study on applications to the poultry supply chain conducted by McCarthy et al. (2015) confirmed that poultry production sectors of the food supply chain has greater environmental impacts than sectors involved in delivery the poultry product to retail outlets.

It seems evident from LCA applications to the food supply system (including agricultural production) indicate that the major impacts on environmental quality are created by the production of the raw

Table 3.1 Volumetric product water footprints for Dolmio pasta sauce and Peanut M&M's manufactured and consumed in Australia showing the proportions of blue, green, and gray water. From Ridoutt and Pfister (2010).

	Total water footprint (l)	Blue water (%)	Green water (%)	Gray water (%)
Dolmio pasta sauce (575 g)	202	63.3	10.6	26.1
Peanut M&M's (250 g)	1153	10.9	85.7	3.4

materials and ingredients required for the food supply system. To reduce these impacts will require significant changes in agricultural practices, and a potential reduction in the raw materials and ingredients needed to meet the food supply for an increasing world population. An alternative to increasing agricultural production to meet the increasing the demand for high quality and safe food is to reduce food waste during all processes required to deliver these products to the consumers.

A tool with specific applications to estimating water footprint was proposed by Ridoutt and Huang (2012). These tools for quantification of the water footprint or a water footprint network (WFN) have been proposed to distinguish the source of the water (blue, green, or gray) as illustrated by Ridoutt and Pfister (2010) in Table 3.1. The levels of quantification can be used to evaluate water requirements for specific food products, and the applications extend to evaluation of the water to be used for various processes throughout the food supply system beyond raw product production. The WFN provides an advancement in the estimation of the water footprint for individual processes throughout the food supply system. A Food Waste Processor (WFP) unit has been defined by Lundie and Peters (2005) as a tool for estimating the impact of food waste on the environment. The unit expresses impact of food waste on environmental quality parameters, such as energy usage, climate change, and acidification potential. The water consumption quantified by the WFN can become an important input to a LCA.

3.6 Process analysis and design

The challenges to ensure sustainability of the food supply chain require analysis of individual processes from the handling of raw food materials and ingredients, following production, through to the preparation of the safe and high-quality food products for consumption. More specifically, process analysis and design is a tool to be used with an understanding of the chemistry and microbiology of the food, as needed to ensure product safety and maximum retention of quality attributes associated with each food product. Process analysis and design has a focus on minimizing the energy demand for each process, reducing the fresh-water requirements for each process, and ensuring that all processes in the food supply chain contribute to an overall reduction in food waste. The outcomes from the process analysis and design can ensure that the maximum quantity of the raw food materials and ingredients are transformed into the food and beverage products to be consumed. One definition of process design is the integration of kinetics of reactions within the food product during the process with the transport phenomena associated with the process as needed to achieve the optimum process in terms of product quality (Heldman, 2011a) as illustrated in Fig. 3.4.

The basic concepts associated with applications of process analysis and design for food preservation have evolved over period of more than 50 years. Early versions of the concept maintained a focus on evaluating

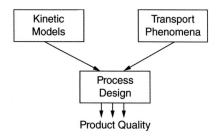

FIG. 3.4

The process analysis and design concept (Heldman, 2011b).

process parameters needed to ensure microbiological safety of the product, while achieving maximum retention of product quality attributes (Teixeira et al., 1969). The same concepts of process analysis and design can be extended to ensuring the most efficient uses of energy and water while minimizing food waste, and the identification of the optimum process parameters. Application of these concepts to processes throughout the food supply chain would contribute to sustainability of the food supply chain.

3.6.1 Applications to process analysis to energy conservation

To accomplish the desired outcomes from process analysis and design of the food supply chain requires a systematic analysis of each process. These processes include handling, preservation, packaging, and distribution of food products between harvest or assembly of the raw material or ingredient and the final preparation of the food for consumption. For each process, the approach involves analysis of energy requirements, as well as alternatives to reduce energy demand for the process. For many food preservation processes, the concepts of thermal energy balances have evolved and become more sophisticated.

The reduction of energy for ultra-high temperature (UHT) pasteurization of a liquid food illustrates process analysis and design. It is well established that the process accomplishes the required product safety and the extension of product shelf-life. The analysis of the process to evaluate the reductions in energy requirements requires a more thorough thermal energy analysis of the process, including a comparison of energy required to increase the temperature to the various levels needed for inactivation of the microbial populations, as well as the thermal energy requirements for cooling the product. A complete analysis would include:

1. Computation or measurement of the thermal energy requirements to increase the product temperature to an established holding temperature. This step would consider the effectiveness and efficiency of different types of heat exchangers, as well as the influence of pressure levels required for the process at higher temperatures.
2. Monitoring of thermal energy loss during the hold step for the process would be recommended. Consideration should be given to various designs of product hold tubes.
3. The third step in the analysis would include computation or measurement of thermal energy requirements for product cooling. The contributions of regeneration (using thermal energy recovered during product cooling as a heating medium for cold product entering the system.

The outcomes from the thermal energy balance become inputs to a process simulation to evaluate the process. Using the kinetics of inactivation for the microbial populations in the raw product, the temperature and time combinations required for product safety and shelf-life can be established. Using the temperature/time combinations for product safety and shelf-life, the process design can focus on product quality retention and energy requirements. The outcomes from the analysis and design reveal the process parameters needed to ensure product safety and shelf-life, while ensuring maximum product quality retention and minimum energy requirements.

An example of process design for minimizing energy was demonstrated by the reduction in refrigeration requirements for individual-quick-freezing (IQF) of food while improving the quality of the frozen product (Heldman and Gorby, 1974). The analysis demonstrated the savings in energy (refrigerant) accomplished by limiting the time-to-freeze to the time required to reduce the mass average temperature of the product to the frozen product storage temperature. The outcome from this process design demonstrated the potential reduction in the refrigeration requirements by approximately 50%. These reductions in refrigeration requirements are accomplished while ensuring the maximum product quality associated with IQF.

The broad applications of process analysis and design for energy reductions to processes within the food supply chain are limited by availability of appropriate data for each of the processes. Although the measurement of data needed for completion of thermal energy balance can be initiated, the lack of appropriate kinetic constants needed for product quality retention in the appropriate temperature ranges associated with product preservation.

3.6.2 Applications of process analysis to water conservation

An analysis of individual processes based on water requirements requires focus on the evaluation all water associated with the process. The basic measurements would include the mass of water in all streams entering the process, as well as the mass of water in all streams leaving the process. The overall goal of the analysis is the identification of the process parameters influencing water requirements, as well as the process parameters with impact on the composition of water streams leaving the process. A complete process analysis with a focus on water conservation would include the following:

1. An estimation or measurement of the water requirements for the process. After defining the process boundaries, all the mass flow rates of water streams entering the process would be identified, followed by measurement or estimation. In addition, the flow rate of all additional streams containing water entering the process would be measured or estimated, along with the amounts of water and other components of those streams.
2. The second step would involve the identification of all streams containing water leaving the process. The analysis of these streams would include measurement or estimation of flow rates, as well as the composition.
3. After establishing the mass flow rates and composition of all input and output water streams, the influence of process parameters on these quantities would be analyzed. Depending on specific goals of the process, the analysis would involve relationships describing the role of water in process. These relationships would estimate composition changes within all the water streams and would focus on water streams leaving the process.
4. The final step of the process analysis is development of a simulation to evaluate the influence of process variables on water requirements for the process. Attention would be given to changes in process parameters needed to reduce the quantities of water required for the process, while ensuring the process meets expectations in terms of product quality and safety, as well as process

efficiency and throughput. The overall goals of product safety, while achieving maximum product quality while minimizing fresh-water requirements would be the ultimate outcomes.

An example of the potential for water conservation is evident when considering the analysis of the water rinse step of the clean-in-place (CIP) process. Fan et al. (2018) have demonstrated that water requirements for cleaning of product-contact surfaces of pipelines could be reduced by up to 50% without impacting the effectiveness of the rinse step. This process analysis required the evaluation of the cleaning process parameters, including the influence of flow rates of rinse water, temperature of the water, and the reactions occurring at the interface between water and product residue on the surface to be cleaned.

3.6.3 Applications to process analysis to waste reduction

The application of process analysis and design to reduction of food waste would be accomplished through analysis of individual processes throughout the food supply chain between production and food consumption. Although there are opportunities to reduce food waste during all processes, the more significant impacts occur during processes near the end of the food supply chain. Many of the processes occurring earlier in the food supply chain influence the potential for food waste during later processes. The analysis must include steps needed to reduce food waste during the process, as well as an evaluation on the impacts of the process on potential food waste occurring at later stages of the food supply chain. The overall goal is to achieve an improvement in the efficiency of converting raw food materials into safe and high-quality foods and beverages with the least food waste.

An analysis of individual processes based on reduction of waste must focus on evaluation all inputs to the process, as well as all waste streams leaving the process. The overall goal of the analysis is the identification of the process parameters needed to reduce food waste and must include process parameters with impact on the waste streams from the process. A complete process analysis with a focus on food waste would include the following:

1. The first step would be estimation or measurement of the raw food materials or ingredients entering the process. Using the same process boundaries as defined for the analysis of energy or water use, the flow rates the various streams carrying components and ingredients would be estimated.
2. The second step of the analysis would be identification of all streams containing product or product components leaving the process and the measurement/estimation of product components in each stream. For most processes, the most important stream will be the primary product to be delivered to the consumer. The analysis of all streams leaving the process must include measurement of composition to be used in the mass balance for the process. A difficult part of this step would be estimating impacts of the process on anticipated food waste occurring during later stages in the food supply chain.
3. After identification of all input and output product streams, the influence of process parameters on those streams is evaluated. The critical part of this step is to ensure
4. Careful analysis of each stream as needed for a mass balance for all product entering and leaving the process. The influence of process parameters on product residuals in waste streams is essential, as well as the parameters needed to ensure minimum product waste are important.
5. The final step in the process analysis to reduce food waste must include a focus on the impact of the process on food waste occurring at later steps in the supply chain. Although these may be the

more challenging parts of the analysis and involve analysis of the food supply chain following the process, the impact on food waste must include these dimensions. Ultimately, the influence of process parameters on food waste would be determined to reduce waste to a minimum.

The concepts of process analysis and design apply equally to minimizing food waste. Maintaining food product safety and shelf-life must be ensured, as well as the quality attributes of the product. Any process parameter with influence on safety and shelf-life must be identified and operated over a range of magnitudes. Any impacts on product quality should be quantified by established kinetic parameters, followed by simulation of the influence of process parameters on food waste. The outcomes from the process analysis and design are the process parameters to ensure product safety and shelf-life, and with maximum quality retention and with minimum food waste.

Most often, packaging is the final process in product manufacturing, and can impact product quality in both positive and negative ways during the storage and distribution stages of the food supply chain. Packaging processes and packaging materials have direct impact on the shelf-life of the product, and severe limits to the shelf-life result in food waste. An analysis of the packaging process must include an analysis of the relationship of the packaging material to the product, with emphasis on the extent the packaging material limits the reactions causing changes in product quality during storage and distribution. The potential for thermal abuse of product during storage and distribution and the subsequent impacts on food waste should be evaluated as part of the process analysis for the packaging process. The role of the packaging material as an interface between the environment and the product during all stages of the food supply chain following packaging must be considered.

A less obvious impact on food waste are date labels or code dates attached to food packages to communicate shelf-life of the product. According to Leib and Gunders (2013), code dates cause premature discard of acceptable quality food products and may represent as much as 40% of the food waste occurring after the product is purchased by the consumer. Since the code date is assigned to the product as part of the packaging process, the process analysis must include an evaluation of the code date and the potential for encouraging food waste (Fig. 3.5).

A process analysis must account for all parameters influencing the product shelf-life during all stages and processes encountered by the product after being placed in the package. Ultimately, process analysis and design should generate information needed for creation of real-time shelf-life indicators to communicate actual shelf-life of the product at all stages of storage and distribution and reduce food waste encouraged by the packaging process.

3.7 Conclusions and recommendations

The challenges to sustainability of the food supply chain can be addressed through applications of process analysis and design. This approach applies the concepts of mass and energy balances to individual processes in the food supply chain and reduces the demand for energy and water, while reducing food waste. The application of these concepts will identify the process parameters to minimize the energy and water requirement for the process, while reducing food waste to a minimum. These challenges are addressed while ensuring product safety and the maximum retention of product quality attributes. Many of the tools associated with LCA contribute to process analysis and design.

In conclusion, the published literature confirms that the food supply chain is energy-intensive, and there are opportunities for energy conservation through application of process analysis and design to

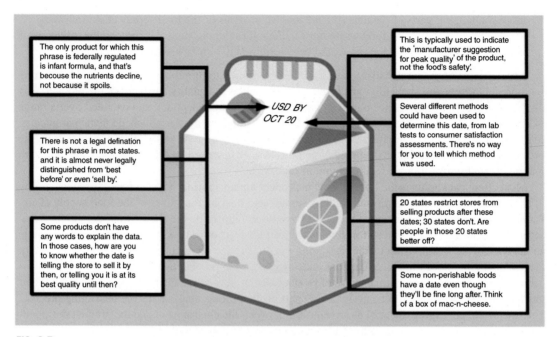

The only product for which this phrase is federally regulated is infant formula, and that's becouse the nutrients decline, not because it spoils.

This is typically used to indicate the 'manufacturer suggestion for peak quality' of the product, not the food's safety'.

USD BY
OCT 20

There is not a legal defination for this phrase in most states. and it is almost never legally distinguished from 'best before' or even 'sell by'.

Several different methods could have been used to determine this date, from lab tests to consumer satisfaction assessments. There's no way for you to tell which method was used.

Some products don't have any words to explain the data. In those cases, how are you to know whether the date is telling the store to sell it by then, or telling you it is at its best quality until then?

20 states restrict stores from selling products after these dates; 30 states don't. Are people in those 20 states better off?

Some non-perishable foods have a date even though they'll be fine long after. Think of a box of mac-n-cheese.

FIG. 3.5

An example of code date assigned to food packages (Leib and Gunders, 2013).

individual processes throughout the food supply chain. The published literature provides additional insights into the significant quantities of fresh water used by the food supply chain. Although water footprints indicate that water demands vary from one product and/or process to another, it is evident that significant overall reductions in water requirements are possible through applications of the concepts of process analysis and design. The application to individual processes throughout the food supply chain will contribute to the overall strategy. Food waste is the most obvious indicator of an inefficient food supply chain. In developed economies, food waste of 25–30% occur after food products are delivered to the final customer. Food wastes at the conclusion of the food supply chain are a significant contributor to the magnitude of energy and water utilized by the entire food supply chain. The reduction of food waste is challenging but can be addressed through applications of the concepts of process analysis and design, including the introduction of sensors to reduce food wastes by the consumer. The process analysis for an individual process must anticipate the impact of the process on food waste occurring after the product is delivered to the consumer. The contributions of food waste to overall energy and water requirements requires unique considerations and can be addressed through applications of process analysis and design.

Acknowledgments

The author wishes to acknowledge the support of the Dale A. Seiberling Endowment during development of the concepts presented in this manuscript.

References

Bakshi, B.R., Fisksel, J., 2003. The quest for sustainability: challenges for process systems engineering. AIChE J. 49 (6), 1350–1359.

Canning, P., Ainsley,C., Huang,S., Polenske, K.R., Walters, A., 2010. Energy use in the U.S. food system. USDA-ERS Report #94. Washington, DC.

Cacurachi, S., Scherer, L., Guinee, J., Tukker, A., 2019. Life cycle assessment of food systems. One Earth 1 (3), 292–297.

Dalgaard, R., Halberg, N., Hermansem, J.E., 2007. Danish pork production: an envirnomental assessment. DJF Anim. Sci. No. 82. https://www.researchgate.net/publication/292309489 ISBN:87-91949-26-2.

Fan, M., Phinney, D.M., Heldman, D.R., 2018. The impact of clean-in-place parameters on rinse water effectiveness and efficiency. J. Food Eng. 222, 276–283.

Food and Agriculture Organization of the United Nations (FAO), 2019. International Fund for Agricultural Development (IFAD), United Nations Children's Fund (UNICEF), World Food Programme (WFP) & World Health Organization (WHO). 2019. The State of Food Security and Nutrition in the World 2019. Safeguarding against economic slowdowns and downturns. Rome, FAO. Available at: http://www.fao.org/3/ca5162en/ca5162en.pdf.

Gleick, P.H., Palaniappan, M., 2010. Peak water limits to freshwater withdrawal and use. Proc. Natl. Acad. Sci. 107 (25), 11155–11162 https://doi.org/10.1073/pnas.1004812107.

Gunders, D., 2012. Wasted: How America is Losing Up to 40 Percent of Its food from Farm to Fork to Landfill. Natural Resources Defense Council, Washington, DC NRDC Issue Paper 12-06-P.

Halberg, N., van der Werf, H., Basset-Mens, C., Dalgaard, R., de Boer, I.J.M., 2005. Environmental assessment tools for the evaluation and improvement of European livestock production systems. Livest. Prod. Sci. 96 (1), 33–50 https://doi.org/10.1016/j.livprodsci.2005.05.013.

Hall, K.D., Guo, J., Dore, M., Chow, C.C., 2009. The progressive increase of food waste in America and its environmental impact. PLoS One 4 (11), e7940 https://doi.org/10.1371/journal.pone.0007940.

Heldman, Dennis R., 2011. Food Process Preservation Design. Academic Press/Elsevier, Inc, Burlington, MA.

Heldman, Dennis R., 2011. Food preservation process design. Food process engineering in a changing world, ICEF11 Proceedings, 685–689.

Heldman, D.R., Gorby, D.P., 1974. Computer Simulation of Ice Cream Freezing; Recent Studies of the Thermophysical Properties of Foodstuffs. Int. Inst. Refr., Paris, p. 29.

Heller, M.C., Koeleian, G.A., 2000. Life Cycle-Based Sustainability Indicators for Assessment of the U.S. Food System. The Center for Sustainable Systems Report no. CSS00-04.

Heller, M.C., Koeleian, G.A., 2003. Assessing the sustainability of the U.S. food system: a life cycle perspective. Agric. Syst. 76, 1007–1041.

Hoekstra, A.Y., Chapagain, A.K., Aldaya, M.M., Mekonnen, M.M., 2011. The Water Footprint Manual. EARTHSCAN, London; Washington, DC.

Institute for Integrated Economic Research, 2019. Energy and resource limits are now inhibiting growth. Energy Stuff. https://www.energyandstuff.org.

Leib E.B., Gunders, D., 2013. The Dating Game: How Confusing Food Date Labels Lead to Food Waste in America. NRDC Report.

Lundie, S., Peters, G.M., 2005. Life cycle assessment of food waste management options. J. Cleaner Prod. 13 (3), 275–286.

McCarthy, D., Matopoulos, A., Davis, P., 2015. Life cycle assessment in the food supply chain: a case study. Int J Logistics: Res. Applic. 18 (2), 140–154.

Mogensen, L., Hermansen, J., Halberg, N., Dalgaard, R., Vis, J., Smith, B., 2012. Life cycle assessment across the food supply chain. In: Baldwin, C. (Ed.), Sustainability in the Food Industry. Wiley-Blackwell, Chapter 5, pp. 115–144, ISBN:9780813808468.

Ridoutt, B.G., Huang, J., 2012. Environmental relevance—the key to understanding water footprints. Proc. Natl. Acad. Sci. U S A 109, E1424.

Ridoutt, B.G., Pfister, S., 2010. A revised approach to water footprinting to make transparent the impacts of consumption and production on global freshwater scarcity. Global Environ. Change. 20 (1), 113–120. doi:10.1016/j.gloenvcha.2009.08.003.

Rosenbaum, M., 1993. Sustainable design strategies. Solar Today 7, 2.

Schubert, Helmar, Schuchmann, H.P., 2011. Food process engineering and innovation in a fast-changing world, Proceedings of the 11th International Congress on Engineering and Food. Athens Greece, 1, 1–2 May 22-26.

Singh, R.P., 1978. Energy accounting in food process operations. Food Technol. 32 (4), 40–46.

Tassou, S.A., 2014. Energy demand and reduction opportunities in the UK food chain. Proc. Inst. Civil Eng. – Energy 167 (3), 162–170. https://doi.org/10.1680/ener.14.00014.

Teixeira, A.A., Dixson, J.R., Zahradnik, J.W, Zinsmeinster, G.E., 1969. Computer optimization of nutrient retention in thermal processing of conduction-heating foods. Food Technol. 23, 137.

Uhlin, H.-E., 1997. Why energy productivity is increasing: an I-O analysis of Swedish agriculture. Agric. Syst. 56 (4), 443–465.

United Nations, Department of Economic and Social Affairs, Populations Division, 2019. World Population Highlights 2019. United Nations, Department of Economic and Social Affairs, Populations Division. Highlights ST/ESA/SER.S/423.

WCED, 1987. Report of the World Commission on Environment and Development: Our Common Future. United Nations General Assembly, New York.

Woods, W., Williams, A., Hughes, J.K., Black, M., Murphy, R., 2010. Energy and the food system. Philos. Trans. R. Soc. Lond. Biol. Sci. 365 (1554), 2991–3006. doi:10.1098/rstb.2010.0172.

Further reading

Food and Agriculture Organization of the United Nations, 2021. Food Waste|Technical Platform on the Measurement and Reduction of Food Loss and Waste. Food and Agriculture Organization of the United Nations. Available at: http://www.fao.org/platform-food-loss-waste/food-waste/definition/en/.

Recovery of high-value compounds from food by-products

4

Jiadai Wu[a,b], Katherine Blackshaw[a,b], Junlae Cho[a,b], Nooshin Koolaji[a,b], Jimmy Yun[c,d], Aaron Schindeler[a,b,e], Peter Valtchev[a,b], Fariba Dehghani[a,b]

[a]*School of Chemical and Biomolecular Engineering, The University of Sydney, Sydney, NSW, Australia*
[b]*Centre for Advanced Food Engineering, The University of Sydney, Sydney, NSW, Australia*
[c]*School of Chemical Engineering, The University of New South Wales, Sydney, NSW, Australia*
[d]*Qingdao International Academician Park Research Institute, Qingdao, Shandong, PR China*
[e]*Bioengineering and Molecular Medicine Laboratory, The Children's Hospital at Westmead and the Westmead Institute for Medical Research, Westmead, NSW, Australia*

4.1 Introduction

Food production has increased in the last few decades to meet the demand of a growing world population. This coincided with an exponential increase in food by-products and waste, particularly resulting from food processing (Sagar et al., 2018). A 2017 report indicated that Australia generates 20 billion dollars' worth of food waste annually (Australian Government. National Food Waste Strategy, 2017). The estimated 1.6 billion tons of food by-products represents a global problem with significant economic costs (approximately 750 billion USD) and substantial environmental impact (Ran et al., 2019; FAO, 2013). To address this issue, there is an increasing impetus to utilize food waste as a valuable source of nutritional and functional compounds.

High-value compounds recovered from food processing waste streams can be classified as macromolecules such as proteins and polysaccharides, or small molecule compounds such as vitamins, antioxidants, and micronutrients (Fig. 4.1; Garcia-Amezquita et al., 2018; Xiong et al., 2019). The nature of food by-products determines what target compounds can be recovered, as well as the best strategy for their recovery. For instance, dietary fiber is commonly recovered from plant-based food by-products such as wheat bran (Yan et al., 2019), while proteins such as collagen are usually recovered from food waste of animal origin (Zanjani et al., 2018).

Common techniques that are used for the recovery of high-value compounds can be summarized in five different categories: drying, homogenization and/or particle size reduction, extraction, isolation, purification, and formation/stabilization (Galanakis, 2012). Emerging technologies are implemented to increase the yield and purity of recovered compounds. Examples include superfine grinding technologies, supercritical fluid extraction, microwave, or ultrasound-assisted extraction, pulsed electric field and plasma assisted extraction, and extraction with ionic liquids (Sagar et al., 2018; Ran et al., 2019). This has led to new opportunities to utilize unconventional food by-products, such as lupin coats (Zhong et al., 2019), coffee sliver skins (Costa et al., 2018), and kimchi by-products (Heo et al., 2019).

Food Engineering Innovations Across the Food Supply Chain. DOI: https://doi.org/10.1016/B978-0-12-821292-9.00002-9

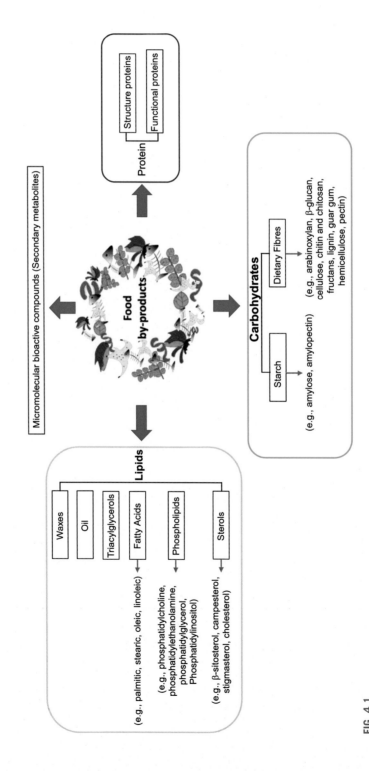

FIG. 4.1

Common high-value bioactive compounds and macromolecules with potential for recovery from food by-products.

This review provides a brief overview on the recovery of high-value-added compounds from food by-products extracted from both plant and animal sources.

4.2 Natural compounds recovered from plant-based by-products

The plant by-products generated during postharvest processing are responsible for 1.4 billion tons of the processing food waste annually, out of the 1.6 billion tons of total wastage (FAO, 2013). The type and composition of plant-based wastage can vary between countries and regions due to multiple factors, such as dietary habits, environment, and the levels of local agricultural development (FAO, 2013). Globally, cereals and vegetables are the major source of food loss (71%), followed by fruit (25%) and oilseed crops and legumes (4%) (Garcia-Amezquita et al., 2018). The amounts of plant-based foods and their by-products are summarized in Table 4.1 based on the most recent available data.

Cereals are a major energy source for humans. The members of the nine Gramineae family species are wheat (*Triticum*), rye (*Secale*), barley (*Hordeum*), oat (*Avena*), rice (*Oryza*), millet (*Pennisetum*), corn (*Zea*), sorghum (*Sorghum*), and triticale (Papageorgiou and Skendi, 2018). The annual production of grain reached 2.58 billion tons in 2016 with associated 1.3 billion tons by-products as listed in Table 4.1 (Papageorgiou and Skendi, 2018). Cereal by-products include bran, germ, and multiple outer layers, are widely used as animal feed (Hemdane et al., 2016). Due to the high concentration of nutrients in cereal by-products, numerous studies have suggested their greater inclusion as dietary supplements. Wheat bran, for example, is used as an additive to increase the fiber content in the production of bread. (Hemdane et al., 2016). Fruit and vegetable by-products generated from juice, canning, and beverage processing are other major alternative sources of high-value-added bioactive compounds. Bioactive compounds with antioxidant, anticancer, and anti-inflammatory properties are widely recovered (Zhou et al., 2019; Santos-Sánchez et al., 2019; Xu et al., 2017; Wilson et al., 2017; Baiano and Del Nobile, 2016).

While increased utilization of traditional cereal by-products has the potential for major global impact, this chapter focuses on more recent research efforts to purify specific compounds such as antioxidants, dietary fiber, proteins, and lipids from alternative plant and animal waste streams.

4.2.1 Antioxidants

Antioxidants are a large group of bioactive compounds that can neutralize free radicals, which are associated with oxidative stress, aging, and the etiology of many chronic diseases (Santos-Sánchez et al., 2019). Compared to endogenous human antioxidants such as glutathione, naturally occurring plant exogenous antioxidants vary significantly in their pharmacokinetic properties (Hwang and Choi, 2015; Lourenco et al., 2019). In addition to having well-recognized health benefits, plant-based antioxidants have also been widely used as natural food preservatives and colorants in the food industry (Xu et al., 2017; Wilson et al., 2017; Khoo et al., 2017). The demand for antioxidants has increased over the last few decades, which encourages the development of new approaches to obtain these antioxidants from low-cost natural resources.

Numerous antioxidant molecules have been identified in the plant by-products, which are grouped based on their solubility in water as hydrophilicity or lipophilicity (Roselló-Soto et al., 2015). The classification of the most well-documented antioxidant molecules is shown in Fig. 4.2, and their mechanism of action in Table 4.2. Generally, hydrophilic antioxidants tend to donate a hydrogen atom to neutralize free radicals. Vitamin C and polyphenols are the abundant water-soluble antioxidants found in plants waste streams (Baiano and Del Nobile, 2016; Roselló-Soto et al., 2015). Carotenoids, lycopene, beta-carotene, and vitamin E are the most widely recovered lipophilic antioxidants that act as quenchers for singlet oxygen (Baiano and Del Nobile, 2016; Roselló-Soto et al., 2015). The physicochemical properties of the targeted plant antioxidants suggested different recovery strategies.

Table 4.1 Annual global production and nature of plant-based by-products.

Commodity	Annual Australian production and by-product			Annual global Production and by-product				Refs.
	Production (tons)	By-product production (tons)	Year	Production (million metric tons)	Year data obtained	By-product production (million metric tons)	Nature of by-products	
Apple	310 k	56 K	2019	86	2020	25.8	Pomace, peel, and seeds	(Sagar et al., 2018; Garcia-Amezquita et al., 2018; Gupta and Joshi, 2000; Oreopoulou and Tzia, 2007; FAOSTAT, 2021)
Banana	372 k	100 k	2018	114	2017	34	Peel	(Sagar et al., 2018; Garcia-Amezquita et al., 2018; FAOSTAT, 2021; Schieber et al., 2001; Safari, 2018)
Beetroot	36 k	10 k	2018	270	2014	61	Pomace, pulp, lime cake	(Garcia-Amezquita et al., 2018; FAOSTAT, 2021; Schieber et al., 2001)
Cereal	-	-	-	2577.85	2017	1330.02	Bran, germ, and seed coat	(Papageorgiou and Skendi, 2018; FAO-AMIS 2017)
Carrot	332 k	93 k	2018	39	2017	23	Pomace, peel	(Garcia-Amezquita et al., 2018; FAOSTAT, 2021)
Citrus	744 k	379 k	2018	124.25	2016	62*	Rag, peel, and seeds	(Sagar et al., 2018; Gupta and Joshi, 2000; FAO, 2017)
Durian	-	-	-	0.9	2018	0.54–0.72	Skin, seeds	(Sagar et al., 2018; Safari, 2018)
Grape	208 k	130 k	2019	75	2016	15	Skin, Stem and Seeds	(Sagar et al., 2018; Garcia-Amezquita et al., 2018; Gupta and Joshi, 2000; Schieber et al., 2001; Food and Agriculture Organization of the United Nations and International Organisation of, 2016)
Mango	74 k	29 k	2018	45	2017	27	Peel, stone	(Sagar et al., 2018; Garcia-Amezquita et al., 2018; Ajila et al., 2008; Cheok et al., 2018; Altendorf, 2019)
Pineapple	70 k	20 k	2019	25	2017	8	Core, skin	(Sagar et al., 2018; Garcia-Amezquita et al., 2018; Cheok et al., 2018; Altendorf, 2019)
Potato	1380 k	238 k	2019	423	2019	153	Peel	(Sagar et al., 2018; Garcia-Amezquita et al., 2018; Singh, 2020)
Tomato	469 k	153 k	2019	182	2018	45.5*	Core, skin, seed	(Sagar et al., 2018; Garcia-Amezquita et al., 2018; Brasesco, 2019)

*Estimated based on the percentage obtained from the reference.

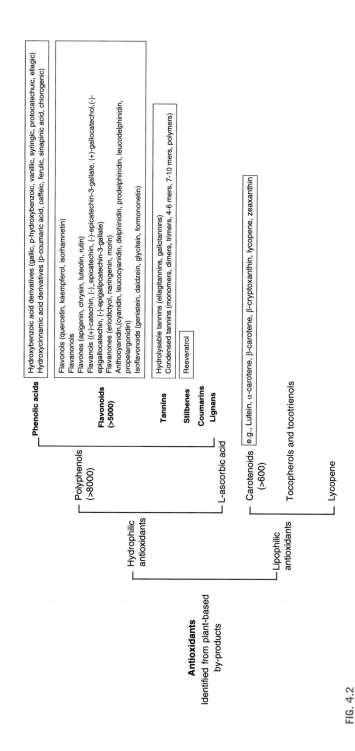

FIG. 4.2

Antioxidants identified in plant-based by-products.

Table 4.2 Classes of antioxidants of plant origin and their mechanisms of action (Sagar et al., 2018; Santos-Sánchez et al., 2019; Xu et al., 2017; Wilson et al., 2017; Baiano and Del Nobile, 2016; Lobo et al., 2010).

Antioxidant	Group	Mechanism of action
Anthocyanins and anthocyanidin	Hydrosoluble	Metal chelation Donation of hydrogen to oxygen radicals
Ascorbic acid (vitamin C)	Hydrosoluble	Chain-breaking scavenger for peroxyl radicals Donation of hydrogen atom to the vitamin E-derived phenolate radicals
Flavonoids	Hydrosoluble	Scavenging free radicals and lipid peroxide radicals Metal chelation
Polyphenolics	Hydrosoluble	Donation of hydrogen atom to free radicals
Carotenoids (vitamin A)	Liposoluble	Quenching of singlet oxygen
Tocopherols and tocotrienols (vitamin E)	Liposoluble	Quenching of singlet oxygen

Antioxidants from plant by-products are commonly recovered via extraction, purification, and drying as summarized in several reviews (Ran et al., 2019; Roselló-Soto et al., 2015; Dzah et al., 2020; Majerska et al., 2019). A selection of recent improvements or optimizations focusing on these procedures is listed in Table 4.3. There is a trend to reduce and replace the usage of toxic organic solvents, which aims to overcome the main drawback of conventional extraction systems. Microwave or ultrasound assisted extractions are the most popular techniques used to reduce solvent usage and shorten the processing time by increasing extraction efficiency as seen in Table 4.3 (Ran et al., 2019). The use of supercritical/subcritical fluids is another approach that replaces toxic solvents, generally with benign substances such as CO_2 or water (Bojanić et al., 2019; Daraee et al., 2019; Putra, 2021; Santana et al., 2019).

Identifying antioxidants from new plants and their by-products is another emerging trend in the field. Antioxidants can now be purified from coffee silver skin (Regazzoni et al., 2016) and chili plants (Iqbal et al., 2017). Another notable trend is the implementation of mathematical modeling to predict the impact of processing parameters, such as pressure and temperature, on the yield of desirable antioxidants (Roselló-Soto et al., 2015; Bojanić et al., 2019; Daraee et al., 2019; Putra, 2021; Santana et al., 2019; Belwal et al., 2019). This aims to facilitate the optimization and scale-up of the recovery process in the future.

4.2.1.1 Vitamin C
Vitamin C, the L-enantiomer of ascorbic acid, is a natural metabolite with the strong antioxidant activity that is present in all living parts of plants (Baiano and Del Nobile, 2016). Citrus fruit by-products, such as peels or pomace, have been found to contain a high amount of vitamin C and are considered promising sources of natural ascorbic acid for the food industry (Quirós et al., 2019). One latest study suggested to reuse the vitamin C in the citrus peels and pomace by directly drying and powdering the by-products under optimized condition. These final product could be used as a new type of food supplement without significantly compromising the high vitamin C content compared with the fresh fruits (>2mg/g) (Bozkir et al., 2020).

Conventionally, ethanol has been used to recover vitamin C from plant by-products, yet the instability of vitamin C remained a challenge for this method (Baiano and Del Nobile, 2016; Nunes et al., 2018; Nizori et al., 2018). Encapsulation of vitamin C with edible proteins, such as soybean protein, has

Table 4.3 **Recent lab-scale improvements for the recovery of antioxidants from plant-based by-products.**

Target compounds	Source	Method		Improvement	Refs.
		Approaches	**Solvent**		
Anthocyanin	Bilberry press cake	Microwave hydrolysis for 30 min at 140 °C	Water	• Extracted multiple compounds with single step • Replaced organic solvent with water • Highest conversion rate (44.2%) achieved for 30 min treatment • Pilot scale (3000 ml) established and tested	(Zhou et al., 2018)
	Blackberry, blueberry and *Grumixama* residues	Ultrasound assisted extraction at 37 kHz, 580 W for 90 min at room temperature	Acidified water (pH = 2.0) ethanol (50%, 70%, and 100%	• Highest yield (1.96 mg/g DM) of anthocyanin was achieved with UAE and 70% ethanol.	(Machado et al., 2017)
	Grape pomace	Ultrasound assisted extraction 59 kHz for 10 min at room temperature	Methanol with 2% formic acid	• The first single-step procedure to extract 14 anthocyanin from grape pomace with high purity (>90%) • Highest yield was 1.13 mg/g	(Zhao et al., 2020)
	Mulberry by-products	Pectinase-assisted	Water	• The highest yield was 6.04 mg/g DM following the conditions: 58 min, solid–liquid ratio (1:20), pH = 5.9, 45°C	(Li et al., 2020)
	Pyrus communis peel	Ultrasonic-assisted extraction at 20 kHz, 162 W for 11 min at 71 °C	Trifluoroacetic acid = 3%, ethanol = 57%,	• The highest yield of Cy3-gal (0.343 ± 0.05 mg/g DM) was obtained with the sample and solvent ratio of 1:30. • The optimized condition prevented the degradation of anthocyanin. • Pilot scale (3000 mL) batch and continuous processing were established and tested • Cy3-gal yield from the pilot scale was 0.285 mg/g DM	(Belwal et al., 2019)
Carotenoids	Tucuma and peach palm peel	Organic solvent extraction	30% of ethyl acetate in acetone	• 7.8 and 7.3 mg/100 g DM of the major carotenoid (β-carotene) • Optimized sample and solvent ratio 1:7	(Matos et al., 2019)
	Tomato peel and seeds	Ohmic heating with voltage ranging from 60 to 280V and different temperatures (0–100 °C) for up to 30 min at 25 kHz.	70% aqueous ethanol	• The highest yield, 2.821 ± 0.211 mg gallic acid equivalents/g fresh weight, was obtained at 70 °C for 15 min with ethanol	(Coelho et al., 2019)

(continued)

Table 4.3 Recent lab-scale improvements for the recovery of antioxidants from plant-based by-products. *Continued*

Target compounds	Source	Method		Improvement	Refs.
		Approaches	**Solvent**		
		Supercritical fluid extraction at the flow rate of 10ml/min· 340 bar and 60 °C	CO_2	• Obtained oleoresin with high carotenoids contents. β-carotene and γ-tocopherols were found at 75 mg/100 g and 575 mg/100 g of the oleoresin	(Romano et al., 2020)
	Jabuticaba peel and seeds	Grilling	Ethanol extraction	• First investigation of using Jabuticaba by-products as the source of carotenoids	(Albuquerque et al., 2020)
Total phenolics	Tropical highland blackberry pomace	3.4 solvent/blackberry by-products extracted with, at 60 °C for 120 min	57.1% aqueous ethanol	• Optimized yield of ellagitannins and anthocyanin are 6.96 mg/g and 1.20 mg/g of the DM, respectively	(Quirós et al., 2019)
	Beet leaves	Solvent extraction	Ethanol	• Optimized yield of bioactive compounds is obtained under the condition: 61 °C, 80% ethanol and solid: solvent ratio of 1:10, with the recovery of 31.7 mg/g (Total phenolics), 0.78 m/g (betacyanins), 2.17 mg/g (betaxanthins)	(Bengardino et al., 2019)
	Mango peel	Microwave-assisted extraction at (400–800 W) for up to 120 sec	60% aqueous ethanol	• The highest yield, 52.08 mg gallic acid eq./g DM, was obtained at 800 W, solid-liquid ratio (1:50), for 90 s	(del Pilar Sánchez-Camargo et al., 2020)
Vitamin C	Sweet cherry petioles, leaves	HPLC analysis	1% ascorbic acid mixed with methanol	• The leaves and petioles had a higher concentration of dietary fiber, vitamin C, carotenoids, and polyphenols as well as an antioxidant activity than the fruit	(Dziadek et al., 2019)
Vitamin E	Bilberry seeds	Supercritical CO_2 treatment at 20, 35, and 50 MPa and temperature (at 40°, 50°, and 60°C) for 80 min	Supercritical CO_2	• The best recovery rate of vitamin E, (14.5 mg/100g DM), was obtained at 20 MPa and 60 °C	(Gustinelli et al., 2018)
	Coffee silver skin	Soxhlet extraction during 2.5 h	N-hexane	• α-tocopherol (2.25 mg/100 g), β-tocotrienol (0.95 mg/100 g)	(Costa et al., 2018)
	Olive pomace	Multifrequency Modulated ultrasonic technique at 20kHz, 90 to 160 W for 10 min	Water	• MMM technique is a faster method to yield 2.63 mg/100 g DM of the Vitamin E • Solvent-free extraction method	(Nunes et al., 2018)

emerged as a method to minimize degradation and increase yields (Papoutsis et al., 2018). Moreover, other food by-products such as the placenta, pericarp, and seeds of pepper, and kiwi peel have been identified as rich sources of vitamin C (Dziadek et al., 2019; Marangi et al., 2018; Baenas et al., 2019). Table 4.3 summarizes examples of recovering vitamin C from these alternative sources at a laboratory scale.

4.2.1.2 Polyphenols

Polyphenols are the main class of water-soluble antioxidants and encompasses more than 8000 molecules (Shahidi and Ambigaipalan, 2015, Pandey and Rizvi, 2009). These bioactive molecules feature one or more aromatic rings (Gligor et al., 2019). Polyphenols are found in a variety of food by-products including orange peel (Shammugasamy et al., 2019), waste streams from wine production (Aliakbarian et al., 2012), potato peels (Sagar et al., 2018), fruit pomace (Sagar et al., 2018; Majerska et al., 2019) and tomato leaves (Arab et al., 2019). For some plants, the uneaten by-products contain a higher concentration of polyphenols than the edible parts, such as apple peel, date seed, and citrus fruit peel (Sagar et al., 2018).

It has been well established that polyphenols have a number of health benefits such as having cardioprotective (Pandey and Rizvi, 2009; Santhakumar et al., 2018), anticancer (Pandey and Rizvi, 2009, Shahidi and Yeo, 2018), antidiabetic, antiaging (Shahidi and Yeo, 2018), antiobesity (Cory et al., 2018), neuroprotective (Cory et al., 2018; Brglez Mojzer et al., 2016), antibacterial (Shahidi and Yeo, 2018), antiviral (Shahidi and Yeo, 2018), anti-inflammatory, and immune-modulatory effects (Shahidi and Yeo, 2018; Cory et al., 2018; Crasci et al., 2018). Polyphenols can also have biologically important interactions with beneficial gut bacteria (Cory et al., 2018). Attempts to recycle phenolic compounds from food by-products and residues emerged in the last couple of decades (Sagar et al., 2018). Of the thousands of bioactive compounds in the polyphenol family, the flavonoids are the most abundant antioxidants found in plant by-products and have well-understood mechanisms of biological activity (Tan et al., 2019; Jatav et al., 2019; Oteiza et al., 2018).

Flavonoids are the largest group of plant polyphenols that includes more than 5000 compounds as illustrated in Fig. 4.2. They are grouped into 13 classes: chalcones, dihydrochalcone, aurons, flavones, flavonols, dihydroflavonol, flavanones, flavanols, flavandioles, anthocyanidins, isoflavones, flavonoids, and condensed tannins (Baiano and Del Nobile, 2016). Among them, anthocyanidins have an exceptionally high antioxidant capacity, contributing to the red, purple, and blue color in some fruits and vegetables (Khoo et al., 2017). Conventionally, anthocyanins have been extracted from fruit seeds, such as the grape seeds (TaracTechnologies, 2016b), or fruit pomace via ethanol or water extraction (Zdybel et al., 2019). Recovery with ultrasonic or microwave-assisted extraction is the commonly applied novel techniques to obtain a high yields and purity of the recovered anthocyanidins. The scale-up of these processes up to a pilot-scale (~30 L) and simplification of the process to a single-step treatment are summarized in Table 4.3.

4.2.1.3 Carotenoids

Carotenoids are essential bioactive compounds that are not biosynthesized by humans and therefore require a dietary intake. More than 600 different carotenoids have been identified, with the most common being α-carotene, β-carotene, β-cryptoxanthin, lutein, zeaxanthin, and lycopene (Baiano and Del Nobile, 2016). As the most abundant lipophilic antioxidants, carotenoids have been conventionally extracted from plant by-products via solid-liquid or solid-phase extraction. These extractions involve multiple organic solvents, including methanol, acetone, ethanol, and ethyl acetate (Martins, 2017). Most of the innovations and improvements for carotenoid recovery have focused on reducing or replacing these organic solvents (Martins, 2017). Besides microwave- and ultrasound-assisted extractions, the recently emerged ohmic heating assisted extraction was able to recover carotenoids from tomato peels using an innocuous 70% ethanol solution (Table 4.3; TaracTechnologies, 2016a; Coelho et al., 2019). Supercritical fluids have also been used to extract carotenoids from various plant waste matrices, such as banana peel, grape peel, and pumpkin peels and seeds (Crasci et al., 2018; Tan et al., 2019). This technique has recently been used with a simplified system to directly produce carotenoid-enriched oleoresin from tomato peels for food, pharmaceutical, and cosmetic formulations (Romano et al., 2020).

4.2.1.4 Vitamin E

There are eight forms of vitamin E: four tocopherols and four tocotrienols (Baiano and Del Nobile, 2016). As an important lipid-soluble antioxidant, vitamin E is typically present in extracted edible oils, such as olive oil (Shahidi and de Camargo, 2016). Other plants, such as beetroot, are also rich in vitamin E (Battistella Lasta et al., 2019). Wheat germ, a common by-product generated in the milling industry, has been found to possess 16 times higher amount of α-tocopherol compared to other fractions of the wheat kernel (Boukid et al., 2018). Therefore, establishing, and optimizing vitamin E from wheat germ was the focus of many studies (Boukid et al., 2018; Ghafoor et al., 2017). Among numerous extraction methods, supercritical/subcritical carbon dioxide ($ScCO_2$) has been suggested to be the most favorable approach for vitamin E recovery from the wheat waste. It has shown the highest yield compared to other solvents and other nonconventional extraction methods (Teslić et al., 2020). Vitamin E has also been successfully recovered from by-products nontraditionally used for recovering bioactives, such as bilberry seeds and coffee silver skin.

4.2.1.5 Solanesol

Solanesol is noncyclic terpene alcohol with nine isoprene units. It has high antioxidant capacity due to nonconjugated double bonds in its structure. Solanesol is difficult to synthesize, but it has been successfully extracted from tobacco plants via saponification with alkaline solutions, and heat reflux extraction with organic solvents (Banožić et al., 2020). Besides tobacco, solanesol is also found and recovered on laboratory scale from other agricultural plant leaves, such as tomato and eggplant (Arab et al., 2019; Yan et al., 2015). In recent decades, microwave and ultrasound-assisted extraction, as well as the $ScCO_2$ have been investigated intensively (Arab et al., 2019; Banožić et al., 2020; Wang and Gu, 2018). Scale-up and optimization may be further improved based on mathematical modeling (Banožić et al., 2020; Wang and Gu, 2018).

4.2.2 Dietary fibers

Dietary fiber (DF) is a term used to encompass polysaccharides with three or more monomeric units. Many dietary fibers are indigestible by human gut enzymes and are not absorbed in the small intestine. However, these fibers can act as a substrate or energy source for beneficial gut microbiota in the colon (Sharma et al., 2018). Beneficial coliform bacterial communities afford desirable metabolites, such as short-chain fatty acids (SCFA), vitamins, and succinate. They have proven benefits to the host's health, and can mitigate the risks of obesity and diabetes (Yan et al., 2019; Sharma et al., 2018; Dalile et al., 2019; Canfora et al., 2019). In addition to prebiotic functionality, DF also acts as a bulking agent in the food industry to reduce caloric intake without adjusting volume by making use of their water and oil retention capacities (Garcia-Amezquita et al., 2018).

Due to the increasing awareness of the health benefits of DF, the demand for DF has also been increasing in the last few decades. Fruit pomace or vegetable matrices contain higher concentrations of soluble dietary fiber (SDF) than cereal by-products (Garcia-Amezquita et al., 2018; Quiles et al., 2018). Certain low utility plant parts, such as leaves and stalks, have higher concentrations of DF than edible fruits (Dziadek et al.,

2019). A growing body of publications has explored the feasibility of extracting DF from cereal, fruits, and vegetable by-products (Garcia-Amezquita et al., 2018; Sharma et al., 2018; Quiles et al., 2018). The commercialized DF products derived from wheat bran, green banana, and artichokes, are good examples of converting these by-products to valuable products to fulfill the emerging DF market (NaturalEvolution 2019).

More recently, nontraditional by-products have been considered as sources of DF, such as lupin seed coats (Zhong et al., 2019), mushroom waste (Xue et al., 2019), foxtail millet bran (Dong et al., 2019), and coffee silver skins (Ates and Elmaci, 2019; Benitez et al., 2019; Gocmen et al., 2019). High DF grade plant by-products have been used in for food engineering. As examples, apple pomace has been added to stirred yogurt to improve the structure and texture *via* the extra dietary fiber (Wang et al., 2020). DF extracted from maize and kimchi by-products has been added to biscuits and muffins to improve their baking and physico-chemical properties (Heo et al., 2019; Paraskevopoulou et al., 2019).

4.2.3 Plant-based proteins

Proteins are one of the major components of food products with significant nutritional value. In recent years, both consumers and scientists have been raising the issue of finding innovative good-quality plant-based protein sources. Traditional livestock farming is associated with considerable disadvantages, including excessive water usage and contamination, greenhouse gas emissions, and animal welfare issues (Richter et al., 2015). Besides the use of the plant-derived protein in the food industry, recent technological breakthroughs suggest that there are also major alternative applications for plant-derived proteins in spontaneously degradable bioplastics, biocomposites, and wastewater treatment (Reinert and Carbone, 2008; Heuer et al., 2001; Moura et al., 2017). Thus, there are ethical, ecological, and economic incentives to increase the sustainable supply of plant-based protein.

Potential under-utilized sources of plant-based protein include by-products from biofuel production, wineries, breweries, the fruit industry, vegetable processing, oil, and the cereal industries (Dhillon, 2016). Zein is a well-understood plant protein found in maize that can be manufactured as a water-insoluble, tasteless powder with a variety of industrial and food engineering uses (Xu et al., 2007). By-products from wine production contain various bioactive nutrients, including 11% -4% of protein with potential health benefits (Teixeira et al., 2014). Every year, millions of tons of apple pomace from the apple juice industry are disposed or used as animal feed. A current study demonstrated an advanced application of apple pomace as a renewable source of protein as apple pomace contains up to 5.7% w/w protein by dry mass (Dhillon et al., 2013). Sweet potato is an ingredient to produce traditional Japanese liquor, and the marc of the sweet potato was generally discarded as industrial waste. Another study suggested that the dietary crude protein and amino acids in the sweet potato marc could be a potential source of protein (Wu and Bagby, 1987). Flaxseed meal, industrial waste from the flaxseed oil industry, is high in protein content (reaching 40%) and a balanced amino-acid composition (Wu et al., 2019). Similarly barley, the fourth most commonly grown grain world-wide (Yalçın et al., 2007), is rich in protein ranging from 7–13% (Pomeranz and Shands, 1974).

4.2.4 Other bioactive compounds

Even though essential fatty acids are considered as one of the necessary nutritional factors, the health benefits of consuming monounsaturated fatty acids or polyunsaturated fatty acids is still unclear (Nettleton et al., 2016). However, previous reports have suggested that the intake of monounsaturated fatty acids or polyunsaturated fatty acids can be beneficial for reducing the risk of metabolic syndrome

(Julibert et al., 2019; Montserrat-de la Paz et al., 2019), the symptoms of rheumatoid arthritis (Veselinovic et al., 2017; Matsumoto et al., 2018), and for promoting healthy aging (Tessier and Chevalier, 2018; Han et al., 2017). Fish by-products, algae/microalgae, and other marine by-products are rich sources of fatty acids (Wu and Bechtel, 2008); Herrero et al., 2006). However, there are some risk factors associated with the use of these sources, including toxins such as heavy metals introduced by environmental pollution as well as potentially unidentified natural toxins (Jennings et al., 2016).

Notably, several recent reports suggest that plant by-products might be a good source of fatty acids (Lazos, 1986; Gustinelli et al., 2018; Antonia Nunes et al., 2018; Pasten et al., 2017). Less traditional crops such as algae, bilberry, pumpkin, and melon seeds have been identified as candidates to produce edible oils. Linoleic acids, α-linoleic acid, oleic acid, omega-3, and triacylglycerol are the major fatty acids from bilberry seeds (Gustinelli et al., 2018). Researchers suggested that oleic acid was the most abundant fatty acid: oleic acid made up to 75% and accounted for up to 5.4 g/100 g of dry mass in bilberry seeds. Less abundant fatty acids such as palmitic acid (10.4%) and linoleic acid (8.46%) are also found in olive seed cake and olive pomace (Antonia Nunes et al., 2018; Pasten et al., 2017). Linoleic acid was the most abundant fatty acid, and the portion of oleic acid was also the second largest fatty acid in melon seeds: 64.6% and 20.1% of dry weight, respectively (Lazos, 1986). These reports clearly demonstrate the feasibility of utilizing by-products from less traditional crops such as algae, bilberry, pumpkin, and melon seeds to produce

4.3 High-value-added compounds from animal-based by-products

The definition of what constitutes waste from animal foods is dependent on cultural perceptions and is therefore highly variable around the world. The most commonly consumed portion of animal products is the muscle, but may also include other tissues, such as blood, entrails, and internal organs (Nollet and Toldrá, 2011). Nevertheless, blood and internal organs are considered by-products in most countries and when the skin, bone, fatty tissue, feet, horns and/or hooves are included this can comprise up to two-thirds of the animal body mass (Irshad and Sharma, 2015). In Europe, it is estimated that every year over 18 million tons of by-products are produced at slaughterhouses and farms (Jędrejek et al., 2016). A report from 2010 found that only 16.5% (24 million of 145 million tons) of the annual catch of fish is utilized as meat, the rest becoming by-products or waste (Sila and Bougatef, 2016).

The use of animal by-products can increase slaughterhouse revenues, which is an important consideration for the business model of a meat industry that typically operates with low margins (Chatli et al., 2005; Shen et al., 2019). Due to a combination of regulatory issues, practices in the meat industry, and cultural attitudes, a significant volume of animal by-products are currently used for the production of low-value products (e.g., biodiesel, fertilizers, or animal/pet feed) or simply disposed of as waste (Mora et al., 2014). Commonly used animal by-products in food engineering are summarized in Table 4.4.

The direct use of animal by-products has an important impact on both the economy and the environment (Tonini et al., 2018). The need to reduce waste meat in landfills has led to an increase in the use of animal by-products for animal feed and energy production (Bujak, 2015). The use of animal by-products for energy production is not be profitable, and therefore it may be advantageous to consider extracting high-value products instead (Ghosh et al., 2019). Therefore, there is an increasing focus on isolating bioactive peptides from animal by-products for high-value-added products such as medicines, nutritional products, and cosmetics.

Table 4.4 Commonly used animal-based by-products in food engineering.

Target	By-product	Extraction method	Function/ application	Refs.
Blood plasma proteins	Animal blood	Fractionation	Food additive for gelation and emulsification, meat binder	(Cofrades et al., 2000; Lennon et al.; 2010)
Blood cellular fraction	Animal blood	Fractionation	Color enhancer	(Hsieh and Ofori, 2011)
Hemoglobin	Animal blood	Enzymatic digestion	Improving iron absorption	(Nissenson et al., 2003)
Bioactive peptides and free amino acids	Various bovine and porcine sources, e.g., trimmings, blood, organs, collagen, hemoglobin	Various methods, e.g., enzymatic hydrolysis, acid, and alkaline hydrolysis, microbial fermentation	Antihypertensive, antioxidant, opioid, antimicrobial, antidiabetic, analgesic, antifatigue, hypocholesterolemic activities	(Vercruysse et al., 2005; Chang et al., 2007)
Antimicrobial peptides	Various, e.g., bovine hemoglobin	Peptic digestion	Antibacterial action	(Nedjar-Arroume et al., 2006)
Gelatin	Collagen from skin, bones, tendons, and ossein	Extraction, controlled hydrolysis of collagen, recovery, and drying processes	Clarifying agent, stabilizer, and protective coating, cosmetic applications	(Djagny et al., 2001)
Collagen	Livestock skin, fish scales	Acid or enzymatic methods (undenatured collaged) or heat (denatured collagen)	Food additive and emulsifier to improve quality and calcium supplement	(Gelse et al., 2003)
Elastin	Poultry hides	Solvent extraction	Incorporated into functional foods for its antioxidant properties	(Nadalian et al., 2013)
Lard, tallow and butter	Clean tissue of healthy pigs, cattle, and sheep	Wet extraction (for edible products) and dry extraction	Nutrition	(Jayathilakan et al., 2012)
Fish oil	Fish offal	Various including organic solvent extraction, high performance liquid chromatography, urea inclusion method, distillation, and supercritical CO_2 fluid extraction	Health food supplement	(Rubio-Rodríguez et al., 2012)
Chondroitin sulfate	Fish bones	Various including alkali method, enzymatic method and ultrasonic method	Broad applications in health foods	(He et al., 2014)
Chitosan and chitin	Shrimp and crab shells	Thermochemical processes: soaking in HCl acid to remove calcium, then in sodium hydroxide to remove protein residues	Strong antibacterial properties for food preservatives. Many applications	(Shen et al., 2019)

4.3.1 Bioactive peptides and polysaccharides

The extraction/generation of bioactive peptides from by-products is the main motivation to manage animal-based food waste. Bioactive peptides are defined as sequences of 2–20 amino acids that have a biological function beyond nutrition. They are usually inactive when still part of the parent protein, but become bioactive after they are hydrolyzed (Najafian and Babji, 2014; Kumar et al., 2015). Such peptides have been found to have promising therapeutic applications.

Among these, collagen has been identified as a precursor for many bioactive peptides, and is highly abundant in discarded tissues such as skin and bones (Fu et al., 2016). The extraction of collagen from fish scales (*Prionace glauca*, *Scyliorhinus canicula*, *Xiphias gladius*, and *Thunnus albacares*) with pepsin yields valuable peptides with therapeutic properties (Blanco et al., 2017; Banerjee and Shanthi, 2016). Angiotensin-converting enzyme (ACE) inhibitory peptides can also be sourced from collagen-containing by-products of animal origin. This activity is attributed to peptides consisting of short, hydrophobic amino acids (Aluko, 2015). The peptides derived from the viscera of the giant catfish, porcine skin, and bovine-derived collagen demonstrated potent ACE-inhibitory activity (Fu et al., 2016; Hong et al., 2019; Ketnawa et al., 2017).

Varun et al. (2017) found that chitosan, chemically extracted from shrimp shell waste, can be depolymerized into chitooligomers using nitrous acid. Antimicrobial activity was found in certain chitooligomers against pathogenic gut bacteria and could potentially be used as a feed additive for gut health or may replace antibiotics (Varun et al., 2017).

4.3.2 New trends in recovery of valuable compounds from animal by-products

Table 4.5 summaries recent reports of new technologies for improving extraction from animal by-products. Recently emerged techniques, such as pulse electric field, which is the commonly used non-thermal pasteurization approach, has been used for protein recovery from chicken meat wastes (Ghosh et al., 2019). This method has also been found to enhance extraction of antioxidants in a chemical-free manner (Gómez et al., 2019). Besides the novel techniques, the latest improvements that focus on the extraction of well-known nutraceutical compounds, for instance collagen, are also reviewed in this section.

4.3.2.1 New techniques for collagen recovery

Collagen is a high-value product, but its production is limited because of high production costs (Bhagwat and Dandge, 2016). Fish scales are a good source of collagen, and Bhagwat found that when demineralized carp scales were acid-treated, acid-soluble collagen was extracted at a rate of 9.79% (Bhagwat and Dandge, 2016). This collagen was successfully included as additive in the milk product paneer. A semipurified form of pepsin from the stomach of sea catfish was also found to have milk clotting activity, which can also be useful for making paneer (Osuna-Ruiz et al., 2019).

Recently, there has been a particular focus on enzyme-assisted extraction of peptides from collagen (Vieira and Ferreira, 2017). Different enzymes have been investigated for both the purpose of increasing the product yield or producing a better-quality product. A new method for extraction of gelatine from Salmon skin used a trypsin assisted process to achieve a twofold higher yield of gelatine (Fan et al., 2017).

Different enzymatic activities were compared in a novel aqueous enzymatic extraction method for processing lard from pig fatback and found that alcalase achieved the best oil extraction, producing a higher yield and quality of lard in respect to the color, acid value, peroxide value, phospholipids,

Table 4.5 Recent studies on recovering valuable bioactive compounds from animal-based by-products.

Target	By-product	Extraction/treatment method	Function/ application	Refs.
Milk fat globule membrane and phospholipids	The β stream from anhydrous milk fat	Zinc and calcium acetate and mild heat treatment and pH adjustment, followed by ethanol extraction of protein pellet	Health benefits and functional properties	(Price et al., 2020)
Antifungal peptides	Whey from goats milk	Trypsin hydrolysis of whey	Shelf-extension of food products	(Luz et al., 2020)
Milk fat globule membrane	Dairy waste products	By-product from different stages of cream processing were compared for profile/composition	Nutrition	(Qu et al., 2019)
Pepsin from the stomach of sea catfish	Fish viscera	Proteolysis	Milk coagulant activity	(Osuna-Ruiz et al., 2019)
Whey	Caprine milk	Ultrafiltration	Nutrition	
Paneer whey	Dairy waste	Fermentation with *Pediococcus pentasaceus*	Shelf-life extension of paneer	(Pandey et al., 2019)
Collagen hydrolysates	Porcine skin	Enzymatic hydrolysis with alcalase and fractionation	Antioxidant and antiaging ingredients in food, cosmetics, and pharmaceuticals, supplement	(Hong et al., 2019)
Chicken meat from off-cuts	Waste chicken meat	Pulse electric field	Human nutrition	(Ghosh et al., 2019)
Protein hydrolysates	Porcine skin	New technology of hydrothermal treatment of porcine skin using high heat and pressure		(Min et al., 2017)
Analgesics	Organ meat	Partial or complete dissembling followed by hot or cold-set binding into cohesive mass to provide acceptable sensory texture	Exert opioid bioactivity (to improve quality of life in situations where there is food insecurity and malnutrition, e.g., internally displaced people	(Fayemi et al., 2018)
Chitin and chitosan	Shellfish	Depolymerization using nitrous acid	Antimicrobial activities against gut bacteria	(Varun et al., 2017)
Gelatin hydrolysates	Fish scales	Visceral peptidase and bovine trypsin for extraction	ACE inhibitory and antioxidant peptides	(Ketnawa et al., 2017)

(continued)

Table 4.5 Recent studies on recovering valuable bioactive compounds from animal-based by-products. *Continued*

Target	By-product	Extraction/treatment method	Function/ application	Refs.
Gelatin with high molecular weight protein chains	Salmon scales	Trypsin enzymatic method optimized with response surface methodology	Higher yield of gelatin	(Fan et al., 2017)
Pepsin soluble collagen	Fish scales	Pepsin extraction	Antioxidant capabilities	(Blanco et al., 2017)
Protein hydrolysate	Skipjack Tuna viscera	Hydrolyzed with alcalase	Nutrition and reduction of waste	(Klomklao and Benjakul, 2017)
Denatured collagen	Fish scales	Heating in water of decalcified fish scales	Optimization of collagen extraction	(Wu and Kang, 2017)
Chelating peptides	Alaska Pollack skin	Stability of peptide was tested using in vitro gastrointestinal enzymatic digestion, and binding site and mineral transport was investigated using caco-2 cells	Prevention of mineral deficiency	(Chen et al., 2017)
Acid soluble Collagen	Salmon scales	Demineralization with EDTA then acid extraction	Enrichment of foods (e.g., paneer) with collagen	(Bhagwat and Dandge, 2016)
Collagen	Telapia scales	Novel extrusion process	Optimization of collagen extraction	(Huang et al., 2016)
High quality, lower cholesterol lard	Pig fatback	Aqueous enzymatic extraction method	Applications in baking and healthier deep fat frying	(Wang et al., 2016)
ACE inhibitory peptide	Collagen	Enzymatic process with papain and bromelain	Valorization of collagen by deriving bioactive peptides	(Fu et al., 2016)

cholesterol, and oxidation stability than the standard method (Wang et al., 2016). The application of peptidase from the viscera of the giant catfish allowed greater yield of gelatine from fish scales with greater antioxidant activities (Ketnawa et al., 2017). Alcalase was used to recover collagen from porcine skin, and the produced peptides showed a high level of antioxidant activity (Hong et al., 2019).

4.3.2.2 Recovery other proteins from animal by-products

Organs are a common animal by-product in slaughterhouse, which is a promising source for edible iron. Processing and repurposing of this meat to be directly consumed can improve both the acceptability of the product and its utilization through conversion to food supplements (Fayemi et al., 2018). Other by-products, such as chicken gizzard inner lining, were found as a valuable source of amino acids, and additionally was found to contain a bioactive compound that has gastroprotective activities (Ni et al., 2018). The viscera of skipjack tuna was hydrolyzed most efficiently with the enzyme alcalase to form an important source of protein hydrolysate for nutritional supplementation purposes (Klomklao and Benjakul, 2017).

4.3.2.3 Recovery valuable compounds from dairy by-products

Whey is a by-product from cheese making and is produced in large quantities. The multifunctional lactoferrin with potent antimicrobial properties are extracted from whey (Jenssen and Hancock, 2009; Caccavo et al., 2002; Smithers et al., 1996). There has been a recent focus on improving the yield of lactoferrin using n-heptanol/cetrimonium bromide (CTAB) by the selected extraction approach (Pawar et al., 2019). Meanwhile, there has been an increased interest in the extraction of milk fat globule membranes (MFGM) which are considered of high nutritional value. Recent studies have focused on the most efficient stage in milk and cream processing to extract MFGM for nutritional supplementation (Price et al., 2020; Qu et al., 2019). Another example of the potentially valuable product derived from dairy waste is caprine whey. Ayunta et al. found that it has similar physicochemical properties to bovine whey, which is a much more culturally accepted and therefore, highly valued product (Ayunta et al., 2019). Peptides with antifungal activities in caprine whey were proposed as a high-value product that has an important application in shelf-life extension (Luz et al., 2020). Pandey et al. found that shelf-life of paneer was extended by the use of fermented whey waste (*Pediococcus pentasaceus)* in the paneer making process (Pandey et al., 2019).

4.4 Antiviral compounds from food by-products

Viruses are human pathogens that are diverse and complex. Unlike bacteria and unicellular eukaryotic parasites, viruses lack cellular structures. Viruses are commonly classified based on the genome (either RNA or DNA) and this nucleic acid is protected by a protein capsid and/or a lipid-containing envelope (Lodish et al., 2000). Most viruses proliferate by invading the hosts' cells and hijacking the cellular apparatus to replicate themselves (Cohen, 2016). This underlies the challenge of treating viral infections, which are poorly treated using antibiotic strategies (Akram et al., 2018). Antiviral compounds can thus act by promoting the hosts' own native immunity, or by specifically blocking an individual virus (such as by impairing docking to a cell receptor used for virus uptake). Safety of antiviral compounds is crucial, and can be a major impediment to drug development (Denaro et al., 2020).

Compared with other human pathogens, viruses can spread easily and rapidly and can feature a period of asymptomatic transmissibility. The 2020 outbreak of coronavirus (SARS-CoV-2) demonstrated the tremendous impact of a highly transmissible viral infection on global health and economies (Liu and Saif, 2020). This pandemic raised public interest in the discovery of natural antiviral compounds that could limit the spread of the virus, particularly before vaccines became available (Denaro et al., 2020). Seeking nonconventional antiviral compounds is the new research trend to fulfill this emerging needs. The previous overlooked food by-products might provide an alternative source for the novel compounds with promising potentials. Unlike compounds synthesized from expensive substrates, antiviral constituents derived from food waste are sustainable and low-cost.

Many plants have been found to possess antiviral properties and are regularly used in the farming industry to control infectious diseases. For instance, seaweed and algae are widely utilized in aquaculture as the diet supplement of salmon to prevent viral outbreaks (Munoz et al., 2019). Most plant extracts' antiviral activity is attributed to the high content of ubiquitous bioactive compounds, such as phenolic acids, flavonoids, terpenes, lignans, coumarins, alkaloids (Daglia, 2012). These plant-derived phytochemicals exert their antiviral actives by aggregating viral particles or binding to viral proteins and preventing viral entry (Munekata et al., 2019; Lagana et al., 2019). Other plant-derived bioactive compounds, such as chrysophanol (Xie et al., 2019), fucoidans, and carrageenan (Mirzadeh et al., 2020) demonstrate broad-spectrum antiviral activity by inactivating key viral structures to prevent viral replication. Similar to antioxidants, these antiviral molecules are also abundant in plant-based by-products (Zhu et al., 2019), and this includes peel (Abdul Ahmad et al., 2017) and fruit pomace (Bekhit et al., 2019). Antiviral herbal extractions can target a range of common viral pathogens, such as Herpes virus, the influenza virus, the Human Immunodeficiency Virus, and Hepatitis C (Denaro et al., 2020). Besides the plants processing waste stream, animal-based food by-products are also rich in diverse antiviral proteins or peptides. Hemocyanin, the protein purified from abalone waste, has been found to prevent infection of HSV-1 in vitro (Zanjani et al., 2018; Talaei Zanjani et al., 2016; Zanjani et al., 2014).

Emerging research on antivirals from plant and animal derived waste streams have tended to focus on investigating the antiviral activity of crude extracts rather than individual compounds (Akram et al., 2018; Denaro et al., 2020; Rouf et al., 2020). However, limited studies have been performed to test purified molecules and reveal their mechanism of actions on the cell level. Examples are summarized in Table 4.6. The lack of detailed mechanistic testing is partially due to the difficulty of establishing a proper models (Denaro et al., 2020). Effective collaboration between virologists and food engineering experts will be paramount to overcome these obstacles. Another novel research trend in improving the discovery rate and hence targeted recovery of antiviral compounds from the food by-product is in silico modeling. This approach allows rapid screening of natural compounds based on putative structural interaction with viral proteins essential for viral entry or replication (Pendyala and Patras, 2021).

Although recovering antiviral compounds from food by-products is a promising area, experimental outcomes need to be carefully evaluated. Most antiviral properties are concluded based on cell-based models, yet the concentration of the purified compounds that tested in vitro may be challenging to achieve in vivo model due to low bioavailability of many plant-derived compounds (Denaro et al., 2020). Besides proper validation, other hurdles to overcome for using food by-product to produce antiviral compounds, are as the process optimization, stability of the final products, and the lengthy and costly regulatory approval.

4.5 Concluding remarks

The importance of recovering valuable compounds from food by-products has been widely recognized. Many studies seek to raise awareness of the utilization of food by-products as sustainable and low-cost sources to produce nutraceutical and pharmaceutical compounds. Meanwhile, optimizing recovery to improve the yield and purity of the valuable ingredients has already been achieved for many compounds at a laboratory scale. Nevertheless, translation to full industrial-scale production often remains a challenge. Novel processing methods has been developed to increase the yield of extraction and attempted to reduce energy consumption, minimize operating temperatures with less environmental impact. The scale-up of those procedures has been progressed to pilot scale for some compounds, which is about 30–40 L. Simplification and substitution of multistep, costly procedures will be a crucial requirement in designing advanced recovery techniques at industrial scale. The widely applied production of dietary fiber from wheat bran serves an example of the successful commercialization of food by-product. The integration of new technologies and processing methods into existing manufacturing systems may

Table 4.6 New antiviral compounds identified from food by-product.

Source	By-product	Active compounds	Virus	Mechanism of action	Refs.
Elderberry (*Sambucus nigra*)	Pulp	Cyandin-3-glucoside	Influenza A	Blocking the function of the hemagglutinin of the virus	(Torabian et al., 2019)
Abalone (*Haliotis rubra*)	Hemolymph	Hemocyanin	Herpes simplex virus type 1 (HSV-1)	Binding to viral glycoprotein B to block viral entry	(Zanjani et al., 2018; Talaei Zanjani et al., 2016; Zanjani et al., 2014; Wu et al., 2016)
Rambutan (*Nephelium lappaceum* L)	Peel	Geraniin	Dengue virus type 2 (DENV-2)	Blocking the envelope protein E of DENV-2	(Abdul Ahmad et al., 2017)
Longan fruit	Pericarp and seed	Pinocembrin	Zika virus (ZIKV) chikungunya virus (CHIKV) DENV2	Inhibition of ZIKV RNA and protein synthesis	(Lee et al., 2019)
Vietnamese cinnamon	Bark	Cinnamomum	Human immunodeficiency virus type 1 (HIV-1)	Inhibition of reverse transcriptase activities	(Silprasit et al., 2011; Kumar et al., 2019)
Dianthus superbus	Whole plant without root	Quercetin-3-O-rutinoside	Influenza A	Inhibition of viral neuraminidase activity	(Kim et al., 2019)
Grape	Wine production residue	Total phenolic compounds	Influenza A	-	(Bekhit et al., 2019)

allow for other by-products to be similarly utilized for societal, environmental, and industrial benefit. While these may require significant capital investment, cost-effective technologies are likely to make use of these by-products profitable enough in the long term, particularly when considering the costs associated with conventional disposal.

Regulation of ingredients derived from food by-product streams are currently inconsistently or poorly defined. As these ingredients are derived from food sources, it is expected that a simplified pathway would apply for the recovered high-value compounds to obtain regulatory approval. A final hurdle to the acceptance of products generated from food by-products is overcoming cultural or personal values associated with consuming products derived from waste. Marketing, education, and a re-alignment of cultural priorities to maximize economic and environmental outcomes will be necessary for some products to become acceptable to consumers, such as the nutraceutical compounds recovered from animal by-product. Nevertheless, utilizing food by-products to produce nutraceutical and pharmaceutical products represents a promising field that is expected to expand with coordination and collaboration among research, industry, and government stakeholders.

Acknowledgments

The authors acknowledge the financial support from the Australian Research Council and the University of Sydney. The authors also acknowledge the contribution of Dr. Diana Bogueva in edition of this article.

References

Abdul Ahmad, S.A.A., Palanisamy, U.D., Tejo, B.A., Chew, M.F., Tham, H.W., Hassan, S.S., et al., 2017. Geraniin extracted from the rind of Nephelium lappaceum binds to dengue virus type-2 envelope protein and inhibits early stage of virus replication. Virol. J. 14 (1), 1–13, 229. doi:https://doi.org/10.1186/s12985-017-0895-1.

Ajila, C., Leelavathi, K., Rao, U.P., 2008. Improvement of dietary fiber content and antioxidant properties in soft dough biscuits with the incorporation of mango peel powder. J. Cereal Sci. 48 (2), 319–326.

Akram, M., et al., 2018. Antiviral potential of medicinal plants against HIV, HSV, influenza, hepatitis, and coxsackievirus: a systematic review. Phytother. Res. 32 (5), 811–822.

Albuquerque, B.R., et al., 2020. Jabuticaba residues (Myrciaria jaboticaba (Vell.) Berg) are rich sources of valuable compounds with bioactive properties. Food Chem. 309.

Aliakbarian, B., et al., 2012. Extraction of antioxidants from winery wastes using subcritical water. J. Supercrit. Fluids 65, 18–24.

Altendorf, S., 2019. Major tropical fruits market review 2017. FAO, Rome, pp. 1–12.

Aluko, R.E., 2015. Antihypertensive peptides from food proteins. Annu. Rev. Food Sci. Technol. 6, 235–262.

Antonia Nunes, M., et al., 2018. Olive pomace as a valuable source of bioactive compounds: a study regarding its lipid- and water-soluble components. Sci. Total Environ. 644, 229–236.

Arab, M., et al., 2019. A benign process for the recovery of solanesol from tomato leaf waste. Heliyon 5 (4), e01523.

Ates, G., Elmaci, Y., 2019. Physical, chemical and sensory characteristics of fiber-enriched cakes prepared with coffee silverskin as wheat flour substitution. J. Food Measur. Charact. 13 (1), 755–763.

Australian Government. National Food Waste Strategy, 2017. Halving Australia's food waste by 2030. Australia.

Ayunta, C.A., et al., 2019. Physicochemical properties of caprine and commercial bovine whey protein concentrate. J. Food Measur. Charact. 13 (4), 2729–2739.

Baenas, N., et al., 2019. Industrial use of pepper (Capsicum annum L.) derived products: technological benefits and biological advantages. Food Chem. 274, 872–885.

Baiano, A., Del Nobile, M.A., 2016. Antioxidant compounds from vegetable matrices: biosynthesis, occurrence, and extraction systems. Crit. Rev. Food Sci. Nutr. 56 (12), 2053–2068.

Banerjee, P., Shanthi, C., 2016. Cryptic peptides from collagen: a critical review. Protein Pept. Lett. 23 (7), 664–672.

Banožić, M., Babić, J., Jokić, S., 2020. Recent advances in extraction of bioactive compounds from tobacco industrial waste—a review. Ind. Crops Prod. 144, 112009.

Battistella Lasta, H.F., et al., 2019. Pressurized liquid extraction applied for the recovery of phenolic compounds from beetroot waste. Biocatal. Agric. Biotechnol. 21, 101353.

Bekhit, A.E.A., et al., 2019. Effect of extraction system and grape variety on anti-influenza compounds from wine production residue. Food Control 99, 180–189.

Belwal, T., et al., 2019. Optimization model for ultrasonic-assisted and scale-up extraction of anthocyanins from Pyrus communis 'Starkrimson' fruit peel. Food Chem., 297.

Belwal, T., Huang, H., Duan, Z., Zhang, X., Aalim, H., Luo, Z., et al., 2019. Optimization model for ultrasonic-assisted and scale-up extraction of anthocyanins from *Pyrus communis* 'Starkrimson' fruit peel. Food Chem. 297, 1–12, 124993. doi:https://doi.org/10.1016/j.foodchem.2019.124993.

Bengardino, M.B., et al., 2019. Recovery of bioactive compounds from beet leaves through simultaneous extraction: modelling and process optimization. Food Bioprod. Process. 118, 227–236.

Benitez, V., et al., 2019. Coffee parchment as a new dietary fiber ingredient: functional and physiological characterization. Food Res. Int. 122, 105–113.

Bhagwat, P.K., Dandge, P.B., 2016. Isolation, characterization and valorizable applications of fish scale collagen in food and agriculture industries. Biocatal. Agric. Biotechnol. 7, 234–240.

Blanco, M., et al., 2017. Hydrolysates of fish skin collagen: an opportunity for valorizing fish industry byproducts. Mar. Drugs 15 (5), 131.

Bojanić, N., Teslić, N., Rakić, D., Brdar, M., Fišteš, A., Zeković, Z., Bodroža-Solarov, M., Pavlić, B., et al., 2019. Extraction kinetics modeling of wheat germ oil supercritical fluid extraction. J. Food Process. Preserv. 43 (9), 1–12, e14098. doi:https://doi.org/10.1111/jfpp.14098.

Boukid, F., et al., 2018. A compendium of wheat germ: separation, stabilization and food applications. Trends Food Sci. Technol. 78, 120–133.

Bozkir, H., Tekgül, Y., Erten, E.S., 2020. Effects of tray drying, vacuum infrared drying, and vacuum microwave drying techniques on quality characteristics and aroma profile of orange peels. J. Food Process Eng., e13611 n/a(n/a).

Brasesco, F., Asgedom, D., and Casari, G., Strategic analysis and intervention plan for fresh and industrial tomato in the agro-commodities procurement zone of the pilot integrated agro-industrial park in Central-Eastern Oromia, Ethiopia. 2019: Addis Ababa. 2019.

Brglez Mojzer, E.B., Knez Hrnčič, M., Škerget, M., Knez, Ž., Bren, U., et al., 2016. Polyphenols: extraction methods, antioxidative action, bioavailability and anticarcinogenic effects. Molecules 21 (7), 1–38, 901. doi:10.3390/molecules21070901. PMC6273793.

Bujak, J.W., 2015. New insights into waste management–meat industry. Renewable Energy 83, 1174–1186.

Caccavo, D., et al., 2002. Antimicrobial and immunoregulatory functions of lactoferrin and its potential therapeutic application. J. Endotoxin Res. 8 (6), 403–417.

Canfora, E.E., et al., 2019. Gut microbial metabolites in obesity, NAFLD and T2DM. Nat. Rev. Endocrinol. 15 (5), 261–273.

Chang, C.-Y., Wu, K.-C., Chiang, S.-H., 2007. Antioxidant properties and protein compositions of porcine haemoglobin hydrolysates. Food Chem. 100 (4), 1537–1543.

Chatli, M.K., Padda, G., Devatkal, S.K., 2005. Augmentation of animal by-products processing for the sustainability of meat industry. Indian Food Industry 24 (5), 69–73.

Chen, Q., et al., 2017. The chelating peptide (GPAGPHGPPG) derived from Alaska pollock skin enhances calcium, zinc and iron transport in Caco-2 cells. Int. J. Food Sci. Technol. 52 (5), 1283–1290.

Cheok, C.Y., et al., 2018. Current trends of tropical fruit waste utilization. Crit. Rev. Food Sci. Nutr. 58 (3), 335–361.

Coelho, M., et al., 2019. Extraction of tomato by-products' bioactive compounds using ohmic technology. Food Bioprod. Process. 117, 329–339.

Cofrades, S., et al., 2000. Plasma protein and soy fiber content effect on bologna sausage properties as influenced by fat level. J. Food Sci. 65 (2), 281–287.

Cohen, F.S., 2016. How viruses invade cells. Biophys. J. 110 (5), 1028–1032.

Cory, H., Passarelli, S., Szeto, J., Tamez, M., Mattei, J., et al., 2018. The role of polyphenols in human health and food systems: a mini-review. Front. Nutr. 5, 1–9, 87. doi:10.3389/fnut.2018.00087.

Costa, A.S.G., et al., 2018. Nutritional, chemical and antioxidant/pro-oxidant profiles of silverskin, a coffee roasting by-product. Food Chem. 267, 28–35.

Crasci, L., et al., 2018. Natural antioxidant polyphenols on inflammation management: anti-glycation activity vs metalloproteinases inhibition. Crit. Rev. Food Sci. Nutr. 58 (6), 893–904.

Daglia, M., 2012. Polyphenols as antimicrobial agents. Curr. Opin. Biotechnol. 23 (2), 174–181.

Dalile, B., et al., 2019. The role of short-chain fatty acids in microbiota-gut-brain communication. Nat. Rev. Gastroenterol. Hepatol. 16 (8), 461–478.

Daraee, A., Ghoreishi, S.M., Hedayati, A., 2019. Supercritical CO_2 extraction of chlorogenic acid from sunflower (Helianthus annuus) seed kernels: modeling and optimization by response surface methodology. J. Supercrit. Fluids 144, 19–27.

del Pilar Sánchez-Camargo, A., Ballesteros-Vivasb, D., Buelvas-Puello, L.M., Martinez-Correa, H.A., Parada-Alfonso, F., Cifuentes, A., Ferreira, S.R.S., Gutiérrez, L.F., et al., 2020. Microwave-assisted extraction of phenolic compounds with antioxidant and anti-proliferative activities from supercritical CO_2 pre-extracted mango peel as valorization strategy. LWT 137, 110414. doi:https://doi.org/10.1016/j.lwt.2020.110414.

Denaro, M., et al., 2020. Antiviral activity of plants and their isolated bioactive compounds: an update. Phytother. Res. 34 (4), 742–768.

Dhillon, G.S., et al., 2016. Chapter 2 - agricultural-based protein by-products: characterization and applications. In: Dhillon, G.S. (Ed.), Protein Byproducts. Academic Press, Canada, pp. 21–36.

Dhillon, G.S., Kaur, S., Brar, S.K., 2013. Perspective of apple processing wastes as low-cost substrates for bioproduction of high value products: a review. Renew. Sustain. Energy Rev. 27, 789–805.

Djagny, K.B., Wang, Z., Xu, S., 2001. Gelatin: a valuable protein for food and pharmaceutical industries. Crit. Rev. Food Sci. Nutr. 41 (6), 481–492.

Dong, J.L., et al., 2019. Structural, antioxidant and adsorption properties of dietary fiber from foxtail millet (Setaria italica) bran. J. Sci. Food Agric. 99 (8), 3886–3894.

Dzah, C.S., et al., 2020. Latest developments in polyphenol recovery and purification from plant by-products: a reviewTrends Food Sci. Technol.99, 375–388.

Dziadek, K., Kopec, A., Tabaszewska, M., 2019. Potential of sweet cherry (Prunus avium L.) by-products: bioactive compounds and antioxidant activity of leaves and petioles. Eur. Food Res. Technol. 245 (3), 763–772.

Fan, H., Dumont, M.-J., Simpson, B.K., 2017. Extraction of gelatin from salmon (Salmo salar) fish skin using trypsin-aided process: optimization by Plackett–Burman and response surface methodological approaches. J. Food Sci. Technol. 54 (12), 4000–4008.

FAO, 2013. Food Wastage Footprint. Impacts on Natural Resources. FAO, Rome.

FAO, 2017. Citrus Fruit Fresh and Processed: Statistical Bulletin 2016. FAO, Rome.

FAO. *Banana facts and figures*. 2018 [cited 2020 16-Jan]; Available from: http://www.fao.org/economic/est/est-commodities/bananas/bananafacts/en/#.Xh-vhGkzaM8. 2020.

FAO-AMIS, 2017. Database. FAO, Rome.

FAOSTAT, 2021. Food and Agriculture Data. Food and Agriculture Organization of the United Nations, Rome.

Fayemi, P.O., et al., 2018. Targeting the pains of food insecurity and malnutrition among internally displaced persons with nutrient synergy and analgesics in organ meat. Food Res. Int. 104, 48–58.

Food and Agriculture Organization of the United Nations and International Organisation of, 2016. Vine and Wine Intergovernmental Organisation, Table and dried grapes. Paris, FAO-Oiv.

Fu, Y., et al., 2016. Revalorisation of bovine collagen as a potential precursor of angiotensin I-converting enzyme (ACE) inhibitory peptides based on in silico and in vitro protein digestions. J. Funct. Foods 24, 196–206.

Galanakis, C.M., 2012. Recovery of high added-value components from food wastes: conventional, emerging technologies and commercialized applications. Trends Food Sci. Technol. 26 (2), 68–87.

Garcia-Amezquita, L.E., et al., 2018. Dietary fiber concentrates from fruit and vegetable by-products: processing, modification, and application as functional ingredients. Food Bioprocess Technol. 11 (8), 1439–1463.

Gelse, K., Pöschl, E., Aigner, T., 2003. Collagens—structure, function, and biosynthesis. Adv. Drug. Deliv. Rev. 55 (12), 1531–1546.

Ghafoor, K., et al., 2017. Nutritional composition, extraction, and utilization of wheat germ oil: a review. Eur. J. Lipid Sci. Technol. 119 (7), 1600160.

Ghosh, S., et al., 2019. Towards waste meat biorefinery: extraction of proteins from waste chicken meat with non-thermal pulsed electric fields and mechanical pressing. J. Cleaner Prod. 208, 220–231.

Gligor, O., et al., 2019. Enzyme-assisted extractions of polyphenols – a comprehensive review. Trends Food Sci. Technol. 88, 302–315.

Gocmen, D., et al., 2019. Use of coffee silverskin to improve the functional properties of cookies. J. Food Sci. Technol.-Mysore 56 (6), 2979–2988.

Gómez, B., Munekata, P.E.S., Gavahian, M., Barba, F.J., Martí-Quijal, F.J., Bolumar, T., Campagnol, P.C.B., Tomasevic, I., Lorenzo, J.M., et al., 2019. Application of pulsed electric fields in meat and fish processing industries: an overview. Food Res. Int 123, 95–105. doi:https://doi.org/10.1016/j.foodres.2019.04.047.

Gupta, K., Joshi, V.K., 2000. Fermentative utilization of waste from food processing industry. In: Verma, L.R., Joshi, V.K. (Eds.), Postharvest Technology of Fruits and Vegetables: Handling Processing Fermentation and Waste Management, 2. Indus Pub Co, New Delhi, pp. 1171–1193.

Gustinelli, G., et al., 2018. Supercritical CO_2 extraction of bilberry (Vaccinium myrtillus L.) seed oil: fatty acid composition and antioxidant activity. J. Supercrit. Fluids 135, 91–97.

Han, S., et al., 2017. Mono-unsaturated fatty acids link H3K4me3 modifiers to C. elegans lifespan. Nature 544 (7649), 185–190.

He, G., et al., 2014. Optimisation extraction of chondroitin sulfate from fish bone by high intensity pulsed electric fields. Food Chem. 164, 205–210.

Hemdane, S., et al., 2016. Wheat (Triticum aestivum L.) bran in bread making: a critical review. Comprehensive Rev. Food Sci. Food Saf. 15 (1), 28–42.

Heo, Y., et al., 2019. Muffins enriched with dietary fiber from kimchi by-product: Baking properties, physical-chemical properties, and consumer acceptance. Food Sci Nutr. 7 (5), 1778–1785.

Herrero, M., Cifuentes, A., Ibañez, E., 2006. Sub- and supercritical fluid extraction of functional ingredients from different natural sources: Plants, food-by-products, algae and microalgae: a review. Food Chem. 98 (1), 136–148.

Heuer, A.H., et al., 2001. Shell: properties. In: Buschow, K.H.J. et al (Ed.), Encyclopedia of Materials: Science and Technology. Elsevier, Oxford, pp. 8462–8469.

Hong, G.-P., Min, S.-G., Jo, Y.-J., 2019. Anti-oxidative and anti-aging activities of porcine by-product collagen hydrolysates produced by commercial proteases: effect of hydrolysis and ultrafiltration. Molecules 24 (6), 1104.

Hsieh, Y.-H.P., Ofori, J.A., 2011. Blood-derived products for human consumption. Revelation Sci. 1 (01), 1–14. doi:https://journals.iium.edu.my/revival/index.php/revival/article/view/15.

Huang, C.-Y., et al., 2016. Isolation and characterization of fish scale collagen from tilapia (Oreochromis sp.) by a novel extrusion–hydro-extraction process. Food Chem. 190, 997–1006.

Hwang, K.A., Choi, K.C., 2015. Anticarcinogenic effects of dietary phytoestrogens and their chemopreventive mechanisms. Nutr. Cancer 67 (5), 796–803.

Iqbal, S., et al., 2017. Potassium and waste water interaction in the regulation of photosynthetic capacity, ascorbic acid and capsaicin in chilli (Capsicum annuum L.) plant. Agric. Water Manage. 184, 201–210.

Irshad, A., Sharma, B., 2015. Abattoir by-product utilization for sustainable meat industry: a review. J. Anim. Prod. Adv. 5 (6), 681–696.

Jatav, S., et al., 2019. Isolation of a new flavonoid and waste to wealth recovery of 6-O-ascorbyl esters from seeds of Aegle marmelos (family- Rutaceae). Nat. Prod. Res. 33 (15), 2236–2242.

Jayathilakan, K., et al., 2012. Utilization of byproducts and waste materials from meat, poultry and fish processing industries: a review. J. Food Sci. Technol. 49 (3), 278–293.

Jędrejek, D., et al., 2016. Characteristics, European regulatory framework, and potential impacts on human and animal health and the environment. J. Anim. Feed Sci. 25, 189–202.

Jennings, S., et al., 2016. Aquatic food security: insights into challenges and solutions from an analysis of interactions between fisheries, aquaculture, food safety, human health, fish and human welfare, economy and environment. Fish Fish. 17 (4), 893–938.

Jenssen, H., Hancock, R.E., 2009. Antimicrobial properties of lactoferrin. Biochimie 91 (1), 19–29.

Julibert, A., Bibiloni, M.D.M., Tur, J.A., 2019. Dietary fat intake and metabolic syndrome in adults: a systematic review. Nutr. Metab Cardiovasc. Dis. 29 (9), 887–905.

Ketnawa, S., et al., 2017. Fish skin gelatin hydrolysates produced by visceral peptidase and bovine trypsin: bioactivity and stability. Food Chem. 215, 383–390.

Khoo, H.E., et al., 2017. Anthocyanidins and anthocyanins: colored pigments as food, pharmaceutical ingredients, and the potential health benefits. Food Nutr. Res. 61 (1), 1361779.

Kim, D.H., et al., 2019. Utilization of Dianthus superbus L and its bioactive compounds for antioxidant, anti-influenza and toxicological effects. Food Chem. Toxicol. 125, 313–321.

Klomklao, S., Benjakul, S., 2017. Utilization of tuna processing byproducts: protein hydrolysate from skipjack tuna (*Katsuwonus pelamis*) viscera. J. Food Process. Preserv. 41 (3), e12970.

Kumar, S., Kumari, R., Mishra, S., 2019. Pharmacological properties and their medicinal uses of Cinnamomum: a review. J. Pharm. Pharmacol. 71 (12), 1735–1761.

Kumar, Y., et al., 2015. Recent trends in the use of natural antioxidants for meat and meat products. Comprehens. Rev. Food Sci. Food Saf. 14 (6), 796–812.

Lagana, P., et al., 2019. Phenolic substances in foods: health effects as anti-inflammatory and antimicrobial agents. J. AOAC Int. 102 (5), 1378–1387.

Lazos, E.S., 1986. Nutritional, fatty acid, and oil characteristics of pumpkin and melon seeds. J. Food Sci. 51 (5), 1382–1383.

Lee, J.L., et al., 2019. Antiviral activity of pinocembrin against Zika virus replication. Antiviral Res. 167, 13–24.

Lennon, A., et al., 2010. Performance of cold-set binding agents in re-formed beef steaks. Meat Sci. 85 (4), 620–624.

Li, Y., et al., 2020. Process optimization for enzymatic assisted extraction of anthocyanins from the mulberry wine residue. IOP Conf. Ser.: Earth Environ. Sci. 559, 012011.

Liu, S.-L., Saif, L., 2020. Emerging Viruses Without Borders: The Wuhan Coronavirus. Multidisciplinary Digital Publishing Institute, America.

Lobo, V., et al., 2010. Free radicals, antioxidants and functional foods: impact on human health. Pharmacogn. Rev. 4 (8), 118–126.

Lodish, H., Berk, A., Zipursky, S.L., 2000. Section 6.3 Viruses: Structure, Function, and Uses. In: Lodish, H. (Ed.), et al., Molecular Cell Biology, 4th ed. W. H. Freeman, New York.

Lourenco, S.C., Moldao-Martins, M., Alves, V.D., 2019. Antioxidants of natural plant origins: from sources to food industry applications. Molecules 24 (22), 1–25, 4132. doi:10.3390/molecules24224132.

Luz, C., et al., 2020. Antifungal and antimycotoxigenic activity of hydrolyzed goat whey on Penicillium spp: an application as biopreservation agent in pita bread. LWT 118, 108717.

Machado, A.P.D., et al., 2017. Recovery of anthocyanins from residues of *Rubus fruticosus*, *Vaccinium myrtillus* and *Eugenia brasiliensis* by ultrasound assisted extraction, pressurized liquid extraction and their combination. Food Chem. 231, 1–10.

Majerska, J., Michalska, A., Figiel, A., 2019. A review of new directions in managing fruit and vegetable processing by-products. Trends Food Sci. Technol. 88, 207–219.

Marangi, F., et al., 2018. Hardy kiwi leaves extracted by multi-frequency multimode modulated technology: a sustainable and promising by-product for industry. Food Res. Int. 112, 184–191.

Martins, N., Ferreira, I.C.F.R., 2017. Wastes and by-products: upcoming sources of carotenoids for biotechnological purposes and health-related applications. Trends Food Sci. Technol. 62, 33–48.

Matos, K.A.N., et al., 2019. Peels of tucuma (Astrocaryum vulgare) and peach palm (Bactris gasipaes) are by-products classified as very high carotenoid sources. Food Chem. 272, 216–221.

Matsumoto, Y., et al., 2018. Monounsaturated fatty acids might be key factors in the Mediterranean diet that suppress rheumatoid arthritis disease activity: the TOMORROW study. Clin. Nutr. 37 (2), 675–680.

Min, S.-G., Jo, Y.-J., Park, S.H., 2017. Potential application of static hydrothermal processing to produce the protein hydrolysates from porcine skin by-products. LWT-Food Sci. Technol. 83, 18–25.

Mirzadeh, M., Arianejad, M.R., Khedmat, L., 2020. Antioxidant, antiradical, and antimicrobial activities of polysaccharides obtained by microwave-assisted extraction method: a review. Carbohydr. Polym., 229, 115421. doi:https://doi.org/10.1016/j.carbpol.2019.115421.

Montserrat-de la Paz, S., et al., 2019. Monounsaturated fatty acids in a high-fat diet and niacin protect from white fat dysfunction in the metabolic syndrome. Mol. Nutr. Food Res. 63 (19), e1900425.

Mora, L., Reig, M., Toldrá, F., 2014. Bioactive peptides generated from meat industry by-products. Food Res. Int. 65, 344–349.

Moura, I.G., Sá, A.V., Abreu, A.S.L.M., Machado, A.V.A., et al., 2017. 7 – Bioplastics from agro-wastes for food packaging applications. In: Grumezescu, A.M. (Ed.), Food Packaging. Academic Press, United Kingdom, pp. 223–263.

Munekata, P.E.S., Alcántara, C., Collado, M.C., Garcia-Perez, J.V., Saraiva, J.A., Lopes, R.P., Barba, F.J., Silva, L.P., Sant'Ana, A.S., Fierro, E.M., Lorenzo, J.M., et al., 2019. Ethnopharmacology, phytochemistry and biological activity of Erodium species: a review. Food Res. Int., 126, 108659. doi:https://doi.org/10.1016/j.foodres.2019.108659.

Munoz, I.L., et al., 2019. Diets enriched in red seaweed (Pyropia columbina and Gracilaria chilensis) cryo concentrates modulate the immune-relevant gene encoding the Mx antiviral protein in salmon (Salmo salar) white blood cells. J. Appl. Phycol. 31 (2), 1415–1424.

Nadalian, M., et al., 2013. Extraction and characterization of elastin from poultry skin, AIP Conference Proceedings. American Institute of Physics.

Najafian, L., Babji, A.S., 2014. Production of bioactive peptides using enzymatic hydrolysis and identification antioxidative peptides from patin (Pangasius sutchi) sarcoplasmic protein hydolysate. J. Funct. Foods 9, 280–289.

NaturalEvolution. *Natural & Organic superfoods*. 2019 [cited 2020; Available from: https://www.naturalevolutionfoods.com.au/.

Nedjar-Arroume, N., et al., 2006. Isolation and characterization of four antibacterial peptides from bovine hemoglobin. Peptides 27 (9), 2082–2089.

Nettleton, J.A., et al., 2016. Dietary fatty acids: is it time to change the recommendations? Ann. Nutr. Metab. 68 (4), 249–257.

Ni, C.-H., et al., 2018. Investigation of the chemical composition and functional proteins of chicken gizzard inner lining. Food Sci. Technol. Res. 24 (5), 893–901.

Nissenson, A.R., et al., 2003. Clinical evaluation of heme iron polypeptide: sustaining a response to rHuEPO in hemodialysis patients. Am. J. Kidney Dis. 42 (2), 325–330.

Nizori, A., et al., 2018. Impact of varying hydrocolloid proportions on encapsulation of ascorbic acid by spray drying. Int. J. Food Sci. Technol. 53 (6), 1363–1370.

Nollet, L., Toldrá, F., 2011. Introduction. Offal meat: definitions, regions, cultures, generalities. In: Nollet, L.M.L., Toldrá, F. (Eds.), Handbook of Analysis of Edible Animal By-Products. CRC Press, Boca Raton, FL, USA, pp. 3–11.

Nunes, M.A., et al., 2018. Olive pomace as a valuable source of bioactive compounds: a study regarding its lipid- and water-soluble components. Sci. Total Environ. 644, 229–236.

Oreopoulou, V., Tzia, C., 2007. Utilization of plant by-products for the recovery of proteins, dietary fibers, antioxidants, and colorants. In: Oreopoulou, V., Russ, W. (Eds.), Utilization of By-Products and Treatment of Waste in the Food Industry. Springer, Boston, MA, pp. 209–232.

Osuna-Ruiz, I., et al., 2019. Biochemical characterization of a semi-purified aspartic protease from sea catfish Bagre panamensis with milk-clotting activity. Food Sci. Biotechnol. 28 (6), 1785–1793.

Oteiza, P.I., et al., 2018. Flavonoids and the gastrointestinal tract: local and systemic effects. Mol. Aspects Med. 61, 41–49.

Pandey, K.B., Rizvi, S.I., 2009. Plant polyphenols as dietary antioxidants in human health and disease. Oxidative Med. Cell. Longevity 2 (5), 270–278.

Pandey, K.K., et al., 2019. Bioutilization of paneer whey waste for production of paneer making powder containing pediocin pa-1 as a biopreservative to enhance shelf life of paneer. LWT 113, 108243.

Papageorgiou, M., Skendi, A., 2018. 1 - Introduction to cereal processing and by-products. In: Galanakis, C.M. (Ed.), 1 - Introduction to cereal processing and by-products. Sustainable Recovery and Reutilization of Cereal Processing By-Products, 1–25.

Papoutsis, K., et al., 2018. Encapsulation of citrus by-product extracts by spray-drying and freeze-drying using combinations of maltodextrin with soybean protein and ι-Carrageenan. Foods 7 (7), 115.

Paraskevopoulou, A., Rizou, T., Kiosseoglou, V., 2019. Biscuits enriched with dietary fibre powder obtained from the water-extraction residue of maize milling by-product. Plant Foods Hum. Nutr. 74 (3), 391–398.

Pasten, A., Uribe, E., Stucken, K., Rodríguez, A., Vega-Gálvez, A., et al., 2017. Influence of drying on the recoverable high-value products from olive (cv. *Arbequina*) waste cake. Waste Biomass Valorization 10, 1627–1638. doi:https://doi.org/10.1007/s12649-017-0187-4.

Pawar, S.S., Iyyaswami, R., Belur, P.D., 2019. Selective extraction of lactoferrin from acidic whey using CTAB/n-heptanol reverse micellar system. J. Food Sci. Technol. 56 (5), 2553–2562.

Pendyala, B., Patras, A., 2021. In silico screening of food bioactive compounds to predict potential inhibitors of COVID-19 main protease (Mpro) and RNA-dependent RNA polymerase (RdRp). J. Agric. Food Chem.

Pomeranz, Y., Shands, H.L., 1974. Food uses of barley. C R C Crit. Rev. Food Technol. 4 (3), 377–394.

Price, N., et al., 2020. Application of zinc and calcium acetate to precipitate milk fat globule membrane components from a dairy by-product. J. Dairy Sci. 103 (2), 1303–1314.

Putra, N.R., et al., 2021. Recovery of valuable compounds from palm-pressed fiber by using supercritical CO_2 assisted by ethanol: modeling and optimization. Sep. Sci. Technol. 55 (17), 3126–3139.

Qu, X., et al., 2019. Proteomics analysis of milk fat globule membrane enriched materials derived from by-products during different stages of milk-fat processing. LWT 116, 108531.

Quiles, A., et al., 2018. Fiber from fruit pomace: a review of applications in cereal-based products. Food Rev. Int. 34 (2), 162–181.

Quirós, A.M., Acosta, O.G., Thompson, E., Soto, M., et al., 2019. Effect of ethanolic extraction, thermal vacuum concentration, ultrafiltration, and spray drying on polyphenolic compounds of tropical highland blackberry (Rubus adenotrichos Schltdl.) by-product. J. Food Process Eng. 42 (4), 1–12, e13051. doi:https://doi.org/10.1111/jfpe.13051.

Ran, X.L., et al., 2019. Novel technologies applied for recovery and value addition of high value compounds from plant byproducts: a review. Crit. Rev. Food Sci. Nutr. 59 (3), 450–461.

Regazzoni, L., et al., 2016. Coffee silver skin as a source of polyphenols: high resolution mass spectrometric profiling of components and antioxidant activity. J. Funct. Foods 20, 472–485.

Reinert, K.H., Carbone, J.P., 2008. Synthetic polymers. In: Jørgensen, S.E., Fath, B.D. (Eds.), Encyclopedia of Ecology. Academic Press, Oxford, pp. 3461–3472.

Richter, C., et al., 2015. Plant protein and animal proteins: do they differentially affect cardiovascular disease risk? Adv. Nutr. 6, 712–728.

Romano, R., et al., 2020. Characterisation of oleoresins extracted from tomato waste by liquid and supercritical carbon dioxide. Int. J. Food Sci. Technol. 55 (10), 3334–3342.

Roselló-Soto, E., et al., 2015. Clean recovery of antioxidant compounds from plant foods, by-products and algae assisted by ultrasounds processing. Modeling approaches to optimize processing conditions. Trends Food Sci. Technol. 42 (2), 134–149.

Rouf, R., et al., 2020. Antiviral potential of garlic (Allium sativum) and its organosulfur compounds: a systematic update of pre-clinical and clinical data. Trends Food Sci. Technol. 104, 219–234.

Rubio-Rodríguez, N., et al., 2012. Supercritical fluid extraction of fish oil from fish by-products: a comparison with other extraction methods. J. Food Eng. 109 (2), 238–248.

Suhana Safari, et al. Durian As New Source of Malaysia's Agricultural Wealth. 2018 10-2018 [cited 2020 16-Jan]; Available from: http://ap.fftc.agnet.org/ap_db.php?id=904&print=1. 2018.

Sagar, N.A., et al., 2018. Fruit and vegetable waste: bioactive compounds, their extraction, and possible utilization. Comprehens. Rev. Food Sci. Food Saf. 17 (3), 512–531.

Santana, A.L., et al., 2019. Pressurized liquid- and supercritical fluid extraction of crude and waste seeds of guarana (Paullinia cupana): obtaining of bioactive compounds and mathematical modeling. Food Bioprod. Process. 117, 194–202.

Santhakumar, A.B., Battino, M., Alvarez-Suarez, J.M., 2018. Dietary polyphenols: structures, bioavailability and protective effects against atherosclerosis. Food Chem. Toxicol. 113, 49–65.

Santos-Sánchez, N.F., Salas-Coronado, R., Villanueva-Cañongo, C., Hernández-Carlos, B., et al., 2019. Antioxidant compounds and their antioxidant mechanism, antioxidants. In: Shalaby, Emad (Ed.), et al., Antioxidant. IntechOpen. Digital published, pp. 1–29.

Schieber, A., Stintzing, F.C., Carle, R., 2001. By-products of plant food processing as a source of functional compounds—recent developments. Trends Food Sci. Technol. 12 (11), 401–413.

Shahidi, F., Ambigaipalan, P., 2015. Phenolics and polyphenolics in foods, beverages and spices: antioxidant activity and health effects – a review. J. Funct. Foods 18, 820–897.

Shahidi, F., de Camargo, A.C., 2016. Tocopherols and tocotrienols in common and emerging dietary sources: occurrence, applications, and health benefits. Int. J. Mol. Sci. 17 (10), 1–29, 1745. doi:10.3390/ijms17101745.

Shahidi, F., Yeo, J., 2018. Bioactivities of phenolics by focusing on suppression of chronic diseases: a review. Int. J. Mol. Sci. 19 (6), 1–16, 1573. doi:10.3390/ijms19061573. PMC6032343.

Shammugasamy, B., et al., 2019. Effect of citrus peel extracts on the cellular quiescence of prostate cancer cells. Food Funct. 10 (6), 3727–3737.

Sharma, P., et al., 2018. Chapter 11 – Dietary fibers: a way to a healthy microbiome. In: Holban, A.M., Grumezescu, A.M. (Eds.), Chapter 11 – Dietary fibers: a way to a healthy microbiome. Diet, Microbiome and Health, 299–345.

Shen, X., et al., 2019. Novel technologies in utilization of byproducts of animal food processing: a review. Crit. Rev. Food Sci. Nutr. 59 (21), 3420–3430.

Sila, A., Bougatef, A., 2016. *Antioxidant peptides from marine by-products: isolation, identification and application in food systems.* A review. J. Funct. Foods 21, 10–26.

Silprasit, K., et al., 2011. Anti-HIV-1 reverse transcriptase activities of hexane extracts from some Asian medicinal plants. J. Med. Plants Res 5, 4194–4201.

Singh, H.P., 2020. Policies and strategies conductive to potato development in Asia and the Pacific region. FAO, Delhi, India.

Smithers, G.W., et al., 1996. New opportunities from the isolation and utilization of whey proteins. J. Dairy Sci. 79 (8), 1454–1459.

Talaei Zanjani, N., et al., 2016. Abalone hemocyanin blocks the entry of herpes simplex virus 1 into cells: a potential new antiviral strategy. Antimicrob. Agents Chemother. 60 (2), 1003–1012.

Tan, L., et al., 2019. Simultaneous identification and quantification of five flavonoids in the seeds of Rheum palmatum L. by using accelerated solvent extraction and HPLC-PDA- ESI/MSn. Arab. J. Chem. 12 (7), 1345–1352.

TaracTechnologies. *List Of Permitted Carrier Or Extraction Solvents (Lists Of Permitted Food Additives).* 2016 [cited 2020; Available from: https://www.canada.ca/en/health-canada/services/food-nutrition/food-safety/food-additives/lists-permitted/15-carrier-extraction-solvents.html.

TaracTechnologies. *GrapEX Grape Seed Tannin.* 2016 [cited 2020 14-12]; Available from: https://www.tarac.com.au/assets/product-attachments/Grapex-Grape-Seed-Tannin-Information-Sheet-26May2016.pdf.

Teixeira, A., et al., 2014. Natural bioactive compounds from winery by-products as health promoters: a review. Int. J. Mol. Sci. 15 (9), 15638–15678.

Teslić, N., Bojanićb, N., Čolovića, D., Fištešb, A., Rakićb, D., Solarov, M.B., Zekovićb, Z., Pavlićb, B., et al., 2020. Conventional versus novel extraction techniques for wheat germ oil recovery: multi-response optimization of supercritical fluid extraction. Sep. Sci. Technol. 56 (9), 1546–1561. doi:10.1080/01496395.2020.1784941.

Tessier, A.J., Chevalier, S., 2018. An update on protein, leucine, omega-3 fatty acids, and vitamin D in the prevention and treatment of Sarcopenia and functional decline. Nutrients 10 (8), 1–17, 1099. doi:10.3390/nu10081099.

Tonini, D., Albizzati, P.F., Astrup, T.F., 2018. Environmental impacts of food waste: learnings and challenges from a case study on UK. Waste Manage. (Oxford) 76, 744–766.

Torabian, G., et al., 2019. Anti-influenza activity of elderberry (Sambucus nigra). J. Funct. Foods 54, 353–360.

Varun, T.K., et al., 2017. Extraction of chitosan and its oligomers from shrimp shell waste, their characterization and antimicrobial effect. Vet. World 10 (2), 170.

Vercruysse, L., Van Camp, J., Smagghe, G., 2005. ACE inhibitory peptides derived from enzymatic hydrolysates of animal muscle protein: a review. J. Agric. Food Chem. 53 (21), 8106–8115.

Veselinovic, M., Vasiljevic, D., Vucic, V., Petrovic, S., Tomic-Lucic, A., Savic, M., Zivanovic, S., Stojic, V., Jakovljevic, V., et al., 2017. Clinical benefits of n-3 PUFA and -linolenic acid in patients with rheumatoid arthritis. Nutrients 9 (4), 1–11, 325. doi:10.3390/nu9040325.

Vieira, E.F., Ferreira, I.M., 2017. Antioxidant and antihypertensive hydrolysates obtained from by-products of cannery sardine and brewing industries. Int. J. Food Prop. 20 (3), 662–673.

Wang, Q.L., et al., 2016. High quality lard with low cholesterol content produced by aqueous enzymatic extraction and β-cyclodextrin treatment. Eur. J. Lipid Sci. Technol. 118 (4), 553–563.

Wang, X.Y., Kristo, E., LaPointe, G., 2020. Adding apple pomace as a functional ingredient in stirred-type yogurt and yogurt drinks. Food Hydrocolloids 100, 10.

Wang, Y., Gu, W., 2018. Study on supercritical fluid extraction of solanesol from industrial tobacco waste. J. Supercrit. Fluids 138, 228–237.

Wilson, D.W., et al., 2017. The role of food antioxidants, benefits of functional foods, and influence of feeding habits on the health of the older person: an overview. Antioxidants (Basel, Switzerland) 6 (4), 81.

Wu, J., et al., 2016. Comparison of Haliotis rubra hemocyanin isoforms 1 and 2. Gene Rep. 4, 123–130.

Wu, S.F., et al., 2019. Bioactive protein/peptides of flaxseed: a review. Trends Food Sci. Technol. 92, 184–193.

Wu, T.H., Bechtel, P.J., 2008. Salmon by-product storage and oil extraction. Food Chem. 111 (4), 868–871.

Wu, Y.V., Bagby, M.O., 1987. Recovery of protein-rich by-products from sweet potato stillage following alcohol distillation. J. Agric. Food Chem. 35 (3), 321–325.

Wu, S., Kang, H., 2017. Advances in research and application of fish scale collagen. Agric. Sci. Technol. 18 (12), 2543–2553.

Xie, L., et al., 2019. Chrysophanol: a review of its pharmacology, toxicity and pharmacokinetics. J. Pharm. Pharmacol. 71 (10), 1475–1487.

Xiong, X.N., Yu, I.K.M., Tsang, D.C.W., Bolan, N., Ok, Y.S., Igalavithana, A.D., Kirkham, M.B., Kim, K.-H., Vikrant, K., et al., 2019. Value-added chemicals from food supply chain wastes: state-of-the-art review and future prospects. Chem. Eng. J., 375, 121983, https://doi.org/10.1016/j.cej.2019.121983.

Xu, D.-P., et al., 2017. Natural antioxidants in foods and medicinal plants: extraction, assessment and resources. Int. J. Mol. Sci. 18 (1), 96.

Xu, W., Reddy, N., Yang, Y., 2007. An acidic method of Zein extraction from DDGS. J. Agric. Food Chem. 55 (15), 6279–6284.

Xue, Z.H., et al., 2019. Structure, thermal and rheological properties of different soluble dietary fiber fractions from mushroom Lentinula edodes (Berk.) Pegler residues. Food Hydrocolloids 95, 10–18.

Yalçın, E., Çelik, S., İbanoğlu, E., 2007. Foaming properties of barley protein isolates and hydrolysates. Eur. Food Res. Technol. 226 (5), 967.

Yan, J.K., et al., 2019. Subcritical water extraction-based methods affect the physicochemical and functional properties of soluble dietary fibers from wheat bran. Food Chem., 298.

Yan, N., et al., 2015. Solanesol: a review of its resources, derivatives, bioactivities, medicinal applications, and biosynthesis. Phytochem. Rev. 14 (3), 403–417.

Zanjani, N.T., et al., 2014. Formulation of abalone hemocyanin with high antiviral activity and stability. Eur. J. Pharm. Sci. 53, 77–85.

Zanjani, N.T., et al., 2018. From ocean to bedside: the therapeutic potential of Molluscan hemocyanins. Curr. Med. Chem. 25 (20), 2292–2303.

Zdybel, B., Rozylo, R., Sagan, A., 2019. Use of a waste product from the pressing of chia seed oil in wheat and gluten-free bread processing. J. Food Process. Preserv. 43 (8), e14002. doi:10.1111/jfpp.14002.

Zhao, X., Zhang, S.-S., Zhang, X.-K., He, F., Duan, C.-Q., et al., 2020. An effective method for the semi-preparative isolation of high-purity anthocyanin monomers from grape pomace. Food Chem., 310, 125830. doi:https://doi.org/10.1016/j.foodchem.2019.125830.

Zhong, L.Z., et al., 2019. Extrusion cooking increases soluble dietary fibre of lupin seed coat. LWT-Food Sci. Technol. 99, 547–554.

Zhou, L., et al., 2018. Natural product recovery from bilberry (Vaccinium myrtillus L.) Presscake via microwave hydrolysis. ACS Sustain. Chem. Eng. 6 (3), 3676–3685.

Zhou, Y., Xu, X.-Y., Gan, R.-Y., Zheng, J., Li, Y., Zhang, J.-J., Xu, D.-P., Li, H.-B., et al., 2019. Optimization of ultrasound-assisted extraction of antioxidant polyphenols from the seed coats of red sword bean (Canavalia gladiate (Jacq.) DC.). Antioxidants 8 (7), 1–13, 200. doi:10.3390/antiox8070200.

Zhu, X.-r., et al., 2019. Pericarp and seed of litchi and longan fruits: constituent, extraction, bioactive activity, and potential utilization. J. Zhejiang Univ.-Sci. B 20 (6), 503–512.

Recent developments in fermentation technology: toward the next revolution in food production

5

Netsanet Shiferaw Terefe

CSIRO Agriculture and Food, Werribee, VIC, Australia

5.1 Introduction

Fermentation is an ancient technology as old as human civilization. Humans invented fermentation primarily as a way of preserving perishable food materials such as fruit, vegetable, and meat for periods of food scarcity (Steinkraus, 2004). Over the years, humans discovered empirically that fermentation not only preserves their food but converts it to foods and beverages with desirable sensory attributes. Fermented foods still make up a large proportion of the diets in the developing world and East Asia, although their importance had declined until recently in the West due to the introduction of modern food preservation techniques such as refrigeration. Yet, fermented foods including bread, sausage, and yogurt have remained essential foods in the western diet; a testament to the importance of fermentation not only as a preservation technology but also a technology for transforming food materials into products with desirable texture and flavor attributes (Terefe, 2016). Currently, there is a renewed interest in fermented foods in the West, driven mainly by the perceived health benefits of fermented foods (Hugenholtz, 2013), a perception shared by many traditional societies.

Today, food fermentation is undergoing a significant transformation. Developments in systems biology, "omics," metabolic engineering and synthetic biology are enabling rational design of starter cultures (Shiferaw Terefe and Augustin, 2020) for sustainable production of food ingredients traditionally sourced from agricultural raw materials. These developments coupled with the significant advances in fermenter design and engineering will potentially have a profound impact on our food production system. This chapter gives an overview of fermentation process design, industrial fermentation processes, and current trends in food fermentation and the impact of advances in synthetic biology on fermentation and its implication to sustainable manufacture of food ingredients.

5.2 Fermentation process engineering
5.2.1 Introduction

Fermentation processes can be broadly divided into submerged and solid-state fermentation processes. Submerged fermentation processes are conducted in the presence of free water in a liquid media whereby the fermenting organisms grow submerged feeding on soluble substrates with total solid content ranging from 5% to 10%. Submerged fermentation processes are by far the most common in biotechnological applications and the most advanced in terms of design and operation (Kaur and Vohra, 2013). Production of yogurt, beer, and condiments such as vinegar are conducted under submerged fermentation conditions. Solid-state fermentation on the other hand involves the cultivation of microorganisms in the absence or near absence of free water but sufficient moisture content and water activity that enables growth of the fermenting organisms (Rodríguez_Leon et al., 2013; Singhania et al., 2009), with total solid content ranging from 60% to 80%. Solid-state fermentation is more common in traditional fermentation of food products such as bread, cheese, miso, tempeh, and koji fermentation for soy sauce. Solid-state fermentation enables high substrate and product concentration, high productivity and the utilization of low-value substrates such as agri-food waste (Singhania et al., 2009). Nevertheless, the process is highly heterogeneous with gradients in temperature, gas, and matrix composition and biomass (Rodríguez_Leon et al., 2013) making the kinetic and thermodynamic characterization of the system difficult. There have been several progresses in modeling of solid-state fermentation processes, reactor design, and application studies (production of enzymes, pigments, organic acids, etc.) especially over the last 30 years (Singhania et al., 2009; Soccol et al., 2017). However, industrial application of the technology so far is rare except in traditional food fermentation, composting and ensiling due to challenges related to scale-up, process monitoring and control, and availability of fermenter designs that enable efficient heat removal and mass transfer at a larger scale (Arora et al., 2018).

In general, fermentation processes are operated in batch, semibatch, or continuous mode. Most industrial fermentation including food fermentation processes are conducted in batch with a few exceptions such as the production of vinegar, baker's yeast, and the mycelial meat analogue Quorn. Continuous fermentation processes were employed by the brewing industry for a period of time using two approaches: a cascade and a tower fermentation system. The cascade system involved the use of three vessels, two for fermentation and one for biomass separation and could be operated continuously for up to 3 months with wort residence time of 19 h. However, it was discontinued due to excessive biomass production. In the tower system, the bottom part of the tower was packed with flocculated yeast while the upper part of the tower was a yeast separation zone with diameter twice that of the bottom part. The wort was introduced at the bottom of the tower and progressively fermented as it passes through the tower leaving the fermenter through the yeast separation zone. The tower process reduced the wort fermentation time from 1 week in a traditional batch fermentation process to 4–8 h. However, it suffers from a number of drawbacks such as long start-up time, the complexity of the plant, the need for a more skilled workforce, and the difficulty to match the flavor of the traditionally fermented beer. Thus, it was discontinued with the introduction of the cylinderoconical batch fermentation vessel that shortened the batch fermentation time to 48 h (Stanbury et al., 1995).

Batch operations are common in general due to the reduced risk of contaminations and cell mutations and the lower capital investment compared to an equivalent continuous process of the same volume although they suffer from lower productivity due to the time required for filling, sterilization, cooling prior to fermentation, and for emptying and cleaning after fermentation. A good compromise between a continuous and a batch process is a semibatch process which involves a continuous supply of a substrate(s) during certain periods of a batch process allowing process optimization and improvement in productivity. For instance, the fermentation may be conducted as a batch process until the limiting substrate is depleted and then a defined batch of the substrate is supplied continuously or intermittently at a specific rate required to extend the exponential growth phase. For a detailed discussion on the various fermentation operation modes, one may refer to Stanbury et al. (1995) and Williams (2002).

5.2.2 Fermentation process design

Fermentation engineering, especially with respect to submerged fermentation, is a mature science with standardized approaches for the design and optimization of fermentation processes and fermenter design. The initial step of the process design is usually the screening, identification, characterization, and selection of microbial strains for target applications. Recent developments in systems biology, metabolic engineering, and synthetic biology techniques are enabling the construction of efficient metabolic pathways for target products and their insertion in chassis organisms such as *E. coli*, reducing the efforts for screening and identification of suitable starters. This is followed by screening of growth media, kinetic studies to determine growth rate and the rate of substrate uptake and product formation via shaking flask experiments. Subsequently, laboratory scale fermentation studies are conducted to determine the optimal processing conditions for the highest titer (g/L), yield (g/g substrate), and productivity (g/L h) followed by a step-wise scale-up to pilot and commercial scales. Yield is especially important for low-value/high volume products, which include whole cells, primary metabolites and many secondary metabolites, where the cost of the substrate comprises a significant proportion of the total production cost (Villadsen et al., 2011b). During scale-up of submerged fermentation processes, factors such as specific power input (P/V), the volumetric mass transfer coefficient (K_La), shear rate, and dissolved oxygen concentration are used as scale-up parameters (Villadsen et al., 2011a). The design of submerged fermentation processes involves the use of the laboratory scale stirred tank reactor (STR) equipped with instrumentation and feedback control system for monitoring and control of parameters such as pH, temperature, stirring, and aeration rates that enables kinetic, thermodynamic, and hydrodynamic characterization of the system for rational process optimization and scale-up.

Such standardized design approach coupled with advances in fermenter design and process control has significantly benefited the biopharmaceutical and the biotechnology industries. Nevertheless, the current feedback process control systems are designed mainly to maintain processing parameters (e.g., pH, temperature…) at designed set values and are not able to make reactive adjustment to processing parameters based on real-time process performance data. As such, they are not adequate to prevent product rework (i.e., reprocessing) and at extreme situations product rejection or recall that may arise due to slight changes in raw material properties or the metabolic activities of the starter organisms (Panikuttira et al., 2018; Rathore et al., 2010). This has led to the introduction of the Process Analytical Technology (PAT) initiative, which is defined by FDA as "a system for designing, analyzing, and controlling manufacturing through timely measurements (i.e., during processing) of critical quality and performance attributes of raw and in-process materials and processes with the goal of ensuring final product quality" (Junker and Wang, 2006). PAT involves identification of critical quality attributes (CQAs) of the target product and critical process parameters (CPPs) that impact them, development of analytic or monitoring technique for CQAs and CPPs (on-line, at-line, or off-line) and understanding the interdependence of CQAs and CPPs as the bases for designing appropriate feedback control system that enables real time process decisions based on analytic data (Rathore et al., 2010). One of the challenges in this respect is establishing accurate mechanistic models that describe the relationship between CPPs and CQAs as the basis for process control and optimization due to insufficient understanding of complex and dynamic biological processes (Junker and Wang, 2006; Mears et al., 2017). In this regard, the use of design of experiments (DOE) methodologies such as response surface (RSM) (Mandenius and Brundin, 2008) and hybrid modeling (mechanistic modeling coupled with data-driven modeling) (Mears et al., 2017) are suggested as possible options. The other issue is the lack of reliable in-line sensors for biomass, substrate, and product concentrations. Fluorescence spectroscopy coupled with chemometrics (Faassen and Hitzmann, 2015), NIR spectroscopy (Grassi and Alamprese, 2018), and attenuated total reflectance-mid-infrared (ATR-MIR) spectroscopy coupled with chemometrics (Schorn-García et al., 2021) have been evaluated for this application. Clearly, in order to fully implement PAT, more research is required to develop robust models describing the relationship between CPPs and CQAs and reliable in-line sensors for real time measurement of CQAs.

In the case of solid-state fermentation processes, there are no standardized process design approaches. The design methodologies developed for submerged fermentation cannot be applied due to differences in the biochemical and hydrodynamic characteristics of the two processes (Soccol et al., 2017). Some of the challenges with scale-up of solid-state fermentation processes include the limited reproducibility of experimental results, the heterogeneity in temperature and substrate concentration that arises from heat generation during fermentation and the resulting evaporation of water, and the inadequacy of the mathematical models developed so far to describe the complex physical and biochemical changes that occur during solid-state fermentation (Rodríguez_Leon et al., 2013; Soccol et al., 2017).

5.2.3 Fermenter design
5.2.3.1 Submerged fermenters
The main types of bioreactors for submerged fermentation processes can be roughly divided into stirred tank (STR) and plug flow reactors. STRs are by far the most used because of their flexibility. These bioreactors are equipped with devices for efficient mixing and cooling of the medium, sensors for continuous monitoring of temperature, pH, dissolved oxygen, and level coupled with feedback control loops to control these parameters at the desired level. They are also equipped with feed ports for continuous supply of air and other gases, nutrients in the case of feed-batch operations, alkali and acid solutions for maintaining pH at the desired value, and antifoaming agents to maintain the level at desired value. Plug flow reactors are used only in applications such as remediation of very toxic wastewater where the complete conversion of the substrate is required in which case a STR is used for pretreatment followed by a plug flow reactor. There are also some variations of the stirred tank fermenter including tower fermenter, airlift fermenter, bubble column reactor and fluidized bed reactor which are pneumatically agitated instead of mechanical agitation. These fermenters are industrially used in applications such as mold fermentation where mechanical agitation induced shear stress may cause cell damage. For instance, the airlift fermenter is used in the production of Quorn, the mycelial meat analogue that is made from the biomass of *Fusarium graminearum*, which is cultivated in a continuous fermentation process (Kaur and Vohra, 2013). For more detailed discussion on submerged fermenters, one may refer to Williams (2002).

5.2.3.2 Solid-state fermenters
There are four types of commonly used solid-state fermenters; tray, packed bed, horizontal drum, and fluidized bed reactors, which vary based on the aeration and agitation strategies (Berovic, 2019) (Fig. 5.1). Tray fermenters are the most basic solid-state fermenters that are commonly used in traditional solid-state fermentation of foods in Asia where the substrates are thinly spread (5–10 cm) over trays made of stainless steel, plastic, or wood, which are stacked in a shelf on top of one another and kept in a room or a cabinet with continuous air circulation to maintain constant temperature and humidity (Fig. 5.2). The trays are typically perforated at the bottom and on the sides to facilitate air flow (Soccol et al., 2017; Berovic, 2019). Most of the other solid-state fermenters are still at the experimental

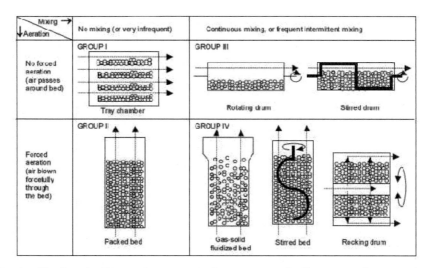

FIG. 5.1 The classification of solid-state fermenters design based on agitation and aeration strategies.

Adapted with permission from Mitchell et al. (2000).

stage except the drum fermenter, which has been developed by the Japanese firm Fujiwara and used for koji fermentation in many parts of Asia (Soccol et al., 2017). The drum fermenter typically consists of a cylindrical drum lying horizontally, which is operated partially filled with a bed of the substrate and air blown over the headspace. In rotary drum fermenters, the drum rotates around its central axis to mix the bed. In stirred drum fermenters, paddles mounted on a shaft along the central axis of the drum rotate and mix the bed while the drum body remains stationary (Fig. 5.3) (Mitchell et al., 2006).

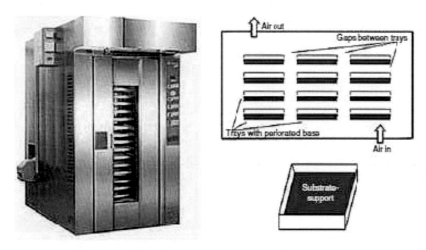

FIG. 5.2 A tray fermenter and schematic diagrams showing a group of trays and an individual tray.

Adapted with permission from Berovic (2019).

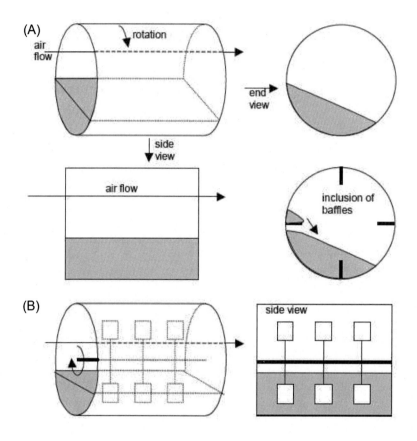

FIG. 5.3

Schematic diagrams showing the basic features of (A) rotary drum solid-state fermenter and (B) stirred drum solid-state fermenter.

Adapted with permission from Mitchell et al. (2006).

5.3 Industrial food fermentation

Food fermentation processes have significantly benefited from developments in microbiology, biotechnology and process engineering over the last century. The production of several fermented foods and beverages including wine, beer, and spirits as well as bread, sausages, sauerkraut, pickles, yogurt, and cheese have been industrialized. However, the design and operation of industrial food fermentation processes, except those for food ingredients and processing aids such as flavorings and enzymes, differ from that of standard submerged fermentation processes described in the previous sections.

5.3.1 Advances in industrial vegetable fermentation

The industrial production of fermented vegetables of commercial importance including sauerkraut, fermented cucumber (pickles), and fermented olives are little different from the traditional fermentation of these products, except the automation of some processing steps, a degree of standardization due to better understanding of the microbiology and biochemistry of the processes and the introduction of process monitoring (periodic measurement of pH, acidity, etc.). For instance, the industrial production of fermented cucumber, often referred to as pickles, is conducted in a similar way as the traditional small-scale process, with some degree of automation dependent on the scale of the operation (Franco et al., 2016). The process involves preparation of brine solution, partial filling of the open fermentation tank with brine for cushioning the cucumbers during loading, sorting, culling out, and loading of cucumbers into the fermentation tank, closing the tank with wooden boards to keep the cucumbers submerged in the brine, final salt addition to adjust the salt content to 6 to 7% which is followed by spontaneous fermentation for 3–4 weeks depending on the temperature until all the fermentable sugars are depleted (Franco et al., 2016; Fleming, 1995). The tanks are periodically sparged with air to remove carbon dioxide especially during the first 14 days of fermentation to prevent bloating (Franco et al., 2016). The brine is usually acidified with acetic acid or vinegar to pH 4.5 or slightly lower to facilitate carbon dioxide sparging and inhibit the growth of undesirable microorganisms and prevent bloating (Fleming, 1995). A small amount of calcium chloride is also added to the brine to improve the texture of the product (Franco et al., 2016, Medina-Pradas et al., 2017). The salt concentration, the pH of the brine, acidity, and sugar are periodically monitored during the various stages of fermentation (Franco et al., 2016).

In the past, open wooden or concrete tanks with flat bottoms were used as fermentation tanks. However, they are now largely replaced by fiberglass reinforced plastic tanks of similar design with capacity ranging from 15 to 20 tons. Wooden planks secured to the tanks with brackets or semicircular fiberglass covers with holes to allow free movement of brine are used to keep the cucumbers submerged inside the brine especially during the early stages of fermentation (Franco et al., 2016). The loading of fresh cucumbers into the tanks and unloading of fermented cucumbers are done manually in small-scale operations while they are largely automated in large-scale operations. Once fermentation is completed, the pickles are either stored in a tank with added salt for preservation for up to 1 year or further processed into finished products. The pickles are typically desalted by flushing with tap water and immersed in water for up to 24 h to reduce their salt content to acceptable level. This is followed by canning in a reformulated brine containing different condiments such as sugar, cinnamon, garlic, etc. depending on the target product (Franco et al., 2016).

5.3.2 Advances in other fermentation processes

As in the case of vegetable fermentation, the basic unit operations of other traditional fermentation processes such as wine (Pérez-Torrado et al., 2018) and beer (Bamforth, 2017) fermentation did not substantially change with industrialization. Perhaps, dairy fermentation has been the most affected by technological advances and industrialization (Medina-Pradas et al., 2017; Aryana and Olson, 2017). Even then, the changes are not of profound nature. Overall, advances in microbiology and biotechnology over the last century have enabled the development of pure starter cultures for the manufacture of products such as yogurt, cheese, and beer. Moreover, better understanding of the microbiology and the biochemistry of the respective fermentation processes has led to the establishment of procedures for rigorous control of raw material properties, standardized pretreatment of substrates, and control and continuous monitoring of processing conditions (e.g., temperature, pH) that improved product safety, quality, stability, and consistency. Some degree of automation of the upstream and the downstream processes (e.g., introduction of industrial centrifuges for beer separation) and changes in the materials of construction and design of fermentation tanks (e.g., changes from earthenware or wooden tanks to stainless steel or fiberglass tanks, introduction of the cylinderoconical vessels for beer and yogurt fermentation) have been introduced with industrialization. There have also been researches toward implementation of PAT in food fermentation processes (Panikuttira et al., 2018; Schorn-García et al., 2021). However, attempts to fundamentally alter traditional processes have met with limited success. For instance, in brewing, alternative processing approaches such as the use of enzymes as alternative to malt, continuous fermentation as opposed to the traditional batch fermentation (refer to previous section), the use of immobilized cells, and synthesizing beer from an alcoholic base and flavorings were not convincing enough to be adopted by industrial brewers (Bamforth, 2017).

Overall, industrial food and beverage fermentation processes still combine science and some element of art peculiar to the specific fermented product, with the design process and degree of automation dependent on the process.

This could be due to several factors including

i. The nature of food substrates such as milk, grains, vegetables, and fruits, which are often complex in structure and composition, unlike the substrates in most biotechnological processes which are relatively simple sterile media composed of carbon and nitrogen sources, vitamins, and minerals designed for the target process.
ii. The complexity of the fermentation processes themselves, which often combine both catabolic biotransformation of macromolecules as well as biosynthesis of new metabolites that determine the safety, functional, and sensory attributes of the final product, unlike standard biotechnological processes designed for the biosynthesis of target primary or secondary metabolites, which are extracted and purified in downstream processing.
iii. The legacy from traditional processes and the expected sensory attributes, which may not be replicated if pure cultures are used, the substrate is sterilized or the process is fully automated, which necessitate a combination of both art and scientific approaches.

5.4 Recent developments in food fermentation

The last decade has seen a significant surge of interest in fermented foods and fermentation technology both by consumers and the food industry. The main drivers being the perceived health benefits of fermented foods, sustainability and consumer's increasing interest in "natural" and preservative free products, and changes in the dietary pattern of consumers, such as the rising prevalence of veganism, vegetarianism, and flexitarianism (Hugenholtz, 2013; Shiferaw Terefe and Augustin, 2020), which are fueling the plant-based trend. The world economic forum in November 2020 described fermentation as "an opportunity to fundamentally change the way the world eats and improve global human and environmental health and the economy" (GFI, 2020). Fermented foods are positioned at the center of mega trends such as healthy living (e.g., natural, clean label, functional...), ethical living (e.g., plant-based, sustainable...), and shifting market frontiers (e.g., exoticism: growing experimentation with exotic flavors especially by millennials ...) that are shaping consumer markets worldwide (International, 2018; Askew, 2018). Among growth drivers, gut health accounted for the highest proportion (1.0 relative score), followed by sugar reduction (0.22), prebiotic/probiotic delivery (0.15), and the plant-based trend (0.07) in that order. Nevertheless, interest in plant based foods increased by 532.8%

compared to 110.3% for gut health foods over the last 4 years (2016–2020) (Ravi, 2020) and it is likely that the plant-based trend will overtake gut health as the main driver for growth in fermentation over the coming years.

5.4.1 Innovations in traditional or "natural" food fermentation

The global fermented food and drink market including fermented alcoholic and nonalcoholic beverages (e.g., wine, beer, kombucha), fermented dairy (yogurt, cheese, kefir), bakery (e.g., sourdough bread, yeast raised doughnuts), and other fermented food products (e.g., sauerkraut, kimchi, pickles) was $1036.38 billion in 2017 and is estimated to reach $1376.42 billion by 2022 growing at a compound annual growth rate (CAGR) of 5.84% (Technavio.com, 2017). Most of the product innovation over the last 10 years has been on plant-based fermented food products. According to Dupont (Dupont, 2019), the growth in the launch of fermented dairy alternatives between 2011 and 2018 was 43% compared to 16% in fermented dairy products.

The growing interest in fermented foods has led to popularization and commercialization of traditional fermented products such as kefir and kombucha out of their home regions with a rapid expansion in their market size. For instance, the global kefir (a fermented dairy product originally from the Caucasus) market tripled between 2011 and 2016 as it became popular worldwide (Technavio.com, 2017). The global kefir market was valued at $1.23 billion in 2019 and is forecasted to reach 1.84 billion by 2027 growing at a CAGR of 5.4% (Insight, 2020a). Similarly, the global market size of kombucha (a fermented sweet tea product originally from China), grew from $0.5 billion to $1.84 billion in 2019, and is forecasted to reach $10.45 billion by 2027 growing at a CAGR of 23.2% (MarkestandMarkets, 2015; Insights, 2020b). The starter cultures of these traditional products are also being increasingly used for development of novel products. Examples include Bionade: fruit flavored malt beverages fermented using gluconic acid bacteria, one of the bacteria in the starter culture of kombucha, Rythem, coconut milk-based beverages with or without fruit juice fermented using kefir starter cultures (kefir grains), and kefir water products with various fruit juice flavors fermented using kefir grains (e.g., TheGoodSeed, Nexba). Traditional fermentation is also being employed for developing novel food ingredients and products with desirable technological or health functionality, enrichment with bioactive compounds or reducing the level of undesirable components such as sugar, undesirable flavor compounds or antinutrients (Shiferaw Terefe and Augustin, 2020). Recently, novel fermentation-based technologies were developed to enhance the yield of the bioactive isothiocynates in brassica products (Augustin and Terefe, 2019; Cai et al., 2020), reduce sugar levels in sugar rich plant-based products via conversion into oligo- and polysaccharides and sugar alcohols (Augustin et al., 2019; Xu et al., 2020), reduction of bitter flavor notes in coffee, (Watson, 2016c) and production of a pea protein hydrolysate ingredient for enhancing salty taste of food products (Bhomik et al., 2017).

From technological innovation perspective, the application of nonthermal processing technologies such as ultrasound, high pressure (HPP) and pulsed electric field (PEF) processing are being explored for enhancing the efficiency of food fermentation processes and modulate the nutritional and sensory attributes of fermented foods with promising outcomes (Shokri et al., 2021). For instance, ultrasonic pretreatment was reported to increase the growth kinetics and metabolic activities of fermentation starter cultures (Shokri et al., 2020, 2021; Ojha et al., 2018), enhance lactose hydrolysis and conversion to galacto-oligosaccharides during milk fermentation (Nguyen et al., 2009), and shorten the fermentation time of sweet whey (Barukčić et al., 2015), and enhance the bioconversion of isoflavone glycosides into aglycones during fermentation of prebiotic supplemented soymilk (Yeo and Liong, 2011). Similarly, HPP at sublethal level (<50 MPa) has been shown to increase the fermentative activities of starter microorganisms (Gálvez et al., 2011; Mota et al., 2018; Ferreira et al., 2019). On the other hand, HPP at higher pressures (>100 MPa) enhanced the ability of nonsaccharomyces yeast to outcompete grape wild yeast and dominate wine fermentation (Bañuelos et al., 2016), improved the acidification rate of milk during fermentation, lowered coagulation pH and improved gel strength (Tsevdou et al., 2013) and improved the texture and reduced syneresis of low-fat set yogurt (Harte et al., 2003). PEF processing has also been shown to stimulate growth and metabolic activities of fermentation starter cultures (Grosse et al., 1988; Fologea et al., 1998; Ohba et al., 2016), modulate the production of metabolites during kimchi fermentation (Joo et al., 2013), increase the production of bioactive aglycones during fermentation of mannitol supplemented soymilk (Yeo and Liong, 2013) and enhance the fermentation rate of apple cider and reduce ethanol content (Al Daccache et al., 2020). Indeed, all these technologies have potentials to enhance fermentation rate and reduce fermentation time and modulate metabolite profile enabling the manufacturing of fermented foods and beverages with differentiated sensory and nutritional attributes. However, the results so far are based on laboratory scale investigations and further research is required to integrate these technologies into standard fermentation processes at larger scale without compromising the robustness of the fermentation processes (Shokri et al., 2021).

5.4.2 Precision fermentation for production of food products ingredients

A pivotal development in food fermentation is the emergence of precision fermentation as a technology for sustainable manufacture of food ingredients that were traditionally derived from plant or animal sources as an offshoot of developments in synthetic biology. Precision fermentation is the use of genetically engineered microorganisms as starters for production of target molecules via fermentation. It is increasingly being used for the manufacture of a variety of food ingredients that were conventionally sourced from animals and plants. Examples of such products include whey proteins by Perfect Day (Watson, 2016b), production of gelatin by Geltor (Watson, 2016a), vanilla, saffron, and steviol glycosides by Evolva-Cargil (Watson, 2015, 2018), and human breast milk lactoferrin by CONAGEN (GlobeNewswire, 2020).

Precision fermentation per se is not new and genetically engineered microorganisms (recombinant DNA technology) have been used since the 1970s for fermentative production of enzymes and other biomolecules for various applications (Katz et al., 2018; Teng et al., 2021). A classic example is chymosin, that has been produced by recombinant DNA technology since the 1980s for use in cheese manufacturing as an alternative to rennet traditionally derived from calf stomach (Teng et al., 2021). Synthetic biology on the other hand has developed over the last two decades as an extension of genetic engineering by incorporating engineering design principles such as standardization, abstraction of modular parts and decoupling of design from fabrication into genetic engineering to streamline the design, build and test of synthetic gene networks and facilitate the construction of biological systems for target applications (Katz et al., 2018; Agapakis, 2014). Synthetic biology is a multidisciplinary approach that encompasses genetic programming, pathway engineering, metabolic engineering, DNA synthesis and assembly and computer aided design among other disciplines (Katz et al., 2018). It goes beyond the mere alteration of certain genes as in genetic engineering to the creation of new genetic constructs that fulfill target functionality (Boldt and Müller, 2008). The current expansion in the application of precision fermentation to the production of diverse food ingredients is a direct outcome of advances in synthetic biology, which was facilitated by reduction in DNA sequencing and synthesis cost and development of genome editing tools such as CRISPR/Cas9 (Katz et al., 2018). Further advances would enable progress toward the design of "synthetic cell factories where all available resources are diverted to produce the desired compound" with minimal formation of by-products (Teng et al., 2021). In fact, synthetic biology and precision fermentation are believed to reduce our dependence on traditional agriculture and animal husbandry, which are constrained due to limited arable land, climate change and population growth, and enable us to sustainably feed the world and ensure food security (Teng et al., 2021; Lv et al., 2021). Nevertheless, realizing that goal requires more than an efficient cell factory. Rather, a multidisciplinary approach from synthetic biology through process engineering, public policy and social studies are required (Fig. 5.4).

Some of the current challenges with respect to precision fermentation for food production include:

i. Scalability, economic feasibility and sustainability: most of the precision fermentation start-ups for food ingredient manufacturing are at an early stage and it is not clear whether they are economically feasible or sustainable when operated at scale (GFI, 2020; Wilcox, 2019).

ii. Consumer perception: some consumers are concerned that synthetic biology may introduce new proteins and other molecules into foods that may be unsafe or allergenic. GMO labeling does not apply to precision fermentation manufactured ingredients in many jurisdictions (USA, Australia) as long as the molecules are nature equivalent, which makes it difficult for consumers to make informed choices (Wilcox, 2019). Appropriate labeling and transparency coupled with consumer education will be required to enhance the consumer trust and acceptance of precision fermentation ingredients (Lv et al., 2021).

iii. Ethical concerns: synthetic biology in its boldest form is not a mere genetic alteration of an organism rather a creation of "a new life form" that fulfills a target vision/application, raising

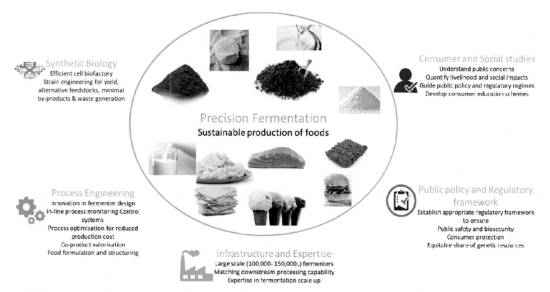

FIG. 5.4 **A schematic representation of the multidisciplinary effort required to unlock the potential of precision fermentation for sustainable food production.**

ethical concerns, and biosafety and biosecurity risks including escape and uncontrolled replications of these organisms to the environment and deliberate misuse by terrorists (Boldt and Müller, 2008; RRDC, 2016). In addition, manufacture of traditionally agriculture derived food ingredients like vanilla and steviol glycosides by precision fermentation affects the livelihood of farmers dependent on these commodities especially in developing countries (Lv et al., 2021; Wilcox, 2019) with societal and ethical implications. All these require concerted effort by academia, governments, and industry to ensure biosecurity, ethical use of the technology and equitable sharing of profits from genetic resources.

iv. Regulatory aspects: policy and regulatory frameworks governing synthetic biology and precision fermentation are still being established worldwide and there are concerns that innovation in the field may outpace the development of appropriate regulation (Wilcox, 2019; RRDC, 2016).

v. Capability in large-scale fermentation and downstream processing: Lack of expertise in fermentation process engineering at commercial scale fermentation and downstream processing facilities has been identified as the main bottlenecks for unlocking the full potential of precision fermentation. The available food fermentation capacity will be consumed in 12–24 months (Warner, 2021), highlighting the need for investment on fermentation facilities.

Overall, more research is required on discovery of better chassis organisms, design of better genetic componentry and artificial intelligence guided metabolic and pathway engineering for higher yield, titer and productivity of target molecules and reduction in by-products, strain engineering for use of alternative feedstocks other than refined sugar including agri-food waste and CO_2, innovations in bioreactors optimized for food grade fermentation, innovation in downstream processing to reduce cost, and co-product valorization, that is, value adding to spent media from fermentation processes (GFI, 2020; Lv et al., 2021) so as to ensure the economic feasibility and sustainability of precision fermentation processes. Further innovations in continuous fermentation processes and the use of immobilized microbial cells may also contribute toward ensuring the cost effectiveness of precision fermentation processes.

5.4.3 Fermentation for valorization of food waste

A significant amount of organic waste is generated in the agri-food supply chain from on-farm production through postharvest handling, processing, distribution, and consumption. Edible food waste alone accounts for over one third of the edible food produced globally, equivalent to about 1.3 billion tons (Uçkun Kiran et al., 2014). In Australia, 7.6 million tons of food waste are generated per year. Approximately 70% of this waste is edible and the cost to the Australian economy is estimated at 36.6 billion per year, which is quite substantial (Li et al., 2014). Currently, most food waste streams worldwide are used for low-value animal feed, compost, anaerobic digestion and incineration with other combustible municipal waste for energy generation (Lin et al., 2014). Clearly, food waste is a valuable resource that needs to be minimized and the unavoidable part converted to food ingredients and other valuable food products (Terefe, 2016). In this regard, fermentation is being explored as one of the enabling technologies for the conversion of food waste into high-value food ingredients and other value-added products (Lin et al., 2014; Koutinas et al., 2014). Some of the applications investigated include fermentation pretreatment of fruit processing waste for enhancing the extractability and bioactivity of phenolic antioxidants (Martinez-Avila et al., 2014; Madeira Junior et al., 2015), oil pressing cakes, fruit, vegetable, and bakery wastes as feedstocks for production of food flavorings (Laufenberg et al., 2004) enzymes (Uçkun Kiran et al., 2014), organic acids (Zhang et al., 2013), and microbial biomass

(Stabnikova et al., 2005; Di Donato et al., 2014). The food industry is also progressing toward integrated processing for efficient utilization of raw materials. An interesting example in Australia is the Manildra group that produces gluten, starch, and glucose syrup from wheat grain and ferment the waste generated in the manufacture of the three products to ethanol for food and beverage, pharmaceutical, and personal care applications enabling 100% utilization of the wheat grain (Manildra, 2021). With respect to food waste valorization, biomass fermentation, i.e. production of microbial biomass using food waste as feedstock is particularly interesting as it enables the efficient conversion of a variety of agri-food waste into high protein microbial biomass for food application (Teng et al., 2021).

5.4.4 Fermentation and the alternative protein trend

Fermentation is considered as one of the key three technologies that are enabling the alternative protein revolution with the other two being plant-based proteins and cultivated proteins (GFI, 2020). In the context of the alternative protein industry, fermentation is being explored for:

i. Transforming plant materials into alternative protein food products like Tempeh, i.e. traditional fermentation
ii. Cultivating microorganisms for use as a primary protein source, i.e. biomass fermentation as in the case of Quorn or
iii. Producing specialized ingredients such as proteins, fats, enzymes, and flavoring for use in plant-based products and cultivated meat mainly via precision fermentation, an example being soy leghemoglobin that is being used in the plant based "Impossible" burger.

The focus in this respect is progressively shifting toward precision fermentation. Of the 14 newly launched alternative protein start-ups in 2019, eight were on biomass fermentation whereas of the 13 new start-ups in 2020, nine were on precision fermentation, three on biomass, and one on traditional fermentation. Fermentation companies in the alternative protein sector raised a total of $587 million investment in 2020, which was a twofold increase from the previous year (GFI, 2020). Clearly, there is a huge market driven interest in fermentation technology for manufacture of ingredients for the alternative protein sector. Nevertheless, most of the start-ups are at a very early stage and they face most of the challenges highlighted earlier for precision fermentation including cost-effective production of the protein ingredients and therefore cost competitiveness with agriculture derived ingredients, regulatory hurdles, and limited availability of large-scale fermentation and downstream processing infrastructure and trained manpower. Perhaps, the biomass fermentation start-ups may have less hurdles in terms of economic feasibility and regulatory approval compared to the precision fermentation-based ones.

In fact, some of the biomass fermentation start-ups (e.g., Air Protein, Solar Foods) are based on the cultivation of hydrogenotrophs on potentially readily available feedstocks (carbon dioxide and hydrogen) (Linder, 2019; Alloul et al., 2021), and may contribute to greenhouse gas emission reduction.

5.5 Conclusion and future perspectives

The 20th century has witnessed the transition of fermentation from a technology for household preservation of primary produce to a sophisticated technology for manufacture of pharmaceuticals, biochemicals, food ingredients, foods, and beverages at industrial scales at the back of advances in microbiology, biotechnology and process engineering.

Fermentation is currently experiencing a resurgence and transformation. This is propelled by changes in consumer attitudes and dietary patterns, concern for the environment, sustainability, and food security on one hand and scientific progress in systems biology and metabolic engineering spawning precision fermentation enabling "designed" microbial cell factories that are able to produce biomolecules of interest using nature's genetic blueprint. Consumer's growing awareness of the relationship between diet and health and interest in "natural," clean label' and sustainably and ethically sourced food products is driving growth and innovation in traditional/natural fermented foods and beverages.

The plant-based trend, which is being fueled by similar issues, is driving growth especially in precision fermentation. Fermentation is currently in a growth trajectory that is set to stay for the foreseeable future. Indeed, precision fermentation, in particular, has a huge potential to improve sustainability, food security, human and environmental health, and the global economy. Nevertheless, further research is required to ensure the economic feasibility and scalability of precision fermentation processes including metabolic engineering and strain optimization for productivity and use of alternative feedstocks, innovations in fermenter and downstream equipment design, real-time process and product monitoring, and control system design and co-product valorization. Social science research is also needed in parallel in order to understand the ethical and social dimensions of the issues surrounding synthetic biology and precision fermentation and guide government policy and regulatory framework to ensure biosecurity and biosafety, consumer protection, and equitable share of economic dividends from genetic resources. Investment on large-scale fermentation and downstream facilities dedicated to food ingredient manufacturing is also urgently needed since lack of infrastructure and expertise at large-scale fermentation is becoming an impediment to unlocking the full potential of precision fermentation processes.

References

Agapakis, C.M., 2014. Designing synthetic biology. ACS Synth. Biol. 3 (3), 121–128.

Al Daccache, M., et al., 2020. Control of the sugar/ethanol conversion rate during moderate pulsed electric field-assisted fermentation of a *Hanseniaspora* sp. strain to produce low-alcohol cider. Innov. Food Sci. Emerg. Technol. 59, 102258.

Alloul, A., et al., 2021. Unlocking the genomic potential of aerobes and phototrophs for the production of nutritious and palatable microbial food without arable land or fossil fuels. Microb. Biotechnol 0 (0), 1–7. https://doi.org/10.1111/1751-7915.13747.

Arora, S., Rani, R., Ghosh, S., 2018. Bioreactors in solid state fermentation technology: Design, applications and engineering aspects. J. Biotechnol. 269, 16–34.

Aryana, K.J., Olson, D.W., 2017. A 100-year review: yogurt and other cultured dairy products. J. Dairy Sci. 100 (12), 9987–10013.

Askew, K., 2018. There is a mega-trend around fermentation': the rising star of fermented foods. Food Navigator.

Augustin, M.A., Terefe, N.S., 2019. Preparing Isothiocyanate Containing Product i.e. Probiotic Used as Feed for Aquatic Animal e.g. Fish by Pre-treating Brassicaceae Material to Improve Access of Myrosinase to Glucosinolate, and Fermenting Material With Lactic Acid Bacteria. CSIRO.

Augustin, M.A., Terefe, N.S., Hlaing, M., 2019. Sugar Reduced Products and Method of Producing Thereof. CSIRO.

Bamforth, C.W., 2017. Progress in brewing science and beer production. Annu. Rev. Chem. Biomol. Eng. 8, 161–176.

Bañuelos, M.A., et al., 2016. Grape processing by high hydrostatic pressure: effect on use of non-saccharomyces in must fermentation. Food Bioprocess Technol. 9 (10), 1769–1778.

Barukčić, I., et al., 2015. Influence of high intensity ultrasound on microbial reduction, physico-chemical characteristics and fermentation of sweet whey. Innov. Food Sci. Emerg. Technol. 27, 94–101.

Berovic, M., 2019. Cultivation of medicinal mushroom biomass by solid-state bioprocessing in bioreactors. In: Steudler, S., Werner, A., Cheng, J.J. (Eds.), Solid State Fermentation: Research and Industrial Applications. Springer, Switzerland, pp. 5–22.

Bhomik, T., C.T. Simmons, and S.I. Myaka, Fermented ingredient. 2017.

Boldt, J., Müller, O., 2008. Newtons of the leaves of grass. Nat. Biotechnol. 26 (4), 387–389.

Cai, Y.X., et al., 2020. Mild heat combined with lactic acid fermentation: a novel approach for enhancing sulforaphane yield in broccoli puree. Food Funct. 11 (1), 779–786.

Di Donato, P., et al., 2014. Biomass and biopolymer production using vegetable wastes as cheap substrates for extremophiles. In: Bardone, E., Bravi, M., Keshavarz, T. (Eds.), Ibic 2014: Fourth International Conference on Industrial Biotechnology, 163–168.

Dupont, The rise of plant-based fermented foods: improving quality, extending shelf-life, and reducing foodwaste. 2019.

Faassen, S.M., Hitzmann, B., 2015. Fluorescence spectroscopy and chemometric modeling for bioprocess monitoring. Sensors (Basel) 15 (5), 10271–10291.

Ferreira, R.M., et al., 2019. Adaptation of *Saccharomyces cerevisiae* to high pressure (15, 25 and 35 MPa) to enhance the production of bioethanol. Food Res. Int. 115, 352–359.

Fleming, H.P., 1995. Vegetable fermentations. In: Rehm, H.-J., Reed, G. (Eds.), Biotechnology. VCH, Weinheim, pp. 629–661.

Fologea, D., et al., 1998. Increase of *Saccharomyces cerevisiae* plating efficiency after treatment with bipolar electric pulses. Bioelectrochem. Bioenerg. 46 (2), 285–287.

Franco, W., et al., 2016. Cucumber fermentation. Lactic Acid Ferment. Fruits Vegetables 2016, 107–155.

Gálvez, A., et al., 2011. 16 – Biological control of pathogens and post-processing spoilage microorganisms in fresh and processed fruit and vegetables A2.. In: Lacroix, C. (Ed.), Protective Cultures, Antimicrobial Metabolites and Bacteriophages for Food and Beverage Biopreservation. Woodhead Publishing, Sawston, UK, pp. 403–432.

GFI, 2020. 2020 State of the industry report, Fermentation: Eats, Egg and Dairy, 57.

GlobeNewswire, I., 2020. Infant Formula Evolves and Advances in Nutrition.

Grassi, S., Alamprese, C., 2018. Advances in NIR spectroscopy applied to process analytical technology in food industries. Curr. Opin. Food Sci. 22, 17–21.

Grosse, H.H., Bauer, E., Berg, H., 1988. Electrostimulation during fermentation. Bioelectrochem. Bioenerg. 20 (1), 279–285.

Harte, F., et al., 2003. Low-fat set yogurt made from milk subjected to combinations of high hydrostatic pressure and thermal processing. J. Dairy Sci. 86 (4), 1074–1082.

Hugenholtz, J., 2013. Traditional biotechnology for new foods and beverages. Curr. Opin. Biotechnol. 24 (2), 155–159.

Insight, F.B., The global kefir market size was $1.23 billion in 2019 and is projected to reach $1.84 billion by 2027, exhibiting a CAGR of 5.4% during the forecast period. Read More at: https://www.fortunebusinessinsights.com/kefir-market-102463. 2020a.

Insights, F.B., The global kombucha market size stood at USD 1.84 billion in 2019 and is projected to reach USD 10.45 billion by 2027, exhibiting a CAGR of 23.2% during the forecast period. Read More at: https://www.fortunebusinessinsights.com/industry-reports/kombucha-market-100230 2020b.

International, E., 8 Food Trends for 2018. 2018.

Johanningsmeier, S.D., 2011. Biochemical characterization of fermented cucumber spoilage using non-targeted, comprehensive, two-dimensional gas chromatography-time-of-flight mass spectrometry: anaerobic lactic acid utilization by lactic acid bacteriaFood Science. North Carolina State University, Raleigh, North Carolina, p. 202.

Joo, D.H., Jeon, B.Y., Park, D.H., 2013. Effects of an electric pulse on variation of bacterial community and metabolite production in kimchi-making culture. Biotechnol. Bioprocess Eng.: BBE 18 (5), 909–917.

Junker, B.H., Wang, H.Y., 2006. Bioprocess monitoring and computer control: key roots of the current PAT initiative. Biotechnol. Bioeng. 95 (2), 226–261.

Katz, L., et al., 2018. Synthetic biology advances and applications in the biotechnology industry: a perspective. J. Ind. Microbiol. Biotechnol. 45 (7), 449–461.

Kaur, P., Vohra, A.Satyanarayana, 2013. Laboratory and industrial scale bioreactors for submerged fermentation. In: Soccol, C.R., Pandey, A., Larroche, C. (Eds.), Fermentation Processes Engineering in the Food Industry. CRC Press, Boca Raton, Florida, USA, pp. 165–179.

Koutinas, A.A., et al., 2014. Valorization of industrial waste and by-product streams via fermentation for the production of chemicals and biopolymers. Chem. Soc. Rev. 43 (8), 2587–2627.

Laufenberg, G., Rosato, P., Kunz, B., 2004. Adding value to vegetable waste: oil press cakes as substrates for microbial decalactone production. Eur. J. Lipid Sci. Technol. 106 (4), 207–217.

Li, T., et al., 2014. Open and continuous fermentation: products, conditions and bioprocess economy. Biotechnol. J. 9 (12), 1503–1511.

Lin, C.S.K., et al., 2014. Current and future trends in food waste valorization for the production of chemicals, materials and fuels: a global perspective. Biofuels Bioprod. Biorefining 8 (5), 686–715.

Linder, T., 2019. Making the case for edible microorganisms as an integral part of a more sustainable and resilient food production system. Food Security 11 (2), 265–278.

Lv, X., et al., 2021. Synthetic biology for future food: research progress and future directions. Fut. Foods 3, 100025.

Madeira Junior, J.V., Teixeira, C.B., Macedo, G.A., 2015. Biotransformation and bioconversion of phenolic compounds obtainment: an overview. Crit. Rev. Biotechnol. 35 (1), 75–81.

Mandenius, C.F., Brundin, A., 2008. Bioprocess optimization using design-of-experiments methodology. Biotechnol. Prog. 24 (6), 1191–1203.

Manildra. 2021.

MarkestandMarkets, Kombucha Market by Types (Bacteria, Yeast, Mold, Others), Flavors (Herbs & Spices, Citrus, Berries, Apple, Coconut & Mangoes, Flowers, Others), & by Region - Forecasts to 2020. 2015.

Martinez-Avila, G.C.G., et al., 2014. Fruit wastes fermentation for phenolic antioxidants production and their application in manufacture of edible coatings and films. Crit. Rev. Food Sci. Nutr. 54 (3), 303–311.

Mears, L., et al., 2017. Mechanistic fermentation models for process design, monitoring, and control. Trends Biotechnol. 35 (10), 914–924.

Medina-Pradas, E., et al., 2017. Chapter 9 - Review of vegetable fermentations with particular emphasis on processing modifications, microbial ecology, and spoilage. In: Bevilacqua, A., Corbo, M.R., Sinigaglia, M. (Eds.), The Microbiological Quality of Food. Woodhead Publishing, Sawston, UK, pp. 211–236.

Mitchell, D.A., et al., Group III: rotating-drum and stirred-drum bioreactors. In: Solid-State Fermentation Bioreactors: Fundamentals of Design and Operation, D.A. Mitchell, M. Berovič, and N. Krieger (Eds.), 2006, Springer Berlin Heidelberg: Berlin, Heidelberg. pp. 95-114.

Mitchell, D.A., Berovic, M., Krieger, N., 2000. Biochemical engineering aspects of solid state bioprocessing. New Products and New Areas of Bioprocess Engineering. Springer Berlin Heidelberg, Berlin, Heidelberg, pp. 61–138.

Mitchell, D.A., M. Berovič, and N. Krieger, Introduction to solid-state fermentation bioreactors. In: Solid-State Fermentation Bioreactors: Fundamentals of Design and Operation, D.A. Mitchell, M. Berovič, and N. Krieger, (Eds.), 2006, Springer Berlin Heidelberg: Berlin, Heidelberg. pp. 33-44.

Mota, M.J., et al., 2018. *Lactobacillus reuteri* growth and fermentation under high pressure towards the production of 1,3-propanediol. Food Res. Int. 113, 424–432.

Nguyen, T.M.P., Lee, Y.K., Zhou, W., 2009. Stimulating fermentative activities of bifidobacteria in milk by highintensity ultrasound. Int. Dairy J. 19 (6), 410–416.

Ohba, T., Uemura, K., Nabetani, H., 2016. Moderate pulsed electric field treatment enhances exopolysaccharide production by *Lactococcus lactis* subspecies cremoris. Process Biochem. 51 (9), 1120–1128.

Ojha, K.S., et al., 2018. Integrated phenotypic-genotypic approach to understand the influence of ultrasound on metabolic response of *Lactobacillus sakei*. PLoS One 13 (1), e0191053.

Panikuttira, B., et al., 2018. Process analytical technology for cheese manufacture. Int. J. Food Sci. Technol. 53 (8), 1803–1815.

Pérez-Torrado, R., Barrio, E., Querol, A., 2018. Alternative yeasts for winemaking: *Saccharomyces* non-cerevisiae and its hybrids. Crit. Rev. Food Sci. Nutr. 58 (11), 1780–1790.

Rathore, A.S., Bhambure, R., Ghare, V., 2010. Process analytical technology (PAT) for biopharmaceutical products. Anal. Bioanal. Chem. 398 (1), 137–154.

Ravi, A., 2020. A microscopic look at the fermentation. Food Trend SPOONSHOT.

Rodríguez_Leon, J.A., Rodríguez-Fernandez, D.E., Soccol, C.R., 2013. Laboratory and industrial bioreactors for solid-state fermentation. In: Soccol, C.R., Pandey, A., Larroche, C. (Eds.), Fermentation Processes Engineering in the Food Industry. CRC Press, Boca Raton, Florida, USA, pp. 181–199.

RRDC, A.G.R.I.R.D.C., Transformative Technologies: Synthetic Biology. 2016.

Schorn-García, D., et al., 2021. ATR-MIR spectroscopy as a process analytical technology in wine alcoholic fermentation – a tutorial. Microchem. J. 166, 106215.

Shiferaw Terefe, N., Augustin, M.A., 2020. Fermentation for tailoring the technological and health related functionality of food products. Crit. Rev. Food Sci. Nutr. 60 (17), 2887–2913.

Shokri, S., et al., 2021. Ultrasound-assisted fermentation for enhancing metabolic and probiotic activities of *Lactobacillus brevis*. Chem. Eng. Process. Process Intensif. Woodhead publishing, Swanston, UK 166, 108470.

Shokri, S., Shekarforoush, S.S., Hosseinzadeh, S., 2020. Stimulatory effects of low intensity ultrasound on the growth kinetics and metabolic activity of *Lactococcus* lactis subsp. Lactis. Process Biochem. 89, 1–8.

Shokri, S., Terefe, N.S., Manzari, M., 2021. Advances in food fermentation: potential application of novel processing technologies for enhancing fermentation kinetics and product yield. In: Knoerzer, K., Muthukumarappan, K. (Eds.), Innovative Food Processing Technologies: A Comprehensive Review. Elsevier, pp. 135–156.

Singhania, R.R., et al., 2009. Recent advances in solid-state fermentation. Biochem. Eng. J. 44 (1), 13–18.

Soccol, C.R., et al., 2017. Recent developments and innovations in solid state fermentation. Biotechnol. Res. Innov. 1 (1), 52–71.

Stabnikova, O., et al., 2005. Biotransformation of vegetable and fruit processing wastes into yeast biomass enriched with selenium. Bioresour. Technol. 96 (6), 747–751.

Stanbury, P.F., Whitaker, A., Hall, S.J., 1995. Chapter 2 – Microbial growth kinetics. In: Stanbury, P.F., Whitaker, A., Hall, S.J. (Eds.), Principles of Fermentation Technology second ed. Pergamon, Amsterdam, pp. 13–33.

Steinkraus, K., 2004. Origin and history of food fermentations. In: Hui, Y.H., Meunier-Goddik, Lisbeth., Josephsen, Jytte. (Eds.), Handbook of Food and Beverage Fermentation Technology. CRC Press, pp. 1–8.

Technavio.com, Global fermented foods and drink market 2018-2022. 2017. p. 146.

Teng, T.S., et al., 2021. Fermentation for future food systems. EMBO Rep. 22 (5), e52680.

Terefe, N.S., 2016. Emerging trends and opportunities in food fermentation. Reference Module in Food Science. Elsevier, pp. 1–9. http://dx.doi.org/10.1016/B978-0-08-100596-5.21087-1.

Tsevdou, M., et al., 2013. Monitoring the effect of high pressure and transglutaminase treatment of milk on the evolution of flavour compounds during lactic acid fermentation using PTR-ToF-MS. Food Chem. 138 (4), 2159–2167.

Uçkun Kiran, E., et al., 2014. Enzyme production from food wastes using a biorefinery concept. Waste Biomass Valorization 5 (6), 903–917.

Villadsen, J., Nielsen, J., Lidén, G., 2011. Scale-up of bioprocesses, Bioreaction Engineering Principles. Springer US, Boston, MA, pp. 497–546.

Villadsen, J., Nielsen, J., Lidén, G., 2011. Design of fermentation processes, Bioreaction Engineering Principles. Springer US, Boston, MA, pp. 383–458.

Warner, M., 2021. Commercial Fermentation Opportunities and Bottlenecks. Warner Advisors.

Watson, E., 2015. Synthetic biology is cheaper, faster, and more sustainable, says Evolva CEO: 'We're proud of what we do'. Food Navigator.

Watson, E., 2016a. Geltor seeks to disrupt the gelatin market with potentially game-changing animal-free alternative. Food Navigator.

Watson, E., 2016b. Don't have a cow? Perfect Day animal-free milk bids for slice of multibillion-dollar global dairy market. Food Navigator.

Watson, E., 2016c. Cultured coffee… the beginning of a new fermented foods revolution?. Food Navigator.

Watson, E., 2018. Cargill launches EverSweet fermented steviol glycosides. Food Navigator.

Wilcox, M., 2019. Synthetic Biology Is Changing What We Eat. Here's What You Need to Know. Civileats.

Williams, J., 2002. Keys to bioreactor selections. Chem. Eng. Prog. 98, 34–41.

Xu, Y., et al., 2020. Fermentation by probiotic *Lactobacillus gasseri* strains enhances the carotenoid and fibre contents of carrot juice. Foods 9 (12), 1–15, 1803.

Yeo, S.K., Liong, M.T., 2011. Effect of ultrasound on the growth of probiotics and bioconversion of isoflavones in prebiotic-supplemented soymilk. J. Agric. Food Chem. 59 (3), 885–897.

Yeo, S.K., Liong, M.T., 2013. Effect of electroporation on viability and bioconversion of isoflavones in mannitol-soymilk fermented by lactobacilli and bifidobacteria. J. Sci. Food Agric. 93 (2), 396–409.

Zhang, A.Y.Z., et al., 2013. Valorisation of bakery waste for succinic acid production. Green Chem. 15 (3), 690–695.

Strategies to mitigate protein deficit

Periaswamy Sivagnanam Saravana[a], Viruja Ummat[a,b], Colm O'Donnell[b], Brijesh Tiwari[a]
[a]*Department of Food Chemistry & Technology, Teagasc Food Research Centre, Ashtown, Dublin, Ireland*
[b]*School of Biosystems and Food Engineering, University College Dublin, Dublin, Ireland*

6.1 Introduction

A sustainable food and agriculture system is required to meet the needs of an increasing global population. The world population now stands at 7.8 billion people, with a growth rate of 1.07%/year and it is expected to reach 10 billion by 2050 (Fasolin et al., 2019). An increase in human population will require an expansion in agricultural and food industries, leading to increases in greenhouse gases, deforestation, water consumption, climate change and other adverse impacts (FAO and UNICEF, 2017). Furthermore, the expansion of food production has resulted in the increased consumption of animal protein. The consumption of meat and dairy products from 2010 to 2050 is likely to increase by 173% and 158%, respectively. The resource requirement to convert plant sources into animal-derived protein is inefficient; that is, circa 7 kg of plant-based product is needed to produce 1 kg of milk or meat for human consumption (Nadathur et al., 2017). In order to meet the objective of eliminating malnutrition in all its forms while developing a sustainable, resilient food system for healthy diets as specified in a report by the Food and Agricultural Organisation of the United Nations (FAO, 2014), food scientists have been investigating alternative foods without compromising on food safety and sustainability. According to a report, from Paris-based consultancies BIPE (Bureau d'informations et de prévisions économiques) and Sofiprotéol, the demand for plant protein will increase by 43% from 2010 to 2030 (Chardigny and Walrand, 2016). This increase will initially be driven mainly by sub-Saharan Africa and India. Meanwhile, the requirement for animal protein is estimated to increase by 33% during this period, mainly driven by increased demand from China (FAO, 2017).

Recent market trends are generating data for purified protein obtained from various well established plant sources such as pea and soy as well as new plant sources such as oats, hemp, flax, rice, and chia. Therefore, in the future protein supply is likely to be from a variety of sources. Current and new sources of protein are shown in Fig. 6.1. Soy is considered as the first-generation source of protein while canola, pea, and rice are considered as the second generation. In the third generation, moringa and duckweed are included. It is anticipated that the demand for proteins obtained from insects, algae, agricultural waste (Ag waste) and generated via new biotechnological processes (Syn bio), will increase globally in the next two to three decades (Fig. 6.1). In vitro proteins and cultured beef product are being developed by companies such as Modern Meadow and Mosa Meat and will be discussed in Section 6.3.7.

It is necessary to produce plant- and animal-based proteins sustainably so that the global protein demand can be met while contributing to the sustainability of biodiversity and food systems (Henchion et al., 2017). Globally, as diets based on plant proteins are widely consumed daily, there has been a

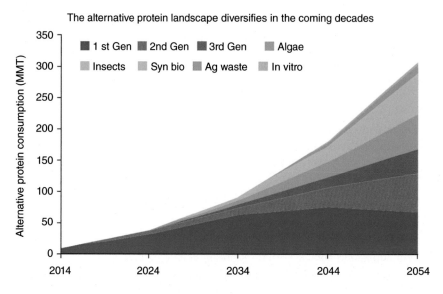

The alternative protein landscape diversifies in the coming decades

FIG. 6.1

Forecast for plant and alternative protein consumption from 2014 to 2054. Adapted from Lux Research.

discussion for some time on finding alternative sources of proteins (AP) and studying their impact on human health as a result of pressure from government and nongovernmental policies (Niva et al., 2017). There are a number of foods that are considered as emerging sources of proteins and these include grains (wheat and zein), pulses (beans, peas, lentils), seeds (chia), leaves (moringa), fungi (mycoproteins), algae (microalgae and seaweed), and insects.

Animal and vegetable protein fractions have good functional properties such as the formation of gels, emulsions, hydrogel, etc. Therefore, these functional properties could be used as a food delivery system to protect and deliver the functional molecules to the human body. For example, utilization of diary waste such as whey serum from milk which is rich in protein also has functionality (enhanced digestibility, gelation, foaming, and emulsifying capacities) and bioactive (antimicrobial, antiviral, and anticarcinogenic) properties and nutritional value (balanced amino acid profile; Ramos et al., 2017). In the food processing industry, proteins are affected by heating and enzymatic treatments, and these processes will influence human gastrointestinal digestion and the immune system. A thorough investigation is required for considering emerging proteins as a source of food and the processing conditions should be optimized to maximize the retention of nutrients and reduce allergenicity (Ramos et al., 2017).

Recent research has identified three innovative goals for future food systems: (1) better efficiency— "mass production at the lowest possible price"; (2) innovation opportunity, driven by consumer trends and the need to balance the use of animal-origin proteins with plant proteins—development of new perceptions/sensations and niche products with high value; and (3) development of functional foods and ingredients targeting health and wellbeing upon consumption (De Vries et al., 2018). Therefore, to utilize AP in future food systems, they should be produced cost effectively and sustainably using innovative processing technologies.

This chapter aims to present emerging alternative protein (AP) sources and discusses sustainable processing approaches that could add value and present new opportunities to undervalued protein rich fractions. In addition, the key determinants for consumer acceptance and health considerations of AP will be critically discussed.

6.2 Protein demand

Globally protein demand has increased due to population growth, higher incomes, urbanization and aging (Popkin et al., 2012). Negative impacts associated with animal-derived protein include 12% of greenhouse gas emissions associated with livestock production and 30% of human-induced terrestrial biodiversity loss caused by animal production (Henchion et al., 2017). In the European Union (EU), land use is also a concern because of two thirds of the agriculture area is used for livestock production and 75% of protein rich feed is imported from South America (Henchion et al., 2017). Ethical issues in animal production and changes in consumer dietary preferences may also reduce future demand for animal-derived protein products.

Currently, global protein demand is 202 million tons for a population of 7.4 billion inhabitants, and animal protein demand is mainly met by pork and poultry (Kim et al., 2019). Poultry demand is projected to increase by 750% by 2030, especially in the Southeast and South Asian countries (Kim et al., 2019). The increase in the demand for pork and poultry protein products will result in increased demand for protein feed, as most of the animals diet contains protein rich feeds (Council, 2012, 2000). Animal feed production was estimated to be around 1 billion tons in 2016, which is a 4% increase over the 2015 data (Alltech, 2018). In addition, the seventh annual Alltech global feed survey suggested that swine, poultry and ruminant feed comprise 26%, 44%, and 22% of total feed production, respectively. Pet and aquaculture feeds comprise only 2% and 4% of total feed production. The survey also suggested that Asia-Pacific, EU, and North America regions produce 36%, 24%, and 19% of the total feed production, respectively (Alltech, 2018).

It was reported that 321 million tons of oilseed meal protein feed was the main source utilized in the year 2016/2017, other sources of feed protein came from fish, animal proteins, and by-products from biofuel (Alltech, 2018). Soybean meal ranks the number one oil seed meal with 226 million tons (70% of total), followed by rapeseed with 39 million tons (12% of total) (Westcott and Hansen, 2021). Production of these feeds in the future may be affected by several factors such as availability of land and water, currency, climate change, yield deficit, energy cost, government policies, and fuel policy (Westcott and Hansen, 2021). Further, the demand for protein feed may be alleviated by developing the feed efficacy nutritional value of feed ingredients and animals (Kim et al., 2019).

6.3 Sustainability of alternative proteins sources

Nutrition is the main function of food, and good nutrition is a dimension of food security and sustainability. A summary of the environmental, social aspects and financial costs of AP vital for assessing long-term sustainability is presented in Fig. 6.2. According to the FAO definition (Burlingame and Dernini, 2010), sustainable food systems are those that deliver food security and nutrition

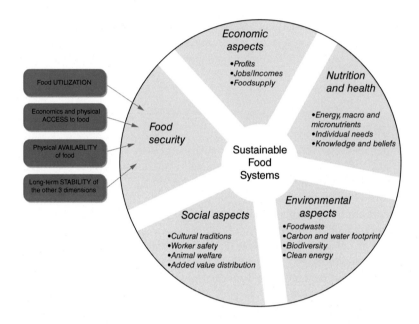

FIG. 6.2

Schematic framework of food safety and sustainable food systems dimensions (Source: Adopted from Fasolin et al., 2019 with permission).

for all in a way that environmental, economic, and social sustainability is not compromised for future generations.

6.3.1 Plant

Plant-based proteins are potential natural substitutes for conventional proteins such as eggs and meat. They have low production costs along with health and environmental benefits (Rubio et al., 2020). In addition, plant-based foods have lower greenhouse gas emissions and much reduced effects on the environment compared to animal-derived foods (Tian et al., 2016). Substituting or reducing animal protein by plant protein could have a significant impact on biodiversity loss and climate change (Stoll-Kleemann and Schmidt, 2017). However plant-based food production can cause undesirable effects such as decreased soil fertility, deforestation, desertification, and pollution of water resources with agrochemicals.

Plant-based protein sources such as vegetables, nuts, seeds, fruits, and legumes are rich in vitamins, fibers, minerals, and antioxidants which are important for human health (Kojima et al., 2018). The high protein contents of some vegetable sources are listed in Table 6.1. Generally, the quantity and quality of amino acids is higher in products obtained from animal sources compared to plants. However the nutritional benefit and quality of protein should be considered in the context of bioavailability and digestibility. Combining plant and animal protein sources in a well-balanced diet is recommended for the intake of adequate essential amino acids (Melina et al., 2016).

Table 6.1 Examples of alternative protein sources and their protein content.

Source	Name	Protein content (% w/w)	References
Grains and legumes	Amaranth (*Amaranthus* spp.)	12.5–17.6	(Caselato-Sousa and Amaya-Farfán, 2012)
	Lupin (*Lupinus* spp.)	38–55	(Bähr et al., 2014)
	Navy bean (*Phaseolus vulgaris*)	26.0	(Tabtabaei et al., 2019)
Meat by-products	Mechanically deboned chicken meat	17.4	(Nuckles et al., 1990)
	Beef heart (cap on)	15.4	(Nuckles et al., 1990)
	Beef lung (lobes only)	17.7	(Nuckles et al., 1990)
	Beef spleen	15.3	(Nuckles et al., 1990)
	Pork liver	22.1	(Nuckles et al., 1990)
	Pork lung (lobes only)	15.5	(Nuckles et al., 1990)
Fish by-product (wet/weight)	Atlantic salmon (head)	11–13	(Sandnes et al., 2003)
	Atlantic salmon (viscera)	5–7	(Sandnes et al., 2003)
	Atlantic salmon (backbone/frames)	10–15	(Sandnes et al., 2003)
	Cod (cut-offs)	13–23	(Søvik, 2005)
	Haddock (cut-offs)	15–18	(Søvik, 2005)
	Ling (cut-offs)	15–20	(Søvik, 2005)
Microbial	*Aspergillus niger*	10.3–61.2	(Kamal et al., 2019)
	Fusarium venenatum	41.8–46.4	(Hosseini and Khosravi-Darani, 2010)
	Saccharomyces cerevisiae	15.3–49.3	(Gervasi et al., 2018)
	Torula utilis (*Candida utilis*)	28.4–48.9	(Kurcz et al., 2018)
	Yarrowia lipolytica	45–55	(EFSA Panel on Nutrition, N.F., 2019)
	Rhodopseudomonas sp.	54–92	(Yang et al., 2017)
Insects	Cricket (*Gryllodes sigillatus*)	56.8	(Hall et al., 2017)
	Grasshopper (*Schistocerca gregaria*)	76.0	(Mishyna et al., 2019)
	Honey bee brood (*Apis mellifera*)	22.1	(Mishyna et al., 2019)
	Mealworm (*Tenebrio molitor*)	51.0	(Zhao et al., 2016)
Algae	Arthrospira platensis (*Spirulina platensis*)	53.5	(Benelhadj et al., 2016)
	Chlorella vulgaris	12.7–53.0	(Laurens et al., 2017)
	Dunaliella salina	51.2–82.2	(Sui et al., 2019)
	Palmaria palmata	25.5	(Galland-Irmouli et al., 1999)
	Ulva reticulata	21.06	(Ratana-arporn and Chirapart, 2006)

Furthermore, the consumption of a mixed diet with animal and vegetable protein will have a favorable effect on gut microbiota and also help in reducing cardiovascular, diabetic risk, and other health-related diseases (Glick-Bauer and Yeh, 2014; Richter et al., 2015). Over the last few decades, several studies have reported that vegetable protein has several advantages over animal protein due to their nutritional and functional properties (Busnelli et al., 2018). These proteins can be varied into globulins, albumins, glutelins, and prolamins, each with their own functional properties. Rapeseed has good emulsion stability and foaming functionalities due to the availability of napin and cruciferin. Several legumes such as pea, lupin, lentil, and nuts have strong foaming, gelling and emulsion properties (Tabilo-Munizaga et al., 2019; Liu et al., 2018; Ettoumi et al., 2016). Also, some vegetables by-products such as oat bran, apple pomace, orange pulp, and sugar beet may be used to isolate protein (Huc-Mathis et al., 2019; Wallecan et al., 2015; Tenorio et al., 2017).

6.3.2 Meat and fish by products

Meat by products such as bones, blood, skin, fat tissue, hoofs, horns, viscera, and skull are produced in large quantities and are costly to treat in waste management facilities (Toldrá et al., 2016). Through innovative technologies these products can be easily processed and valorized for other applications. The European Commission Regulation (EC) 1069/2009 regulates animal by-products and derived products not intended for human consumption. The Food and Drug Administration (FDA) in 2004 introduced legislation to prevent the establishment and spread of bovine spongiform encephalopathy (BSE) in the United States, including a prohibition on the use of high-risk, cattle-derived materials that can carry the BSE agent which are defined as specified risk material.

Meat by products are mainly used as start up materials for the production of bioactive molecules of interest such as protein hydrolysate (Lasekan et al., 2013), products rich in functional activity (Chernukha et al., 2015), or bioactive peptides (Martínez-Alvarez et al., 2015). In addition, by-products such as bone meal and fat are also used as biodiesel, fertilizer or a feed/ingredient for other industries. Meat by products such as deboned chicken, beef heart, spleen and lung, pork liver - lung also have high protein contents (15–22%; Table 6.1) and are widely used in research to improve its product value (Nuckles et al., 1990)

Fish consumption has increased globally and it contributes up to 15% of the total animal protein. Among which fish fillets, krill, and shrimp are considered as low protein sources of 30–40%, 15%, and 10%, respectively (Gehring et al., 2011). The high consumption of seafood and seafood products leads to an increase in fish processing by-products and waste. Fish by-products include head, bone, skin, carcass, and viscera. The by-products obtained in commercial fishing equates to approximately 60–70% of the fresh fish caught (Ananey-Obiri and Tahergorabi, 2018). Fish by-products can be used as fertilizer, feed, and valuable foods such as fish oil and these products can be obtained by clean and green techniques and utilized as dietary supplements. However, a large amount of by-products obtained from fish caught is discarded back to the sea. Fish by-products are a good source of protein, soluble vitamins, bio actives and phospholipids (Villamil et al., 2017). The protein content of selected fish by-products are displayed in Table 6.1.

Various types of protein fractions can be obtained from fish by-products. Flesh stuck to the fish skeletal structure and cut-offs can be used for making fish surimi, mince, or other products based on surimi (Taylor et al., 2007). Minced fish by products can be used as a readymade product but require a stable and high quality raw material supply. However, the minced product market is quite small but resolving quality issues could increase its market size. Mince can be used in various products in natural or altered forms (Kim and Park, 2007). SuspenTec uses fish trimmings and these products are micro sized particles which are stored at a low temperature around −4 to 6°C and fused into brines for injection into fish fillets (Thorkelsson et al., 2008).

Surimi is prepared by repeatedly washing the fish mince to remove sarcoplasmic proteins and other contents that will help in denaturation of protein during freezing (Kim and Park, 2007). Most of the time it is mixed with cryoprotectant and it exhibits good functional properties such as water holding, emulsifying and gelling. The production of surimi gel requires large amounts of water, that is, 20 times the amount for deboned meat. This water is high in biological oxygen demand and must be treated prior to discharge. White fish is mostly used in the production of high quality surimi. Several research studies investigated manufacture of surimi from fish by products, but the surimi obtained in this manner had lower gelation properties (Gehring et al., 2011).

Fish protein hydrolysates obtained from by-products are used in several applications because of their excellent functional properties (gelling, water holding, foaming, emulsifying, and texture). These hydrolysates also have good nutritional and bioactive properties such as antioxidant, antithrombotic, antihypertensive, and immunomodulatory activities, etc. (Kim and Mendis, 2006). Peptides and their chain lengths are widely researched because of their relationship to organoleptic and functional properties such as emulsifying, bitterness, and solubility. The final product of the hydrolysate depends on the type of raw materials, properties of raw material (state of degradation, enzyme activity), type of enzyme, and treatment conditions (Šližytė et al., 2009). In large-scale production, hydrolysis processes need to be carefully controlled as consistent properties of the products are necessary and is quite challenging one to achieve (Rustad et al., 2011). A knowledge of raw material composition and quality including information of cutting/mincing, type of enzyme, inactivation of endogenous enzymes, conditions of hydrolysis process (pH, temperature, time, and water requirement), and the variety and activity of endogenous enzymes are required. These features will impact the hydrolysis process and end product quality and yield.

6.3.3 Microbial

Microbial or single cell protein is obtained from unicellular or even multicellular organisms, mostly bacteria and fungi. Utilizing microbial protein as a protein supplement in animal feeds and human diet is not a new idea and during World War I, yeast was employed as a protein source. However large-scale production of microbial protein has been hindered by challenges such as product quality, costs limitations, high level of nucleic acid, protein recovery, and other practical issues (Reihani and Khosravi-Darani, 2019).

Microbes cultivation does not require a large area like animal and crop husbandry, since microbes can grow in reactor or tanks (Laurens et al., 2017). Microbial protein can be produced using waste water treatment as a media (Reihani and Khosravi-Darani, 2019) and it is an eco-friendly process with a great advantage over other APs. *Rhodopseudomonas* sp. can produce up to 92% protein through biogas slurry treatment with high ammonia and high salinity (Yang et al., 2017). Microbial protein can also be obtained as a co-product from certain industrial processes such as spent yeast from breweries and biorefineries (Pietrzak and Kawa-Rygielska, 2013; Vieira et al., 2019).

The protein content of the microorganisms (Table 6.1) and its amino acid content can change according to species and also substrate, nutrient sources, environment condition, and cell growth stage (Reihani and Khosravi-Darani, 2019). Microbial-derived proteins are considered as high quality because they can meet the FAO/WHO reference value of 40% (Organization and University, 2007) of essential amino acids. In brewery spent yeast (*Saccharomyces pastorianus*) the main essential amino acids in the protein were valine (6.7%) and leucine (8.2%; Vieira et al., 2019). In *Fusarium venenatum*, the protein content was around 33% and phenylalanine, valine and lysine were the main essential amino acids present (Hosseini and Khosravi-Darani, 2010).

Besides protein, these microorganisms also contain lipids, nucleic acids, fats, vitamins, pigments, and carbohydrates (Kurcz et al., 2018). Microbial-derived protein can be categorized as a feed/food or flavorings, texture modifiers, colorants and preservatives to improve food/feed functionalities (Ritala et al., 2017). Recently, Ritala et al. (2017) reviewed the industrial utilization of microbial protein and relevant patents published over recent years. They outlined that fungi are used as an edible ingredient for human consumption either as supplements or food ingredients. Some examples of fungi that are used in commercial sectors are *Fusarium Saccharomyces, Aphanizomenon, Torulopsis, Torula*, etc. and in the case of bacteria, it has been mostly used as animal feed (Ritala et al., 2017).

6.3.4 Insects

Insect-based proteins are considered to be the most debated alternative to animal protein sources due to their conflicts with cultural issues and traditional lifestyles in some countries. There are nearly 2000 insect species categorized as edible and their consumption is common in Africa and some Eastern and Latin American countries. However, insect-based proteins are not widely accepted in Western countries (Ritala et al., 2017) due to their unappealing nature and possibility of allergens. Processed insect powder is used as an ingredient in other food products and has higher consumer acceptance (Woolf et al., 2019). Insect-based protein is used in small quantities to make chips, pasta and bread. It has also been evaluated for structural, sensory and nutritional values (Duda et al., 2019; Roncolini et al., 2019). Insects are considered to be a viable food because of their nutritional value and in addition, insect production has positive economic and environmental benefits. Lepidoptera (18%), Coleoptera (31%), Orthoptera (13%), Hymenoptera (14%), and Hemiptera (10%) are the more widely consumed insect groups (Sun-Waterhouse et al., 2016).

Insect production has numerous advantages over animal and crop husbandry. They can be grown in small land areas, they do not affect soil fertility or require deforestation (Ooniex, 2017). Insect production requires less water and also emits very low amounts of ammonia and greenhouse gases compared to farmed poultry, cattle, and aquaculture. Insect production also benefits from high reproduction rates, short life-cycles and high feed conversion efficiency. Furthermore, insects can be cultivated with foods and by-products obtained from waste streams and food processing (Sun-Waterhouse et al., 2016). Insects have a high content of fat and protein along with a quantity of vitamins and minerals (de Castro et al., 2018). Minerals such as zinc, sodium, iron, potassium, phosphorus, copper, manganese, magnesium, and vitamins such as folic acid, biotin pantothenic acid, and riboflavin are found in insects. Interestingly, some insects have a high content of iron and zinc which can be similar to beef and greater than pork and chicken (de Castro et al., 2018).

Generally, insects are rich in protein content (above 60%), which can vary among species by around 7–91% (dry weight; Table 6.1). Other factors which influence the protein content include sex or developmental stage of the insects (Mishyna et al., 2019). Generally, the high quality and protein content of insects is comparable to plant and meat sources. However, the nutritional properties of insect protein are better than other sources because they include all essential amino acids (Zielińska et al., 2015). Lately, studies have reported that functional properties such as foamability, solubility, emulsifying, and gelling ability of insect proteins are similar to some commercial proteins (Mishyna et al., 2019). Currently, insect protein powders are added as an ingredient in food applications. However further information on their functionality, technological feasibility and consumer acceptance is required for the widespread use of insects as a source of protein. Recently, one study investigated the acceptance factors of several edible insects and processed food. They studied different products such as snacks, protein bars, pasta, granola, and burgers (Fig. 6.3; Orsi et al., 2019). They concluded that consumers had a very low level awareness of the health benefits of the insect-based products. In addition, the results showed that a prevalence of psychological, low willingness, and personality barriers to consumption which included food neophobia and disgust.

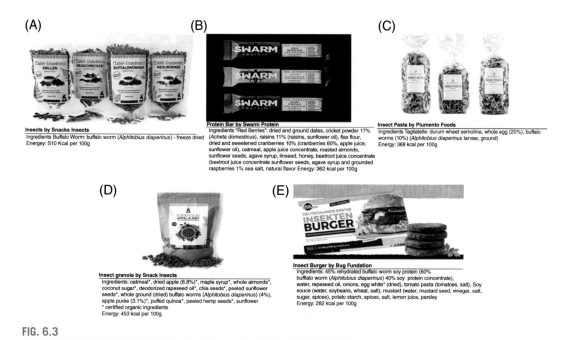

FIG. 6.3

(A) Whole insects by snack insects. (B) Protein bars by swarm protein. (C) Insect pasta by plumento foods. (D) Insect granola by snack insects. (E) Insect burger by bug foundation (Source: Adopted from Orsi et al., 2019 with permission).

6.3.5 Algae

Marine plants such as seaweed and microalgae are a strong potential source of protein. They are generally called algae. Seaweeds are made of multicellular components that are raised in a marine environment or saltwater, while microalgae are single cell organisms that can grow in a wide array of atmospheric conditions. Around 24 million tons of algae, rich in protein are cultivated globally with a value of €6.4 billion, and could offer a partial solution to close the so-called "protein gap" (Henchion et al., 2017). Microalgae such as *Chlorella* sp., *Arthrospira* sp., and *Dunaliella salina* are used for human consumption and in the EU, *Chlorella* sp. and *Spirulina* sp. are primarily used. In China and Japan however, a wide range of microalgae are used, which may contain microbial contaminants, toxins and other drainage components which can be of concern for human use. These algae are generally cultivated in an open stream or outdoor ponds (Becker, 2007).

Algal protein are rich in nutrients with a sufficient amount of essential amino acids and can be compared with vegetable protein. However, disadvantages include high costs, complex refining and extraction processes and sensory issues related to use in food products. Microalgal products are rich in EPA/DHA and products in the market include oils, pasta, cosmetics and animal feed (Fleurence, 1999). About 30% of the algal meal produced commercially is used as animal. Some de-oiled algae products are fed to poultry and mixed with soybean as pig feed, thus replacing soybean meal in animal feed (Rosegrant et al., 2001). A large USA-based biotechnology company, TerraVia (formerly Solazyme) produces a protein rich algal product named AlgaVia which has 65% protein and they have a production capacity of 100,000 mt/year in the USA and Brazil (Henchion et al., 2017).

Seaweeds are cultivated in the ocean and are a natural source of several bioactive compounds. They are classified as brown, red, and green algae. Green and red algae are rich in protein content, especially *Porphrya* sp. which has 47% (dry weight) protein. Generally, red or green seaweed have 10–47% of the dry weight while brown seaweed has a low content of 3–15% of the dry weight (Table 6.1; Sampels, 2014). Protein content may vary according to the species, seasonal, and also geographic location. Seaweed-derived protein has a similar protein composition to soybean or eggs. Glutamic acids and aspartic acid (22–44% of total amino acids) are the major amino acids in brown seaweeds. Researchers have found that the extraction technique influenced the seaweed protein content and influenced its digestibility because seaweed is a complex material and during the extraction process phlorotannins and carbohydrates are extracted together which may lead to digestion problems (Sampels, 2014). There are also reported concerns regarding the consumption of seaweed because of high iodine content, heavy metals, and other containments such as pesticides and dioxins (Van der Spiegel et al., 2013).

Seaweeds are widely used in animal feed and aquaculture diets. For example, *Porphyra* sp. is rich in protein content and is used as a feed for sea bream (Sampels, 2014). Seaweed protein and derived peptides can be used as food ingredients because of the fifth taste, umami. Monosodium glutamate (MSG), the sodium salt of glutamic acid found in major seaweeds provides the umami taste. Brown algae, especially *Saccharina japonica* is rich in MSG, which become available when seaweed is softened or heated. In addition, *Porphyra* sp., commonly known as Nori, also gives an umami taste because nucleic acids present within their cells (Henchion et al., 2017). The commercialization of algal products as an under explored protein source is likely to be delayed because of nutrition, food safety and health issues.

6.3.6 In vitro meat

In vitro meat, which is also called lab grown meat, clean meat or cell-culture meat is obtained by cell identification (adult or embryo cells) and isolation, tissue engineering and cell culture procedures (Langelaan et al., 2010). Proponents of in vitro meat claim that a specially designed bioreactor with a size of a typical swimming pool can nourish 40,000 people for a year and would decrease the necessity for farm sourced meat products. In 2013, Mark Post (Maastricht University) created a cultured hamburger which cost £200,000 and required 2 years of research time (Futures, 2021). Later, a Dutch company (Mosa Meat) tried to improve the flavor of the lab cultured meat with fat cells and reduce the production costs. Currently several companies have started to investigate the production of in vitro meat and are working to similar objectives.

In addition to sensory effects, economic, ethical, societal, and environmental factors play a key role in the production of in vitro meat. In vitro meat can cut greenhouse gas emissions by 96%, avoid slaughter of animals and does not require large-scale land farming. However, disadvantages of in vitro meat includes complex production systems that consumes more energy and chemical usage (nutrients, hormones; Futures, 2021). Current advances in stem cell research, biomedical engineering, and regenerative medicine suggest that in vitro meat production systems are likely to be feasible in the future. A possible system to produce in vitro meat fabrication is shown in Fig. 6.4.

The challenges faced by in vitro meat production are culturing of cells, that is, adipose and muscle cells have possible genetic uncertainty owing to rapid growth which could lead to cancer cell formation. The culturing medium which grows the cells needs to be food grade and large-scale bioreactors

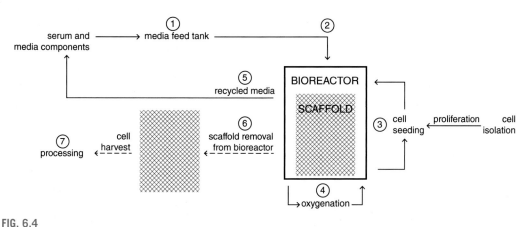

FIG. 6.4

A potential scheme for the production of cultured meat (Source: Derived from Datar and Betti, 2010 with permission).

are required for commercialization. Another challenge is matching the final product quality to that of real meat. Cultured meat requires testing of several issues such as energy utilization, glycolysis, rigor onset, calcium release, enzyme effects, and denaturation and oxidation, since these factors affect the consumption quality. Finally, a complete ecological investigation of in vitro meat is necessary and must take these features into account (Mattick et al., 2015).

6.4 Alternative protein extraction techniques

In this section, various processes to extract protein (acid and alkaline based, enzyme, ultrasound, pulsed electric field, and microwave assisted) are described. Table 6.2 summarizes the extraction methods and protein recovery/content of the alternative protein sources. However, some authors have failed to report some of these values/content as a percentage, making it difficult to compare the results.

6.4.1 Acid-based extraction

Acid-based extraction is usually carried out at a pH < 4 using HCl, citric acid, or acetic acid, followed by isoelectric precipitation. A low pH can solubilize proteins and is followed by pH shift or cryo-precipitation. For example, kidney bean (*Phaseolus vulgaris*) protein extracted with a citric acid solution (0.4 N, pH 4.0) followed by cryo-precipitation (4 °C, 18 h) had a protein content of 95% and a carbohydrate content 3% (Alli et al., 1993). Using minor modifications to this method, alternative proteins were extracted from sources such as beef heart (10.34% protein; DeWitt et al., 2002) and seaweed *Ascophyllum nodosum* (57% protein; Kadam et al., 2017). However, the seaweed protein was extracted using a combination of acid/alkali extraction which was found to be the most efficient method. In addition, the authors suggested that ultrasound assisted extraction reduced the time required.

Table 6.2 Different protein extraction methods and resulting protein content.

Source	Protein extraction method	Protein content	References
Quinoa (*Chenopodium quinoa*)	Alkaline extraction	62.1% (9.06 g of protein/100 g defatted Quinoa seed meal)	(Guerreo-Ochoa et al., 2015)
Chenopodium quinoa	Alkaline extraction	72.70 g/100 g	(Takao et al., 2005)
Beef heart	Acid solubilization	10.34%	(DeWitt et al., 2002)
Beef lung (lobes only)	Alkaline extraction	17.7%	(Nuckles et al., 1990)
Porcine myocardium	Ultrasound assisted alkaline extraction	90%	(Kim et al., 2017)
Salmon by-product	Enzymatic treatment	85.70%	(Gbogouri et al., 2004)
Rainbow trout by-products	Isoelectric solubilization/ precipitation	71.5%	(Chen and Jaczynski, 2007)
Palmaria palmata	Enzymatic pretreatment	54.9%	(Bjarnadóttir et al., 2018)
Ascophyllum nodosum	Acid, alkali, ultrasound	57%	(Kadam et al., 2017)
Grateloupia turuturu	Ultrasound assisted enzymatic hydrolysis	3.6 ± 0.3 mg/g dw	(Le Guillard et al., 2016)
Papaya seeds	PEF	100 mg/L	(Parniakov et al., 2015)
Lupinus mutabilis	Ultrasound	14%	(Aguilar-Acosta et al., 2020)
Lobster shells	Microwave-intensified enzymatic deproteinization	85.8%	(Nguyen et al., 2016)
Amaranthus seeds	Isoelectric precipitation	61.0%	(Salcedo-Chávez et al., 2002)

6.4.2 Alkaline-based extraction

Alkali-based extraction is the most widely used method to extract the protein at laboratory or industry scale (Qiaoyun et al., 2017). NaOH/KOH are the most frequently used bases for this process and they result in a higher extraction yield compared to acid- or chemical-based extraction. The key mechanism is that the alkaline process helps to break the disulfide bonds in protein which improves protein recovery and resultant yield (del Mar Contreras et al., 2019). Alkaline-based extraction of proteins from plant sources enhances the digestibility/bioavailability of the extracted plant protein. However it also degrades the protein quality by denaturation, crosslinking, lysinoalanine formation, loss of amino acids and protein hydrolysis (del Mar Contreras et al., 2019).

Quinoa is a popular Andean pseudo-cereal that has sparked a lot attention due to its excellent nutritional value. Recently the best conditions to extract protein from quinoa meal were found to be a solvent/meal ratio of 19.6/1 (v/w), pH 11, temperature 36.2 °C, an extraction time of 150 min, and a particle size of 500 μm. Under these optimal conditions, a protein yield of 62.1% was obtained (Guerreo-Ochoa et al., 2015).

6.4.3 Enzyme assisted extraction

Enzyme assisted extraction (EAE) is a reliable method for the large-scale production of high quality alternative proteins. EAE results from the enzymatic degradation of the cell wall such as hemicellulose, cellulose and pectin of the protein source. (Kumar et al., 2021). Specific activities of carbohydrates/ pectinases in the cell wall lead to dissolution and the release of cellular proteins from the materials such as cereals, oilseeds, and legumes, etc. (Rommi et al., 2014). Protease enzymes can help to break unbound proteins from the polysaccharide matrix and improve the protein yield. Enzymes need a range of acidic and alkaline environments to function at their optimum catalytic efficiency. Generally, carbohydrates function under mild acidic environments while several proteases, function under mild alkaline conditions. Proteases working in an alkaline environment require an optimum pH (8–10) and temperature (45–60 °C), respectively (Sari et al., 2015).

Press cakes made from dehulled rapeseeds are rich in proteins (36–40%) and carbohydrates (35%), and were treated with cellulolytic, pectinolytic, and xylanolytic enzyme preparations for protein extraction. Proteolytic enzymes disintegrated the cell walls by hydrolyzing glucans and pectic polysaccharides and increased the protein extraction yield by over 70%. Extraction rates of 74% and 56% for proteins were reported from dehulled and intact rapeseed press cakes, respectively (Rommi et al., 2014). Similarly, salmon by-products treated with Alcalase enzyme gave a protein yield of 85.70% (Gbogouri et al., 2004). *Palmaria palmata* seaweed gave a protein yield of 54.9% when treated with xylanase (Bjarnadóttir et al., 2018).

6.4.4 Ultrasound assisted extraction

In ultrasound assisted extraction, ultrasonic waves lead to cavitation which in turn generates hotspots of higher pressure and temperature. This mechanism helps to extract proteins from targeted cell walls. Rapid formation and collapse of gas bubbles (cavitation) generated by ultrasonic waves on the cell surface of the sample, microstreaming, and shockwaves exert high shear and mechanical forces, causing membrane and cell wall disruption (Lupatini et al., 2017). The surrounding solvent can then effectively penetrate into the cells and release the intracellular proteins into the solvent (Ummat et al., 2021).

Rapeseed proteins can be extracted in 80 min using ultrasound, the optimum extraction conditions were found to be 450 W, 1:20 sample to solvent ratio and a temperature from 25 to 35 °C. It has been shown that the temperature has a strong effect on the extraction process and the protein yield decreased above 35–55 °C (Dong et al., 2011). A study investigated the use of ultrasound to extract protein from porcine myocardium using 0.2 M NaCl, pH 8.0, extraction temperature <40 °C and an amplitude of 60–80% for extracting protein and reported a yield of 90% (Kim et al., 2017).

Another study showed that the use of ultrasound assisted extraction on *Lupinus mutabilis* gave a higher protein yield compared to conventional extraction, with a 14% yield increase after 10 min treatment, whereas for *L. angustifolius* isolates, no significant difference with the use of ultrasound technology was observed (Aguilar-Acosta et al., 2020). An interesting result was that in vitro digestibility was not affected by ultrasound treatment, although SDS-PAGE showed an increase in protein agglomerates due to ultrasound application during the alkaline extraction procedure. Another recent study reported that ultrasound treatment combined with enzymes such as carbohydrates, improves protein extraction from *Grateloupia turuturu,* the authors found that 3.6 ± 0.3 mg/g dw can be obtained with this synergistic process in 2 h (Le Guillard et al., 2016).

6.4.5 Pulsed electric field assisted extraction

Pulsed electric field (PEF) assisted extraction involves exposing the plant material to several pulses of high electric field intensity (10–80 kV/cm) for short durations. During this process, the sample is positioned between two electrodes and a transmembrane voltage, which is a function of the amplitude of the electric field, cell radius, and location of the membrane with respect to the electric field direction vector, is induced across the cell membrane (Golberg et al., 2016). The primary extraction mechanism is caused by the electroporation process, which can lead to the mass transport of intracellular components to the surrounding matrix, hence improving the extraction yields of bioactive compounds (Kotnik et al., 2015).

Rapeseed biomass was studied for protein and polyphenols extraction using PEF as a pretreatment step followed by use of a hydraulic press, this gave a better protein yield compared to unpretreated raw material (Yu et al., 2016). However another study reported a relatively low effect of PEF treatment on the yield of different components from papaya seeds (proteins, total phenolic compounds, carbohydrates, and isothiocyanates) when compared to aqueous extraction (Parniakov et al., 2015). An advantage of this technique compared to other conventional and nonconventional techniques is the extraction of proteins can be obtained at higher purity.

6.4.6 Microwave assisted extraction

Microwave assisted extraction (MAE) has been demonstrated for bioactive extraction from a wide range of matrices. Microwaves are electromagnetic radiation emitted in the range of 300 MHz–300 GHz. Two main frequencies (915 MHz and 2.45 GHz) are employed for microwave processing. Microwave heating is generated by ionic conduction of dissolved ions and dipole rotation of polar solvent. Rapid internal heating leads to effective cell rupture which releases the target compounds into a solvent (Ummat et al., 2021).

MAE yielded 1.54-fold more protein than alkaline solvents-based extraction from rice bran (Phongthai et al., 2016). MAE has several advantages compared to conventional thermal extraction, including rapid heating, improved extraction rate, lesser solvent consumption and shorter extraction time, making it suitable for solid-liquid extraction (Kumar et al., 2021). Recently, microwave-intensified enzymatic protein extraction was reported to significantly enhance the yield (96.4 mg/g) from Australian rock lobster shells. The protein hydrolysate produced using this process had an excellent techno functionality and high nutritional quality (Nguyen et al., 2016).

6.5 Key determinants for the acceptance of alternative proteins

The consumer acceptance or rejection of AP from novel sources such as algae, microbes, insects, byproducts etc. depend not only on product quality attributes (such as taste) or price, but also on other factors such as cultural beliefs and emotional aspects.

6.5.1 Food neophobia

Food neophobia is the expression used to describe an individual's tendency to reject new, unknown foods. Therefore, products made of AP, such as those listed above, could result in food neophobia. A recent study indicates that the probability to eat insects as a food ingredient or a dried insect is low as it showed a negative correlation between consumer's food preferences and neophobic tendencies. Food neophobia can be an important factor to predict consumer's readiness to eat beetles in Western countries, when it is related to a further descriptive variables such as gender, perceived healthiness, prior consumption, disgust sensitivity and convenience (Schlup and Brunner, 2018).

6.5.2 Disgust

EU culture consider that eating AP sources of insect, algae, and microbes as dangerous, dirty, and disgusting. It is also considered as a sheer nuisance and a disease vector as a substitute for their food choices. A recent study showed a significant negative impact of disgust on the willingness to eat insects, which resulted in rejection even before they tasted it. The impression of disgust associated with AP foods (insect, microbes, algae) generally comes from traditional beliefs in rejection and adverse expectations of taste (Hartmann and Siegrist, 2016). Another study revealed that the impact of eating insects leads to nausea, vomiting, and disgust and these findings suggest that consumers are strongly prejudiced even before tasting the products (Hartmann et al., 2015). On a positive note, some consumers are willing to eat insects as a result of a tasting activity (Sogari et al., 2017). These results indicate that an encouraging sampling experience can help to diminish the nous of disgust toward APs in countries where they are not part of their gastronomic heritage.

6.5.3 Environmental awareness

There is potential to produce AP as an eco-friendly and sustainable food resource, and consumers should have an awareness of the environmental impact on the production of such foods. Alternatively, governments should create awareness about the environmental impact of AP to consumers. It has been shown from some surveys that consumers who are concerned about the ecological impacts of the foods that they eat and have awareness of the benefits of APs are more likely to consuming APs (Schiemer et al., 2018). A study in Italy stated that consumers consider eating APs as a part of their contribution to prevent ecological disasters and the study also noted that consumers have a positive attitude (Menozzi et al., 2017). Lombardi et al. (2019) showed that consumers are eager to pay attention to various insect-based diets when facts on the advantage of consuming of insect-based protein (such as environmental factors) were given (Lombardi et al., 2019). Though the literature includes the significance of environmental awareness for accepting APs in their diets, this is not adequate to provide effective strategies and to stimulate actual behaviors for the intake of APs.

6.5.4 Health consciousness

Due to their valuable nutritional profile, edible insects are undoubtedly a healthy and highly nutritional food source. Ruby et al. (2015) observed that consuming insect-based food would enable nutritive food and along with that it provides the benefits of a healthier lifestyle. Therefore, these factors could change the future food habits by consuming APs (Ruby et al., 2015). In the modern era, athletes are looking for functional, balanced, and healthy diets and are likely to accept APs. Likewise, a study found that the main factor of consumer acceptance of APs is for health and wellbeing purposes (Gere et al., 2017).

6.5.5 Risk assessment

Apparent risks play a vital role in the approval of novel foods. Feelings of hesitation, distaste, and disgust to try APs are regularly based on the mistake of sources being dangerous, harmful, dirty to human health (Orsi et al., 2019). Baker et al. (2018) inspected the various features of risk awareness and its impact on willingness to try consuming APs. In an insect-based food study, it was found that several customers observed these sources can have a very high risk to safety, which eventually lead to refusal (Baker et al., 2018). Therefore, consumers risk assessment is necessary to predict the acceptance of APs.

6.5.6 Personal experiences

Most studies focusing on APs as human food has defined previous experiences of eating of the source is a key factor in the approval of APs. When AP products satisfy consumers in terms of flavor, taste and willingness to eat, can gain attention to consume those foods again and a study shows that participants liked the insect-based food product after tasting it (Orsi et al., 2019). Verbeke et al. (2015) disclosed that customers could not accept insect-based APs as a substitute for meat because it had a negative influence (Verbeke et al., 2015).

6.5.7 Familiarity

Familiarity and experience are closely related. Consumer awareness of the familiarity with APs can provide them a clear idea of why a particular AP is needed. If the consumers have tasted the AP, they will have familiarity with it before the product comes into the market. However, the consumer evaluation is not completely associated with the sensory experience after the first tasting. Studies have shown that familiarity with traditional foods prepared with APs could boost consumer acceptability. For example, the preparation of familiar foods such as cookies and meat balls with insect-based ingredients have increased the consumer likelihood to taste the food (Tan et al., 2015). A recent study found that a positive level of sensory liking was obtained for a hamburger made of insect-based ingredients (Megido et al., 2016).

6.5.8 Socio demographic factors

Gender issue has an effect on the acceptance of AP and studies have shown that males have a positive approach than females (Orsi et al., 2019). However, controversially, De Boer et al. (2013) and Hartmann et al. (2015) did not find an impact of gender on the satisfactoriness of insect-based foods. However, another author claimed that younger people are keener to eat APs than elder people (Hartmann et al., 2015). This shows that the level of education will play a key factor in the acceptance of and utilizing APs. Like meat, insects, algae, microbes and in vitro meat can be an issue related to nutritional limitations based on dietary or animal safety constraints. In particular, people who are consuming vegan or vegetarian diets are unlikely to accept APs and convincing them to consume APs will be difficult (Hartmann et al., 2015).

6.6 Health considerations

When introducing APs from new resources into the human diet, it is vital to take into account their bio-availability and other functional characteristics through the gastrointestinal tract and to evaluate their potential toxicity effects on human health.

6.6.1 Digestibility

APs or their fractions can be digested in the human gut and intestinal epithelium and absorbed. These absorbed products can affect the health of humans either in a positive (nutraceutical) or negative way (i.e., allergy or antinutritional; Ribeiro et al., 2017). Generally, novel food products can have allergenic proteins and their effects on intestinal digestion should be evaluated frequently. Therefore, in vitro digestion tests should be done to evaluate the allergenic properties of APs (i.e., testing AP for their resistance to gastric fluids; Fasolin et al., 2019). A study used lupin seed globulins for the in vitro digestion; they found that pepsin enzyme hydrolyzed the globulins completely and consecutively other enzymes such as chymotrypsin or pancreatin were treated but those enzymes did not hydrolyze it. γ-conglutin, which is present in the protein structure and its antigenic properties might have led to the improper hydrolysis of the enzymes (Czubiński et al., 2016). Another study identified and characterized chickpea seed proteins and found that these proteins are resistant to in vitro digestion. They reported that 7S vicilin and 11S legumin seed storage protein class was the reason for this characteristic and in addition, these proteins are bioactive. (Ribeiro et al., 2017). The outcomes of protein digestibility studies should be examined with caution. Conversely, these proteins might have specific allergens that cannot be solubilized in digestive juices. This shows that although digestion stability data can provide very useful information, this will not be enough to predict the allergenic potential of a protein (Untersmayr and Jensen-Jarolim, 2008). APs produced by *Corynebacterium ammoniagenes* and soybean proteins were studied for their potential digestibility properties in pigs. They found that protein from microbes gave lower digestibility than the soy protein and it is due to a higher amount of nonprotein nitrogen content (e.g., nucleotide; Wang et al., 2013).

6.6.2 Cytotoxicity

Numerous novel peptides and proteins have great potential toward nutrient supplements or as functional food components. Therefore, these products should be assessed for their availability, quality and digestibility. Most of the APs obtained from by products, microbes, insect, and algae should undergo a safety assessment before it comes to the market (Loveday, 2019). Recently, several insects such as *Gryllodes sigillatus*, *Schistocerca gregaria*, and *Tenebrio molitor* were checked for their cytotoxicity to human skin fibroblasts CRL-2522 using different products such as raw, cooked and protein hydrolysate. They found that *G. sigillatus* and *T. molitor* hydrolysate did not show any cytotoxic effects (before and after heat treatment), but *S. gregaria* hydrolysate showed cytotoxicity toward fibroblast CRL-2522 (up to 40% cell death; Zielińska et al., 2015). Moreover, cytotoxicity and antioxidant properties of peptides were obtained from microalgae *Navicula incerta* against HepG2/CYP2E1 cells and the peptides of microalgae decreased the ethanol-induced cytotoxicity of HepG2 cells (Kang et al., 2012). Another report stated that sunflower and rapeseed protein hydrolysates exhibit no toxicity effects on mice and no changes were found in food intake organ weight and blood biochemical and growth parameters (Canistro et al., 2017).

6.6.3 Allergenicity

Food allergies are increasing steadily, especially in infants and they occur due to an immune system reaction after consuming certain foods. A small amount of food (containing allergens) can activate signs and indications such as gastrointestinal difficulties, rashes, or inflammations. For example anaphylaxis, which is a serious fatal allergic reaction, and it can happen within seconds due to an allergic reaction, for instance, due to dairy proteins, peanuts, etc. (Verhoeckx et al., 2015). In general, allergic food reactions result in immunoglobulins (IgE) and non-IgE facilitated primary immunological sensitivities, secondary sensitivities, and nonimmunological food intolerances. The chief form of immunological allergens by foods is due to IgE formation. Some examples of food sources which are associated with IgE reactions are milk, egg, wheat, fish, shellfish, fruits, vegetables, and nuts (Valenta et al., 2015). These antibodies migrate to a different part of the body and react with allergen chemicals (ingested proteins), causing an allergic reaction. The symptoms can occur in mouth, throat, lungs, nose or skin (Valenta et al., 2015). Furthermore, studies have shown that nonthermal treatments, like microwave, ultrasonication, and high pressure processing can alter the food allergens (Pojić et al., 2018).

Moreover, a whole allergenic potential risk assessment is needed (according to the European Food Safety Authority (EFSA) legislation) to agree with APs as possible future food, taking into account several traits such as oral treatment and dosage of protein content, how a protein reacts to each individual's immune system and protein properties (e.g., physicochemical properties; Parenti et al., 2019). Various approaches can be used to evaluate allergenic potential risk. One potential approach is "weight-of-evidence approach" used for foods derived from genetically modified plants. There is a potential risk of an AP being an allergen and this emphasizes on cross reactivity. However, APs can also present a threat of de novo sensitization, which could lead to novel food allergies (Pali-Schöll et al., 2019).

6.7 Conclusions

APs such as vegetables, meat, and fish by-products, microbes, insect, and algae can be used in the food system for food sustainability and security. Novel ways of farming can contribute to preserving biodiversity and natural resources and reduce climate change and environmental damage. APs are considered as a source of macro and micronutrients which include high quality proteins when used as part of a healthy diet. Novel extraction processes based on enzymes, subcritical water, ultrasound, microwaves, and PEF, could improve the extraction rates and the energy efficiency, resulting in a more sustainable process to produce shelf stable proteins of specific functionalities. As there are some concerns that these novel processes could affect the biological, nutritional, and functional properties of the proteins as well as their allergenicity, and digestibility, further research is needed to validate the suitability of these novel technologies to extract and improve the functionalities of AP rich fractions.

The functional properties of most APs are still not known. Although vegetable proteins have been explored during the last few years, insect, algae, and microbial proteins have not been explored much. Recent research has investigated the effect of extraction parameters on the functional properties such as gelling, water holding capacity, and aggregation of APs and these findings reveal that APs can be used as a substitute for conventional proteins in the food industry. Furthermore, APs containing bioactive compounds with micro and nanostructures obtained from novel sources could be used in a wide range of innovative food products in specific applications that conforms to the environmental sustainability in the food industry.

It is important that the effect of APs in the human gastrointestinal tract (e.g., digestion and bioavailability) and their role in food allergens is studied. Furthermore, as several proteins appear to have allergenic or cytotoxic effects on human cells, more research is needed in this area to find out whether the proteins from these alternative sources will lead to benefits or risks. As there is limited data on the utilization of these proteins in nanotechnology and their safety aspects, additional risk assessment studies are required, especially on the impact on allergenicity and toxicity with their long-term usage.

Acknowledgments

This research was/is supported by Biorbic SFI Bioeconomy Research Centre, which is funded by Ireland's European Structural and Investment Programmes, Science Foundation Ireland (16/RC/3889) and the European Regional Development Fund.

References

Aguilar-Acosta, L.A., et al., 2020. Effect of ultrasound application on protein yield and fate of alkaloids during lupin alkaline extraction process. Biomolecules 10 (2), 292.

Alli, I., et al., 1993. Identification and characterization of phaseolin polypeptides in a crystalline protein isolated from white kidney beans (Phaseolus vulgaris). J. Agric. Food Chem. 41 (11), 1830–1834.

Alltech, *7th Annual Alltech Global Feed Survey*. Alltech Global, USA - California, 2018.

Ananey-Obiri, D., Tahergorabi, R., 2018. Development and characterization of fish-based superfoods. Curr. Top. Superfoods, 33.

Bähr, M., et al., 2014. Chemical composition of dehulled seeds of selected lupin cultivars in comparison to pea and soya bean. LWT-Food Sci. Technol. 59 (1), 587–590.

Baker, M.A., Shin, J.T., Kim, Y.W., 2018. Customer acceptance, barriers, and preferences in the US. In: Eilenberg, J., van Loon, J.J.A. (Eds.), Edible Insects in Sustainable Food Systems. Springer, Germany, pp. 387–399.

Becker, E., 2007. Micro-algae as a source of protein. Biotechnol. Adv. 25 (2), 207–210.

Benelhadj, S., et al., 2016. Effect of pH on the functional properties of *Arthrospira* (Spirulina) platensis protein isolate. Food Chem. 194, 1056–1063.

Bjarnadóttir, M., et al., 2018. Palmaria palmata as an alternative protein source: enzymatic protein extraction, amino acid composition, and nitrogen-to-protein conversion factor. J. Appl. Phycol. 30 (3), 2061–2070.

Burlingame, B., Dernini, S., 2010. Biodiversity and Sustainable Diets United Against Hunger, Rome.

Busnelli, M., et al., 2018. Effects of vegetable proteins on hypercholesterolemia and gut microbiota modulation. Nutrients 10 (9), 1249.

Canistro, D., et al., 2017. Digestibility, toxicity and metabolic effects of rapeseed and sunflower protein hydrolysates in mice. Ital. J. Anim. Sci. 16 (3), 462–473.

Caselato-Sousa, V.M., Amaya-Farfán, J., 2012. State of knowledge on amaranth grain: a comprehensive review. J. Food Sci. 77 (4), R93–R104.

Chardigny, J., Walrand, S., 2016. How might oil seeds help meet the protein challenge. EDP Sci, 23–24.

Chen, Y.-C., Jaczynski, J., 2007. Protein recovery from rainbow trout (Oncorhynchus mykiss) processing byproducts via isoelectric solubilization/precipitation and its gelation properties as affected by functional additives. J. Agric. Food Chem. 55 (22), 9079–9088.

Chernukha, I.M., Fedulova, L.V., Kotenkova, E.A., 2015. Meat by-product is a source of tissue-specific bioactive proteins and peptides against cardio-vascular diseases. Proc. Food Sci. 5, 50–53.

Council, N.R., 2000. Nutrient Requirements of Beef Cattle. National Research Council, *Washington,* DC.

Council, N.R., 2012. Nutrient Requirements of Swine. National Academies Press, uganda.

Czubiński, J., Siger, A., Lampart-Szczapa, E., 2016. Digestion susceptibility of seed globulins isolated from different lupin species. Eur. Food Res. Technol. 242 (3), 391–403.

Datar, I., Betti, M., 2010. Possibilities for an in vitro meat production system. Innovative Food Sci. Emerg. Technol. 11, 13–22.

De Boer, J., Schösler, H., Boersema, J.J., 2013. Climate change and meat eating: an inconvenient couple? J. Environ. Psychol. 33, 1–8.

de Castro, R.J.S., et al., 2018. Nutritional, functional and biological properties of insect proteins: processes for obtaining, consumption and future challenges. Trends Food Sci. Technol. 76, 82–89.

De Vries, H., et al., 2018. Small-scale food process engineering—challenges and perspectives. Innovative Food Sci. Emerg. Technol. 46, 122–130.

del Mar Contreras, M., et al., 2019. Protein extraction from agri-food residues for integration in biorefinery: Potential techniques and current status. Bioresour. Technol. 280, 459–477.

DeWitt, C.M., Gomez, G., James, J., 2002. Protein extraction from beef heart using acid solubilization. J. Food Sci. 67 (9), 3335–3341.

Dong, X.Y., et al., 2011. Some characteristics and functional properties of rapeseed protein prepared by ultrasonication, ultrafiltration and isoelectric precipitation. J. Sci. Food Agric. 91 (8), 1488–1498.

Duda, A., et al., 2019. Quality and nutritional/textural properties of durum wheat pasta enriched with cricket powder. Foods 8 (2), 46.

EFSA Panel on Nutrition, N.F., 2019. Safety of Yarrowia lipolytica yeast biomass as a novel food pursuant to Regulation (EU) 2015/2283. EFSA J. 17 (2), e05594.

Ettoumi, Y.L., Chibane, M., Romero, A., 2016. Emulsifying properties of legume proteins at acidic conditions: effect of protein concentration and ionic strength. LWT-Food Sci. Technol. 66, 260–266.

FAO, F., *The Future of Food and Agriculture – Trends and Challenges.* Annual Report, 2017.

FAO, I., UNICEF, W., 2017. The state of food security and nutrition in the world 2017Building Resilience for Peace and Food Security. FAO, Rome.

FAO, W., 2014. Conference outcome document: Rome declaration on nutrition, Second International Conference on Nutrition. FAO and WHO, Rome.

Fasolin, L.H., et al., 2019. Emergent food proteins – towards sustainability, health and innovation. Food Res. Int., 108586.

Fleurence, J., 1999. Seaweed proteins: biochemical, nutritional aspects and potential uses. Trends Food Sci. Technol. 10 (1), 25–28.

Futures, W.F., 2021. From Business as Usual to Business Unusual, World Economic Forum, Switzerland.

Galland-Irmouli, A.-V., et al., 1999. Nutritional value of proteins from edible seaweed *Palmaria palmata* (Dulse). J. Nutr. Biochem. 10 (6), 353–359.

Gbogouri, G., et al., 2004. Influence of hydrolysis degree on the functional properties of salmon byproducts hydrolysates. J. Food Sci. 69 (8), C615–C622.

Gehring, C., et al., 2011. Functional and nutritional characteristics of proteins and lipids recovered by isoelectric processing of fish by-products and low-value fish: a review. Food Chem. 124 (2), 422–431.

Gere, A., et al., 2017. Readiness to adopt insects in Hungary: a case study. Food Qual. Preference 59, 81–86.

Gervasi, T., et al., 2018. Production of single cell protein (SCP) from food and agricultural waste by using Saccharomyces cerevisiae. Nat. Prod. Res. 32 (6), 648–653.

Glick-Bauer, M., Yeh, M.-C., 2014. The health advantage of a vegan diet: exploring the gut microbiota connection. Nutrients 6 (11), 4822–4838.

Golberg, A., et al., 2016. Energy-efficient biomass processing with pulsed electric fields for bioeconomy and sustainable development. Biotechnol. Biofuels 9 (1), 1–22.

Guerreo-Ochoa, M.R., Pedreschi, R., Chirinos, R., 2015. Optimised methodology for the extraction of protein from quinoa (Chenopodium quinoa Willd. Int. J. Food Sci. Technol. 50 (8), 1815–1822.

Hall, F.G., et al., 2017. Functional properties of tropical banded cricket (*Gryllodes sigillatus*) protein hydrolysates. Food Chem. 224, 414–422.

Hartmann, C., et al., 2015. The psychology of eating insects: a cross-cultural comparison between Germany and China. Food Qual. Preference 44, 148–156.

Hartmann, C., Siegrist, M., 2016. Becoming an insectivore: results of an experiment. Food Qual. Preference 51, 118–122.

Henchion, M., et al., 2017. Future protein supply and demand: strategies and factors influencing a sustainable equilibrium. Foods 6 (7), 53.

Hosseini, S., Khosravi-Darani, K., 2010. Response surface methodology for mycoprotein production by Fusarium venenatum ATCC 20334. J. Bioprocess Biotechniq. 1, 1–5.

Huc-Mathis, D., et al., 2019. Emulsifying properties of food by-products: valorizing apple pomace and oat bran. Colloids Surf. A 568, 84–91.

Kadam, S.U., et al., 2017. Extraction and characterization of protein from Irish brown seaweed Ascophyllum nodosum. Food Res. Int. 99, 1021–1027.

Kamal, M.M., et al., 2019. Optimization of process parameters for improved production of biomass protein from Aspergillus niger using banana peel as a substrate. Food Sci. Biotechnol. 28 (6), 1693–1702.

Kang, K.-H., et al., 2012. Protective effects of protein hydrolysate from marine microalgae *Navicula incerta* on ethanol-induced toxicity in HepG2/CYP2E1 cells. Food Chem. 132 (2), 677–685.

Kim, H.K., et al., 2017. Protein extraction from porcine myocardium using ultrasonication. J. Food Sci. 82 (5), 1059–1065.

Kim, J.-S., Park, J.W., 2007. Mince from seafood processing by-product and Surimi as food ingredients. In: Kim, J.-S., (Eds.), Maximising the Value of Marine By-Products. Elsevier, Amsterdam, pp. 196–228.

Kim, S.-K., Mendis, E., 2006. Bioactive compounds from marine processing byproducts—a review. Food Res. Int. 39 (4), 383–393.

Kim, S.W., et al., 2019. Meeting global feed protein demand: challenge, opportunity, and strategy. Annu. Rev. Anim. Biosci.

Kojima, G., et al., 2018. Fruit and vegetable consumption and frailty: a systematic review. J. Nutr. Health Aging 22 (8), 1010–1017.

Kotnik, T., et al., 2015. Electroporation-based applications in biotechnology. Trends Biotechnol. 33 (8), 480–488.

Kumar, M., et al., 2021. Advances in the plant protein extraction: mechanism and recommendations. Food Hydrocolloids 115, 106595.

Kurcz, A., et al., 2018. Application of industrial wastes for the production of microbial single-cell protein by fodder yeast Candida utilis. Waste Biomass Valorization 9 (1), 57–64.

Langelaan, M.L., et al., 2010. Meet the new meat: tissue engineered skeletal muscle. Trends Food Sci. Technol. 21 (2), 59–66.

Lasekan, A., Bakar, F.A., Hashim, D., 2013. Potential of chicken by-products as sources of useful biological resources. Waste Manage. (Oxford) 33 (3), 552–565.

Laurens, L.M., et al., 2017. Development of algae biorefinery concepts for biofuels and bioproducts; a perspective on process-compatible products and their impact on cost-reduction. Energy Environ. Sci. 10 (8), 1716–1738.

Le Guillard, C., et al., 2016. Soft liquefaction of the red seaweed Grateloupia turuturu Yamada by ultrasound-assisted enzymatic hydrolysis process. J. Appl. Phycol. 28 (4), 2575–2585.

Liu, C.-m., et al., 2018. Molecular and functional properties of protein fractions and isolate from cashew nut (Anacardium occidentale L.). Molecules 23 (2), 393.

Lombardi, A., et al., 2019. Willingness to pay for insect-based food: the role of information and carrier. Food Qual. Preference 72, 177–187.

Loveday, S.M., 2019. Food proteins: technological, nutritional, and sustainability attributes of traditional and emerging proteins. Annu. Rev. Food Sci. Technol. 10, 311–339.

Lupatini, A.L., et al., 2017. Protein and carbohydrate extraction from S. platensis biomass by ultrasound and mechanical agitation. Food Res. Int. 99, 1028–1035.

Martínez-Alvarez, O., Chamorro, S., Brenes, A., 2015. Protein hydrolysates from animal processing by-products as a source of bioactive molecules with interest in animal feeding: a review. Food Res. Int. 73, 204–212.

Mattick, C.S., Landis, A.E., Allenby, B.R., 2015. A case for systemic environmental analysis of cultured meat. J. Integrative Agric. 14 (2), 249–254.

Megido, R.C., et al., 2016. Consumer acceptance of insect-based alternative meat products in Western countries. Food Qual. Preference 52, 237–243.

Melina, V., Craig, W., Levin, S., 2016. Position of the academy of nutrition and dietetics: vegetarian diets. J. Acad. Nutr. Dietetics 116 (12), 1970–1980.

Menozzi, D., et al., 2017. Eating novel foods: an application of the theory of planned behaviour to predict the consumption of an insect-based product. Food Qual. Preference 59, 27–34.

Mishyna, M., et al., 2019. Extraction, characterization and functional properties of soluble proteins from edible grasshopper (Schistocerca gregaria) and honey bee (Apis mellifera). Food Res. Int. 116, 697–706.

Mishyna, M., et al., 2019. Heat-induced aggregation and gelation of proteins from edible honey bee brood (Apis mellifera) as a function of temperature and pH. Food Hydrocolloids. 91, 117–126.

Nadathur, S., Wanasundara, J., Scanlin, L., 2017. Proteins in the diet: challenges in feeding the global populationSustainable Protein Sources. Elsevier, pp. 1–19.

Nguyen, T.T., et al., 2016. Microwave-intensified enzymatic deproteinization of Australian rock lobster shells (*Jasus edwardsii*) for the efficient recovery of protein hydrolysate as food functional nutrients. Food Bioprocess Technol. 9 (4), 628–636.

Niva, M., Vainio, A., Jallinoja, P., 2017. Barriers to increasing plant protein consumption in Western populations. In: Mariotti, F., (Ed.), Vegetarian and Plant-Based Diets in Health and Disease Prevention. Elsevier, amsterdam, pp. 157–171.

Nuckles, R., Smith, D., Merkel, R., 1990. Meat by-product protein composition and functional properties in model systems. J. Food Sci. 55 (3), 640–643.

Ooniex, D., 2017. Environmental impact of insect production. In: Ooniex, D., (Ed.), Insects as Food and Feed: From Production to Consumption. Wageningen Academic Publishers, Wageningen, Netherlands, pp. 79–93.

Organization, W.H., University, U.N., 2007. Protein and Amino Acid Requirements in Human Nutrition, 935. World Health Organization, Geneva, Switzerland.

Orsi, L., Voege, L.L., Stranieri, S., 2019. Eating edible insects as sustainable food? Exploring the determinants of consumer acceptance in Germany. Food Res. Int. 125, 108573.

Pali-Schöll, I., et al., 2019. Allergenic and novel food proteins: State of the art and challenges in the allergenicity assessment. Trends Food Sci. Technol. 84, 45–48.

Parenti, M.D., et al., 2019Literature Review in Support of Adjuvanticity/Immunogenicity Assessment of Proteins16. EFSA Supporting Publications, p. 1551E.

Parniakov, O., et al., 2015. New approaches for the effective valorization of papaya seeds: extraction of proteins, phenolic compounds, carbohydrates, and isothiocyanates assisted by pulsed electric energy. Food Res. Int. 77, 711–717.

Phongthai, S., Lim, S.-T., Rawdkuen, S., 2016. Optimization of microwave-assisted extraction of rice bran protein and its hydrolysates properties. J. Cereal Sci. 70, 146–154.

Pietrzak, W., Kawa-Rygielska, J., 2013. Utilization of spent brewer's yeast for supplementation of distillery corn mashes. Polish J. Chem. Technol. 15 (4), 102–106.

Pojić, M., Mišan, A., Tiwari, B., 2018. Eco-innovative technologies for extraction of proteins for human consumption from renewable protein sources of plant origin. Trends Food Sci. Technol. 75, 93–104.

Popkin, B.M., Adair, L.S., Ng, S.W., 2012. Global nutrition transition and the pandemic of obesity in developing countries. Nutr. Rev. 70 (1), 3–21.

Qiaoyun, C., et al., 2017. Optimization of protein extraction and decoloration conditions for tea residues. Hortic. Plant J. 3 (4), 172–176.

Ramos, O.L., et al., 2017. Design of whey protein nanostructures for incorporation and release of nutraceutical compounds in food. Crit. Rev. Food Sci. Nutr. 57 (7), 1377–1393.

Ratana-arporn, P., Chirapart, A., 2006. Nutritional evaluation of tropical green seaweeds Caulerpa lentillifera and Ulva reticulata. Kasetsart J. (Nat. Sci.) 40 (Suppl), 75–83.

Reihani, S.F.S., Khosravi-Darani, K., 2019. Influencing factors on single-cell protein production by submerged fermentation: a review. Electron. J. Biotechnol. 37, 34–40.

Ribeiro, I., et al., 2017. Identification of chickpea seed proteins resistant to simulated in vitro human digestion. J. Proteomics 169, 143–152.

Richter, C.K., et al., 2015. Plant protein and animal proteins: do they differentially affect cardiovascular disease risk? Adv. Nutr. 6 (6), 712–728.

Ritala, A., et al., 2017. Single cell protein—state-of-the-art, industrial landscape and patents 2001–2016. Front. Microbiol. 8, 2009.

Rommi, K., et al., 2014. Effect of enzyme-aided cell wall disintegration on protein extractability from intact and dehulled rapeseed (Brassica rapa L. and Brassica napus L.) press cakes. J. Agric. Food Chem. 62 (32), 7989–7997.

Roncolini, A., et al., 2019. Protein fortification with mealworm (Tenebrio molitor L.) powder: effect on textural, microbiological, nutritional and sensory features of bread. PLoS One 14 (2), 21–26.

Rosegrant, M.W., et al., 2001. Global Food Projections to 2020: Emerging Trends and Alternative Futures. International Food Policy Research Institute, UK.

Rubio, N.R., Xiang, N., Kaplan, D.L., 2020. Plant-based and cell-based approaches to meat production. Nat. Commun. 11 (1), 1–11.

Ruby, M.B., Rozin, P., Chan, C., 2015. Determinants of willingness to eat insects in the USA and India. J. Insects Food Feed 1 (3), 215–225.

Rustad, T., Storrø, I., Slizyte, R., 2011. Possibilities for the utilisation of marine by-products. Int. J. Food Sci. Technol. 46 (10), 2001–2014.

Salcedo-Chávez, B., et al., 2002. Optimization of the isoelectric precipitation method to obtain protein isolates from amaranth (Amaranthus cruentus) seeds. J. Agric. Food Chem. 50 (22), 6515–6520.

Sampels, S., 2014. Towards a more sustainable production of fish as an important protein source for human nutrition. J. Fisheries Livest Prod 2, 119. doi:10.4172/2332-2608.1000119 *Page 2 of 2 9. NordgardenUF, OppedalF, TarangerGL, HemreGI, HansenT (2003) Seasonally changing metabolism in Atlantic salmon (Salmo salar L.) I–Growt h and feed conversion ratio. Aquacult. Nutr., 2014. 9: p. 287-293.*

Sandnes, K., Pedersen, K., Hagen, H., 2003. Prosessering av fiskeraÊstoff ved hjelp av industrielle enzymer Final report, RUBIN.

Sari, Y.W., et al., 2015. Towards plant protein refinery: review on protein extraction using alkali and potential enzymatic assistance. Biotechnol. J. 10 (8), 1138–1157.

Schiemer, C., et al., 2018. Marketing insects: superfood or solution-food? In: Schiemer, C., (Ed.), Edible Insects in Sustainable Food Systems. Springer, Germany, pp. 213–236.

Schlup, Y., Brunner, T., 2018. Prospects for insects as food in Switzerland: a tobit regression. Food Qual. Preference 64, 37–46.

Šližytė, R., et al., 2009. Functional, bioactive and antioxidative properties of hydrolysates obtained from cod (*Gadus morhua*) backbones. Process Biochem. 44 (6), 668–677.

Sogari, G., Menozzi, D., Mora, C., 2017. Exploring young foodies' knowledge and attitude regarding entomophagy: a qualitative study in Italy. Int. J. Gastronom. Food Sci. 7, 16–19.

Søvik, S.L., 2005. Characterisation of Enzymatic Activities in By-Products From Cod Species: Effect of Species, Season and Fishing Ground. Elsevier, Amsterdam.

Stoll-Kleemann, S., Schmidt, U.J., 2017. Reducing meat consumption in developed and transition countries to counter climate change and biodiversity loss: a review of influence factors. Reg. Environ. Change 17 (5), 1261–1277.

Sui, Y., et al., 2019. Light regime and growth phase affect the microalgal production of protein quantity and quality with Dunaliella salina. Bioresour. Technol. 275, 145–152.

Sun-Waterhouse, D., et al., 2016. Transforming insect biomass into consumer wellness foods: a review. Food Res. Int. 89, 129–151.

Tabilo-Munizaga, G., et al., 2019. Physicochemical properties of high-pressure treated lentil protein-based nanoemulsions. LWT 101, 590–598.

Tabtabaei, S., et al., 2019. Functional properties of navy bean (*Phaseolus vulgaris*) protein concentrates obtained by pneumatic tribo-electrostatic separation. Food Chem. 283, 101–110.

Takao, T., et al., 2005. Hypocholesterolemic effect of protein isolated from quinoa (*Chenopodium quinoa* Willd.) seeds. Food Sci. Technol. Res. 11 (2), 161–167.

Tan, H.S.G., et al., 2015. Insects as food: exploring cultural exposure and individual experience as determinants of acceptance. Food Qual. Preference 42, 78–89.

Taylor, K., Himonides, A., Alasalvar, C., 2007. Increased processed flesh yield by recovery from marine by-products In: Taylor, K., (Ed.), Maximising the Value of Marine By-products. Elsevier, Amsterdam, pp. 91–106.

Tenorio, A.T., et al., 2017. Interfacial properties of green leaf cellulosic particles. Food Hydrocolloids. 71, 8–16.

Thorkelsson, G., et al., 2008. Mild Processing Techniques and Development of Functional Marine Protein and Peptide Ingredients. Woodhead Publishing, United Kingdom.

Tian, J., Bryksa, B.C., Yada, R.Y., 2016. Feeding the world into the future—food and nutrition security: the role of food science and technology. Front. Life Sci. 9 (3), 155–166.

Toldrá, F., Mora, L., Reig, M., 2016. New insights into meat by-product utilization. Meat Sci. 120, 54–59.

Ummat, V., et al., 2021. Advances in pre-treatment techniques and green extraction technologies for bioactives from seaweeds. Trends Food Sci. Technol.

Untersmayr, E., Jensen-Jarolim, E., 2008. The role of protein digestibility and antacids on food allergy outcomes. J. Allergy Clin. Immunol. 121 (6), 1301–1308.

Valenta, R., et al., 2015. Food allergies: the basics. Gastroenterology 148 (6), 1120–1131 e4.

Van der Spiegel, M., Noordam, M., Van der Fels-Klerx, H., 2013. Safety of novel protein sources (insects, microalgae, seaweed, duckweed, and rapeseed) and legislative aspects for their application in food and feed production. Comprehens. Rev. Food Sci. Food Safety 12 (6), 662–678.

Verbeke, W., Sans, P., Van Loo, E.J., 2015. Challenges and Prospects for Consumer Acceptance of Cultured Meat. Elsevier, Amsterdam.

Verhoeckx, K.C., et al., 2015. Food processing and allergenicity. Food Chem. Toxicol. 80, 223–240.

Vieira, E., Cunha, S.C., Ferreira, I.M., 2019. Characterization of a potential bioactive food ingredient from inner cellular content of brewer's spent yeast. Waste Biomass Valorization 10 (11), 3235–3242.

Villamil, O., Váquiro, H., Solanilla, J.F., 2017. Fish viscera protein hydrolysates: production, potential applications and functional and bioactive properties. Food Chem. 224, 160–171.

Wallecan, J., et al., 2015. Emulsifying and stabilizing properties of functionalized orange pulp fibers. Food Hydrocolloids. 47, 115–123.

Wang, J., et al., 2013. Amino acid digestibility of single cell protein from *Corynebacterium ammoniagenes* in growing pigs. Anim. Feed Sci. Technol. 180 (1-4), 111–114.

Westcott, P. and J. Hansen, 2021 *USDA Agricultural Projections to 2025*.No. 2016, OCE-2016-1). Office of the Chief Economist, World Agricultural Outlook Board.

Woolf, E., et al., 2019. Willingness to consume insect-containing foods: a survey in the United States. LWT 102, 100–105.

Yang, A., et al., 2017. Enhancing protein to extremely high content in photosynthetic bacteria during biogas slurry treatment. Bioresour. Technol. 245, 1277–1281.

Yu, X., et al., 2016. Pulsed electric field pretreatment of rapeseed green biomass (stems) to enhance pressing and extractives recovery. Bioresour. Technol. 199, 194–201.

Zhao, X., et al., 2016. Yellow mealworm protein for food purposes-extraction and functional properties. PLoS One 11 (2), 22–26.

Zielińska, E., et al., 2015. Selected species of edible insects as a source of nutrient composition. Food Res. Int. 77, 460–466.

CHAPTER

7

Key technological advances of extrusion processing

M. Azad Emin

Chair of Food Process Engineering, Karlsruhe Institute of Technology, Karlsruhe, Germany

7.1 Introduction

Extrusion is one of the few food technologies that has been continuously developed since its invention, and recently, it has also gained a significant attention with respect to the design of sustainable food products. This is due to its high degree of flexibility concerning the selection of raw materials and process conditions which are necessary to produce a wide variety of food products. With this, extrusion has been offering concrete solutions to one of the major problems of current food research, which is the design of sustainable food products without sacrificing their appeal and convenience character, and in a manner that they will indeed promote health and well-being of the consumers.

In almost all extrusion applications, however, there is a major challenge in the determination and characterization of the critical parameters required to control structural changes. Extrusion is a closed continuous process operating in dynamic steady state equilibrium. In most studies, extruders are therefore considered as black boxes, as the influence of process parameters on resulting extrusion conditions and their effect on structural changes are unknown. This results in the control of this process and design of new extruded products based on empirical knowledge. The empirical approach links independent process variables (e.g., screw speed, barrel temperature, water content, ingredients) to final product structure and characteristics. The empirical approach has been shown to be successful for many conventional food products. However, it does not allow determining the mechanisms responsible for structural changes. With increased complexity of the food systems, the conventional empirical approach generally fails to provide fundamental understanding and is not efficient in finding a rational compromise between the large numbers of parameters influencing the structure and functionality of the food product. Furthermore, product range extension and process scale-up remain challenging, as the information gained is highly material and machine dependent. Therefore, the ever-increasing demand for an in-depth analysis and better control of the extrusion processing remains unrealized and pertinent to this day.

Already in the 1980s, Meuser and van Lengerich et al. addressed the limitations of the empirical approach in their publications (Klingler et al., 1986; Meuser et al., 1982; van Lengerich, 1990). They have also presented a so-called "system analytical model," which can be considered as the initial step toward improved process understanding at mechanistic level. The system analytical model attempts to link the structural changes directly with the process characterized by the specific mechanical energy input (SME) and specific thermal energy input (STE), calculated from motor torque and product temperature at the die, respectively (Klingler et al., 1986; Meuser et al., 1982; van Lengerich, 1990). This approach

Food Engineering Innovations Across the Food Supply Chain. DOI: https://doi.org/10.1016/B978-0-12-821292-9.00005-4

has considerably improved the analysis of the structural changes in extrusion processing (Bouvier and Campanella, 2014). However, the parameters used (i.e., SME, STE) to characterize the process are insufficient for determining and controlling the decisive mechanisms behind the structural changes in complex food systems (Emin et al., 2015; Alig et al., 2010; Barnes et al., 2007). SME is strongly dependent on rheological, and therefore, the structural changes. Hence, it almost always correlates with the product properties, even if it is not the underlying cause of these changes. This severely restricts its application for process analysis. Moreover, SME and STE are integral values, and supply no information on local thermal and mechanical stresses, varying strongly along the extruder screws. Such local temperature peaks and distributions are often the primary cause of the many structural changes (e.g., reactions and morphological changes) that occur during extrusion processing (Emin and Schuchmann, 2013a; Liu et al., 2010; van Den Einde et al., 2004).

Since 2013, the research group "Extrusion Biopolymeric Materials" at the Karlsruhe Institute of Technology aims to overcome these limitations by developing a research approach and methods to analyze extrusion processing at the mechanistic level. This contribution gives a brief overview to the main outcomes of these efforts.

7.2 Research approach

In general, the extrusion process can be analyzed as two coherent sections: the screw section and the die section (Fig. 7.1) (Emin, 2015). The first section can be considered as a bioreactor, where the food material is exposed to thermal and mechanical stresses generated by the rotation of the screws and the heating of the cylinders. According to the extruded raw material (e.g., starch or protein), these stresses can alter the physical and chemical structure of the material through various mechanisms such as gelatinization, melting, polymerization, depolymerization, etc. (Camire et al., 1990; Lai and Kokini, 1991; Emin et al., 2012). The kinetics of these reactions strongly correlate with the extent of thermal and mechanical treatment in the screw section (Pietsch et al., 2019). These stresses therefore play a fundamental role in the resulting structure and properties of the plasticized material, which is then forced to flow through the extruder die.

The die section is mainly used to give the product the desired product shape and texture. Directly expanded ready-to-eat cereals often require small die diameters and short lengths at high temperatures (>100 °C) to achieve high and sudden pressure drops (Bouvier and Campanella, 2014). For meat analogues extruded at high moisture content (>40%), long cooling slit dies (<50 °C) are required (Pietsch et al., 2019). However, regardless of the product, the structure formation in the die section depends not only on the die geometry or temperature, but also on the material properties (Della Valle et al., 1997; Horvat et al., 2013; Moraru and Kokini, 2003). Consequently, the control of the structuring in the die section is only possible if the control of the reactions in the screw section is also possible. Such control of the reactions is also essential for the design of novel components by reactive extrusion or for the retention of labile components during the fortification of extrusion products.

During extrusion, there are strong interactions between mass, energy, and momentum transfer, coupled with complex physicochemical transformations of the material, which govern its material properties. In other words, the food structure is not only the outcome but also an integral part of the

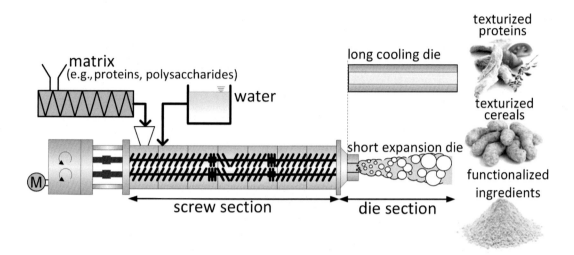

FIG. 7.1

A schematic illustration of twin-screw extrusion processing and its principal sections.

process, as it affects the evolution of the material properties and thus the processing conditions along the extruder. For instance, mechanical stresses generated during extrusion processing are a direct function of rheological properties, which again depend on the material structure prone to change because of mechanical stresses along the extruder. This dynamic interrelation is depicted in Fig. 7.2 (Emin and Schuchmann, 2017).

For understanding and control of structural changes in extrusion processing, therefore, an in-depth analysis of the processing conditions as well as material design properties are necessary. Here, we define the material design properties as the process-related material properties, such as rheological and reaction properties, which play a decisive role on the structure formation during extrusion processing. Therefore, these properties, as well as their dynamic dependency on the processing conditions and material structure, must be known and controlled to move away from purely empirical approaches, and to design functional and sustainable food systems in a targeted manner. The next sections focus on the methods proposed for the analysis of these two major groups of in-process variables and their dynamic interrelation.

7.3 Analysis of material design properties
7.3.1 Reaction properties
Reaction properties include but are not limited to the types of reactions and molecular interactions involved, the onset temperatures of these reactions as well as their kinetics as a function of processing conditions (e.g., temperature, time, shear stress) and formulation (e.g., components, mixing ratio, water

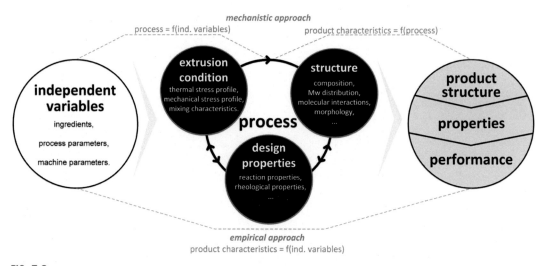

FIG. 7.2

Mechanistic interrelation between independent variables, extrusion conditions, material properties, and structure in extrusion processing (Emin and Schuchmann, 2017).

content). Reaction properties are one of the most important but somehow least investigated material properties for extrusion applications. Although there is vast amount of information about diluted systems, the reaction properties of highly concentrated biopolymers at extrusion conditions (i.e., polymer concentrations > 40%, temperature > 100 °C, and shear stresses >50 kPa) are a subject of only very few studies. This is probably because of the undefined nature of extrusion making it not suitable for such analysis, as here the material is exposed to both thermal and mechanical stresses simultaneously. Moreover, these stresses extensively vary along the screw due to its complex geometry.

For this purpose, we propose to use a model process, in which near real extrusion conditions can be generated in a defined manner (Emin et al., 2017). In our research, we use a closed cavity rheometer (TA Instruments, RPA Elite) which was originally developed in the early 1990s for the characterization of rubber applications (e.g., vulcanization). With such rheometers, it is possible to apply defined thermomechanical stresses on materials at elevated pressure (as necessary to plasticize the matrix) without moisture loss due to evaporation (Madeka and Kokini, 1996; Morales and Kokini, 1997; Pommet et al., 2004). Furthermore, high torque levels (up to 25 N m) necessary for processing of highly viscous biopolymeric melts can be realized. As it is a rheometer, the dynamic rheological properties of the melt can be measured inline during or after a thermomechanical treatment to monitor the structural changes and reaction behavior (Dotsch et al., 2003; Leblanc and Cartault, 2001; Pommet et al., 2004). At the end of the treatment, the samples can be further analyzed by offline analysis to deduce detailed information about the role of individual processing conditions (e.g., time, temperature, shear) on the reaction behavior and structural changes of the material. This way, the kinetic models necessary for the process control can be developed (Quevedo et al., 2020; Pietsch et al., 2018, 2019). As the conditions can be kept constant, the difference in the reaction behavior of various ingredients as well as the role of individual components in mixtures can also be analyzed. In our previous studies, we used this method to perform the analysis of the reaction properties of wheat, soy, and whey proteins, as well as their mixtures with polysaccharides at various conditions related to the texturization and functionalization of these proteins by extrusion processing. Fig. 7.3 shows selected results from this study to demonstrate the capabilities of the proposed method. A comprehensive analysis of these results with complementary analyses of molecular interactions and techno-functional properties can be found in the relevant studies (Emin et al., 2017; Koch et al., 2017a and 2017b).

Fig. 7.3A and B shows exemplary analyses of reaction onset temperatures for different protein concentrations, protein types, and mixtures. Fig. 7.3C demonstrates the analyses of reaction behavior as a function of time, temperature, and shear, whereas Fig. 7.3D depicts a schematic illustration of the closed cavity rheometer (CCR) used for these measurements. In the CCR, the sample (approximately

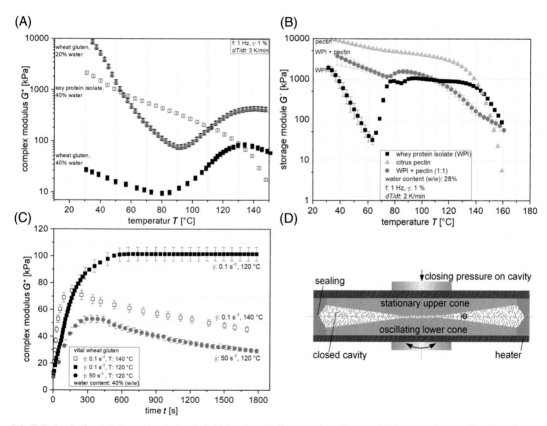

FIG. 7.3 Analysis of the reaction behavior of biopolymers at extrusion-like conditions by closed cavity rheometry.

(A) Temperature sweep analysis showing the effect of protein concentration and protein type, exemplified using wheat gluten at two different water content of 20% and 40% and soy protein isolate with a water content of 40% (Emin et al., 2017). (B) Temperature sweep analysis showing the effect of polysaccharides on the reaction behavior of proteins, exemplified using whey protein isolate and citrus pectin at a water content of 28% (Koch et al., 2017a). (C) Time sweep analysis showing the difference in reaction kinetics as a function of temperature and shear rate, exemplified by using wheat gluten at a water content of 40% (Emin et al., 2017). (D) Schematic illustration of the closed cavity rheometer used for these measurements.

4.5 g) is loaded between two cones. The lower cone is driven by a motor in an oscillatory movement at a user-defined frequency and shear strain amplitude. The sample in this apparatus is kept in a sealed and pressurized cavity (up to 8500 kPa). The bi-conical cavity geometry transmits the shear stress on the sample volume homogeneously. The temperature of the cavity can be varied from 25 °C to 170 °C.

In Fig. 7.3A, temperature sweep analyses for wheat gluten (87% protein) at two different water contents of 20%, as well as for soy protein isolate (90 % protein) with water content of 40% (w/w), are shown. For the wheat gluten samples, the results show three distinctive regions: decrease in complex modulus at mild temperature conditions; followed by steep increase; and final decrease in complex modulus at elevated temperatures. The initial decrease in the complex modulus is related to the higher molecular mobility due to increased temperature. The difference in the absolute values is associated with the higher viscosity of the samples with lower water content, as expected. A steep increase at elevated temperatures can be considered as onset temperature for the polymerization/aggregation reactions leading to the formation of a crosslinked network structure (Lagrain et al., 2010; Hayta and Schofield, 2005; Madeka and Kokini, 1994). This onset temperature is of strong interest, as such a dramatic change in the rheological properties does not only significantly affect the material structure but also the conditions generated during processing. The results also show that increasing water content leads to lower onset temperature, which is probably due to higher molecular mobility leading to lower activation energy necessary for the reactions. The increase in the complex modulus continues with increasing temperature and reaches a peak at around 130 °C and 140 °C for the wheat gluten samples with 40% and 20% water content, respectively. These maxima are related to the maximum structure build-up followed by a decrease in the complex modulus due to end of polymerization/aggregation reactions and/ or the onset of degradation reactions (Kokini et al., 1994; Pommet et al., 2004).

In comparison to wheat gluten with 40% water, soy protein isolate (SPI) shows a different behavior. At $T < 130$ °C, it shows significantly higher matrix viscosity. Furthermore, it shows no increase in G* with temperature, indicating that no aggregation reaction takes place. These results are verified by offline analysis suggesting that this specific SPI is already completely denatured during its commercial extraction/production step prior to extrusion.

Interestingly, the results from Fig 7.3A show that SPI reaches similar values as those of wheat gluten with 40% water at $T > 130$ °C (e.g., typical conditions for protein texturization), but in this case through an increased molecular mobility and not aggregation.

During many extrusion applications (e.g., protein functionalization), proteins are often extruded with polysaccharides. Therefore, a possible effect of the polysaccharides on the reaction behavior of proteins are of high interest. Fig. 7.3B gives an example for such a case at which whey proteins are mixed with citrus pectin and temperature sweep measurements are performed. Here, the results show a distinct difference in the reaction behaviors of each component. Pectin shows degradation reaction at $T > 120$ °C, whereas WPI shows aggregation onset temperature at around 62 °C. The mixture of these components leads to completely different behavior showing a shift in the aggregation temperature of WPI to 72 °C, whereas the degradation temperature of pectin seems to be reduced. However, it must be noted that these results also include the time effect with a large extent due to the heating rate of 2 K/min. To deduce more detailed information about the reaction behavior of biopolymers, an isothermal time sweep analyses can be performed.

Fig. 7.3C depicts an example for such an analysis showing the effect of constant temperature and shear as a function of time. The complex modulus of the samples measured at 120 °C and a shear rate of 0.1 s^{-1} show a linear increase until a maximum is reached at around 600 s, whereas the sample measured at 140 °C shows a faster increase in the complex modulus reaching its maximum at 150 s followed by a continuous decrease. This change in rheological behavior is directly related to the kinetics of aggregation taking place faster at elevated temperatures. The decrease in the complex modulus of the sample measured at 140 °C after the maximum can be linked to depolymerization and/or dissociation of the gluten proteins.

In addition to these results, the sample treated with increased shear (i.e., 50 s^{-1}) shows a different behavior. First, the results show that increasing shear rate leads to lower absolute values of complex modulus at the same temperature (i.e., 120 s^{-1}), as expected, because the plasticized gluten proteins show a shear thinning behavior. The results further show that the reaction behavior is also influenced by shear. The sample treated at minimum shear shows no reduction in complex modulus in the time range investigated, whereas a treatment at elevated shear rate of 50 s^{-1}, causes a reduction of the complex viscosity after reaching a maximum value. These results suggest that the depolymerization reaction took place at an earlier stage with the application of shear.

These measurements allow to gain insight into the reaction properties of biopolymers under combined thermal and mechanical stress under extrusion-like conditions. Stresses triggering significant changes in the biopolymer structure can be identified. Moreover, the effect of individual parameters (e.g., process variables or formulation) on the reaction kinetics of biopolymers can be investigated in detail through targeted sampling and a subsequent analysis of the samples using offline methods.

7.3.2 Rheological properties

The rheological properties of biopolymers during extrusion is not only a marker for changes in structure, but also performs a decisive role in almost all critical process variables such as mixing properties, thermal and mechanical stress profile in the screw section or expansion and texturizing in the die section (Della Valle et al., 1997; Emin and Schuchmann, 2013A; van Lengerich, 1990).

Biopolymers, such as plasticized starch or protein, subjected to extrusion-type conditions, show non-Newtonian rheological behavior, depending on different parameters such as screw speed, temperature, and moisture content (Lai and Kokini, 1991; Xie et al., 2012). Measurement of their rheological behavior under extrusion conditions is nontrivial due to the requirement of elevated temperature and pressure. Such conditions are required to preserve the water content in the sample to measure rheological data accurately. Moreover, the rheometer must be capable of handling very high torque values resulting from very high matrix viscosity. Such requirements cannot be fulfilled by conventional offline rheometers. For this purpose, we propose to use specific rheometers like the closed cavity rheometers, which can ensure such conditions and give detailed information about the viscous and elastic properties of material at low to very large deformations which are relevant for the extrusion processing of biopolymers (Pietsch et al., 2019; Wittek et al., 2020). Fig. 7.4A shows an example for a frequency sweep analysis giving information about the complex rheological properties of biopolymers in the linear viscoelastic region. Here, the results show the effect of methylcellulose on the elastic and shear modulus as well as complex viscosity of soy protein isolate with water content of 52%. Methylcellulose (MC) is often added to the formulation of high-moisture extrusion applications to improve the texture formation in the meat analogues. Here, the result shows that the addition of 10% MC leads to significant increase in the elasticity as well as complex viscosity of soy proteins. Such measurements are often trivial and a standard practice in the field of rheology, which however is not the case for the highly concentrated biopolymers at extrusion-like conditions. Our results show that the CCR is capable of filling this important

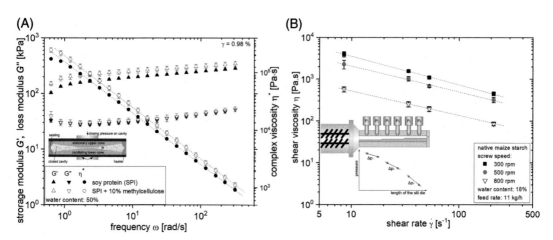

FIG. 7.4 Analysis of the rheological properties by using offline closed cavity rheometer and inline multiple steps slit-die rheometer.

(A) Frequency sweep analysis by CCR showing the complex rheological properties of biopolymers in the linear viscoelastic region, exemplified by using soy protein isolate and methylcellulose as model system (Wittek et al., 2020). (B) Inline analysis of the shear viscosity by slit-die rheometry, exemplified by using maize starch as model system (Emin and Schuchmann, 2013a).

gap and allow gaining detailed information about viscous and elastic behavior of the biopolymers for the extrusion applications. These results are especially useful for the constitutive modeling necessary for the simulation of viscoelastic biopolymer flow during extrusion processing.

However, despite its extensive capabilities, CCR offers no information about the shear viscosity of the matrix at high shear rates. Such information is necessary for deducing the viscosity function used for modeling purposes (e.g., computational fluid dynamics). The complex viscosity depicted in Fig. 7.4B can be used for this, if the Cox-Merz relations is valid. The Cox-Merz rule is an empirical rule which states that the dependence of the steady shear viscosity on the shear rate can be estimated from the complex viscosity as a function of frequency as the two curves are approximately identical. This is however often not the case.

For the measurement of shear viscosity of plasticized biopolymers, inline capillary rheometry can be accepted as most accurate method (Padmanabhan and Bhattacharya, 1991; Della Valle et al., 1996). Inline slit die rheometers are often used (van Lengerich, 1990; Vergnes et al., 1993). For this, often a slit die with known dimensions is mounted at the end of the extruder running at constant flow rate. The pressure gradient in the fully developed flow region of the slit die is measured and then used to determine the shear viscosity by using the Hagen and Poiseuille equation.

For such measurements, it is important to avoid a change in the backpressure in the extruder to avoid any difference in mechanical and thermal stress profile. This can be achieved by using a modular single channel inline rheometer. The rheometer has a modular design, with exchangeable inner geometry by using different slide-in modules (Emin and Schuchmann, 2013a; Horvat et al., 2013).

The inner geometry of the slit die can be designed in such a way that backpressure and therefore thermomechanical history of the material can be kept constant. The slit die rheometer is also equipped with a heating/cooling system keeping the material at desired temperature. This is required for describing the temperature dependency of rheological properties, for example, by using the Arrhenius equation (Horvat et al., 2013). Fig. 7.4B shows the schematic illustration of this rheometer together with an exemplary measurement showing the viscosity of plasticized starch at various screw speeds (Emin and Schuchmann, 2013a). These results show the shear thinning behavior of starch as well as the decrease in the matrix viscosity at higher screw speeds due to an increase in starch fragmentation.

7.4 Analysis of processing conditions
7.4.1 Analysis of thermal stress profile

Control of the structural changes during extrusion is only possible if the thermal stress profile of the processed material is known, as both reaction and rheological properties depend on the temperature (Cheftel, 1986; Emin et al., 2012; Horvat et al., 2013; Van Den Einde et al., 2004). This requires information about the material temperature and residence time of the material at this temperature.

Temperature measurements during extrusion processing are usually made using thermocouples that are often flush-mounted to the extruder barrel and die head as depicted in Fig. 7.5C (Bouvier and Campanella, 2014; Rauwendaal, 2014). Such measurements are, however, highly susceptible to the thermal conduction from the barrel wall to the thermocouple junction (Davis, 1988; Karwe and Godavarti, 1997; Maier, 1996). A possible difference in the barrel and material temperature, which is often the case, affects the accuracy of the measurements significantly. Furthermore, thermocouples have slow response times, typically 1 s, which makes them not suitable to measure the local temperature peaks, which occur due to maximum viscous dissipation generated between the screw tips and barrel wall (Emin and Schuchmann, 2013a; Ishikawa et al., 2001; Kohlgrüber, 2007).

Infrared sensors offer more accurate and sensitive alternatives to the thermocouples (Chen, 1992; Emin et al., 2015; Karwe and Godavarti, 1997; Maier, 1996). They offer the advantage of quick response time in the form of a noninvasive probe to measure the temperature distribution of extruded biopolymers (Emin et al., 2015; Karwe and Godavarti, 1997; Maier, 1996).

Despite their advantages, the application of infrared sensors in food extrusion has only been investigated by very few studies (Emin et al., 2015; Karwe and Godavarti, 1997). One of the main challenges concerning such a study on biopolymeric melts is the necessity of a calibration method to determine the material specific emissivity coefficients at processing conditions. For this, we recently developed an inline method based on the use of a specific channel mounted at the end of the extruder (Emin et al., 2015). The calibrated infrared sensor can then be used in the transient nonisothermal conditions of the extruder. Fig. 7.5A shows exemplary results for such measurements compared to the measurements of conventional thermocouples. The measurements are performed in the fifth barrel of a seven barrel twin screw extruder (Coperion ZSK 26 Mc, Coperion GmbH, Stuttgart, Germany) with a screw diameter of 26 mm and length to diameter ratio (L/D) of 29.

The material temperature measured by the infrared sensor shows strong variations due to the transient conditions generated by the rotation of the screws. These variations increased with screw speed as

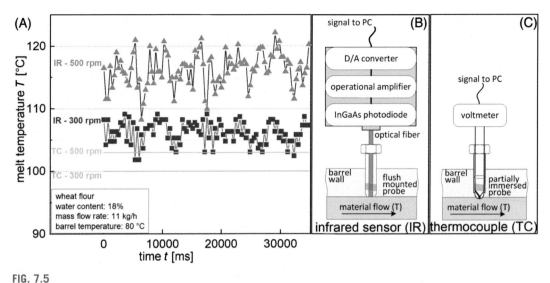

FIG. 7.5

Analysis of local temperature variations during extrusion measured by infrared sensor in comparison to the measurements of thermocouples, exemplified by using wheat starch as model system (Emin et al., 2015).

the viscous dissipation, and therefore, the heat generation increases. A comparison to the results from thermocouples shows a significant difference (up to 10 K for mean values and up to 20 K for the peak values) in the range of screw speeds investigated. This shows that thermocouples are not suitable for the measurement of material temperature along the extruder. Our results further showed that, even if accurate, the time-averaged measurement of the material temperature alone (i.e., mixture temperature) can be insufficient for controlling the structural changes during the applications at which viscous dissipation can lead to significantly higher temperature peaks. For this, infrared sensors can offer a solution and capture the true melt temperature as well as temperature variations and peaks, which are expected to play a decisive role on structural changes.

For the measurement of residence time distribution, the so-called tracer methods can be used. These methods are based on instant incorporation of a certain tracer (e.g., dye or fluorescence tracer) at the first extruder barrel and the measurements of the tracer intensity at different locations along the extruder barrel as a function of time by using various methods (e.g., UV–Vis; Lee et al., 2009), fluorescence detector (Gerstorfer et al., 2013) depending on the tracer. For food extrusion applications, a food grade coloring is often used as a tracer, whose intensity can be monitored through the processing of images recorded by a camera at the extruder die exit (Lee et al., 2009). However, this method is limited to the residence time in the whole extruder. As the temperature along the extruder is not constant, local measurements at different sections of the extruder are also necessary. Even then, the measurement of local temperature and residence time distribution can only give limited information regarding the thermal stress profile, as the stress distribution along the extruded material, or the amount of material which is exposed to maximum stresses, cannot be deduced from this information. Nevertheless, the gained information offers a more reliable basis for process design and can also be used to improve the accuracy of numerical simulations focusing on the thermal stress profile in an extruder.

7.4.2 Analysis of thermomechanical stress profile and mixing characteristics

The measurement of the thermal and mechanical stress profile distribution in an extruder is a very complicated task, and to the best of our knowledge, the experimental methods alone are not capable of monitoring these parameters. For this, numerical analysis can be performed, which offers an access to local processing conditions as well as to the stress profile within an extruder (Alsteens et al., 2004; Bravo et al., 2000; Cheng and Manas-Zloczower, 1997; Emin and Schuchmann, 2013a; Ishikawa et al., 2001).

Among others, 3-dimensional (3D) numerical simulations offer the most comprehensive analysis of the flow in extruders. For this, 3D models are developed using the exact extruder geometry with experimentally determined material data. The flow field is then calculated by solving the according continuity, Navier–Stokes, and energy equations. If performed by using reliable experimental data and mathematical models, it can deliver highly realistic flow information with respect to the generated thermal and mechanical stresses in an extruder. With this, the influence of individual process or material parameters on the resulting flow characteristics can be analyzed. In addition, by using particle tracking simulations, the stress profile and stress distribution experienced by the material during its flow through the extruder, both in axial and radial direction, can be calculated (Kalyon et al., 1988; Lawal et al., 1993; Lawal and Kalyon, 1995; Avalosse and Rubin, 2000; Ishikawa et al., 2001; Cleary and Robinson, 2011; Sarhangi Fard and Anderson, 2013). Such information is essential to have a control on the process, as our experimental analysis showed that the temperature in an extruder can vary locally to a great extent (Emin et al., 2015).

Finite element method (FEM) is the most commonly used method for the solution of mass, momentum, and energy equations, and the constitutive equations of highly viscous non-Newtonian fluid flows in extruders (Alsteens et al., 2004; Avalosse and Rubin, 2000; Dhanasekharan and Kokini, 2003; Emin and Schuchmann, 2013a). In our research, we have been using a commercially available FEM code ANSYS POLFLOW (Ansys Inc., Canonsburg, PA, USA). The flow in a co-rotating twin-screw extruder is three dimensional and unsteady due to moving and intermeshing parts (two rotating intermeshing screws). To simplify the set-up of such simulation and to avoid the use of a remeshing algorithm, the mesh superimposition technique introduced by Avalosse et al. (1996) can be used. For mesh superposition technique, a static mesh is created for the extruder barrel (representing the flow domain) and a dynamic mesh to describe the moving screws (Fig. 7.6E). In each time step, a procedure identifies the elements of the static mesh, which contain elements of the dynamic mesh. The velocity of the moving part is then imposed on the nodes of these static elements, using a penalty technique that modifies the equation of motion. For this purpose, a penalty force term, $H(v - v_p)$, has been introduced where H is either zero outside the moving part or 1 within the moving part and v_p is the velocity of the moving part. The term is used by modifying the equation of motion as follows:

$$H\left(v - v_p\right) + \left(1 - H\right)\left(-\nabla p + \nabla \cdot \bar{\bar{\tau}} + \rho g - \rho \frac{Dv}{Dt}\right) = 0 \qquad (7.1)$$

For $H = 0$, Eq. (7.1) is reduced to the normal Navier–Stokes equations, but for $H = 1$ the equation degenerates into $v = v_p$. The value of H is determined by the generation of an "inside" field that depends on the position of the moving part. When an element or node is greater than 60% within the moving part, H is given the value of 1. To ensure the mass conservation and physically meaningful pressure in the zones where geometrical penetration occurs, the mass conservation equation is modified to become

$$\nabla \cdot v + \frac{\beta}{\eta} \nabla p = 0 \tag{7.2}$$

where β is a relative compression factor, η is the shear viscosity, and p is the pressure.

By using these methods, we successfully performed a time-dependent simulation of a nonisothermal non-Newtonian flow of biopolymers in a co-rotating twin-screw extruder (Emin et al., 2013, 2020). To the best of our knowledge, there is no other study focusing on biopolymer extrusion at such conditions, which is however required to analyze the stress history of the material during the processing realistically. To perform the simulation as close as possible to the real experimental conditions, experiments with a lab scale co-rotating twin screw extruder equipped with an inline slit-die

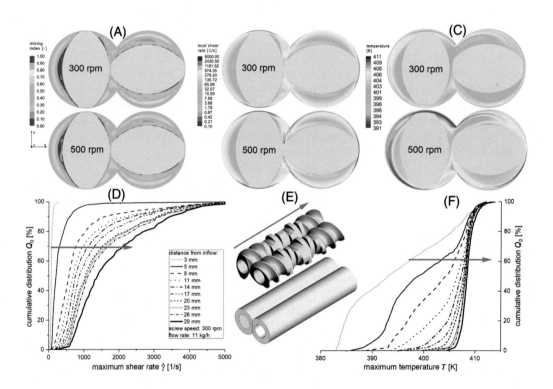

FIG. 7.6 Analysis of thermomechanical stress profile during extrusion by computational fluid dynamics and particle tracking simulations.

(A–C) Mixing index, shear rate, and temperature distribution calculated for the cross section from the middle of the geometry depicted in E. (D and F) Change in the maximum shear rate and temperature distribution of the particles tracked along the screws (only for 300 rpm). These exemplary results are from the simulation performed for maize starch with a water content of 18% at screw speeds of 300 and 500 rpm (Emin et al., 2013, 2020).

rheometer were performed (see Fig. 7.4A). Fig. 7.6 shows selected results from such simulation performed for maize starch extruded at two different processing conditions with a screw speed of 300 rpm and 500 rpm.

Fig. 7.6B and C shows that the maximum thermal and mechanical stresses are generated at the tip of the screws, as expected. Fig. 7.6A shows that these high stress regions are mainly dominated by simple shear flow calculated by mixing index. The mixing index, λ, is defined as:

$$\lambda = \frac{\dot{\gamma}}{\dot{\gamma} + \omega} \tag{7.3}$$

where $\dot{\gamma}$ is the rate of strain tensor and ω is the vorticity tensor, which are the symmetric and asymmetric components of the velocity gradient tensor. For a mixing index of 0, the system is undergoing purely rotational flow. A mixing index of 0.5 denotes simple shear flow, while a value of 1.0 denotes pure elongational flow. At the conditions investigated, the temperature increase at these high shear regions can be significantly higher (up to 10 K) than low shear regions. These results are in accordance with the experimental data obtained in our previous study (see Fig. 7.5). At the conditions investigated, increasing screw speed does not only lead to higher temperatures but also to spread of these hot regions to higher extruder volume due to higher frequency of the screw rotation. This results in higher amount of material experiencing these temperature peaks (results not shown).

To calculate the distribution of the mechanical and thermal stresses along the material, that is, shear stress and temperature profile, particle tracking analysis can be performed (Alsteens et al., 2004; Emin and Schuchmann, 2013a). In particle tracking analysis, it is assumed that marker particles are massless, volumeless, and noninteracting with each other, and therefore have no effect on the flow field. Once the particle trajectories are known, the distribution of the stresses along the particles can be monitored as a function of time or coordinates. Fig. 7.6D and F are examples for such results for the maximum shear and temperature experienced by the particles along the screws depicted in Fig. 7.6E. The result shows that the maximum stress generated remains constant, but the number of particles subjected to these stresses increase as they flow along the screw. Such analysis allows to analyze the effect of important process parameters such as screw speed, screw geometry or flow rate on the stress distribution which is otherwise not accessible through experimental methods.

Moreover, the particle tracking analysis and postprocessing can be further used to get information on more specific measures such as dispersive or distributive mixing efficiency, which are necessary to control the mixing during the extrusion of the multicomponent systems. For this purpose, theoretical or semiempirical measures of mixing, such as capillary number, mixing index, or scale of segregation, can be calculated (Connelly and Kokini, 2006; Emin and Schuchmann, 2013b). More detailed information about these analyses can be found in Emin et al. (2013a) and Emin et al. (2020).

7.5 Concluding remarks

The understanding of structural changes in food extrusion processes is still a major challenge because of the interactions between transient processing conditions, extensive structural changes, and material properties. To advance the state-of-the-art knowledge in this field, we presented an approach based on the analysis of the process at the mechanistic level. This approach involves the splitting of the process into coherent sections and the in-depth analysis of these sections using various numerical, rheological, and optical methods that we developed.

While the proposed methods still have several limitations, they offer enormous technological advancements and are therefore well suited for the characterization and control of the extrusion process. To support and facilitate the progress in this field, it is important to further improve the reliability of the methods available by critically applying them to a broad range of well-designed extrusion applications. Both standardization of current methods and the development of robust and sensitive new methods are also highly desirable. Such methods can provide essential information necessary to model complex flow conditions and structural changes of biopolymers in extruders. The main benefit of the mechanistic approach is that the focus remains on the critical processing mechanisms and not on the independent extrusion variables. As a result, the true interrelation between the independent variables and the final product characteristics becomes visible. Such information can then be efficiently used to adapt the process to perform a targeted product design or to transfer the process to different scales or materials more successfully. We therefore strongly encourage that future research efforts be directed toward the analysis of local processing conditions and material design properties in relation to these conditions by applying an interdisciplinary scientific approach at the interfaces between numerical modeling, rheology, sensor technology, and biopolymer science.

References

Alig, I., Steinhoff, B., Lellinger, D., 2010. Monitoring of polymer melt processing. Meas. Sci. Technol. 21 (6), 062001.

Alsteens, B., Legat, V., Avalosse, T., 2004. Parametric study of the mixing efficiency in a kneading block section of a twin-screw extruder. Int. Polym. Process. 3, 207–217.

Avalosse, T., 1996. Macromol. Symp. 112 (1), 91–98.

Avalosse, T., Rubin, Y., 2000. Analysis of mixing in co-rotating twin screw extruders through numerical simulation. Int. Polym. Process. 15 (2), 117–123.

Barnes, S.E., Sibley, M.G., Edwards, H.G.M., Coates, P.D, 2007. Process monitoring of polymer melts using in-line spectroscopy. Trans. Inst. Meas. Control 29 (5), 453–465.

Bouvier, J., Campanella, O.H., 2014. Extrusion Processing Technology: Food and Non-Food Biomaterials. Wiley-Blackwell, New Jersey.

Bravo, V.L., Hrymak, A.N., Wright, J.D., 2000. Numerical simulation of pressure and velocity profiles in kneading elements of a co-rotating twin screw extruder. Polym. Eng. Sci. 40 (2), 525–541.

Camire, M.E., Camire, A., Krumhar, K., 1990. Chemical and nutritional changes in foods during extrusion. Crit. Rev. Food Sci. Nutr. 29 (1), 35–57.

Cheftel, J.C., 1986. Nutritional effects of extrusion-cooking. Food Chem. 20 (4), 263–283.

Chen, C.C., 1992. Melt temperature measurement in a twin-screw extruder with infrared fiber optic probes, ANTEC '92, 1931–1936.

Cheng, H., Manas-Zloczower, I., 1997. Study of mixing efficiency in kneading discs of co-rotating twin-screw extruders. Polym. Eng. Sci. 37 (6), 1082–1090.

Cleary, P.W., Robinson, M., 2011. Understanding viscous fluid transport and mixing in a twin screw extruder, Eighth International Conference on CFD in Oil & Gas, Metallurgical and Process Industries. Trondheim, Norway.

Connelly, R.K., Kokini, J.L., 2006. Mixing simulation of a viscous Newtonian liquid in a twin sigma blade mixer. AIChE J. 52 (10), 3383–3393.

Davis, W.M., 1988. Heat transfer in extruder reactors. Chem. Eng. Prog. 84 (11), 32.

Della Valle, G., Colonna, P., Patria, A., Vergnes, B, 1996. Influence of amylose content on the viscous behavior of low hydrated molten starches. J. Rheol. 40 (3), 347.

Della Valle, G., Vergnes, B., Colonna, P., Patria, A, 1997. Relations between rheological properties of molten starches and their expansion behaviour in extrusion. J. Food Eng. 31 (3), 277–295.

Dhanasekharan, K., Kokini, J., 2003. Design and scaling of wheat dough extrusion by numerical simulation of flow and heat transfer. J. Food Eng. 60 (4), 421–430.

Dotsch, T., Pollard, M., Wilhelm, M., 2003. Kinetics of isothermal crystallization in isotactic polypropylene monitored with rheology and Fourier-transform rheology. J. Phys. Condens. Matter 15 (11), S923–S931.

Emin, M.A., 2015. Modeling extrusion processes. In: Bakalis, S., Knoerzer, K., Fryer, P.J. (Eds.), Modeling Food Processing Operations. Woodhead Publishing. United Kingdom, pp. 235–253.

Emin, M.A., Mayer-Miebach, E., Schuchmann, H.P., 2012. Retention of β-carotene as a model substance for lipophilic phytochemicals during extrusion cooking. LWT - Food Sci. Technol. 48 (2), 302–307.

Emin, M.A., Quevedo, M., Wilhelm, M., Karbstein, H.P., 2017. Analysis of the reaction behavior of highly concentrated plant proteins in extrusion-like conditions. Innovative Food Sci. Emerg. Technol. 44, 15–20.

Emin, M.A., Schuchmann, H.P., 2013a. Droplet breakup and coalescence in a twin-screw extrusion processing of starch based matrix. J. Food Eng. 116 (1), 118–129.

Emin, M.A., Schuchmann, H.P., 2013b. Analysis of the dispersive mixing efficiency in a twin-screw extrusion processing of starch based matrix. J. Food Eng. 115 (1), 132–143.

Emin, M.A., Schuchmann, H.P., 2017. A mechanistic approach to analyze extrusion processing of biopolymers by numerical, rheological, and optical methods. Trends Food Sci. Technol. 60, 88–95.

Emin, M.A., Schwegeler, Y., Wittek, P., 2020. Numerical analysis of thermal and mechanical stress profile during the extrusion processing of plasticized starch by non-isothermal flow simulation. J. Food Eng. 294, 110407.

Emin, M.A., Teumer, T., Schmitt, W., Rädle, M., Schuchmann, H.P., 2015. Measurement of the true melt temperature in a twin-screw extrusion processing of starch based matrices via infrared sensor. J. Food Eng. 170, 119–124.

Gerstorfer, G., Lepschi, A., Miethlinger, J., Zagar, B.G., 2013. An optical system for measuring the residence time distribution in co-rotating twin-screw extruders. J. Polym. Eng. 33 (8), 683–690.

Hayta, M., Schofield, J.D., 2005. Dynamic rheological behavior of wheat glutens during heating. J. Sci. Food Agric. 85 (12), 1992–1998. doi:10.1002/jsfa.2212.

Horvat, M., Emin, M.A., Hochstein, B., Willenbacher, N., Schuchmann, H.P., 2013. Influence of medium-chain triglycerides on expansion and rheological properties of extruded corn starch. Carbohydr. Polym. 93 (2), 492–498.

Horvat, M., Ladiges, D., Schuchmann, H.P., 2013. Investigation of the nucleation during extrusion cooking of corn starch by a novel nucleation die. Food Bioprocess Technol. 7 (3), 654–660.

Ishikawa, T., Kihara, S.I., Funatsu, K., 2001. 3-D non-isothermal flow field analysis and mixing performance evaluation of kneading blocks in a co-rotating twin srew extruder. Polym. Eng. Sci. 41 (5), 840–849.

Kalyon, D.M., Gotsis, A.D., Yilmazer, U., Gogos, C.G., Sangani, H., Aral, B., Tsenoglou, C., 1988. Development of experimental techniques and simulation methods to analyze mixing in co-rotating twin screw extrusion. Adv. Polym. Technol. 8, 337–353.

Karwe, M.V., Godavarti, S., 1997. Accurate measurement of extrudate temperature and heat loss on a twin-screw extruder. J. Food Sci. 62 (2), 367–372.

Khanal, R.C., Howard, L.R., Brownmiller, C.R., Prior, R.L., 2009. Influence of extrusion processing on procyanidin composition and total anthocyanin contents of blueberry pomace. J. Food Sci. 74 (2), 52–58.

Klingler, R.W., Meuser, F., Niediek, E.a., 1986. Einfluß der Art der Energieübertragung auf strukturelle und funktionelle Merkmale von Stärke. Starch - Stärke 38 (2), 40–44.

Koch, L., Hummel, L., Schuchmann, H.P., Emin, M.A., 2017a. Structural changes and functional properties of highly concentrated whey protein isolate-citrus pectin blends after defined, high temperature treatments. LWT – Food Sci. Technol 84, 634–642.

Koch, L., Hummel, L., Schuchmann, H.P., Emin, M.A., 2017b. Influence of defined shear rates on structural changes and functional properties of highly concentrated whey protein isolate-citrus pectin blends at elevated temperatures. Food Biophys. 12, 309–322.

Kohlgrüber, K., 2007. Der gleichläufige Doppelschneckenextruder: Grundlagen, Technologie, Anwendungen. Hanser Verlag.

Kokini, J.L., Cocero, A.M., Madeka, H., Graaf, E.de., 1994. The development of state diagrams for cereal proteins. Trends Food Sci. Technol. 5 (9), 281–288.

Lagrain, B., Goderis, B., Brijs, K., Delcour, J.A., 2010. Molecular basis of processing wheat gluten toward biobased materials. Biomacromolecules 11 (3), 533–541.

Lai, L., Kokini, J., 1991. Physicochemical changes and rheological properties of starch during extrusion (a review). Biotechnol. Prog. 7 (3), 251–266.

Lawal, A., Kalyon, D.M., 1995. Simulation of intensity of segregation distributions using three-dimensional FEM analysis: application to co-rotating twin screw extrusion processing. J. Appl. Polym. Sci. 58, 1501–1507.

Lawal, A., Kalyon, D.M., Ji, Z., 1993. Computational study of chaotic mixing in co-rotating two-tipped kneading paddles: two-dimensional approach. Polym. Eng. Sci. 33, 140–148.

Leblanc, J.L., Cartault, M., 2001. Advanced torsional dynamic methods to study the morphology of uncured filled rubber compounds. J. Appl. Polym. Sci. 80 (11), 2093–2104.

Lee, S.Y., Hanna, M.A., Jones, D.D., 2009. Residence time distribution determination using on-line digital image processing. Starch - Stärke 61 (3-4), 146–153.

Liu, K., Hsieh, F.-H., 2008. Protein-protein interactions during high-moisture extrusion for fibrous meat analogues and comparison of protein solubility methods using different solvent systems. J. Agric. Food Chem. 56 (8), 2681–2687.

Liu, W.-C., Halley, P.J., Gilbert, R.G., 2010. Mechanism of degradation of starch, a highly branched polymer, during extrusion. Macromolecules 43 (6), 2855–2864.

Madeka, H., Kokini, J.L., 1994. Changes in rheological properties of gliadin as a function of temperature and moisture: development of a state diagram. J. Food Eng. 22 Elsevier, 241–252.

Madeka, H., Kokini, J.L., 1996. Effect of glass transition and cross-linking on rheological properties of zein: development of a preliminary state diagram. Cereal Chem. 73 (4), 433–438.

Maier, C., 1996. Infrared temperature measurement of polymers. Polym. Eng. Sci. 36 (11), 1502–1512.

Meuser, F., van Lengerich, B., Köhler, F., 1982. Einfluss der Extrusionsparameter auf die funktionellen Eigenschaften von Weizenstärke. Starch/Stärke 34 (11), 366–372.

Morales, A., Kokini, J.L., 1997. Glass transition of soy globulins using differential scanning calorimetry and mechanical spectrometry. Biotechnology 13 (1984), 624–629.

Moraru, C.I., Kokini, J.L., 2003. Nucleation and expansion during extrusion and microwave heating of cereal foods. Compr. Rev. Food Sci. Food Safety, 2.

Padmanabhan, M., Bhattacharya, M., 1991. Flow behavior and exit pressures of corn meal under high-shear–high-temperature extrusion conditions using a slit die. J. Rheol. 35 (3), 315.

Pietsch, V.L., Bühler, J.M., Karbstein, H.P., Emin, M.A., 2019. High moisture extrusion of soy protein concentrate: influence of thermomechanical treatment on protein-protein interactions and rheological properties. J. Food Eng. 251, 11–18.

Pietsch, V.L., Karbstein, H.P., Emin, M.A., 2018. Kinetics of wheat gluten polymerization at extrusion-like conditions relevant for the production of meat analog products. Food Hydrocolloids 85, 102–109.

Pietsch, V.L., Schöffel, F., Rädle, M., Karbstein, H.P., Emin, M.A., 2019. High moisture extrusion of wheat gluten: modeling of the polymerization behavior in the screw section of the extrusion process. J. Food Eng. 246, 67–74.

Pommet, M., Morel, M.-H., Redl, A., Guilbert, S., 2004. Aggregation and degradation of plasticized wheat gluten during thermo-mechanical treatments, as monitored by rheological and biochemical changes. Polymer 45 (20), 6853–6860.

Quevedo, M., Kulozik, U., Karbstein, H.P., Emin, M.A., 2020. Kinetics of denaturation and aggregation of highly concentrated β-lactoglobulin under defined thermomechanical treatment. J. Food Eng. 274.

Rauwendaal, C., 2014. Polymer Extrusion. Hanser Verlag.

Sarhangi Fard, A., Anderson, P.D., 2013. Simulation of distributive mixing inside mixing elements of co-rotating twin-screw extruders. Comput. Fluids 87, 79–89.

Stojceska, V., Ainsworth, P., Plunkett, A., Ibanoğlu, S., 2008. The recycling of brewer's processing by-product into ready-to-eat snacks using extrusion technology. J. Cereal Sci. 47 (3), 469–479.

van Den Einde, R.M., Akkermans, C., Van Der Goot, A.J., Boom, R.M., 2004. Molecular breakdown of corn starch by thermal and mechanical effects. Carbohydr. Polym. 56 (4), 415–422.

van Lengerich, B., 1990. Influence of extrusion processing on in-line rheological behavior, structure, and function of wheat starch. In: Faridi, H., Faubion, J.M. (Eds.), Dough Rheology and Baked Product Structure. Van Nostrand Reinhold, New York, pp. 421–472.

Vergnes, B., Della Valle, G., Tayeb, J., 1993. A specific slit die rheometer for extruded starchy products. Design, validation and application to maize starch. Rheol. Acta 32 (5), 465–476.

Wittek, P., Zeiler, N., Karbstein, H.P., Emin, M.A., 2020. Analysis of the complex rheological properties of highly concentrated biopolymers. Appl. Rheol., 64–76, Submitted.

Xie, F., Halley, P.J., Avérous, L., 2012. Rheology to understand and optimize processibility, structures and properties of starch polymeric materials. Prog. Polym. Sci. 37 (4), 595–623.

Key technological advances and industrialization of continuous flow microwave processing for foods and beverages

Josip Simunovic, K.P. Sandeep

Department of Food, Bioprocessing and Nutrition Sciences, North Carolina State University, Raleigh, NC,
United States

8.1 Introduction

For several decades, researchers from the food processing industry, academia, and government institutions have recognized the shortcomings of conventional methods of food preservation for low acid, shelf stable food products.

These methods rely predominately on delivery of thermal treatment to inactivate microorganisms of public health significance and causative agents of spoilage under the temperatures and conditions typical of their storage, transport, and distribution.

In order to deliver appropriately effective thermal treatments for inactivation of microbial targets (thermo-resistant endo-spores of proteolytic strains of *Clostridium botulinum*) and sufficient to achieve commercial sterility for low acid foods, these products need to be exposed to temperatures and pressures exceeding levels achievable under atmospheric conditions. These time-temperature treatment levels have been extensively researched and are consequently very well defined, widely known and extensively published in the scientific and professional literature on thermal food preservation.

This existing knowledge base about the target microbial entities of concern, conditions of their inactivation and methods and techniques of process and time-temperature treatment delivery, monitoring, and validation allow efficient implementation of thermal preservation/sterilization to a very wide range and number of treated products, resulting in a superior track record of safety and an overwhelming dominance in volume and variety of thermally preserved food products in the United States and global markets.

Unfortunately, time and temperature treatment levels required to inactivate microorganisms of public health significance and to render these products safe for consumption throughout their cycle of storage, transport, and distribution, also result in significant degradation of nutritional value (degradation or elimination of thermo-sensitive nutrients such as vitamins and antioxidants) and sensory quality (color, flavor, texture) of most foods preserved by these methods. These negative aspects of thermal preservation are particularly high for food materials with poor heat transfer characteristics such as very viscous, solid, and particle-containing food materials, effectively limiting the size of packages which can be preserved by thermal sterilization if it is implemented by conventional heating methods, while the product is contained in the package.

This major degradation of nutrients and sensory characteristics is the main shortcoming of industrial food preservation by thermal processing and is the cause of wide ranging consequences for the food processing industry, market perception of its products and nutritional well-being of its consumers. Notably, perception of quality of shelf stable, thermally preserved foods in the consumer marketplace is relatively low, leading to a low perception of their value and resulting in relatively low profit margins for this category of products. Comparatively, refrigerated and frozen products are perceived as having high quality and value, but their storage, distribution and shelf life are limited by the need to maintain strictly controlled chilled storage conditions throughout the cycle of distribution from production to consumption.

It can be claimed that most of the significant improvements and developments in the industrial food preservation during the last century have in one way or other resulted from a variety of attempts to address this quality and value-related shortcoming of thermal preservation technology, while maintaining the safety and ambient temperature conditions of the resulting products.

Among the notable developments are low-profile retortable packaging (to reduce the time of thermal exposure necessary for sterilization), and an impressive range of improvements to the autoclave technologies to improve the conditions and efficiency of heat exchange—circulating, static, rotating, agitating, and vibrating versions, condensing steam, and pressurized water versions, etc. Numerous efforts are in progress to implement advanced volumetric heating as a means of sterilization for prepackaged foods such as in-pack microwave and in-pack ohmic sterilization. Some of these technologies are implemented commercially in other parts of the world, but have yet to be commercialized in the North American marketplace.

A variety of nonthermal technologies have also been under research and development for several decades; such as irradiation, ultra-high pressure, pulsed electric field, ultrasound, ultraviolet, and pulsed light treatments etc., all in an effort to address the noted shortcomings of thermal methods of food preservation.

For a variety of reasons, nonthermal technologies and products preserved by their implementation have so far had limited success in the wider commercial marketplace. For most of these technologies, work is still in progress to identify to microbial entities of most significance for consumer safety under the nonthermal conditions of inactivation, their nonhazardous surrogates to be used in process and product development and appropriate and quantifiable means of preservation process design, characterization, monitoring and safety confirmation.

The most significant development in industrial food processing over the last century; which has had also had a major impact in the commercial market place has been the emergence of high temperature, short time thermal processing methods (also known as ultra-high temperature or aseptic processing) which implement separate sterilization of products and packages and subsequent filling and hermetic sealing within an aseptic (microbe-free) environment. This technology enables a faster and more efficient sterilization of the food product prior to packaging and has resulted in numerous products packaged in semiflexible packaging like shelf-stable milk and dairy beverages, fruit juices and beverages and various types of homogeneous, pureed and small-particulate containing soups. However, in order for this technology to advance further to high viscosity, high protein, meat containing and large-particle vegetable soups, stews, sauces, and meals, continuous flow heating technology needs to be advanced to methods capable of rapid, uniform (volumetric) heating of solid and heterogeneous food materials otherwise difficult to heat conventionally (by heat conduction or convection). Only by implementing an advanced continuous flow, volumetric heating method will these product become commercially viable, since the conventional methods of sterilization, even when performed under continuous flow conditions, result in quality degradation comparable to the one imparted on conventionally canned products, if the treated food materials have poor heat transfer properties.

Methods like continuous flow ohmic heating and sterilization and radio frequency heating have been researched and implemented for several decades, while continuous flow microwave sterilization has been investigated rarely and sporadically at best, due to the inability of previous systems to deliver consistent and predictable levels of thermal treatment under these conditions.

Continuous flow microwave processing is an advanced thermal processing technology used for heating and preservation of foods, beverages and biomaterials. In 1996 this technology became the focus of research and development efforts of the Food Process Engineering team at the Department of Food, Bioprocessing and Nutrition Sciences at North Carolina State University in Raleigh, NC.

Also known as microwave assisted aseptic processing, it was identified and developed as the technology with the highest commercial potential for continuous sterilization and aseptic packaging of viscous, poorly conductive, thermos-sensitive, and complex particulate products such as chunky soups, stews, and sauces. Starting with the 1 kW and 5 kW bench top and pilot systems, to 60 kW and 75 kW semi-industrial installations, and continuing to 100 kW, 200 kW, and 400 kW commercial processing installations, recent development efforts have been concentrated on smaller capacity and mobile systems for distributed processing and production of customized and personalized nutrition products. Several novel and unique products have been enabled and commercialized using continuous flow microwave heating followed by aseptic packaging.

This chapter will provide a brief overview of the evolution of continuous flow microwave technologies as initiated by the research, development, and innovations conceptualized and implemented at the Department, start-up businesses, and industrial users of the resulting technologies, rationale for various system designs and the growing list of food, beverage, and biomaterial products introduced by industrial users of the technologies.

8.2 Continuous flow microwave processing prototypes

Three main generations of continuous flow microwave processing system designs have been developed, tested, and implemented at benchtop, pilot, semi-industrial, and commercial production scales, resulting in 4 R&D and 6 industrial facilities, and multiple processing lines for a variety of food, beverage, and biomaterial products.

Commercial packaging formats range from individual serving spouted pouches to multiserving aseptic cartons to flexible packaging for food service intended to replace #10 cans, to 1000 kg bulk bag in box totes containing ingredients for bakery, beverage, and infant food products.

8.2.1 First generation of continuous flow microwave processing technologies

Early models and initial benchtop and pilot capacity installations have been tested using 915 MHz cylindrical microwave applicators (Industrial Microwave Systems, 3000 Perimeter Park Dr, Morrisville, NC 27560) starting with the first benchtop applicator system in 1998. Evolution of the scale and capacity of the processing systems implemented in labs and pilot plants at NC State are illustrated by Fig. 8.1.

All first-generation microwave processing systems were provided by Industrial Microwave Systems and used for studies of high acid and low acid pasteurization and full sterilization of homogeneous and particle containing foods and beverages.

FIG. 8.1

Early evolution of installed R&D continuous flow microwave processing systems at NC State University:
(A) 1998, benchtop, 600 W, 915 MHz, belt conveyer, used for development of surimi and meat paste
products produced by thermal gelation, (B) 2000, pilot plant, 5 kW, 915 MHz, positive displacement pump
conveyance, preliminary testing, cross-sectional temperature profile during continuous flow, dielectric
measurements under continuous flow, feasibility trials, (C) 2002, semi-industrial, 60 kW, 915 MHz, positive
displacement piston pump, coupled to an aseptic Scholle bag filler.

Research and development related to the first generation of technology focused on generation of a
database of previously unavailable dielectric properties of foods and biomaterials at temperatures ranging from ambient to sterilization levels (Brinley et al., 2008; Coronel et al., 2008b; Koskiniemi et al.,
2013; Hanson et al., 2014; Kumar et al., 2008a). Kumar et al. (2008b) developed a new method for
dielectric property measurements under continuous flow conditions, particularly relevant for characterization of multicomponent and multiphase (particulate containing) foods and biomaterials.

Approximately 12 years of intense R&D, primarily supported by funding from the NSF sponsored
Industry/University Collaborative Research Center for Advanced Processing and Packaging Studies (the
only NSF IU/CRC focused on food processing and packaging) led to the first commercial installation
for processing of shelf stable aseptic sweet potato purees at Yamco (Snow Hill, NC), pictured in Fig. 8.2.

North Carolina agribusiness is one of the major producers of agricultural products among all states
in the United States. However, the number and volume of value-added agricultural products and related
NC-based businesses are comparatively low. There was a recognized need to build and add to value of
North Carolina based agribusinesses and North Carolina produced agricultural products. High quality
food and biomaterial products processed, preserved and packaged by advanced and sophisticated new
methodologies could provide the momentum and economic stimulus to establish the competitive advantage needed to achieve these goals.

The researchers working in the laboratory and pilot plant facilities at NCSU Department of Food,
Bioprocessing and Nutrition Sciences have developed and implemented a system of tools, methods,
devices, and procedures for advanced thermal processing of foods and biomaterials utilizing continuous flow microwave pasteurization and sterilization. These methods and devices have been integrated
with other sophisticated for food and biomaterial packaging to enable production and packaging of

FIG. 8.2

Research batch, first commercial batch and the commercial installation of IMS Cylindrical Continuous Flow Microwave Sterilization Reactor at Yamco, LLC (Snow Hill, NC).

high quality and high value-added products. Additionally, the practical environment was provided for education of experts to work on industrial and commercial implementation of developed technologies.

Scientific and technical basis has been established for rapid commercial deployment of technologies developed for food and biomaterials processing by advanced thermal pasteurization and sterilization and integrated with advanced aseptic packaging capabilities. The integrated pilot plant installation at the NCSU Department of Food, Bioprocessing and Nutrition Sciences was the first facility of its kind open to major, intermediate, small, and independent food and bioprocessing companies to test and experiment with own products and formulations and assist with commercialization.

These research and development efforts initially addressed the evolving equipment and processing challenges related to the new technology (Coronel et al., 2008a; Kumar et al., 2008a), methods development for microwave sterilization process monitoring and validation (Coronel et al., 2003; Brinley et al., 2007). These efforts allowed the research to progress to development and implementation of customized processes to a variety of microwave processed and aseptically packaged foods and beverages such as milk (Clare et al., 2005), orange fleshed sweet potato purees (Coronel et al., 2005), salsa con queso (Kumar et al., 2007a; Kumar et al., 2007b), and purple fleshed sweet potato purees (Steed et al., 2008).

The first major commercial application of the new continuous flow microwave sterilization has started production of shelf stable aseptically packaged sweet potato puree in early 2008 and received the first FDA letter of no objection ever for an advanced thermal processing system for sterilization of low acid foods at YamCo LLC of Snow Hill, NC, founded by a consortium of seven sweet potato growers and investors.

The use of microwave assisted aseptic processing equipment and technology allowed sweet potato growers in North Carolina to process and add value to a crop that had previously been ploughed back into the land or sold as animal food. As much as 45% of the annual harvest in the State of North Carolina would not meet the standard for retail sales of freshly harvested sweet potatoes and was effectively discarded. This innovation enabled farmers to virtually eliminate this wastage, and turn a cost into a profit. The first industrial microwave assisted aseptic processing and packaging installation at Yamco was subsequently described by Parrott (2010).

The Yamco commercial production facility in Snow Hill, NC is currently still one of a kind, which at the same time represents:

- the first installation of a continuous flow microwave sterilization system for shelf stable, low acid food products anywhere in the world
- the first commercial installation of a microwave-based sterilization process in the United States
- the first FDA letter of no objection for an advanced heating-based aseptic process for low acid food preservation

Subsequently, Yamco continued to implement continuous flow microwave sterilization to a variety of other fruit and vegetable purees, some of which are pictured in Fig. 8.3.

8.2.2 Second generation of continuous flow microwave processing technologies

Today's consumers are educated and aware of low-quality, nutrient-depleted foods, produced by applications of outdated industrial processing and packaging methods—primarily shelf stable ready to eat

FIG. 8.3

Current selection of shelf stable aseptically packaged fruit and vegetable purees sterilized using continuous flow microwave processing at Yamco: red beets, orange-fleshed sweet potato, cauliflower, butternut squash, carrots, pumpkin, blueberries, and purple-fleshed sweet potato.

products—the category which is traditionally associated with thermally sterilized products packaged in cans.

The continuing controversy associated with the use of Bisphenol-A in can linings also contributes to the erosion of the image and reputation of this product category and by extension the food processing industry in general.

Alternative diets are emerging (vegetarian, vegan, raw, paleo, local, organic, non-GMO), but tend to be expensive, time-consuming, fluctuating in quality, and therefore impractical for most consumers.

Most of the alternative diets have emerged as consumer lifestyle choices, while the active roles of food processors and advanced technologies in these changes have remained limited to niche products with limited shelf life and requiring refrigerated distribution, storage and retail presentation. This results in a limited marketing reach and potential, as well as limited distribution areas and time windows and difficult logistics of transport, distribution, and strict temperature maintenance, monitoring in order to avoid premature product degradation and spoilage. Therefore, there remains a recognized and increasing need and desire for healthy, minimally processed, nutrient-rich products which are affordable and convenient.

These concerns, as well as the limitations of in-pack and batch type processing systems have been some of the main causes in the slow acceptance and implementation of emerging and particularly non-thermal methods of food preservation by most of the major food processors.

Shelf stability at ambient temperatures over prolonged time periods remains a very desirable attribute for all the actors in the food product distribution and consumption chain:

- For the industrial processors, it insures the full safety of commercially sterilized products, quality maintenance, and uniformity over extended periods of time without refrigeration and when achieved using thermal preservation technologies without added preservatives enabling production of products with simple, clean labels. All of these attributes contribute to the geographically extended market reach, yielding nationally and internationally relevant product lines and associated brand value.
- For the food product distributors, food service operations and retailers, shelf stability means less expensive (i.e., nonfrozen and nonrefrigerated) transportation, storage, and point-of sale display space.
- For the consumers it provides convenience, variety and, due to the lower costs of production, transportation and storage, a favorable pricing structure and an ability to plan and budget food purchasing over extended time periods.

Therefore, in order to remain relevant in the changing consumer marketplace, the current dominant challenge for the major global food processors is to deliver new, high quality and nutritionally superior shelf stable products using newer, more environmentally friendly packaging formats and materials.

Aseptic packaging formats and filler technologies have been available to deliver these types of products for several decades, but have been mostly limited to simple liquids, mostly fruit and dairy-based products and more recently functional beverages and simple low viscosity soups. In order for aseptic packaging to expand its application range and market relevance to more difficult and complex food products such as fruit and vegetable purees, complex sauces, soups, and multiphase/particulate products, new advanced methods of thermal sterilization of such products under continuous flow conditions needed to be developed, implemented, validated and applied at full industrial scale.

The need and the motivation to develop the new generations of microwave-assisted aseptic technologies have been based on the experience and recognition of limitations of the previous systems and processes for microwave sterilization of sweet potato and other fruit and vegetable purees.

These second-generation technologies address the issues of expanded application range, improved controls and run times, enabling construction of multigenerator systems and installations and integration with a wider variety of ultra-clean and aseptic filling equipment. Equipment and process modeling software and simulations have been developed and have been used for the development and optimization of several processing and aseptic packaging lines. Proprietary software for process modeling and optimization has been used to generate a variety of processing scenarios for a variety of target products and select the optimal treatments to maximize the quality and nutrient retention in processed products.

The second generation of 915 MHz processing systems (AseptiWave systems) was initiated with the development and testing of a 5 KW prototype illustrated by Fig. 8.4 and continued with the installation of the modular 75 kW traveling wave system coupled with an aseptic pouch filler at NC State Department of Food, Bioprocessing and Nutrition Sciences (Fig. 8.5), and after extensive testing subsequently used in the largest commercial installation of the technology, Wright Foods in Troy, NC. The first industrial installation of continuous microwave processing for low acid particulate foods is depicted in Fig. 8.6. This facility was used to generate the data for the first filing for microwave sterilized low acid particulate product lines (single particle type and multiple particle types) with the United States FDA and received two respective Letters of no Questions (LONQ) in 2015 The validated products were single particle type and multiple/complex particulate low acid vegetable soups. Fig. 8.7 presents the first fully aseptic industrial installation for processing and packaging of consumer package formats (CombiBloc cartons - 0.25, 0.5 and 1 liter volumes) of low acid products using continuous flow

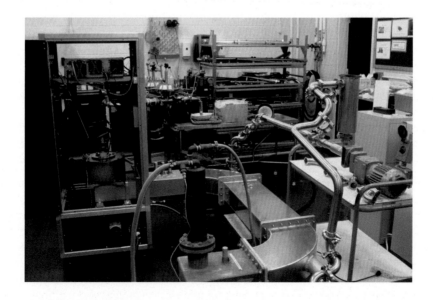

FIG. 8.4

5 kW, 915 MHz prototype of AseptiWave (second-generation continuous flow microwave technology) at NC State University.

FIG. 8.5

75 kW, 915 MHz installation AseptiWave (second-generation continuous flow microwave technology) at NC State University Department of Food, Bioprocessing and Nutrition Sciences, integrated with positive displacement piston pump (Marlen Industries, Riverside, MO) and aseptic pouch filler (Astepo, Alfa-Laval, Richmond, VA).

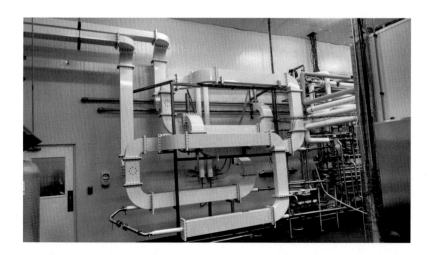

FIG. 8.6

200 kW, 915 MHz installation AseptiWave (second-generation continuous flow microwave technology) at Wright Foods (Troy, NC).

FIG. 8.7

300 kW, 915 MHz installation AseptiWave (second-generation continuous flow microwave technology) at Wright Foods (Troy, NC).

microwave sterilization. Some of the products processed and packaged on this system, and introduced to the commercial markets are presented in Fig. 8.8 (low acid homogeneous soups, high acid fruit purees and acidified fruit and vegetable blend smoothies). Additionally, transparent bags of low acid shelf stable vegetable blends for use as ingredients in restaurants, microwave processed and packaged using the Sealed Air Flavour Mark aseptic packaging system.

Wright Foods has established and put into commercial production the largest installation of advanced thermal processing production lines, with the total microwave generation power installed approaching 2 MW, including the microwave assisted sterilization installations, all first to the commercial market in their respective categories and capacities:

Installation	Number of MW generators	Max. capacity	Integrated filler type	Packaging
Production 1	2 × 100 kW	15 L/min	Effytec Hot Fill SIG Combibloc	90–150 g pouches 500 mL, 750 mL, 1 L cartons
Production 2	3 × 100 kW	50 L/min	Aseptic	
Production 3	2 × 100 kW	15 L/min	Effytec Hot Fill	90–150 g pouches
Production 4	4 × 100 kW	100 L/min	Cryovac Aseptic	5–8 kg bags
R&D/Production 5	2 × 100 kW	15 L/min	Rapak/Intasept Aseptic	1–3 kg

These installations have been used to develop the products, implement and validate the processes, submit the regulatory agency filings, and commercialize multiple lines of microwave sterilized and aseptically packaged products, all first to the retail and food service commercial markets, notably:

- A line of vegetable soups in aseptic cartons
- A line of 100% aseptic fruit purees in cartons

FIG. 8.8

Variety of consumer and food service products commercialized using AseptiWave (second-generation) technology at Wright Foods (Troy, NC).

- A #10 can replacement aseptic bag of banana puree
- A line of 100% aseptic fruit smoothies
- A line of vegetable purees/soup starter blends for food service in aseptic bags.

8.2.3 Third generation of continuous flow microwave processing technologies

The third generation of the technology implemented the established AseptiWave and the novel Nomatic designs using 2450 MHz microwave generators for a broad variety of processing capabilities, with R&D and production installations at SinnovaTek in Raleigh, NC; North Carolina Food Innovation Lab in Kannapolis, NC; First Wave Innovations in Raleigh, NC and Panaceutics Nutrition in Danville, VA (Fig. 8.9).

a. Benchtop—Nomatic—2450 MHz ~1 ton per day
b. Pilot—AseptiWave—915 MHz ~1 ton per hour
c. Commercial production—AseptiWave—915 MHz ~4 tons per hour

All three generations of continuous flow microwave processing technologies depicted on Fig. 8.10, as well as three production scale capacities are in current commercial use and are manufactured, maintained, and supported by SinnovaTek Inc. (Raleigh, NC).

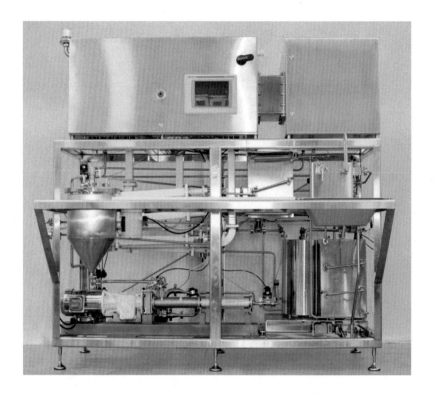

FIG. 8.9

18 kW 2450 MHz SinnovaTek Nomatic processing system installed at the precision scale aseptic processing facility First Wave Innovations in Raleigh, NC, at North Carolina Food Innovation Lab in Kannapolis, NC, Panaceutics Nutrition in Danville, VA and at Burton and Bamber Company aseptic processing facility in Thika, Kenya.

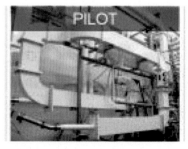

FIG. 8.10

Current range of commercial offerings of continuous flow microwave processing (heating, pasteurization, sterilization) at different scales produced by SinnovaTek (Raleigh, NC).

8.3 Intellectual property

Intellectual property development and protection activities evolved concurrently with the technology and product developments and commercial implementations. Protection was pursued both nationally and internationally. This enabled and encouraged investments in industrial installations and accelerated the extension of applications to new materials and facilities.

The first generation of technology was addressed by Simunovic et al. (2014 and 2017); second and third generation by Simunovic and Drozd (2012), Drozd and Simunovic (2013 and 2017); and Simunovic (2017). Activities in this domain continue with new pending United States and international applications.

8.4 Conclusions

Continuous flow microwave heating, pasteurization, and sterilization technologies for foods, beverages, and biomaterials have been under development for 25 years and in commercial production since 2007.

Subsequently, over 100 new products have been introduced to the consumer, food service, and bulk ingredient commercial markets since the first installation. Different generations of the technology, its developers and the resulting products have been recognized as major industrial innovations by professional, government, and scientific organizations, including two IFT Food Technology Industrial Achievement Awards in 2009 and 2015.

New installations and new applications of continuous flow microwave processing continue to grow in numbers, capacity and variety of outputs. The original objective of commercial production of high-quality shelf stable aseptic particulate/multiphase products using microwave sterilization is still pursued. Recent award of the first FDA clearance for microwave sterilization of particulate aseptic food products marked another significant positive development in these efforts.

Ready availability of an advanced thermal technology and its continued evolution from experimental to commercial scale establishes a new advanced level of capability for the food processing industry to produce high quality, shelf stable high acid, acidified and low acid foods, and beverages.

Microwave processing technologies, in combination with the new aseptic packaging technologies, particularly the recent introduction of fully aseptic spouted pouch formats, opens the opportunity for development and commercial introduction of product lines previously unavailable in aseptically packaged shelf stable formats.

Continuous flow microwave processing technologies provide a commercially tested and viable advanced thermal processing option for the food industry to maximize safety, sensory quality, nutrient retention, and profitability of established and novel food products in shelf stable packaging formats with ambient shelf life of 12–18 months, requiring no frozen or refrigerated storage or distribution.

References

Brinley, T.A., Dock, C.N., Truong, V.D., Coronel, P., Kumar, P., Simunovic, J., Cartwright, G.D., Swartzel, K.R., Jaykus, L.A., 2007. Feasibility of utilizing bioindicators for testing microbial inactivation in sweetpotato purees processed with a continuous-flow microwave system. J. Food Sci. 72 (5), E235–E242.

Brinley, T.A., Truong, V.D., Coronel, P., Simunovic, J., Sandeep, K.P., 2008. Dielectric properties of sweet potato purees at 915 MHz as affected by temperature and chemical composition. Int. J. Food Prop. 11 (1), 158–172.

Clare, D.A., Bang, W.S., Cartwright, G., Drake, M.A., Coronel, P., Simunovic, J., 2005. Comparison of sensory, microbiological, and biochemical parameters of microwave versus indirect UHT fluid skim milk during storage. J. Dairy Sci. 88 (12), 4172–4182.

Coronel, P., Simunovic, J., Sandeep, K.P., 2003. Temperature profiles within milk after heating in a continuous-flow tubular microwave system operating at 915 MHz. J. Food Sci. 68 (6), 1976–1981.

Coronel, P., Truong, V.D., Simunovic, J., Sandeep, K.P., Cartwright, G.D., 2005. Aseptic processing of sweetpotato purees using a continuous flow microwave system. J. Food Sci. 70 (9), E531–E536.

Coronel, P., Simunovic, J., Sandeep, K.P., Cartwright, G.D., Kumar, P., 2008a. Sterilization solutions for aseptic processing using a continuous flow microwave system. J. Food Eng. 85 (4), 528–536.

Coronel, P., Simunovic, J., Sandeep, K.P., Kumar, P., 2008b. Dielectric properties of pumpable food materials at 915 MHz. Int. J. Food Prop. 11 (3), 508–518.

Drozd, J.M., Simunovic, J., (2013). Method for processing materials US Patent 8,574,651.

Drozd, J.M., Simunovic, J., (2017). Electromagnetic system US Patent 9,713,340.

Hanson, D., Kemal, M., Tennant, T., Simunovic, J., 2014. Dielectric properties of selected pork organ meats and experimental pate product formulations for continuous flow microwave processing. Arch. Latinoam. Producción Anim. 22 (5), 303–307.

Koskiniemi, C.B., Truong, V.D., McFeeters, R.F., Simunovic, J., 2013. Effects of acid, salt, and soaking time on the dielectric properties of acidified vegetables. Int. J. Food Prop. 16 (4), 917–927.

Kumar, P., Coronel, P., Simunovic, J., Truong, V.D., Sandeep, K.P., 2007a. Measurement of dielectric properties of pumpable food materials under static and continuous flow conditions. J. Food Sci. 72 (4), E177–E183.

Kumar, P., Coronel, P., Simunovic, J., Sandeep, K.P., 2007b. Feasibility of aseptic processing of a low-acid multiphase food product (salsa con queso) using a continuous flow microwave system. J. Food Sci. 72 (3), E121–E124.

Kumar, P., Coronel, P., Truong, V.D., Simunovic, J., Swartzel, K.R., Sandeep, K.P., 2008a. Overcoming issues associated with the scale-up of a continuous flow microwave system for aseptic processing of vegetable purees. Food Res. Int. 41 (5), 454–461.

Kumar, P., Coronel, P., Simunovic, J., Sandeep, K.P., 2008b. Thermophysical and dielectric properties of salsa con queso and its vegetable ingredients at sterilization temperatures. Int. J. Food Prop. 11 (1), 112–126.

Parrott, D., 2010. Microwave technology sterilizes sweet potato puree. Food Technol. 64 (7), 66–70.

Simunovic, j., Swartzel, K.R., Truong, V.D., Cartwright, Coronel, P., (2014). Methods and apparatuses for thermal treatment of foods and other biomaterials, and products obtained thereby US Patent 8,742,305.

Simunovic, j., Drozd, J.M., (2012) Method for processing biomaterials US Patent 8,337,920.

Simunovic, j., (2017). Modular devices and systems for continuous flow thermal processing using microwaves US Patent App. 15/320,676.

Simunovic, j., Swartzel, K.R., Truong, V.D., Cartwright, G.D., Sandeep, K.P., Coronel, P., (2017). Methods and apparatuses for thermal treatment of foods and other biomaterials, and products obtained thereby US Patent 9,615,593.

Simunovic, j., Drozd, J.M., (2019). Method for processing biomaterials US Patent 10,390,550.

Steed, L.E., Truong, V.D., Simunovic, J., Sandeep, K.P., Kumar, P., Cartwright, G.D., Coronel, P., 2008. Continuous flow microwave-assisted processing and aseptic packaging of purple-fleshed sweetpotato purees. J. Food Sci. 73 (9), E455–E462.

Update on emerging technologies including novel applications: radio frequency

Shaojin Wang[a,b], Yvan Llave[c], Fanbin Kong[d], Francesco Marra[e], Ferruh Erdoğdu[f]

[a]*College of Mechanical and Electronic Engineering, Northwest A&F University, Yangling, Shaanxi, PR China*
[b]*Department of Biological Systems Engineering, Washington State University, Pullman, WA, United States*
[c]*Food Thermal Engineering Laboratory, Department of Food Science and Technology, Tokyo University of Marine Science and Technology, Japan*
[d]*Department of Food Science and Technology, The University of Georgia, Athens, GA, United States*
[e]*Dipartimento di Ingegneria Industriale, Università degli studi di Salerno, Fisciano, SA, Italy*
[f]*Professor of Food Process Engineering, Ankara University, Turkey*

9.1 Introduction

Radio frequency (RF) heating is known as capacitive dielectric heating with a frequency range of 0.3–300 MHz and has been recognized as an innovative emerging technology for food processing. The frequencies at 6.78, 13.56, 27.12, and 40.68 MHz are commonly used for industrial, scientific, and medical purposes while, in another dielectric application, microwave processing, specifically home appliance usage at 2450 MHz, are more common. A frequency of 915 MHz for microwave processing is usually preferred for industrial processes due to longer wavelengths and greater penetration depth. The lower power generation within the product due to the lower frequency and greater penetration depth of the RF energy due to the longer wavelength, compared to microwave processing, make it more suitable for bulk size product processing, especially at industrial scale (Altin et al., 2021). The lower energy dissipation in the product and resulting temperature increase are more controllable and uniform for RF processing.

Industrial applications of RF processes can be mainly found in textile, wood, plastic welding, composites, surface plasma treatment, and food manufacturing (Bernard et al., 2015), and the industrial systems are classified into two categories:

- Self-oscillator RF generators (Fig. 9.1A), and
- 50 Ω RF amplifiers (Fig. 9.1B).

Free running oscillator with a parallel-plate electrode configuration is the most commonly applied one in food and agricultural products processing. RF energy in this system is generated by the standard oscillator circuit formed by the triode tubes, coupled into the product, and its potential can be adjusted by the electrode gap. Moving of the top electrode upward and downward adjusts the required gap and the electrode potential. The products can also be moved through the RF EM field by conveyor belts to allow for continuous processing (Wang et al., 2007a, 2007b; Awuah et al., 2015).

Food Engineering Innovations Across the Food Supply Chain. DOI: https://doi.org/10.1016/B978-0-12-821292-9.00013-3

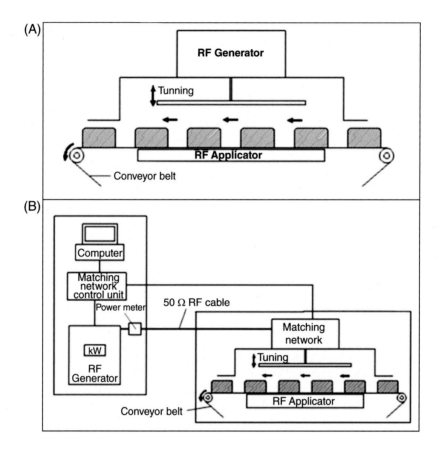

FIG. 9.1

Schematic view of the typical free running oscillator (A) and 50 Ω (B) radio frequency heating systems (Awuah et al., 2015).

About 50 Ω systems are similar to the free running oscillator systems except for the presence of an automatic impedance matching system, which can be automatically turned to keep the overall imped-ance of the working circuit at 50 Ω. These systems can keep the coupling power in products stable throughout the process and provide a fixed frequency with precise power and feedback control com-pared to the free running oscillator systems (Zhou et al., 2017; Jiao et al., 2018). The free running oscil-lator RF systems have higher heating rates and energy efficiency under the same processing conditions. However, control accuracy and heating uniformity of the 50-Ω RF systems are reported to be better (Zhou et al., 2017). Besides these positive remarks for the use of 50-Ω systems, higher manufacture costs limit their use.

The RF applicators, as an interface between RF generators and the product, are designed based on the processed product, and their design is not dependent upon the applied RF generators (Bernard et al., 2015). The RF applicators are an important part of these two RF systems and provide the effective RF energy coupling in the treated product. The RF applicator has to be designed for the particular product dimensions. Although the size and shape of the applicators might vary in a wide range, they are mostly divided into the following electrode configurations:

- the through field electrodes (Fig. 9.2A),
- the fringe field or stray field electrodes (Fig. 9.2B), and
- the staggered through field electrode systems (Fig. 9.2C; Jones and Rowley, 1996).

Conceptually, a through field RF applicator (with a pair of parallel plates) is the simplest and most commonly used design, where the product is placed between two electrodes (electric potential applied and ground electrodes) to form the parallel plate capacitor. This type of applicator is widely applied for heating bulk materials or large and thick products. High potential is applied through the RF generator onto the electrodes. This leads to the formation of an electric field between the electrodes. In conventional configurations, the product is placed between the electrodes, and, the potential on the electrodes is reversed with the frequency.

The fringe-field applicators (also called stray-field electrodes) involve a series of electrodes in the shape of a bar, rod, or narrow plate connected alternatively to either side of the RF power supply. This applicator structure concentrates the high energy density in a sheet-like thin material that passes over or under an array of electrodes. In a Stray-field electrode applicator RF system, charged and ground electrodes (in the form of rods) are placed next to each other below the product, being more suitable for thin layer products (Bernard et al., 2015). However, the generated electric field might have a limited effect on the above placed product. The staggered through-field applicators (also called Garland electrodes) include the rod or tube electrodes staggered on either top or bottom of the conveyor belt. This arrangement can transfer a high power to the thin product on the moving belt (Jones and Rowley, 1996). Staggered through field electrode systems enable thicker materials to be processed (Bedane et al., 2018). Bedane et al. (2018) investigated the electric field distribution in a staggered through-field electrode RF system and thawing temperature uniformity with changes in quality parameters during RF thawing of frozen chicken breast. Higher electric field intensity was observed at the center and middle edges of the bottom. In addition to these three electrode types, inductive applicators, arranged around the material, are another electrode configuration (Bernard et al., 2015).

Another special design for RF applicators is the tubular one, which is used to process liquid or other pump-able products in RF systems. A typical RF tubular applicator is shown in Fig. 9.3 where food product is pumped through a plastic RF – transparent tube (Awuah et al., 2005). Here, the RF energy is applied to the tube by a pair of curved electrodes. The tube can be placed vertically, horizontally or inclined with an angle. These arrangements are designed depending upon the application requirement. The continuous RF heating process with tube applicator has been used for treating sausage emulsions (Houben et al., 1993), CMC solution (Zhong et al., 2003), starch solutions (Awuah et al., 2002), and milk (Awuah et al., 2005).

Electrode configuration is an important factor for designing an RF system while the dielectric properties, affecting the temperature increase and uniformity, are important factors for the process design. Electromagnetic properties of food products in determining the process parameters are permeability and permittivity while the permeability does not significantly affect the absorbed power and evolved heat transfer. Therefore, permittivity (dielectric constant and dielectric loss factors), affecting the interaction with the electromagnetic waves and attenuation—absorption within the product, is the most significant property. In a simple definition, dielectric loss factor indicates the ability of the material to store the electrical energy in an electromagnetic field, and loss factors is about the energy dissipation

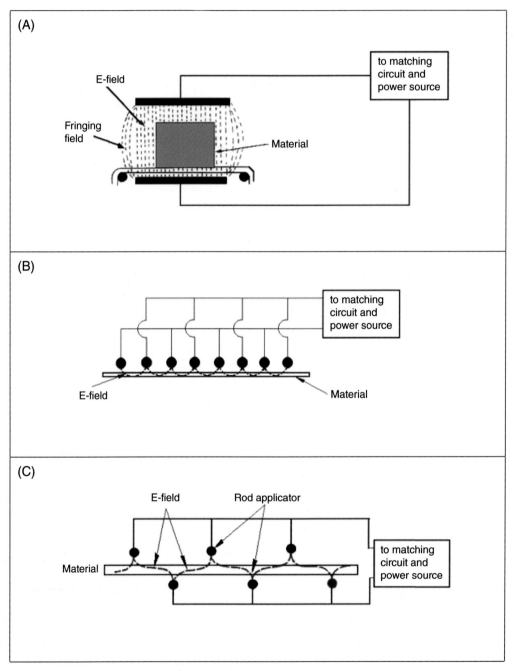

FIG. 9.2

Electrode configurations for RF applicators, (A) Through-field applicator; (B) Fringe-field applicator; and (C) Staggered through-field applicator (Jones and Rowley, 1996).

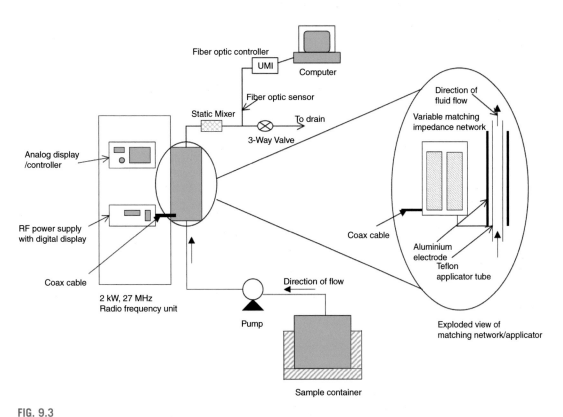

FIG. 9.3

A schematic sketch of the 2-kW, 27.12 MHz radio frequency unit with a tube applicator (Awuah et al., 2005).

and conversion of electrical energy into heat energy. Therefore, besides the effect of the thermo-physical properties on the heat transfer, dielectric properties (dielectric constant and dielectric loss factor) are the major factors affecting the penetration depth, heating rate and resulting process uniformity. Applied frequency, temperature, and composition of the product (moisture and ionic contents) are the main variables affecting the dielectric properties influencing the dielectric heating behavior. Considering that the dielectric loss factor for most of the food products increase with temperature, this might lead to thermal runaway heating (specifically over the surface if the penetration depth is also limited due to the variation of the dielectric properties). This is a significant case to consider for process design purposes. Two different techniques are commonly preferred to measure the dielectric properties of food products: the open-ended coaxial probe method and the parallel plate method. The open-ended coaxial probe is a popular method for measuring the dielectric properties of agricultural products. It can be used for a wide frequency range, but the accuracy is reduced below 200 MHz. It is not suitable for measuring materials with low values of dielectric constant and loss factor, which is the case with low moisture foods including food powders. The parallel plate method involves using an inductance, capacitance, and resistance (LCR) meter or impedance analyzer to test the products placed between the electrodes. The capacitance and resistance values of the foods are measured and used to calculate the dielectric constant and loss factor of the sample (Ozturk et al., 2016). Compared with the open-ended coaxial probe, the parallel plate methods can achieve higher accuracy in the low frequency range, which makes it more suitable for the measurement of dielectric properties of foods subject to RF heating (<100 MHz). In addition to the effects of the applied frequency, temperature, composition, compact density is another variable affecting the dielectric properties of food powders. RF processing of food powders is one of the applications in the food processing industry, and therefore the knowledge of the dielectric properties is required for further process design and optimization purposes. For instance, it was determined that the dielectric properties of broccoli powder increased when the compact density increased, and then decreased after reaching a peak when the compact density continued to increase (Ozturk et al., 2016). The interesting case of the food powders is their rather low dielectric properties (with the dielectric constant less than 10 and the dielectric loss constant below 1.5) due to the low moisture content and low density that make the RF application a challenge.

While there is an increasing trend to the RF processing both in the literature and industry, the applications focus on disinfestation of agricultural commodities, and pasteurization of food products and powders. Microbial inactivation by RF heating is predominantly through thermal effects. High temperature leads to denaturation of enzymes, proteins, nucleic acids, and causes structural disruption of cell membranes (Dwivedi, 2019). An additional significant progress was observed in the thawing-tempering of food products. This latter process has become a rather significant issue considering that all the frozen products, with an exception of ice-cream, need to tempered or thawed for further processing and finally consumption. In addition, the temperature uniformity and process design are additional challenges for all RF applications. Mathematical modeling was introduced as a virtualization approach for the process design and optimization in addition to the experimental studies. Therefore, the chapter will focus on recent RF applications and mathematical modeling approaches for further process design and optimization.

9.2 Radio frequency disinfestation of agricultural products

Applications of RF heating as a postharvest disinfestation method goes back to the 1920s (Headlee and Burdette, 1929). With the continuous improvement of manufacturing technology and reduced cost, RF heating is increasingly used for disinfesting agricultural products with the aim to replace chemical fumigations. Insect control is the most extensive and in-depth research field of RF heating technology as compared to pasteurization and thawing. Since RF treatment is a gentle treatment, reducing the impact on the quality of agricultural commodities and with its selective heating of insects, larvae, etc. (due to significant differences in their dielectric properties compared to the dry products), this technology has been successfully applied for disinfesting low-moisture agricultural products, especially cereals, grains, legumes, and nuts (Table 9.1).

To develop effective RF treatment protocols, it is essential to know the thermal-tolerant characteristics of both target insects or larvae and products themselves over a relatively wide range of time–temperature combinations. The overlap between the area below the quality curve and above the insect mortality curve identifies a range of temperature and exposure times for complete control of the target without damaging the product (Wang et al., 2008). Developing an RF treatment protocol often uses a small- or pilot-scale RF heating system to determine the optimum treatment parameters. RF disinfestation of cereals or grains is currently a hot-topic research area, among which the studies mostly focused on rice, wheat and corn; the three most important cereal crops. For example, based on the thermal death kinetics of adult rice weevil (RWL) and a heating uniformity study, Zhou and Wang (2016a, 2016b)

Table 9.1 Summary of radio frequency disinfestation of agricultural commodities (adapted from Ling et al., 2019).

Agricultural commodity	Target insect for disinfestation	References
Almond	Codling moth	Wang et al. (2013)
Brown rice	Rice weevil	Jiao et al. (2017)
		Zhou and Wang (2016a, 2016b)
Canola seed	Red flour beetle	Yu et al. (2017)
Chestnuts	Yellow peach moth	Hou et al. (2014)
Chickpea	Cowpea weevil	Wang et al. (2010)
Corn seed	Maize weevil	Hassan (2012)
Lentil	Cowpea weevil	Jiao et al. (2012)
Milled rice	Rice moth	Yang et al. (2018)
	Rice weevil	Mirhoseini et al. (2009)
		Jiao et al. (2017)
		Zhou and Wang (2016a, 2016b)
		Vearasilp et al. (2015)
Canola seeds	Red flour beetle	Yu et al. (2017)
Rice flour	Rice weevil	Li et al. (2015)
Rough rice	Rice weevil	Jiao et al. (2017)
		Wangspa et al. (2015)
		Zhou and Wang (2016a, 2016b)
	Grain moths	Lagunas-Solar et al. (2007)
	Lesser grain borers	
Soybean	Indian meal moth	Wang et al. (2010)
Walnuts	Navel orange worm	Wang et al. (2007b)
Wheat seed	Rusty grain beetle	Shrestha et al. (2013)

carried RF disinfestations of adult RWL in brown rice, milled rice, and rough rice both in batch and continuous modes. The results demonstrated that, under continuous treatment, the uniformity of RF heating was better compared to the batch mode treatment. While achieving a complete mortality of RWL, the average throughput of the RF treatments could be as high as 78%, 76%, or 74% and 269, 247, or 225 kg/h for milled, brown, or rough rice, respectively. In other studies, 100% mortality of insects in RF treated cereal and grain products was also achieved (Lagunas-Solar et al., 2007; Hassan, 2012; Shrestha et al., 2013; Li et al., 2015; Yang et al., 2018).

Dried legumes (e.g., soybeans, chickpeas, green peas, or lentils) rank second after cereal grains as the main food grain crop. One of the most common insect pests in legumes around the world is cowpea weevil, and another pest of concern is the Indian meal moth, a common pest of many stored agricultural products. After an RF treatment protocol in batch mode was optimized to achieve 100% mortality of cowpea weevil in lentils (Wang et al., 2010) without product quality degradation, a continuous RF process with an electrode gap and conveyor belt speed of 14.0 cm and 7.5 m/h was also developed to obtain the average heating efficiency and throughput of 76.7% and 209 kg/h (Jiao et al., 2012).

To achieve satisfactory heating uniformity in nuts, the protocol usually consisted of RF heating to a target temperature, holding for a predetermined time with hot air and cooling by forced air circulation at room temperature in a single layer (Wang et al., 2007a; Gao et al., 2010; Hou et al., 2014). An initial study for nuts focused on evaluating the feasibility of RF disinfestation of a small quantity of in-shell walnuts (Wang et al., 2001) followed by differential RF heating between insects and walnut kernels (Wang et al., 2003). After determining the most thermos-tolerant insects in walnuts, the fifth-instar larvae of the navel orange worm were used as a target insect for RF treatment validation studies (Wang et al., 2002). Wang et al. (2007a, 2007b) used 27 MHz, 25 kW industrial RF systems to simulate the industrial processing and evaluate the energy efficiency and cost of the RF process. After RF treatments, average and minimum walnut temperatures were 60 °C and 52 °C, respectively, with 60 °C hot air holding for 5 min. This process resulted in 100% insect mortality without negative effects on product quality stored for 2 years at 4 °C. The average heating efficiency was estimated to be 79.5% for treating walnuts at 1561.7 kg/h. The overall electrical consumption for the treatments was US$ 0.0027/kg, comparable to a unit fumigation cost (US$0.0020–0.0027/kg) for commercial in-shell walnut treatments (Wang et al., 2007a, 2007b).

9.3 Radio frequency pasteurization of food products

Food safety has recently been recognized especially in the postharvest processing of agricultural products. Several recalls and outbreaks have occurred in food products, such as fruits, vegetables, meat, and nuts. *Salmonella* spp., *E. coli, Listeria, Enterobacter sakazakii,* and *Staphylococcus aureus* are the main foodborne pathogens linked to these outbreaks in food products. The survival of these pathogens especially in low moisture products can lead to fast out quickly as soon as suitable environmental conditions are reached. This results in a significant threat to human health. RF pasteurization has been developed to reduce the potential risks associated with these microorganisms and related food products.

Table 9.2 summarizes the recent developments in RF pasteurization of food products. RF heating, due to the given heating rate and temperature increase of the food products, achieves higher inactivation rates of pathogenic microorganisms compared to conventional thermal processing while maintaining product quality. This feature becomes especially significant in products with lower moisture content since microbial inactivation with conventional thermal processing in these products is more difficult due to the lower values of the thermal and physical properties (e.g., thermal conductivity and other properties). In addition, the application of combining RF heating with other technologies or heat transfer media, such as RF application with hot air, can avoid the limitation of the individual methods

Table 9.2 Summary of radio frequency pasteurization in agricultural products (adapted from Zhang et al., 2021).

Food product	Pathogen/log reduction	References
Almond	*S. enteritidis* PT 30/3.7 log *S. typhimurium*/6.0 log *S. senftenberg*/5.6 log *E. coli* ATCC 25922/5 log *E. coli* ATCC 25922/>4 log *E. coli* ATCC 25922/4 log *E. faecium*/2.3 and 3.89 log	Jeong et al. (2017) Li et al. (2017) Li et al. (2018a) Cheng and Wang (2019)
Carrots	Total aerobic microbial/>1 log	Xu et al. (2017)
Chestnuts	Fungi/3–5 log	Hou et al. (2018)
Corn	*Aspergillus parasiticus*/1–6 log *Aspergillus flavus*/3–4 log	Zheng et al. (2017) Jiao et al. (2016)
Nuts	*E. faecium*/3.5–6.7 log	Salazar et al. (2018)
Peach	*Monilinia* spp./not provided *S. typhimurium*/2.41–3.46 log	Sisquella et al. (2014) Hu et al. (2018)
Walnut	*S. aureus* ATCC 25923/>4 log	Zhang et al. (2019)
Wheat	*Aspergillus flavus*/2–3 log	Jiao et al. (2016)
	E. coli O157:H7/0.56–8 log	Choi et al. (2018)
	Staphylococcus aureus/0.87–8 log	

(e.g., temperature nonuniformity of the RF processing and lower heating rates of the conventional thermal processing). Low moisture nature of many agricultural commodities also leads to resistance against microbial inactivation.

Prior to RF processing, increasing water content in agricultural products is used to reduce the thermal resistance of the pathogens and achieve the required microbial reduction. This process is followed by RF drying to reduce the moisture content back to the original level for safe storage (Zheng et al., 2017; Li et al., 2018a). The RF heating can also be combined with other methods, such as controlled atmosphere (Cheng et al., 2020) and cold shock (Ozturk et al., 2019a) to achieve synergistic effects and further improving the efficiency of pasteurization in agricultural products. Future studies should focus on pasteurizing the cocktail of pathogens and transferring RF pasteurization to industrial scale cases, and developing comprehensive and feasible RF postharvest treatment protocols to integrate disinfestation, pasteurization, and drying purposes together.

9.4 Radio frequency pasteurization of food powders

Food powders have advantages of easy handling, transportation, storage, and longer shelf life, therefore they are widely used in the food industry. While RF heating can be used in dehydration and pasteurization, RF pasteurization has emerged as an active research topic in the last decade due to the increasing concern of pathogen contamination in low moisture foods (Dag et al., 2019, 2020). Table 9.3 summarizes the most recent studies:

Table 9.3 Summary of radio frequency pasteurization of food powders (adapted from Zhang et al., 2021).

Food product	Pathogen/log reduction	References
Black pepper	*S. enterica*/>5 log *E. faecium*/>4.8 log	Wei et al. (2018)
	Salmonella spp./3.98 and 5.93 log *E. faecium*/2.3 and 3.89 log	Wei et al. (2019)
Corn flour	*S. enteritidis* PT30/6.59 log	Ozturk et al. (2019b)
Paprika powder	*E. faecium*/4.79 log *Salmonella cocktail*/4.16 log	Ozturk et al. (2020)
White pepper powder	*E. faecium*/1.92 log *Salmonella cocktail*/3.34 log	Ozturk et al. (2020)
Cumin powder	*E. faecium*/1.41 log *Salmonella cocktail*/2.83 log	Ozturk et al. (2020)
Cumin seed	*E. faecium*/0.88 log *S. enterica* cocktail/> 5.8 log	Chen et al. (2019)
Red pepper powder	*E. faecium* NRRL B-2354/> 6.4 log *E. coli* O157:H7/> 5 log	Jeong and Kang (2014)
Wheat flour	*S. typhimurium*/> 5 log *S. typhimurium*/> 3.2 log (at a_w of 0.57) *S. typhimurium*/>4.8 log (at a_w of 0.64) *S. enteritidis* PT 30/> 4.98 log	Hu et al. (2018) Villa-Rojas et al. (2017)
	E. faecium/> 3.24 log *S. enteritidis* and *E. faecium*/3–5 log	Liu et al. (2018)
	E. faecium NRRL B-2354/1.0–4.9 log	Xu et al. (2018)

Current technologies for food powder pasteurization include hot air, steam, irradiation and chemical fumigation. These methods can cause quality deterioration or safety concerns. RF pasteurization has the advantage of fast volumetric heating that can reduce heating time potentially improving food quality, therefore it has been studied as a promising alternative technology. To develop an effective RF process, understanding of RF heating characteristics is crucial including dielectric properties, heating rate and heating uniformity, and microbial validation is eventually required prior to the industrial application (Villa-Rojas et al., 2017; Ozturk et al., 2020). Using surrogate microorganisms (e.g., *Enterococcus faecium* NRRL B-2354 for *Salmonella enteritidis* PT30) is a common way for this purpose as pathogen microorganisms are not permissible for studies in food processing plants (Wei et al., 2019). Microbial validation also requires proper inoculation method and a verified protocol. Considering the inoculation of liquid cultures into a low-moisture matrix, dry inoculum is preferred. Target temperature is usually in the range of 65–85 °C. RF heating rates are dependent on product (dielectric properties, bulk density, moisture content, and chemical composition) and system (applied power, electrode configuration, and shape) properties. Correlations were reported between heating rate and moisture content (and resulting dielectric loss factor; Ozturk et al., 2016). Water activity is an important factor affecting the efficiency of RF process on both heating rate and pasteurization efficiency. Hu et al. (2018) demonstrated the effect of water activity of RF heating with the accompanied hot air process. These conclusions are also supported by Liu et al. (2018), Xu et al. (2018), and Zhang et al. (2019). The heating rate of food is also affected by the size and shape of the food material, and an increased sample size can lead to a slower heating rate (Dag et al., 2019). Tiwari et al. (2011a, 2011b) indicated that larger sample size and lower dielectric properties facilitated lower heating rates with better heating uniformity. Package geometry and dimensions are also important, cubic and cylindrical containers showing more edge heating but center heating was dominant in the ellipsoid container (Huang et al., 2016a). Depending on dielectric properties, product geometry/structure and the RF system design, the cold spot may vary (Xu et al., 2020), for example, geometric center for corn flour (Ozturk et al., 2019a, 2019b) and top surface center for ground black pepper (Wei et al., 2019). It is therefore important to understand the progressive changes of the temperature distribution and the cold spot during the entire heating/cooling process. Computer simulation has been used as a useful tool to predict the complex heat transfer phenomena and temperature variation during RF heating (Huang et al., 2018). Considering that packaging and surrounding material can greatly influence RF heating uniformity, keeping the dielectric constant of the sample and container material similar can result in a more uniform electric field distribution, and subsequently a better temperature distribution. Previous studies have compared the effect of various packaging and surrounding materials (Jiao et al., 2014, 2015; Ozturk et al., 2017).

As summarized above, the recent literature studies have proven the potential of RF heating to be an alternative pasteurization technology for food powders. However, nonuniformity in temperature distribution (especially the location of cold spot) still remain to be a major challenge for the industrial application. Based on the literature studies, combination of RF processing with an additional treatment such as hot air heating seems to be required to eliminate the temperature uniformity problem and achieve the extent of microbial inactivation. Xu et al. (2018) and Ozturk et al. (2020) (the latter also applied an additional post freezing process) demonstrated the significance of hot air holding following the RF process to increase the microbial inactivation in wheat and corn flours.

Further improvements in RF heating uniformity is required with combination of experimental, computer simulation and microbiological validation studies. Computer simulation is helpful tool to provide an understanding and predictions of the heating rate and distribution by solving the equations of the underlying physical phenomena. Innovations in RF systems will also contribute to the successful application of the technology in the food industry. For example, most previous studies utilized free running oscillator systems, while 50 Ω RF systems can also be considered due to their improved power control, heating uniformity and stability. Improvements in electrode shape and configuration are also expected to promote the advancement of RF pasteurization technology applied to food powders.

9.5 Radio frequency tempering and thawing of frozen foods

Frozen storage is a widely applied method for extending shelf life of eventually refrigerated food products, mainly meat, fish and seafood. This makes tempering and thawing required processes before the final consumption of these products or their use in a processing line (e.g., tempering of fish in a canning process line). Inappropriate selection of thawing/tempering methods can result in food quality degradation, especially deterioration in texture, flavor, color, and lipid and protein oxidations with potential microbial spoilage and additional drip loss (Jia et al., 2017). Depending on the final temperature of the food product, tempering (to final temperatures of –5 °C to –2 °C, at thermal center) or thawing (to 0 °C at the thermal center) are preferred approaches for defrosting purposes. Conventional processes use air or water as thawing or tempering medium, and these methods have the long-lasting problem of slow thawing, for example, it may take days depending on the size of the frozen products. These extensive times are related to relatively slow convective heat transfer, and even more so conductive heat transfer taking place inside the product. Llave and Erdogdu (2020) reported that higher temperatures (over +4 °C) are observed over the product surface (higher temperature medium of water or air is preferred to shorten the thawing time in the conventional approaches) in conventional processes, and this leads to possible microbial growth and quality losses. Therefore, the main goal of an optimal thawing/tempering process is to minimize process time while limiting quality losses and microbial growth (Bedane et al., 2018).

In recent years, several new thawing technologies have been extensively studied. These were rapid thawing approaches to avoid problems related to long thawing times and changes in quality attributes. Thermal processes such as dielectric (microwave—MW and radio frequency—RF) or resistive (ohmic) thawing have the advantages of reduced time and less damage to product quality owing to the volumetric heating characteristic of electromagnetic energy. Moreover, RF heating provided the ability to penetrate deep into the frozen food, and opportunities to heat food products within the package itself, due to the longer wavelength and penetration depth. In addition, drip losses are minimized, and microbial safety is not compromised (Li and Sun, 2002). In this section, recent literature on RF tempering and thawing applications of frozen products and approaches to improve RF tempering and thawing uniformity are summarized. Compared to conventional thawing approaches, so-called proper thawing procedures, tempering approaches can be carried out in shorter processing times, reducing the loss of quality attributes (including organoleptic characteristics) do not allow microbial growth while keeping the food product at a desired quality and safety (Llave and Erdogdu, 2020).

In Fig. 9.4, the typical thawing curves for different thawing methods are shown: RF heating, MW heating, and conventional natural-convection air heating (for comparison). In the conventional method, the temperature rise is rapid initially because of the higher thermal conductivity of the frozen layer. Due to the endothermic conversion of ice to water (latent heat of fusion), the heating slows

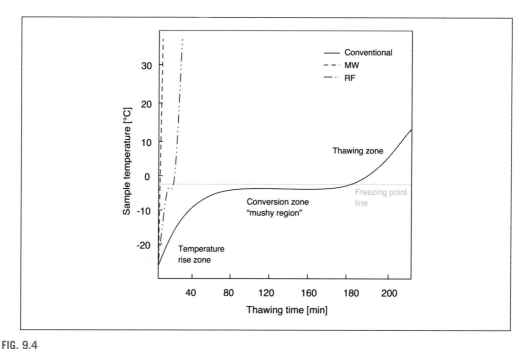

FIG. 9.4

Typical thawing curves for different thawing methods: conventional (natural-convection air), radio frequency (RF) heating, and microwave (MW) heating.

down significantly until all ice is converted to water, which is then followed by a faster temperature increase (Llave and Sakai, 2018). Application of MW yields faster thawing due to the higher frequency (and power absorption) compared to the RF application. However, the control of this high power is difficult, and the process comes at the expense of a significant temperature uniformity. Heating rate for the various thawing methods are MW > RF > convectional method with respect to the thawing time. Choi et al. (2017) demonstrated that the forced air convection approach as a conventional process was a time consuming approach, while a curved electrode RF process, in addition to the improvement in the process time, led to lower drip losses. Considering that tempering is about to the increase of temperature of the product below the thawing front temperature, due to the limited changes in the dielectric properties in the frozen range (before the phase change), temperature uniformity is achievable except at the corners and edges. Due to the improved penetration depth and easier control of the power absorption with the application to the larger sized food products, industrially RF is a preferred process (Altin et al., 2021).

Initial attempts for RF thawing date back to the 1940s where its potential for thawing frozen meat, fish, and vegetables was explored; however, the reported information in the published literature is scarce. Perhaps the earliest RF research on food thawing was reported by Cathcart and Parker (1946) where an RF process (at 14 to 17 MHz) was applied to frozen fish, eggs, fruits and vegetables reducing the thawing time. Bengtsson (1963) reported some of the earliest dielectric property data in the frozen region for lean beef and cod fish at several RF frequencies (10, 35, 100 and 200 MHz) and temperatures

(−25 °C, −10 °C, −5 °C, and 10 °C). Although commercial RF thawing/tempering equipment has been used for decades, rather significant research findings on meat and fish thawing/tempering have been published within the last decade (Tables 9.4–9.6).

For an efficient RF process, it is essential that the electromagnetic field should be uniformly distributed within the RF cavity and the food material to ensure uniform heating. In addition to dielectric properties and electromagnetic field intensity, heat capacity and bulk density of the sample are additional significant factors (Zhang et al., 2010). Due to the intrinsic properties of the dielectric material and interaction with the electromagnetic field over the surface and along the edges of the food product is the cause of nonuniform heating in foods (Margulies, 1983). During processing where the product is moved, this interaction is called magnet-like effect (Erdogdu et al., 2017), and the possible run-away heating caused the temperature nonuniformity. The fringing field (see Fig. 9.2A, field lines at the edge of the plates, which are nonparallel but bet outward) effect (the area extending outside the electrodes and where the field lines are bent) in these processes, becomes also significant. The placement of the product is also a factor to cause this effect, and to balance out the fringing field effect, uniformity improvement methods were developed. The electromagnetic field of RF heating equipment between two parallel electrodes was found to depend on the geometry of, especially, the top electrode. Some studies report on the modification of the geometry leading to a reduction of the fringe effect from the edges of

Table 9.4 Recent applications of RF tempering and thawing of selected frozen seafood products.

Food product	Processing condition	Radio frequency (RF) system	Effect	References
Tuna fish (lean, semifat, and fat muscles)	13.56 MHz; 10 W to 6 kW; upper-electrode projection adjustable to different sizes	Parallel-plate RF system (FRT-1, Yamamoto Vinita Corp. Ltd., Osaka, Japan) with a 50-Ω automatic impedance matching network	Improvement of the temperature uniformity when the top electrode was similar in size to the sample	Llave et al. (2014)
Fish eggs, fish or whale meat, minced fish, and meat and minced meat	100 ± 10 MHz; 100 W to 1000 W	A self-constructed thawing device, including a furnace member (cavity), an amplifier, and a matching device. An antenna is provided within the application furnace member	A more uniform temperature distribution (compared to 2.45 GHz) and rapid thawing (compared to 13.56 MHz) while maintaining the food's quality	Sato et al. (2016)
Shrimp	27.12 MHz; 500 W and 1 kW; gaps: 150 and 160 mm	Laboratory scale free-running oscillator RF system with a parallel plate electrode design	A uniform overall temperature distribution with no local overheating at the surface	Palazoğlu and Miran (2017)
			The most suitable gap for the thawing of industrial blocks was found (16 cm)	
Minced fish	27.12 MHz; 6 kW	A free-running oscillator machine with parallel-plate electrodes	Unaffected cooked gel properties after RF heating	Yang et al. (2019)

Table 9.5 Recent applications of RF tempering and thawing of selected frozen beef meat products.

Food product	Processing condition	Radio frequency (RF) system	Effect	References
Beef meat	27.12 MHz; 400 W and 500 W	A custom built 50-Ω system (Ctech Innovation-Chester, UK) using a RF generator and an impedance matching network and controller	Thawing time and power consumption were reduced. Temperature was uniform	Farag et al. (2008)
Lean beef meat	27.12 MHz; 0.01–20 GHz; target temperature: 5 °C		A significant decrease in drip and micronutrient loss	Farag et al. (2009)
Lean beef meat	27.12 MHz; 6 KW; conveyor belt speed: 3 m/h; electrode gap: 10 cm	Free running oscillator RF equipment	Heating uniformity improved by adjusting the moving speed of conveyor belt and the gap between electrodes. Continuous RF thawing improved heating uniformity. Small electrode gap resulted in nonuniform heating	Bedane et al. (2017a)
Lean minced beef meat	27.12 MHz; 6 kW			Bedane et al. (2017a)
Beef meat	27.12 MHz. Sample thickness (40 mm, 50 mm, 60 mm), base area (small, medium, and large), and shape (cuboid, trapezoidal prism, step)	A parallel-plate RF system	Among all three shapes, the cuboid shape has the best heating uniformity > trapezoidal prism > step shape	Li et al. (2018c)
Lean beef meat	27.12 MHz; 2 kW. Three different vertical positions under moving and stationary conditions	Laboratory scale free-running oscillator RF system with a parallel plate electrode design	Downwardly inclined conveyor configuration may be beneficial in applications where rapid heating at the beginning and a more controlled heating toward the end is desired	Palazoğlu and Miran (2018)
Frozen beef	27.12 MHz; 12 kW with an electrode gap of 115 mm under stationary conditions	50 ohm parallel plate RF system	Heating uniformity was determined using modeling and experimental approach	Li et al. (2018b)

the electrode (Llave et al., 2015; Erdogdu et al., 2017). Kim et al. (2016) investigated the thawing of pork meat using curved shape charged—top electrode considering different temperatures in the cavity. It was reported that the use of an electrode in a geometry resembling the shape of the product geometry together with reduced temperature in the cavity improved the temperature uniformity. Curving the top electrode to conform to the sample geometry reduced the corner and edge heating issues, that is, higher electric field strengths and therefore higher heating at edges and corners. Movement and rotation of sample were also found to have an ability to redistribute the electromagnetic field and heat within the food product, thus improving the RF heating uniformity (Bedane et al., 2017a, 2017b; Erdogdu et al., 2017). Fiore et al. (2013) reported the significance of RF power to avoid the local overheating while

Table 9.6 Recent applications of RF tempering and thawing of selected frozen pork and chicken products.

Food product	Processing condition	Radio frequency (RF) system	Effect	References
Pork sirloin	27.12 MHz; 800 W. Parallel-plate electrodes and curved-electrode were used	RF generator	Heating was uniform and the surface burning was prevented by curving the top flat electrode to fit	Kim et al. (2016)
Pork loin	27.12 MHz; 1000 W and 1500 W. Parallel-plate electrodes and curved-electrode were used 2-kW RF generator, 50-Ω automatic impedance matching network, and a controller set at 27.12 MHz constant frequency	2-kW RF generator, 50-Ω automatic impedance matching network, and a controller set at 27.12 MHz constant frequency	The cylindrical pork samples Drip losses and tempering time reduced by curved-electrodes RF system, and relatively uniform tempering compared to a parallel-plate RF system	Choi et al. (2017)
Lean pork meat	27.12 MHz; 3 kW; electrode gap: 11 cm	Commercial RF heating system	Shorter tempering time than air and water tempering. Tempering temperature (-1 °C and -4 °C) had a significant effect on tempering time, drip loss, cook loss, pH, and had nonsignificant influence on color, TVBN, and TBARS	Zhu et al. (2019)
Chicken breast meat	27.12 MHz; 10 kW; electrode gap: 65, 75, and 85 mm at 2 power levels	Free oscillating staggered through-field electrode RF system	The meat was fully thawed with uniform temperature distribution using 65 mm electrode gap setting	Bedane et al. (2018)

Hansen et al. (2006) and Farag et al. (2011) suggested a pulse mode application to achieve a better heating uniformity compared to a continuous process, due to allowing thermal condition to partially equilibrate temperatures. Using a package or a container that have dielectric properties similar to the sample can also be an efficient way to improve the uniformity of RF heating. Recently, Wang et al. (2014), Jiao et al. (2015), and Huang et al. (2016a) investigated the effects of using polyurethane foams, polyetherimide (PEI), and polystyrene containers, respectively, to improve heating uniformity for low-moisture agricultural products. The package effect was also reported in the above section, and these approaches could be implemented for the RF tempering and thawing applications. In addition, surrounding the food with a tempering medium can reduce the temperature variation in the food and also allows for holding the temperature at a desired level. The selection of the tempering medium depends on the sample properties and applied process.

9.6 Advantages and disadvantages of radio frequency processing

One of the main reasons for using RF heating in food processing (rather than any other thermal technology) is improved food quality with a process efficiency. This is expected due to the volumetric heating process leading to a uniform temperature distribution, uniform heating and increased quality. However, the given penetration depth (≈11 m at 27.12 MHz) does not guarantee the increased depth since the dielectric properties are the main factors affecting this. Therefore, the expected advantage of the process might be a certain disadvantage if the penetration depth is limited. This apparently caused the electromagnetic energy over the surface leading to a significant temperature nonuniformity like in the case of a conventional process. For this reason, a process design study must be planned first to conform the suitability of the RF process for the given product. Once this was carried out and demonstrated, the selective nature of RF heating, with energy being dissipated according to the local loss factor, might be used for improved process efficiency and product quality also resulting in possible shorter process times and increased throughput with process efficiency. Besides the expected temperature uniformity during a process, the improved process control is a tempting feature of an RF application. Considering that, the power dissipated within the product is due to the presence of the electric field (Eq. (9.3)), the product response to that is instantaneous. Hence, changes within the electric field and temperature response of the product might be carried out by on-off cycles to obtain a better heating—temperature uniformity during a process.

While the RF heating mechanism has essentially the same feature with the ohmic-heating, the contactless heating of the RF process is a significant advantage of this innovative process. In the few of the power generation/absorption of the food product during the RF process, an increased rate will be the expected feature of a microwave process while the penetration depth and expected temperature uniformity with lower power absorption are the aspects of the RF application. Currently, the major problem in an RF process lies in the initial capital and energy costs since this demonstrates a significant issue specifically for the SMEs.

9.7 Mathematical modeling

Mathematical modeling of relevant transport phenomena occurring during RF assisted food processing and its analysis based on computer aided engineering were proven to be efficient virtual tools to improve the design and to optimize the operation performances of food processes and products based on the application of RF technology.

Recent review (Huang et al., 2016b, 2018) emphasized the role of mathematical modeling based simulation for obtaining a better uniformity of temperature distribution in foods processed in RF systems and discussed the most important advancements about this subject. A further study used this approach to design an appropriate RF heating uniformity (Jiao et al., 2018; Guo et al., 2019).

Simulation studies focused on electrode applicator position, electrode shape and size, gap between RF electrodes and sample surface, use of translational and rotational movement of the food product, electrode vertical movement, adjustment in electromagnetic potential, and use of two-cavity systems (Chan et al., 2004; Marra et al., 2007; Birla et al., 2008; Romano and Marra, 2008; Wang et al., 2012; Uyar et al., 2015; Llave et al., 2015; Bedane et al., 2017b; Erdogdu et al., 2017; Yang et al., 2019).

In RF assisted food processes the most evident phenomenon is the transfer of energy (particularly, the transfer of the electromagnetic energy dissipated as heat in a "lossy" medium); as discussed earlier, the temperature distribution inside the food product is determined by the process conditions, the nature of the food product being processes, it is the thermal and dielectric properties, which determine the distribution of the electromagnetic field in the food itself and the amount of electromagnetic energy absorbed by the food and dissipated as thermal energy.

The utilized equations' describing the RF heating of foods must include the distribution of electromagnetic field and the heat transfer equations, including a volumetric heat source term, notifying the dissipation of electromagnetic energy in the processed food. Other phenomena that may be of relevance is the case of mass transfer during RF cooking or RF assisted drying. These must be taken into account and must be described mathematically.

The entire set of Maxwell equations (which generally describe the distribution of an electromagnetic field) can be reduced to the Laplace equation for the electric field in the range of 1–100 MHz (Metaxas, 1996):

$$-\vec{\nabla}\cdot\left[\left(\sigma + j\,2\,\pi\,f\,\varepsilon_0\varepsilon_r\right)\vec{\nabla}V\right]=0 \tag{9.1}$$

where V is the electrical potential, σ is the electrical conductivity, f is the frequency, ε_0 is the permittivity of the electromagnetic waves in free space ($8.85E{-}12$ F/m) and ε_r is the relative permittivity of the food product. The transient heat transfer equation can take into consideration the generation term linked to the dissipation of the electromagnetic energy into heat; however, as RF is mostly applied to solid-like foods, the conductive contribution to heat transfer is included too:

$$\rho c_p\frac{\partial T}{\partial t}=\vec{\nabla}k\cdot\vec{\nabla}T+P \tag{9.2}$$

where T is the temperature, t is the time, ρ is density, c_p is specific heat, k is thermal conductivity, and P is the volumetric heat generation (electromagnetic power dissipation) term, defined as:

$$P=2\,\pi\,f\,\varepsilon_0\varepsilon_r\left|\vec{E}\right|^2 \tag{9.3}$$

where $\left|\vec{E}\right|$ is the modulus of the E-field, evaluated in accordance with Eq. (9.1). This term depends on the electric field distribution within the food, the frequency of the electromagnetic field, and the dielectric properties of the food material.

Usually, the heat transfer in the medium around the food product included as boundary conditions (i.e., when an overall convective heat flux at boundaries is considered), the distribution of the electric field must be evaluated for the space between the two electrodes, which includes the sample and the medium around it (Marra et al., 2007). Marra et al. (2009), Uyar et al. (2015), and Huang et al. (2018) provide a comprehensive discussion on modeling aspects, multiphysics approach, and boundary conditions.

9.8 Conclusions

While up until some 20 years ago, the industrial application of RF processes focused mainly on textile, plastic welding, composites, and surface plasma treatment, studies more recently have changed their focus. Industrial applications emerged for disinfestation of agricultural commodities, pasteurization of food products and food powders, and thawing—tempering of frozen food products. Due to the significant change of the dielectric properties within the frozen and thawed stages, a significant temperature nonuniformity was observed for thawing purposes, and tempering has become the process of choice. To design these processes and further develop RF systems, mathematical modeling approaches were introduced, and even rather complex models for moving food products in industrial scale systems were developed to simulate simultaneous temperature increase, phase change for thawing processes and changes in the electric field.

References

Altin, O., Ozturk, S., Topcam, H., Karatas, O., Erdogdu, F., 2021. Process uniformity during electro-thermal applications and modeling approaches. In: Knoerzer, K., Muthukumarappan, K. (Eds.), Innovative Food Processing Technologies – A Comprehensive Review. Elsevier Ltd., Amsterdam, NL. pp. 1–45.

Awuah, G.B., Ramaswamy, H.S., Economides, A, Mallikarjunan, K., 2005. Inactivation of *Escherichia coli* K-12 and *Listeria innocua* in milk using radio frequency (RF) heating. Innovative Food Sci. Emerg. Technol. 6 (4), 396–402.

Awuah, G.B., Ramaswamy, H.S., Piyasena, P., 2002. Radio frequency (RF) heating of starch solutions under continuous flow conditions: effect of system and product parameters on temperature change across the applicator tube. J. Food Process Eng. 25, 201–223.

Awuah, G.B., Koral, T., Guan, D., 2015. Radio-frequency baking and roasting of food products. In: Awuah, G.B., Ramaswamy, H.S., Tang, J. (Eds.), Radio Frequency in Food Processing: Principles and Applications. CRC Press, Boca Raton, FL, pp. 231–243.

Bedane, T.F., Altin, O., Erol, B., Marra, F., Erdogdu, F., 2018. Thawing of frozen food products in a staggered through-field electrode radio frequency system: a case study for frozen chicken breast meat with effects on drip loss and texture. Innovative Food Sci. Emerg. Technol. 50, 139–147.

Bedane, T.F., Chen, L., Marra, F, Wang, S., 2017a. Experimental study of radio frequency (RF) thawing of foods with movement on conveyor belt. J. Food Eng. 201, 17–25.

Bedane, T.F., Marra, F., Wang, S., 2017b. Performance comparison between batch and continuous thawing of food products assisted by radio frequency heating. Chem. Eng. Trans. 57, 2017–2022.

Bengtsson, N., 1963. Electronic defrosting of meat and fish at 35 and 2450 MHz – a laboratory comparison. Food Technol. 17 (10), 1309–1312.

Bernard, J.P., Jacomino, J.M., Radoiu, M., 2015. RF 50 Ω technology versus variable-frequency RF technology. In: Awuah, G.B., Ramaswamy, H.S., Tang, J. (Eds.), Radio-Frequency Heating in Food Processing – Principles and Applications. CRC Press, Boca Raton, FL, USA. Chapter 7, pp. 119–139.

Birla, S.L., Wang, S., Tang, J., 2008. Computer simulation of radio frequency heating of model fruit immersed in water. J. Food Eng. 84, 270–280.

Cathcart, W., Parker, J.J., 1946. Defrosting frozen foods by high-frequency heat. Food Res. 11, 341–344.

Chan, T.V., Tang, J., Younce, F., 2004. 3-Dimensional numerical modeling of an industrial radio frequency heating system using finite elements. J. Microw. Power Electromagn. Energy 39, 87–105.

Chen, L., Wei, X., Irmak, S., Chaves, B.D., Subbiah, J., 2019. Inactivation of *Salmonella enterica* and *Enterococcus faecium* NRRL B-2354 in cumin seeds by radiofrequency heating. Food Control 103, 59–69.

Cheng, T., Ramaswamy, H., Xu, R., Liu, Q., Lan, R., Wang, S., 2020. Controlled atmosphere assisted 50 Ω radio frequency treatments for inactivating *Escherichia coli* ATCC 25922 in almond kernels. LWT-Food Sci. Technol. 123C, 109124.

Cheng, T., Wang, S., 2019. Modified atmosphere packaging pre-storage treatment for thermal control of *E. coli* ATCC 25922 in almond kernels assisted by radio frequency energy. J. Food Eng. 246, 253–260.

Choi, E.J., Park, H.W., Yang, H.S., Kim, J.S., Chun, H.H., 2017. Effects of 27.12 MHz radio frequency on the rapid and uniform tempering of cylindrical frozen pork loin (*Longissimus thoracis et lumborum*). Korean J. Food Sci. Anim. Resour. 37, 518–528.

Choi, E.J., Yang, H.S., Park, H.W., Chun, H.H., 2018. Inactivation of *Escherichia coli* O157:H7 and *Staphylococcus aureus* in red pepper powder using a combination of radio frequency thermal and indirect dielectric barrier discharge plasma non-thermal treatments. LWT-Food Sci. Technol. 93, 477–484.

Dag, D., Singh, R.K., Kong, F., 2019. Dielectric properties, effect of geometry, and quality changes of whole, nonfat milk powder and their mixtures associated with radio frequency heating. J. Food Eng. 261, 40–50.

Dag, D., Singh, R.K., Kong, F., 2020. Developments in radio frequency pasteurization of food powders. Food Rev. Int. doi:10.1080/87559129.2020.1775641.

Dwivedi, M., 2019. Effect of thermal and non-thermal techniques for microbial safety in food powder: recent advances. Food Res. Int. 126, 108654.

Erdogdu, F., Altin, O., Marra, F., Bedane, T.F., 2017. A computational study to design process conditions in industrial radio-frequency tempering/thawing process. J. Food Eng. 213, 99–112.

Farag, K.W., Duggan, E., Morgan, D.J., Cronin, D.A., Lyng, J.G., 2009. A comparison of conventional and radio frequency defrosting of lean beef meats: effects on water binding characteristics. Meat Sci. 83, 278–284.

Farag, K.W., Lyng, J.G., Morgan, D.J., Cronin, D.A., 2008. A comparison of conventional and radio frequency tempering of beef meats: effects on product temperature distribution. Meat Sci. 80, 488–495.

Farag, K.W., Lyng, J.G., Morgan, D.J., Cronin, D.A., 2011. A comparison of conventional and radio frequency tempering of beef meats: effects on product temperature distribution. Food Bioprocess Technol. 4, 1128–1136.

Fiore, A., Di Monaco, R., Cavella, S., Visconti, A., Karneili, O., Bernhardt, S., Fogliano, V., 2013. Chemical profile and sensory properties of different foods cooked by a new radiofrequency oven. Food Chem. 139, 515–520.

Gao, M., Tang, J., Wang, Y., Powers, J., Wang, S., 2010. Almond quality as influenced by radio frequency heat treatments for disinfestation. Postharvest Biol. Technol. 58, 225–231.

Guo, C., Mujumdar, A.S., Zhang, M., 2019. New development in radio frequency heating for fresh food processing: a review. Food Eng. Rev. 11, 29–43.

Hansen, J.D., Drake, S.R., Watkins, M.A., Heidt, M.L., Anderson, P.A., Tang, J., 2006. Radio frequency pulse application for heating uniformity in postharvest codling moth (Lepidoptera: Tortricidae) control of fresh apples (*Malus domestica Borkh*). J. Food Qual. 29, 492–504.

Hassan, A., 2012. Application of Microwave and Radio Frequency Energy to Control *Sitophilus zeamais* (Coleoptera: Curculionidae) in Maize Grains Unpublished Ph.D. Thesis. Georg-August-University Göttingen, Germany.

Headlee, T., Burdette, R., 1929. Some facts relative to the effect of high frequency radio waves on insect activity. J. New York Entomol. Soc. 37, 59–64.

Hou, L., Kou, X., Li, R., Wang, S., 2018. Thermal inactivation of fungi in chestnuts by hot air assisted radio frequency treatments. Food Control 93, 297–304.

Hou, L., Ling, B., Wang, S., 2014. Development of thermal treatment protocol for disinfesting chestnuts using radio frequency energy. Postharvest Biol. Technol. 98, 65–71.

Houben, J.H., Vanroon, P.S., Krol, B., 1993. Continuous radiofrequency pasteurization in sausage manufacturing lines. Fleischwirtschaft 73, 1146–1149.

Hu, S., Zhao, Y., Hayouka, Z., Wang, D., Jiao, S., 2018. Inactivation kinetics for *Salmonella typhimurium* in red pepper powders treated by radio frequency heating. Food Control 85, 437–442.

Huang, Z., Marra, F., Subbiah, J., Wang, S., 2016b. Computer simulation for improving radio frequency (RF) heating uniformity of food products: a review. Crit. Rev. Food Sci. Nutr. 58, 1–25.

Huang, Z., Marra, F., Subbiah, J., Wang, S., 2018. Computer simulation for improving radio frequency (RF) heating uniformity of food products: a review. Crit. Rev. Food Sci. Nutr. 58, 1033–1057.

Huang, Z., Zhang, B., Marra, F., Wang, S., 2016a. Computational modelling of the impact of polystyrene containers on radio frequency heating uniformity improvement for dried soybeans. Innovative Food Sci. Emerg. Technol. 33, 365–380.

Jeong, S.-G., Baik, O.-D., Kang, D.-H., 2017. Evaluation of radio-frequency heating in controlling *Salmonella enterica* in raw shelled almonds. Int. J. Food Microbiol. 254, 54–61.

Jeong, S.G., Kang, D.H., 2014. Influence of moisture content on inactivation of *Escherichia coli* O157:H7 and *Salmonella enterica* serovar *typhimurium* in powdered red and black pepper spices by radio-frequency heating. Int. J. Food Microbiol. 176, 15–22.

Jia, G.L., He, X.L., Nirasawa, S., Tatsumi, E., Liu, H.J., 2017. Effect of high-voltage electrostatic field on the freezing behavior and quality of pork tenderloin. J. Food Eng. 204, 18–26.

Jiao, S., Johnson, J., Tang, J., Wang, S., 2012. Industrial-scale radio frequency treatments for insect control in lentils. J. Stored Prod. Res. 48, 143–148.

Jiao, Y., Shi, H., Tang, J., et al., 2015. Improvement of radio frequency (RF) heating uniformity on low moisture foods with polyetherimide (PEI) blocks. Food Res. Int. 74, 106–114.

Jiao, S., Sun, W., Yang, T., Zou, Y., Zhu, X., Zhao, Y., 2017. Investigation of the feasibility of radio frequency energy for controlling insects in milled rice. Food Bioprocess Technol. 10, 781–788.

Jiao, Y., Tang, J., Wang, S., 2014. A new strategy to improve heating uniformity of low moisture foods in radio frequency treatment for pathogen control. J. Food Eng. 141, 128–138.

Jiao, Y., Tang, J., Wang, Y., Koral, T.L., 2018. Radio-frequency applications for food processing and safety. Annu. Rev. Food Sci. Technol. 9, 105–127.

Jiao, S., Zhong, Y., Deng, Y., 2016. Hot air-assisted radio frequency heating effects on wheat and corn seeds: quality change and fungi inhibition. J. Stored Prod. Res. 69, 265–271.

Jones, P.L., Rowley, A.T., 1996. Dielectric drying. Dry Technol. 14, 1063–1098.

Kim, J., Park, J.W., Park, S., Choi, D.S., Kim, G.H., Lee, S.J., Park, C.W., Han, G.J., Cho, B.-K., et al., 2016. Study of radio frequency thawing for cylindrical pork sirloin. Biosyst. Eng. 41, 108–115.

Lagunas-Solar, M., Pan, Z., Zeng, N., Truong, T., Khir, R., Amaratunga, K., 2007. Application of radio frequency power for non-chemical disinfestation of rough rice with full retention of quality attributes. Appl. Eng. Agric. 23, 647–654.

Li, B., Sun, D.-W., 2002. Novel methods for rapid freezing and thawing of foods – a review. J. Food Eng. 54, 175–182.

Li, R., Kou, X., Cheng, T., Zheng, A., Wang, S., 2017. Verification of radio frequency pasteurization process for in-shell almonds. J. Food Eng. 192, 103–110.

Li, R., Kou, X., Hou, L., Ling, B., Wang, S., 2018a. Developing and validating radio frequency pasteurization processes for almond kernels. Biosyst. Eng. 169, 217–225.

Li, Y., Chen, S., Yao, M., 2015. Effects of radio frequency heating on disinfestation and pasteurization of rice flour. Taiwanese J. Agric. Chem. Food Sci. 53, 125–134.

Li, Y., Li, .F, Tang, J., Zhang, R., Wang, Y., Koral, T., Jiao, Y, 2018b. Radio frequency tempering uniformity investigation of frozen beef with various shapes and sizes. Innovative Food Sci. Emerg. Technol. 48, 42–55.

Li, Y., Li, F., Tang, J., Zhang, R., Wang, Y., Koral, T., Jiao, Y., 2018c. Radio frequency tempering uniformity investigation of frozen beef with various shapes and sizes. Innovative Food Sci. Emerg. Technol. 48, 42–55.

Ling, B., Cheng, T., Wang, S., 2019. Recent developments in applications of radio frequency heating for improving safety and quality of food grains and their products: a review. Crit. Rev. Food Sci. Nutr. 60, 2622–2642.

Liu, S., Ozturk, S., Xu, J., Kong, F., Gray, P., Zhu, MJ., Sablani, S.S., Tang, J., 2018. Microbial validation of radio frequency pasteurization of wheat flour by inoculated pack studies. J. Food Eng. 217, 68–74.

Llave, Y., Erdogdu, F., 2020. Radio frequency processing and recent advances on thawing and tempering of frozen food products – review. Crit. Rev. Food Sci. Nutr. submitted.

Llave, Y., Liu, S., Fukuoka, M., Sakai, N., 2015. Computer simulation of radio frequency defrosting of frozen foods. J. Food Eng. 152, 32–42.

Llave, Y., Sakai, N., 2018. Dielectric defrosting of frozen foods. In: Grumezescu, A.M., Holban, A.M. (Eds.), Handbook of Food Bioengineering (Multi Volume Set—Volume XVIII: Food Processing for Increased Quality and Consumption. Elsevier, London, UK, pp. 383–422.

Llave, Y., Terada, Y., Fukuoka, M., Sakai, N., 2014. Dielectric properties of frozen tuna and analysis of defrosting using a radio-frequency system at low frequencies. J. Food Eng. 139, 1–9.

Margulies, S., 1983. Force on a dielectric slab inserted into a parallel-plate capacitor. Am. Assoc. Phys. Teachers 52, 515–518.

Marra, F., Lyng, J., Romano, V., McKenna, B., 2007. Radio-frequency heating of foodstuff: solution and validation of a mathematical model. J. Food Eng. 79, 998–1006.

Marra, F., Zhang, L., Lyng, J.G., 2009. Radio frequency treatment of foods: review of recent advances. J. Food Eng. 91, 497–508.

Metaxas, A.C., 1996. Foundations of Electroheat: A Unified Approach. Wiley, New York.

Mirhoseini, S., Heydari, M., Shoulaie, A., Seidavi, A., 2009. Investigation on the possibility of foodstuff pest control using radiofrequency based on dielectric heating (case study: rice and wheat flour pests). J. Biol. Sci. 9, 43–51.

Ozturk, S., Kong, F., Singh, R.K., 2017. Radio frequency heating of corn flour: heating rate and uniformity. Innovative Food Sci. Emerg. Technol. 44, 191–201.

Ozturk, S., Kong, F., Singh, R.K., 2020. Evaluation of *Enterococcus faecium* NRRL B-2354 as a potential surrogate of *Salmonella* in packaged paprika, white pepper and cumin powder during radio frequency heating. Food Control 108, 106833.

Ozturk, S., Kong, F., Trabelsi, S., Singh, R.K., 2016. Dielectric properties of dried vegetable powders and their temperature profile during radio frequency heating. J. Food Eng. 169, 91–100.

Ozturk, S., Liu, S., Xu, J., Tang, J., Chen, J., Singh, R.K., Kong, F., 2019a. Inactivation of *Salmonella enteritidis* and *Enterococcus faecium* NRRL B-2354 in corn flour by radio frequency heating with subsequent freezing. LWT-Food Sci. Technol. 111, 782–789.

Ozturk, S., Liu, S., Xu, J., et al., 2019b. Inactivation of *Salmonella enteritidis* and *Enterococcus faecium* NRRL B-2354 in corn flour by radio frequency heating with subsequent freezing. LWT - Food Sci. Technol. 111, 782–789.

Palazoğlu, T.K., Miran, W., 2017. Experimental comparison of microwave and radio frequency tempering of frozen block of shrimp. Innovative Food Sci. Emerg. Technol. 41, 292–300.

Palazoğlu, T.K., Miran, W., 2018. Experimental investigation of the effect of conveyor movement and sample's vertical position on radio frequency tempering of frozen beef. J. Food Eng. 219, 71–80.

Romano, V., Marra, F., 2008. A numerical analysis of radio frequency heating of regular shaped foodstuff. J. Food Eng. 84, 449–457.

Salazar, F., Garcia, S., Lagunas-Solar, M., Pan, Z., Cullor, J., 2018. Effect of a heat-spray and heat-double spray process using radio frequency technology and ethanol on inoculated nuts. J. Food Eng. 227, 51–57.

Sato, M., Yamaguchi, T., Nakano, T., 2016. Method of Thawing Frozen Food. US Patent and Trademark Office, Washington, DC US Patent No. 0192-0667.

Shrestha, B., Yu, D., Baik, O.D., 2013. Elimination of *Cryptolestes ferrungineus* S. in wheat by radio frequency dielectric heating at different moisture contents. Prog. Electromagn. Res. 139, 517–538.

Sisquella, M., Vinas, I., Picouet, P., Torres, R., Usall, J., 2014. Effect of host and *Monilinia* spp. variables on the efficacy of radio frequency treatment on peaches. Postharvest Biol. Technol. 87, 6–12.

Tiwari, G., Wang, S., Tang, J., Birla, S.L., 2011a. Analysis of radio frequency (RF) power distribution in dry food materials. J. Food Eng. 104, 548–556.

Tiwari, G., Wang, S., Tang, J., Birla, S.L., 2011b. Computer simulation model development and validation for radio frequency (RF) heating of dry food materials. J. Food Eng. 105, 48–55.

Uyar, R., Bedane, T.F., Erdogdu, F., Palazoglu, T.K., Farag, K.W., Marra, F., 2015. Radio-frequency thawing of food products—a computational study. J. Food Eng. 146, 163–171.

Vearasilp, S., Thanapornpoonpong, S., Krittigamas, N., Suriyong, S., Akaranuchat, P., von Hoersten, D., 2015. Vertical operating prototype development supported radio frequency heating system in controlling rice weevil in milled rice, First International Conference on Asian Highland Natural Resources Management, vol. 5, 184–192.

Villa-Rojas, R., Zhu, M.J., Marks, B.P., Tang, J.M., 2017. Radio frequency inactivation of *Salmonella enteritidis* PT 30 and *Enterococcus faecium* in wheat flour at different water activities. Biosyst. Eng. 156, 7–16.

Wang, J., Luechapattanaporn, K., Wang, Y., Tang, J., 2012. Radio-frequency heating of heterogeneous food-meat lasagna. J. Food Eng. 108, 183–193.

Wang, S., Ikediala, J., Tang, J., Hansen, J., Mitcham, E., Mao, R., Swanson, B., 2001. Radio frequency treatments to control codling moth in in-shell walnuts. Postharvest Biol. Technol. 22, 29–38.

Wang, S., Monzon, A., Johnson, J.A., Mitcham, E.J., Tang, J., 2007a. Industrial-scale radio frequency treatments for insect control in walnuts I: heating uniformity and energy efficiency. Postharvest Biol. Technol. 45, 240–246.

Wang, S., Monzon, A., Johnson, J.A., Mitcham, E.J., Tang, J., 2007b. Industrial-scale radio frequency treatments for insect control in walnuts II: insect mortality and product quality. Postharvest Biol. Technol. 45, 247–253.

Wang, S., Tang, J., Cavalieri, R.P., Davis, D, 2003. Differential heating of insects in dried nuts and fruits associated with radio frequency and microwave treatments. Trans. ASAE. 46 (4), 1175–1182.

Wang, S., Tang, J., Johnson, J., Mitcham, E., Hansen, J., Cavalieri, R., Bower, J., Biasi, B., 2002. Process protocols based on radio frequency energy to control field and storage pests in in-shell walnuts. Postharvest Biol. Technol. 26, 265–273.

Wang, S., Tiwari, G., Jiao, S., Johnson, J., Tang, J., 2010. Developing postharvest disinfestation treatments for legumes using radio frequency energy. Biosyst. Eng. 105, 341–349.

Wang, S., Yue, J., Chen, B., Tang, J, 2008. Treatment design of radio frequency heating based on insect control and product quality. Postharvest Biol. Technol. 49 (3), 417–423.

Wang, S., Tang, J., Johnson, J.A., Cavalieri, R.P., 2013. Heating uniformity and differential heating of insects in almonds associated with radio frequency energy. J. Stored Prod. Res. 55, 15–20.

Wangspa, W., Chanbang, Y., Vearasilp, S., 2015. Radio frequency heat treatment for controlling rice weevil in rough rice cv. Khao Dawk Mali 105. Chiang Mai Univ. J. Nat. Sci. 14, 189–197.

Wang, Y., Zhang, L., Gao, M., Tang, J., Wang, S., 2014. Evaluating radiofrequency heating uniformity using polyurethane foams. J. Food Eng. 136, 28–33.

Wei, X., Lau, S.K., Stratton, J., Irmak, S., Bianchini, A., Subbiah, J., 2018. Radio-frequency processing for inactivation of *Salmonella enterica* and *Enterococcus faecium* NRRL B-2354 in black peppercorn. J. Food Prot. 81, 1685–1695.

Wei, X., Lau, S.K., Stratton, J., Irmak, S., Subbiah, J., 2019. Radio frequency pasteurization process for inactivation of *Salmonella* spp. and *Enterococcus faecium* NRRL B-2354 on ground black pepper. Food Microbiol. 82, 388–397.

Xu, J., Liu, S., Tang, J., Ozturk, S., Kong, F., Shah, D.H., 2018. Application of freeze-dried *Enterococcus faecium* NRRL B-2354 in radio frequency pasteurization of wheat flour. LWT-Food Sci. Technol. 90, 124–131.

Xu, J., Yang, R., Jin, Y., et al., 2020. Modeling the temperature-dependent microbial reduction of *Enterococcus faecium* NRRL B-2354 in radio-frequency pasteurized wheat flour. Food Control 107, 106778.

Xu, J., Zhang, M., Bhandari, B., Kachele, R., 2017. ZnO nanoparticles combined radio frequency heating: a novel method to control microorganism and improve product quality of prepared carrots. Innovative Food Sci. Emerg. Technol. 44, 46–53.

Yang, C., Zhao, Y., Tang, Y., Yang, R., Yan, W., Zhao, W., 2018. Radio frequency heating as a disinfestation method against *Corcyra cephalonica* and its effect on properties of milled rice. J. Stored Prod. Res. 77, 112–121.

Yang, H., Chen, Q., Cao, H., Fan, D., Huang, J., Zhao, J., Yan, B., Zhou, W., Zhang, W., Zhang, H., 2019. Radiofrequency thawing of frozen minced fish based on the dielectric response mechanism. Innovative Food Sci. Emerg. Technol. 52, 80–88.

Yu, D., Shrestha, B., Baik, O.D., 2017. Thermal death kinetics of adult red flour beetle *Tribolium castaneum* (Herbst) in canola seeds during radio frequency heating. Int. J. Food Prop. 20, 3064–3075.

Zhang, L., Lan, R., Zhang, B., Erdogdu, F., Wang, S., 2021. A comprehensive review on recent development of radio frequency treatment for pasteurizing agricultural products. Crit. Rev. Food Sci. Nutr. 61, 380–394. doi:10.1080/10408398.2020.1733929.

Zhang, L., Lyng, J.G., Xu, R., Zhang, S., Zhou, X., Wang, S., 2019. Influence of radio frequency treatment on in-shell walnut quality and *Staphylococcus aureus* ATCC 25923 survival. Food Control 102, 197–205.

Zhang, M., Jiang, H., Lim, R., 2010. Recent developments in microwave-assisted drying of vegetables, fruits, and aquatic products–drying kinetics and quality considerations. Dry Technol. 28, 1307–1316.

Zheng, A., Zhang, L., Wang, S., 2017. Verification of radio frequency pasteurization treatment for controlling *Aspergillus parasiticus* on corn grains. Int. J. Food Microbiol. 249, 27–34.

Zhong, Q., Sandeep, K.P., Swartzel, K.R., 2003. Continuous flow radio frequency heating of water and carboxymethylcellulose solutions. J. Food Sci. 68, 217–223.

Zhou, H., Guo, C., Wang, S., 2017. Performance comparison between the free running oscillator and 50 Ω radio frequency systems. Innovative Food Sci. Emerg. Technol. 39, 171–178.

Zhou, L., Wang, S., 2016a. Industrial-scale radio frequency treatments to control *Sitophilus oryzae* in rough, brown, and milled rice. J. Stored Prod. Res. 68, 9–18.

Zhou, L., Wang, S., 2016b. Verification of radio frequency heating uniformity and *Sitophilus oryzae* control in rough, brown, and milled rice. J. Stored Prod. Res. 65, 40–47.

Zhu, Y., Li, F., Tang, J., Wang, T.T., Jiao, Y., 2019. Effects of radio frequency, air and water tempering, and different end-point tempering temperatures on pork quality. J. Food Process Eng. 42, e13026.

Recent advances in freezing processes: an overview

10

Piyush Kumar Jha[a,b], Alain Le-Bail[a,b], Soojin Jun[a,b]

[a]ONIRIS, GEPEA CNRS 6144, Nantes, France
[b]Department of Human Nutrition, Food and Animal Sciences, University of Hawaii, Honolulu, HI, United States

10.1 Introduction

Food freezing describes the conversion of water to ice in food, being biological matrices; besides this straightforward definition, the formation of ice crystals results in the increase in the concentration of the aqueous solution contained in the tissue. This highly concentrated solution can be very impacting on the final quality of the frozen foods due to its capability to denature proteins for example, which may affect the biochemistry and the mechanical properties (texture) of the food tissue. What occurs can be compared to a salting process; it is therefore recommended to cross the crystallization zone for two reasons; (a) favoring the apparition of smaller ice crystals that will results in less mechanical damage caused by ice formation and (b) minimize the time of exposure of the tissue to the highly concentrated aqueous solution.

A phase diagram of food, such as the one presented in Fig. 10.1, provides a good overview of the key temperatures to look at for frozen foods. The first ice crystals are expected to appear at point A, corresponding to the initial freezing point temperature (T_{IFP}). In the case of supercooling, a food will be lower than T_{IFP} but without ice formation; this highly unstable state is very challenging to maintain. The temperature interval between 0 °C and T_{IFP} corresponds to the superchilling stage. Between A and B, the ice formation will appear until reaching the maximum cryo-concentration point (B) at temperature T'_m. Below T'_m, no water will reach the ice crystals and at temperature Tg, the concentrated matrice will undergo a glass transition corresponding to state for which the material will become glassy and very fragile. $T'm$ is usually in the range of -30 °C to -40 °C and is therefore lower than the reference frozen storage temperature of -18 °C. As a consequence, most of the "frozen foods" stored at -18 °C (0 °F) are thus not fully frozen, allowing quality change during frozen storage (i.e., ice coarsening, oxidation phenomena, enzymatic reactions, etc.).

An excess of freezing rate may yield mechanical damages caused by the fact that ice contracts at low temperature, whereas water expands during water to ice transition (Le-Bail and Jha, 2019). This differential in terms of thermomechanical expansion may result in severe mechanical damages of the food structure with possible cracks and major disruption as shown by Shi et al. (1999). Therefore, a very high freezing rate can have a highly negative impact on frozen food quality. As a consequence, alternative techniques to control the ice formation and the size of ice crystals should be considered and will be presented in this chapter.

Food Engineering Innovations Across the Food Supply Chain. DOI: https://doi.org/10.1016/B978-0-12-821292-9.00024-8

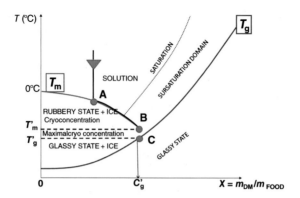

FIG. 10.1 Phase diagram of a food covering the frozen stage.

The blue line is the initial freezing temperature line. The red line is the one of the glass transition. The ice formation starts in A (T_{IFP}) and terminates in B (T_m) with a maximal cryo-concentration. A few Celsius lower, the glass transition of the maximally cryo-concentrated matrix occurs in C (Alain Le-Bail and Jha, 2019).

10.2 Noninvasive innovative freezing methods

10.2.1 Pressure shift freezing and pressure assisted freezing

The phase change temperature of water is depressed under pressure for Ice I until 200 MPa (2000 bar) corresponding to a phase change temperature of around -20 °C (Fig. 10.2; Le-Bail et al., 2002). For higher pressure, the phase change temperature increases. This very specific property is considered for three specific freezing processes;

- Pressure shift freezing (PSF)
- Pressure assisted freezing (PAF)
- High pressure induced crystallization (HPIC)

PAF consists of carrying out the freezing process at a constant and high pressure. This results in a long process with heat transfer driven by natural convection in a high pressure vessel, which yields an expansive process. In the case of PSF, the food product to freeze is placed in a vessel containing a refrigeration fluid at freezing temperature, then the vessel is closed and the pressure is increased above the freezing temperature of the product. The product is then undergoing cooling in the vessel without phase change. Then the pressure is suddenly and quickly released back to atmospheric pressure. During the pressure drop, the sensible heat of the product is converted into ice crystals. During the pressure drop, a shift of the pressure-temperature phase change equilibrium is usually observed, resulting in a high level of supercooling and in a much-refined ice crystals structure. However, the freezing is partial and it is estimated that around 10% of ice can be formed in the best case according to Chevalier et al. (2001). PSF is already a quite long process which requires in addition a final freezing process at atmospheric condition to finalize the full conversion of the partly frozen water to full freezing. In HPIC process, high pressure is applied after conventional freezing to induce solid-solid transition (Ice I/ice III phase transitions). This process has been proposed as a possible low-temperature method for reducing microbial load on foods (James et al., 2015).

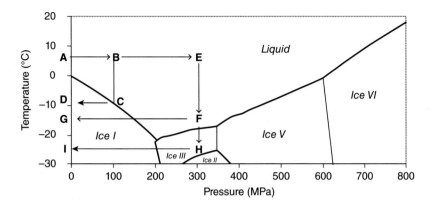

FIG. 10.2

High pressure freezing and thawing processes based on the phase diagram of water (ABCD: pressure-assisted freezing, DCBA: pressure-assisted thawing, ABEFG: pressure shift freezing, GFEBA: pressure-induced thawing, ABEFHI: freezing to ice III, IHFEBA: thawing to ice III).

Multiple research papers are available on this technology, in particular Cheftel et al. (2002), Le-Bail et al. (2002), or Fernández et al. (2006). An important overview on the terminology of the different processes related to high pressure – low temperature processing was proposed by Urrutia Benet et al. (2004). In spite of this technology's performance in terms of refining the size for ice crystals, it did not find interest from the industry. This is because of the cost of equipment and also due to the low temperature that address specific constrains in terms of mechanical resistance of the steel used to construct the high pressure vessel. Also, frosting of fragile mechanical parts does not facilitate the long-term operation of the equipment. However, the size of the ice crystals tend to increase with storage time (ice coarsening effect due to surface tension forces), as shown by Chevalier et al. (2000), so that the initial quality of the frozen foods can only be preserved during low temperature storage. In conclusion, PSF should rather be considered as a reference freezing process which gives access to the finest ice crystals that have been observed in foods so far compared to any other freezing process. This technology is used for example in research with applications in cryo-microscopy for samples preparation for example with TEM (transmission electron microscopy). Pressure assisted freezing (PAF) is another type of application of high pressure. It is more time demanding than PSF, it does not bring major advantages compared to conventional freezing except some possible microbial inactivation due to the combination of high pressure and low temperature and is, therefore, less attractive than PSF.

10.2.2 Static electric and magnetic fields; impact on phase change and freezing

10.2.2.1 Interaction between atoms, molecules, and electric field

An electric field (EF) possesses both the direction as well as the magnitude (or strength) and is a vector quantity. It is measured by the Coulomb force at any point in the space experienced per unit positive test charge at that point. Interactions between EF and atoms or molecules can be classified according to the strengths of electric fields, that is, LOW and HIGH fields. For LOW FIELDS ($E < 0.1$ V/Å or 10^9 V/m) atoms and molecules only get polarized with a change in energy equal to $P.E$ ($P =$ polarizability,

E = electric field strength). For example, considering an ion at position r of charge q, the gain in potential energy will be $q.E(r) \cdot r$. For HIGH FIELDS ($E > 0.1$ V/Å or 10^9 V/m), which are of the order of electrostatic fields inside atoms, will affect the level structure of atoms and thus alter their chemical characteristics making and breaking new bonds in molecules. EF are created by electric charges and can be induced by time-varying magnetic fields. The electric field combines with the magnetic field to form the electromagnetic field. A DC-SEF (direct current-static electric field) can be obtained by applying a DC voltage between two electrodes. The value of the electric field E is given by the ratio between the voltage and the distance between the two electrodes; however, the effective local electric field depends on the dielectric relative permittivity of the matrices located between the electrodes. Computation of the effective EF in a complex system can be done knowing the permittivity's of the different materials in presence. Such modeling has been done in several configuration used in experimental studies with Comsol Multiphysics are as shown in Fig. 10.3 (Havet et al., 2009).

The first scientist to study static electric field (SEF) assisted freezing of supercooled water seems to be Dufour (Dufour, 1861). To date, most of the SEF assisted freezing studies are associated with water and a few on model and real food models. With respect to real food matrix, there exist only three studies, and all of them related to freezing of meat matrix. Based on the experimental results, it seems that SEF can not only influence the freezing parameters but also impact the quality parameters. To be precise, compared with no field conditions, SEF application can promote nucleation at a higher temperature, lessen the induction time, enhance the nucleation and finally prolong the phase transition time. Concerning quality parameters, SEF assisted freezing produces smaller average size ice crystals than compared to no field conditions ones, and thereby offers less freeze damage (Dalvi-isfahan et al., 2017; Jha et al., 2017a; Orlowska et al., 2009). Fig. 10.4 shows the impact of DC-SEF on the size of ice crystals in the case of pork loin (Xanthakis et al., 2013).

Concerning the strength of electric field required to influence the freezing process, at present, it is difficult to mention the exact strength. Nevertheless, based on simulation and experimental studies it seems that electric field between 10^3 and 10^{10} V/m is required manipulate the crystallization process of water. At this point we need to remark that the use of electric field $>10^7$ V/m may lead to a dielectric breakdown of the system, thereby promoting current flow and ohmic heating of the system (Jha et al., 2017a, 2018).

Although this technology has shown promising results in assisting freezing technology and reducing freeze damage of food products but is still limited to processing small samples at a laboratory

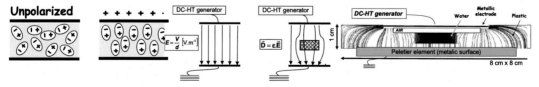

FIG. 10.3

Left: A dielectric medium with polar molecules can have a lower ratio of electric flux to charge (more permittivity) than empty space. Molecules will lock to the field for given DC-SEF. Center: The permittivity of the medium exposed to a DC-SEF must be considered to determine the effective displacement field. Right: Modeling done at ONIRIS-GEPEA for a SEF crystallization prototype; the sample was water as a model food undergoing freezing. DC-HT in the figure refers to direct current-high tension.

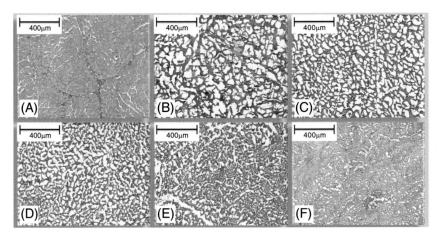

FIG. 10.4 Impact of increasing SEF on the size of ice crystals in pork meat (tenderloin).
(A) Nonfrozen, (B) frozen without SEF, and (C, D, E, F) with increasing SEF of 3, 6, 9, 12 kV, respectively. About 12 kV corresponded to ca. 6×10^6 V/m (Orlowska et al., 2009). The refinement of the ice crystals is clearly visible (Xanthakis et al., 2014).

scale. Tailored optimization of parameters and scaleup is required to make this technology viable at a commercial level. Besides, as this technology involves the use of high voltage, appropriate safety arrangements should be needed to be made to avoid any risk of electric shock.

10.2.3 Possible mechanisms

- *Surface theory*: Pruppacher et al. (1968) and also Fletcher (1970) explained that a semiliquid layer develops around an ice crystal under formation during the freezing of an aqueous solution containing a solute. Within this semiliquid layer, solute ions are moving toward the mother solution, whereas H_3O^+ and OH^- ions are traveling toward the pure ice crystal. A voltage difference of up to 10 V generating a tiny current in the range of 10^{-7} A has been proposed by Fletcher (Fletcher, 1970; Fig. 10.5). Applying an external electric field may interfere with these charge motion. Since SEF is directional, it is likely that crystals growth may be rather favored in selected facets of the crystals under formation.
- *Modification of free energy*: In the presence of a DC-SEF, the expression of the variation of free energy is modified resulting in an increase of the volume term of the variation of free energy Eqs. (10.1) and (10.2) by a term proportional to the E (Marand et al., 1988). The resulting impact on the free energy variation function is shown in Fig. 10.6, yielding in a reduction of the critical radius; smaller crystals are therefore expected.

$$G_n = U - TS + pV - \left[V_c \cdot E \cdot \left(\varepsilon_0 \cdot \varepsilon_r \cdot E + P \right) \right] \tag{10.1}$$

$$\Delta G_n = 4\pi r^2 \gamma - \frac{4}{3}\pi r^3 (\Delta G_v + PE) \tag{10.2}$$

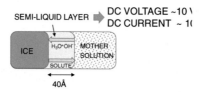

FIG. 10.5

Scheme of the semiliquid layer surrounding ice crystals under formation within an aqueous solution (Fletcher, 1970). A tiny voltage difference is present within this layer and is linked to mass diffusion of chemical species (solute) and ions traveling within this layer. Imposing an external voltage and static charges may affect ice crystallization and as a global approach any crystallization process (Fletcher, 1970).

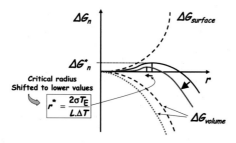

FIG. 10.6

Impact of DC-SEF on crystallization. A reduction of the critical radius is observed thanks to the increase of the volumetric term of the free energy variation function (Eq. (10.2)). The maximum of this function defines the critical radius of ice crystal; in the case of DC-SEF, the critical radius is shifted to lower values resulting in a better stability of smaller ice crystals and also to a higher nucleation rate. This theory has been established by Marand et al. (1988) and applies to freezing but also to any material exposed to DC-SEF (Jha et al., 2017).

where G_n is the free energy for a linear system having permanent polarization P, U is the inner energy (J), T is the temperature (K), p is the pressure (Pa), V is the volume (m^3), V_C is the volume of the system subjected to electrostatic field (m^3), E is the electric field strength (V/m), εr is the relative permittivity of the system, ε_0 is the vacuum permittivity [F/m], P is the permanent polarization (C/m^2), ΔG_n is the free energy of the system with applied electrostatic field (J), r is the radius of the nuclei (m), γ is the surface free energy of the crystal fluid interface (J/m^2), and ΔG_v is the free energy change of the transformation per unit volume [J/m^3].

- *Polarization effect*: SEF aligns the water molecules from random directions to the direction of the electric field vector. This polarization phenomenon makes the hydrogen bond between the water molecules stronger in the direction of the electric field. As a result, water clusters structure can be reordered, which in turn may facilitate in the nucleation process (Shevkunov and Vegiri, 2002; Sun et al., 2006; Vegiri, 2004; Vegiri and Schevkunov, 2001; Wei et al., 2008).
- *Electric double layer formation*: During SEF assisted freezing of water droplets, electric double layer formed at water–dielectric interface and was believed to have enhanced the nucleation rate (Zhang et al., 2016).

Compared to SEF assisted freezing, only a small number of studies have been performed on alternating electric field assisted freezing. These studies are focused mainly on freezing of aqueous solutions and a few on water. Results from these studies reveal that the alternating electric field (AEF) can interfere with the stages of crystallization, that is, supercooling, nucleation of ice crystal in the supercooled water and its subsequent growth. The frequency and intensity of applied AEF determines the effect extent on the crystallization process. For instance, supercooling degree, freezing time, pure ice crystal size, and crystallization fraction of pure ice grains while freezing salty solution (e.g., 0.9% NaCl or K_2MnO_4 water) decreased with an increase in frequency and reached the minimum at a particular frequency. Further increase in the frequency increased the above-mentioned parameters (Ma et al., 2013; Sun and Li, 2010; Sun et al., 2006b, 2006c). On the other hand, with respect to the strength, application of AEF at 1.5×10^6 V/m (60 Hz) elevated the nucleation temperature (reduced supercooling degree) up to 8 °C compared to no electric field condition (−2 °C observed under AEF condition against −10 °C in the case of field-free case; Salt, 1961). In another study, it was observed that AEF (up to $(1.6 \pm 0.4) \times 10^5$ V/m and 3-100 kHz) affected the homogeneous nucleation of ice by less than a factor of 1.5 (Stan et al., 2011). Ma et al. (2013) reported that the grain size and the crystallization fraction of the resulting ice grains while freezing 0.9% NaCl solution decreased with increasing strength of AEF.

To date, no peer-reviewed article is available that is related to freezing of a real food matrix using AEF assisted freezing technology. However, patents claim better food preservations using AEF assisted freezing and/or refrigeration (Kim et al., 2013; Owada, 2007).

The actual mechanism of action of AEF assisted freezing is still not clear, however a few hypotheses have been put forward by the researchers. The torque exerted by an AEF can destabilize the equilibrium state of water clusters, and thus can interfere with both ice nucleation and kinetics of crystal growth (Jackson et al., 1997; Sun et al., 2006a; Woo and Mujumdar, 2010). Another hypothesis states that the vibration and collision induced by the AEF could produce thinner solid–liquid boundaries and decrease heat transfer resistance, similar to acoustic stress (Hu et al., 2013; Mok et al., 2015).

10.2.3.1 Magnetic field assisted freezing; interaction between atoms, molecules, and magnetic field

A magnetic field (MF) is also represented by a vector field. A MF is used for two distinct but closely related fields denoted by the symbols B (in tesla—symbol T) and H (in Ampere/meter or A m^{-1}), where H is measured in units of amperes per meter (symbol: A m^{-1} or A/m) in the SI. B is measured in tesla (symbol: T) and newton per meter per ampere (symbol: N m^{-1} A^{-1} or N/(m A)). B is most commonly defined in terms of the Lorentz force it exerts on moving electric charges. S-MF defines a Static MF stable with time and O-MF defines an oscillating MF. The interaction between MF and molecules or atoms undergoing phase change such as crystallization is still under debate in the literature. The MF on earth is 0.5 Gauss (5×10^{-5} T).

The specific properties of water such as surface tension force and viscosity, optical property (changes in optical feature of water including infrared, Raman, visible, ultraviolet lights, and X-ray spectra), electromagnetic property (refractive index, dielectric constant, and electrical conductivity), thermodynamic property (enthalpy of vaporization), dynamic property (self-diffusion coefficient), and molecular structure of water (hydrogen bond structure) tend to change when subjected to a S-MF (Cai et al., 2009; Chang and Weng, 2006; Deng and Pang, 2007; Hosoda et al., 2004; Martiniano and Galamba, 2013; Pang and Bo, 2008b, 2008a; Toledo et al., 2008; Wang et al., 2013; Zhao et al., 2015; Zhou et al., 2000). However, the exact mechanisms on how S-MF affects water properties are unclear. The conclusions from most of the study indicate that the S-MF application alters the hydrogen-bond network of water, but there exists no agreement on how it is modified. Some of the researchers claim that the S-MF can weaken hydrogen bonds thereby decreases the average number of hydrogen bonds between the water molecules (Wang et al., 2013; Zhou et al., 2000), while other reported that S-MF makes the bonding among the water molecules more robust (Chang and Weng, 2006). The modification of hydrogen bonds structure (either weakening or strengthening) is expected to profoundly affect the properties of water governing the kinetics of processes such as freezing or other related processes. In this context, several studies have shown the impact of magnetic field (both static and oscillating) on the freezing behavior of water and other aqueous solutions. With respect to S-MF assisted freezing, important freezing parameters such as supercooling, freezing temperature and phase transition time were seen affected due to S-MF exposure. In the presence of S-MF, Aleksandrov et al. (2000) reported decrease in the supercooling of water drops, while Zhou et al. (2012) observed an increase in the degree of supercooling of tap water. Inaba et al. (2012) observed that the exposure to S-MFs increased the freezing temperature of both H_2O and D_2O. Meanwhile, Mok et al. (2015) found that the S-MF can affect the phase transition time of 0.9% NaCl solution both positively (reduce the phase transition time) and negatively (increase the phase transition time) depending on the types of S-MF (either attractive or repulsive).

On the other hand, Semikhina and Kiselev (1988) reported that water (bi-distilled water) subjected to a weak O-MF ($B = 0.025\ \mu T$ to $0.88\ mT$ and frequencies between 10^{-2} and $200\ Hz$) for 5 h had a larger supercooling compared to the untreated sample. It was also observed that the supercooling degree was dependent on the strength and frequency of applied O-MF. Moreover, when a particular strength of O-MF was used, the maximum degree of supercooling was obtained at a specific MF frequency. With respect to real food matrix, James et al. (2014) noticed minor additional effect on the degree of supercooling during garlic freezing in the presence of O-MF than compared to similar freezing conditions in the absence of O-MF.

Few of the research works have shown that the MF application can impact the shape and size of the ice crystals. It was observed by Mok et al. (2015) that the ice crystals formed under S-MF influence were more irregular shaped than those obtained without MF. Moreover, the applied external S-MF (attractive or repulsive) governed the pattern of formed ice crystals. For instance, "parting" pattern of ice crystals was obtained under attractive S-MF, while repulsive S-MF yielded a unique pattern of ice crystals. Iwasaka et al. (2011) reported that the freezing of aqueous solution under pulsed-train magnetic fields up to 325 T/s at 6.5 mT produced larger and more uniform ice crystals versus ice crystals obtained without MF.

Some studies such as Watanabe et al. (2011) demonstrated that there was no impact, both from a theoretical and experimental point of view using food systems exposed to freezing with horizontal S-MF 5 to 100 Gauss (G) (10^4 Gauss = 1 Tesla) and also high frequency field O-MF) used in NMR with a vertical high-frequency magnetic field of (1 MHz/1.2 G). Similar to Watanabe et al. (2011), Otero et al. (2018), and Naito et al. (2012) reported no effect of S-MF (attractive S-MF = 107 to 359 mT and repulsive S-MF = 0 to 241 mT) and O-MF (0.5 mT at 30 Hz) on freezing parameters (such as induction time, degrees of supercooling and the phase transition and total freezing times) of saline and distilled/pure water. At this point in time, extensive research is required in this domain to reach a decisive conclusion on the efficacy of MF versus quality of frozen products.

A few patents have emerged on this topic which claims better quality preservation in foods (Owada, 2007; Owada and Kurita, 2001; Owada and Saito, 2010; Sato and Fujita, 2008). The invention from

Owada's group has been commercially sold with the name CAS (cell alive system). A significant number of commercial installations can be found in Japan. A few units are also used in Europe (Spain, France, UK, and other countries). This equipment is made of conventional (compression units) or of cryogenic freezers in which several toroidal coils circulated by AC voltage are installed to obtain an oscillating magnetic field (a priori at 50Hz). The oscillating magnetic field yield an electromagnetic field. Detailed information on patents of magnetic fields assisted freezing can be found in the paper from Jha et al. (2017b).

10.2.4 Microwave (MW) and radio frequency (RF) assisted freezing

Some experimental studies have shown that it is possible to obtain smaller ice crystals by using power electromagnetic waves assisted freezing: Hanyu et al. (1992) reported that MW irradiation (for 50 ms) followed by freezing produced smaller ice crystals in the frozen samples (squid retina, rat liver, and heart muscle) with good repeatability. Chaplin (2021) has proposed that depending on the frequency, water can be affected by different means; for microwaves (2.45 GHz), a heating effect is mainly observed and is also accompanied by an enhancement of water mobility. An enhancement of water mobility is foreseen in the GHz domain, the free water being mainly concerned. In the MHz domain, both free and bound water are concerned whereas in the kHz domain only ice is concerned. Therefore, RF (MHz domain) is a good alternative to MW as RF is less energy dissipating and will be less impacting on the cooling rate.

Focusing on MW assisted freezing, the zone of good freezing extended to a greater depth into the microwaves irradiated sample than compared to the control sample. The zone of freezing can be referred as the area where there is no sign of detectable ice-crystal damage. In other words, MW irradiated sample had a larger ice-free (vitrified) region compared to the untreated sample. Hanyu et al. (1992) have shown a clear reduction of ice crystals size in living tissue (Squid Retina) exposed to freezing under microwave. The explanation proposed is that MW (2.45 GHz) are disrupting water pentamers (five water molecules making ca. 85% of liquid water) in monomers, resulting in a higher supercooling and higher nucleation rate. Jackson et al. (1997) reported that the continuous application of MW (2.45 GHz and 1000 W) during attempted vitrification of ethylene glycol solution (cryo-protectant) caused a significant reduction in ice formation.

More recent studies made by Anese et al. (2012) and Xanthakis et al. (2014) showed a clear reduction of ice crystals size in pork loin exposed to RF waves and MW respectively as shown in Figs. 10.7 and 10.8. Anese et al. (2012) used a pilot-scale RF equipment while liquid nitrogen was flowing in a chamber to provide cryogenic freezing conditions to their samples. Although no data were presented on ice crystal size, the histological micrographs seemed to indicate that the ice crystal size decreased as shown in Fig. 10.7. More recently, studies conducted under "Freezewave Project" have revealed that the MAF (microwaves assisted freezing) process can significantly reduce the texture and drip loss along with reducing the average size of ice crystals in apples and potatoes (Jha et al., 2020).

Two concepts are proposed to explain the observed ice crystal size reduction in frozen systems by using microwaves. The "NITOM" concept (Nucleation Induced by Temperature Oscillation caused by Microwaves) is based on the fact that under specific duty ratio, the temperature of the sample undergoing freezing will oscillate (Xanthakis et al., 2014). The fluctuating temperature is expected to favor

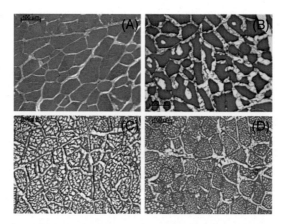

FIG. 10.7 Reduction of the size of ice crystals in pork loin exposed to RF waves (91 MHz) during freezing (Anese et al., 2012).

A (fresh), B (air blast freezing), C (cryogenic freezing), D (cryogenic freezing + RF).

secondary nucleation which in turn suppress the crystal growth. The second concept is called "NIMIW" (Nucleation Induced by constant or pulsed MicroWaves power) and it is based on constant or pulsed (short time) emission. NIMIW can be explained by the impact of MW on the hydrogen bonds between water molecules, which may affect the water cluster structures during freezing. It is expected that MWs could exert a torque and displace the water molecules from their equilibrium relationships in the ice cluster resulting in fragmentation of existing ice crystals when ice crystals are in the form of nuclei. The fragmented ice crystals nuclei may act as new nucleation sites and promote the secondary nucleation, thus, causing ice crystal size reduction (Anese et al., 2012; Cheng et al., 2017; Dalvi-Isfahan et al.,

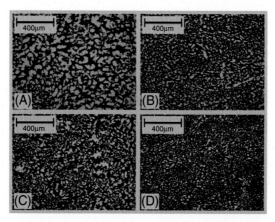

FIG. 10.8 Reduction of the size of ice crystals in pork loin exposed to MW (2.45 GHz) during freezing (Xanthakis et al., 2014).

(A) Static freezing (control condition), (B), (C), (D) increasing MW power irradiation.

2016; Hanyu et al., 1992; Jackson et al., 1997; Xanthakis et al., 2014). Another possibility could be that the electrical disturbances caused by MWs application would alter the mass diffusion in the quasi-liquid layer surrounding the ice crystals under formation (Fletcher, 1970). The delayed growth of ice crystals would, in turn, yield a higher nucleation rate resulting in a refinement of the size of ice crystals in the product (Anese et al., 2012; Dalvi-Isfahan et al., 2016; Hafezparast-Moadab et al., 2018; Hanyu et al., 1992; Jackson et al., 1997). A recent study made on a model gel (methylcellulose) proposed by Sadot et al. (2020b, 2020a) concluded that the NIMIW hypothesis can explain the reduction of the size of ice crystals. These two hypotheses, NITOM and NIMIW may thus be considered to explain the observed reduction of ice crystals size in samples under electromagnetic waves assisted freezing. Further research is needed to better understand the observed reduction of the size of the ice crystals, even though the most recent results tend to confirm that the NIMIW effect should rather be considered as an explanation.

10.3 Ultrasound assisted freezing

Ultrasound assisted freezing is among one of the widely studied novel freezing technologies. This method has not only shown its potential in affecting the freezing of liquid products, but also several solid food products (both model and real food matrix; Tian et al., 2020; Xu et al., 2019; Zhu et al., 2020). There have been evidences that the use of ultrasound during freezing promotes nucleation (both primary and secondary), accelerates the freezing rate, govern the ice crystals size and shape, and ultimately offers lower freeze damage (Tian et al., 2020; Xu et al., 2019; Zhu et al., 2020).

Some of the factors that impact the efficacy of the ultrasound assisted freezing process are: (i) ultrasound power, (ii) exposure time, (iii) use of single frequency versus multiple frequencies, (iv) physical properties of food, (v) pretreatment such as introduction of gas (like CO_2) in the food (mostly liquid foods, also in model solid food) or pre-degassing of the tissues prior to ultrasound freezing (Adhikari et al., 2020; Li and Sun, 2002; Sun and Li, 2003; Tian et al., 2020; Tian et al., 2020; Xu et al., 2019; Zhu et al., 2020).

The exact mechanism on how freezing process is altered by the ultrasound is still unclear. However, all the proposed mechanisms are related to the cavitation effect of ultrasound. Briefly, ultrasound application creates high amounts of cavitation bubbles in the transmissions medium; these cavitation bubbles can promote primary nucleation by acting as crystal nuclei (Chow et al., 2003, 2004, 2005; Zhang et al., 2018). Further, the violent collapse of theses bubbles can release high pressure and temperature at a local level, induce microstreaming and mechanical shocks thereby enhancing the mass transfer and mass and heat transfer efficiency (Kiani et al., 2012; Ye et al., 2019; Zhang et al., 2018). Additionally, microstreaming can break the existing large ice crystals resulting in an overall reduction in the size of ice crystals and also promote secondary nucleation (Chow et al., 2003, 2004, 2005).

Another aspect claimed in most studies is the reduction of the freezing time, which can be difficult to explain as the ultrasound is a powerful wave that tends to heat up the product. On possible explanation could be that the ultrasound provokes an enhanced mass transfer at the surface of the object resulting in an increased superficial rate of heat transfer. The application of this technology to mass production of frozen foods remains a challenge as the exposure of the food products to the wave depends on the view factor between the food and the emitter. Also, a brine should be used to transfer the waves to the products which yield possible cross contamination and mass transfer between the product and the brine. Nevertheless, this technology is used to freeze liquids ("sono-freezing") in particular in the area of biotechnology (Cheng et al., 2015).

10.4 Substances regulating freezing process and final product quality

In this section we will discuss mainly the substances that can impact the freezing process and product quality. In this method, foreign agents like ice-nucleating proteins (INPs), antifreeze proteins (AFPs), natural deep eutectic solvents (NEDs), and other substances (cryoprotectants, pectinmethylesterase (PME), and calcium, etc.) are introduced or brought in contact with the sample to manipulate the freezing process and reduce the freeze damage. With respect to the application of INPs, AFPs and NEDs and their respective mechanism of action, Tian et al. (2020c) has published a very detailed review recently, thus this will not be discussed in detail in this chapter. Briefly, both INPs and AFPs are ice-binding proteins, while NEDs being naturally occurring eutectic solvents, are mixtures of two or more liquid or solid substances. INPs tend to elevate the freezing point and promote nucleation at a higher subzero temperature. Whereas, both AFPs and NEDs suppresses the freezing point. These freezing point regulators (INPs, AFPs, and NEDs) have a great potential to be used in freezing related processes such as food freezing, frozen storage, freeze concentration, freeze drying and cryopreservation. Moreover, freezing point regulators with eco-friendly, green, nontoxic and highly effective characteristics can be sourced from nature, and potentially to satisfy the clean label demand of the consumers (Sutariya and Sunkesula, 2020; Tian et al., 2020).

On other hand, pre-freezing treatments (e.g., incorporation of different substances in the tissue system) can reduce the softening of samples (especially in fruits and vegetables) caused by the freezing-thawing process. Alonso et al. (1995) reported that firmness loss of sweet frozen cherries was less when pretreatment was performed by immersion in 10 and 100 mM $CaCl_2$ solution. The effect was even detected in fruits preserved for 6 months in the frozen state. Suutarinen et al. (2000) found that $CaCl_2$ or sucrose pre-freezing treatments resulted in a firmer product (strawberries) than control sample upon thawing. Van Buggenhout et al. (2006) observed a remarkable texture improvement in rapid and/or cryogenic frozen strawberries when the product was infused with PME and calcium using a vacuum impregnated (VI) process. Xie and Zhao (2003) reported that samples infused with cryoprotectants (high fructose corn syrup and high methoxyl pectin) and minerals (calcium and zinc in the form of calcium gluconal and zinc lactate) using VI process increased the maximum compression force of frozen-thawed Marionberries by 45% to 137% and reduced the drip loss by 28% to 48% depending on the specific VI condition (in terms of infused materials combination). Moreover, they observed that calcium provided additional benefits to the texture quality, and zinc improved color stability of frozen-thawed berries. However, the efficacy of pretreatments depends on product, infused substance and on the specific combination of concentration and temperature of the dipping solution and exposure time (Alonso et al., 1995; Hoeft et al., 1973; Suutarinen et al., 2000; Van Buggenhout et al., 2006).

As a conclusion of this section, it can be said that substances regulating the freezing process require special attention as they are added to the food. A liquid food (i.e., ice cream) is well adapted to these technologies whereas application to solid food requires pre-processing that may extend the overall processing time, which may be detrimental in the case of mass production in the industry.

10.5 Chilling, superchilling, and supercooling

The precise definition of chilling and superchilling is sometimes confused in the literature (Kaale, 2011). This concept was defined and patented by LeDanois (1920). Chilling corresponds to storage of food products in an unfrozen state at negative temperature, that is, between 0°C and the initial freezing temperature. Superchilling involves storage at a temperature below the initial freezing point temperature in a partially frozen state. Finally, supercooling corresponds to storage at a temperature below the initial freezing point in an unfrozen and therefore unstable state (Moore and Molinero, 2011). The implementation of superchilling in industry is still exposed to regulatory constraints.

In the EU regulation on best practice, it is specified that products obtained by a "cold hardening process" as result of partial freezing may be stored up to 48 h in a cold room (in the case of public holidays, this period can reach 72 h, exceptionally 96 h), at the desired target temperature (most often around −10°C). In practice, temperatures between -8°C and -14°C are usual for such objectives. In addition, prolonged storage between the initial freezing point and a temperature of around -12°C (corresponding to the case of "frozen product" in France) or more conventionally -18°C (deep frozen products) is prohibited by the EU Regulation.

10.5.1 Chilling applied to foods

Chilling is the only authorized subzero nonfrozen storage allowed by the EU regulation. It allows a slight reduction of the water activity of foods (for example, the water activity of water or ice is 0.953 at -5 °C or 0.907 at -10 °C according to Bald (1991)) and is, therefore, expected to contribute to a prolonged shelf-life in comparison to a conventional chilled storage usually performed at approximately 4 °C. There is little documentation on this technology, which addresses technical challenges; indeed, keeping the temperature just above the initial freezing temperature is quite difficult from a technical point of view and partial superficial freezing may occur on food surface in the case of negative temperature excursion resulting in possible damage of the food structure and appearance.

10.5.2 Impact of superchilling of food products quality

A recent review from Kaale et al. (2011) showed that superchilling can be a relevant way to mitigate quality losses in food caused by chemical, enzymatic and bacteriological activities. Superchilling allows the suppression of the freezing/thawing stages as the freezing of water is only partial in superchilled food, resulting in energy savings. In practice, the control of the growth of microbial contamination appears to be not as efficient as in the full frozen state. Entering the subzero temperature domain close to the initial freezing temperature does not result in a very significant reduction of the water activity, which is one of the main drivers to reduce microbial growth. More research is, thus, needed to better understand the behavior of spoilage microorganisms in foods undergoing superchilling regime for a more efficient application of superchilling in food applications. Also, quite some references have studied the behavior of the ice crystals during superchilled storage. For example, Kaale and Eikevik (2013) studied the superchilled storage of salmon for 28 days at -1.7 °C and showed that the ice crystal size was stabilized after 1 day of storage. From a technical point of view, the rapid evolution of the ice formation just below the initial freezing point is so that slight temperature oscillations of 1 or 2 °C may result in the formation of a large amount of ice. For example, a food with an initial freezing temperature of -1 °C will have around 50% of the freezable water frozen at -2 °C. Temperature oscillations in a frozen storage are often in the range of a few Celsius, which may yield oscillation of the amount of frozen water at the surface of foods in superchilled regime.

10.5.3 Alternative supercooling technology supported by external magnetic and electric fields

When the internal temperature of a food is below its equilibrium freezing point before ice nucleation has occurred said to be in the supercooled state. The process of supercooling foods is advantageous over superchilling as, unlike superchilling, no ice is present, thus, the food maintains its fresh textural integrity. Though, little is known on how to maintain foods in their supercooled state throughout their storage. The supercooled state is very unstable and stochastic in nature, therefore strict temperature control has been the most utilized technique. Even with strict temperature control and little temperature fluctuations during storage, ice crystallization can occur. Furthermore, any physical vibration can stimulate the onset of ice nucleation. The difficult nature of maintaining the supercooled state of a material is why more research has been reported on superchilling (partial freezing) than supercooling.

Recently, Jun et al. (2018) claimed that a combination treatment of pulsed electric fields (PEF) and oscillating magnetic fields (OMF) has the potential to vibrate water molecules, thus suppressing ice nucleation and maintaining a supercooled state within foods. For persistent water dynamics even in a freezing condition, PEF and OMF are applied within a chamber at a specific power and sequence to inhibit ice crystal nucleation effectively. Water in food is susceptible to be magnetized by a magnetic field since it is a diamagnetic material. In addition, water consists of dipole molecules, indicating that water molecules tend to realign and re-orientate under an applied electric field. Therefore, electric and magnetic fields directly act upon water to prevent ice nucleation and promote supercooling during the freezing process. Furthermore, the combined intermittent emission of PEF and OMF would allow extensive supercooling. Unlike the other supercooling methods which are based on the precision temperature control, the developed supercooling temperature zone is farther below the freezing point (-0.5 to -1.5°C), thus providing better control of food quality and safety. The developed invention has been successfully applied to preserve fish at subzero temperatures without ice crystal formation (Shafel, 2015). Mok et al. (2017) and You et al. (2020a) experimentally demonstrated that a combination of PEF (square-wave pulses) and OMF inhibited sudden ice nucleation and effectively extended the supercooled state within chicken breast fillets and beef steaks, respectively. Fruits (honeydew and pineapple) have also been preserved through supercooling (Her et al., 2019; Kang et al., 2019). Their findings include experimental data showing preservation for up to 4 weeks while the quality of supercooled fruits was kept as fresh as raw materials.

Up to date, there is a tremendous need for the development of a scaled-up system for supercooling of multiple, larger food samples since the prior effective treatment area was often limited to small volumes, which is technically impractical for consumers/users (You et al., 2020b).

10.6 Conclusions

This chapter provides an overview on selected freezing technologies based on invasive technologies using ingredients or means installed in the food system to control the ice nucleation or noninvasive technics based on external means to control ice nucleation. The main objective of the freezing process is to extend the shelf life of foods; however, mitigating the damage caused by ice crystals formation remains a challenge. Means to reduce the size of ice crystals has been the main target of the innovative freezing processes recently developed. More recently, a "hybrid" approach of keeping the foods at low temperature unfrozen (chilling or supercooling) or partly frozen (superchilling) have been considered and seem to be relevant alternatives to full freezing. Research effort is still needed to optimize these processes. Also, some regulatory limitations may occur for the case of supercooling and superchilling, since many regulations on foods worldwide do not permit food materials to be stored at the temperature domain between the initial freezing point and -18°C.

References

Adhikari, B.M., Tung, V.P., Truong, T., Bansal, N., Bhandari, B., 2020. Impact of in-situ CO_2 nano-bubbles generation on freezing parameters of selected liquid foods. Food Biophys. 15 (1), 97–112.

Aleksandrov, V.D., Barannikov, A.A., Dobritsa, N.V, 2000. Effect of magnetic field on the supercooling of water drops. Inorg. Mater. 36 (9), 1072–1075.

Alonso, Jesus, Rodriguez, Teresa, Canet, Wenceslao, 1995. Effect of calcium pretreatments on the texture of frozen cherries. Role of Pectinesterase in the changes in the pectic materials. J. Agric. Food Chem. 43 (4), 1011–1016.

Anese, Monica, Manzocco, Lara, Panozzo, Agnese, Beraldo, Paola, Foschia, Martina, Nicoli, Maria Cristina, 2012. Effect of radiofrequency assisted freezing on meat microstructure and quality. Food Res. Int. 46 (1), 50–54.

Bald, W.B., 1991. Food Freezing: Today and Tomorrow, first ed. Springer-Verlag, London.

Cai, Ran, Yang, Hongwei, He, Jinsong, Zhu, Wanpeng, 2009. The effects of magnetic fields on water molecular hydrogen bonds. J. Mol. Struct. 938 (1–3), 15–19.

Chang, Kai-tai, Weng, Cheng-i, 2006. The effect of an external magnetic field on the structure of liquid water using molecular dynamics simulation. J. Appl. Phys. 100 (4), 43917–43922.

Chaplin, M., 2021. Water structure and behavior. Retrieved on 16 May 2021 from https://sites.science.oregonstate.edu/~hetheriw/astro/rt/info/water/water_dielectric_function_and_microwave_radiation.html#top.

Cheftel, J.C., Thiebaud, M., Dumay, E., 2002. Pressure-assisted freezing and thawing of foods: a review of recent studies. High Pressure Res. 22 (3–4), 601–611.

Cheng, Lina, Sun, Da Wen, Zhu, Zhiwei, Zhang, Zi, 2017. Emerging techniques for assisting and accelerating food freezing processes—a review of recent research progresses. Crit. Rev. Food Sci. Nutr. 57 (4), 769–781.

Cheng, Xinfeng, Zhang, Min, Xu, Baoguo, Adhikari, Benu, Sun, Jincai, 2015. The principles of ultrasound and its application in freezing related processes of food materials: a review. Ultrason. Sonochem. 27, 576–585.

Chevalier, D., Bail, A.Le, Ghoul, M., 2001. Evaluation of the ice ratio formed during quasi-adiabatic pressure shift freezing. High Pressure Res. 21 (5), 227–235.

Chevalier, Dominique, Sequeira-Munoz, Amaral, Bail, Alain Le, Simpson, Benjamin K., Ghoul, Mohamed, 2000. Effect of freezing conditions and storage on ice crystal and drip volume in turbot (*Scophthalmus maximus*) evaluation of pressure shift freezing vs. air-blast freezing. Innov. Food Sci. Emerg. Technol. 1 (3), 193–201.

Chow, R., Blindt, R., Chivers, R., Povey, M., 2005. A study on the primary and secondary nucleation of ice by power ultrasound. Ultrasonics 43 (4), 227–230.

Chow, Rachel, Blindt, Renoo, Chivers, Robert, Povey, Malcolm, 2003. The sonocrystallisation of ice in sucrose solutions: primary and secondary nucleation. Ultrasonics 41 (8), 595–604.

Chow, Rachel, Blindt, Renoo, Kamp, Arnold, Grocutt, Peter, Chivers, Robert, 2004. The microscopic visualisation of the sonocrystallisation of ice using a novel ultrasonic cold stage. Ultrason. Sonochem. 11 (3–4), 245–250.

Dalvi-Isfahan, M., Hamdami, N., Le-Bail, A., 2016. Effect of freezing under electrostatic field on the quality of lamb meat. Innov. Food Sci. Emerg. Technol. 37, 68–73.

Dalvi-Isfahan, M., Hamdami, Nasser, Xanthakis, Epameinondas, Le-Bail, Alain, 2016. Review on the control of ice nucleation by ultrasound waves, electric and magnetic fields. J. Food Eng. 195, 222–234.

Dalvi-isfahan, Mohsen, Hamdami, Nasser, Le-bail, Alain, 2017. Effect of freezing under electrostatic field on selected properties of an agar gel. Innov. Food Sci. Emerg. Technol. 42, 151–156.

Deng, Bo, Pang, XiaoFeng, 2007. Variations of optic properties of water under action of static magnetic field. Chin. Sci. Bull. 52 (23), 3179–3182.

Dufour, L., 1861. Ueber Das Gefrieren Des Wassers Und Über Die Bildung Des Hagels. Ann. Der Phys. 190 (12), 530–554.

Fernández, P.P., Otero, L., Guignon, B., Sanz, P.D., 2006. High-pressure shift freezing versus high-pressure assisted freezing: effects on the microstructure of a food model. Food Hydrocolloids 20 (4), 510–522.

Fletcher, N.H. 1970. *The Chemical Physics of Ice*. A. Herzenberg, M. M. Woolfson, and J.M. Ziman (Eds.). Cambridge: Cambridge University Press.

Hafezparast-Moadab, N., Hamdami, N., Dalvi-Isfahan, M., Farahnaky, A., 2018. Effects of radiofrequency-assisted freezing on microstructure and quality of rainbow trout (*Oncorhynchus mykiss*) fillet. Innov. Food Sci. Emerg. Technol. 47, 81–87.

Hanyu, Y., Ichikawa, M., Matsumoto, G., 1992. An improved cryofixation method – cryoquenching of small tissue blocks during microwave irradiation. J. Microsc.-Oxford 165 (Part 2), 255–271.

Havet, Michel, Orlowska, Marta, Le-Bail, Alain, 2009. Effects of an electrostatic field on ice nucleation, International Conference on Bio and Food Electrotechnologies. Compiègne, France.

Her, J.Y., Kang, T., Hoptowit, R., Wall, M.M., Jun, S., 2019. Oscillating magnetic field (OMF) based supercooling preservation of fresh-cut honeydew melon. Trans. ASABE 62 (3), 779–785.

Hoeft, R., Bates, R.P., Ahmed, E.M., 1973. Cryogenic freezing of tomato slices. J. Food Sci. 38, 362.

Hosoda, Haruki, Mori, Haruki, Sogoshi, Norihito, Nagasawa, Akira, Nakabayashi, Seiichiro, 2004. Refractive indices of water and aqueous electrolyte solutions under high magnetic fields. J. Phys. Chem. A 108 (9), 1461–1464.

Hu, Song Qing, Liu, Guang, Li, Lin, Li, Zhi Xin, Hou, Yi, 2013. An improvement in the immersion freezing process for frozen dough via ultrasound irradiation. J. Food Eng. 114 (1), 22–28.

Inaba, Hideaki, Saitou, Tetsuya, Tozaki, Ken-ichi, Hayashi, Hideko, 2012. Effect of the magnetic field on the melting transition of H_2O and D_2O measured by a high resolution and supersensitive differential scanning calorimeter. J. Appl. Phys. 96 (11), 6127–6132.

Iwasaka, M., Onishi, M., Kurita, S., Owada, N., 2011. Effects of pulsed magnetic fields on the light scattering property of the freezing process of aqueous solutions. J. Appl. Phys. 109 (7), 1–4.

Jackson, T.H., Ungan, A., Critser, J.K., Gao, D., 1997. Novel microwave technology for cryopreservation of biomaterials by suppression of apparent ice formation. Cryobiology 34 (4), 363–372.

James, Christian, Purnell, Graham, James, Stephen J., 2015. A review of novel and innovative food freezing technologies. Food Bioprocess Technol. 8 (8), 1616–1634.

James, Christian, Reitz, Baptiste, James, Stephen J., 2014. The freezing characteristics of garlic bulbs (*Allium sativum* L.) frozen conventionally or with the assistance of an oscillating weak magnetic field. Food Bioprocess Technol. 8 (3), 702–708.

Jha, P.K., Sadot, M., Vino, S.A., Jury, V., Curet-Ploquin, S., Rouaud, O., Havet, M., Le-Bail, A., 2017a. A review on effect of DC voltage on crystallization process in food systems. Innov. Food Sci. Emerg. Technol. 42, 204–219.

Jha, P.K., Xanthakis, E., Jury, V., Havet, M., Le-Bail, A., 2018. Advances of electro-freezing in food processing. Curr. Opin. Food Sci. 23, 85–89.

Jha, P.K., Xanthakis, E., Jury, V., Le-Bail, A., 2017b. An overview on magnetic field and electric field interactions with ice crystallisation; application in the case of frozen food. Crystals 7 (10), 299.

Jha, Piyush Kumar, Chevallier, Sylvie, Xanthakis, Epameinondas, Jury, Vanessa, Le-Bail, Alain, 2020. Effect of innovative microwave assisted freezing (MAF) on the quality attributes of apples and potatoes. Food Chem. 309, 125594.

Jun, S., J. H. Mok, and S. H. Park. 2018. "Method of Supercooling Perishable Materials." U.S. Patent No. 10,111,452.

Kaale, Lilian Daniel, Eikevik, Trygve Magne, 2013. A histological study of the microstructure sizes of the red and white muscles of Atlantic salmon (*Salmo salar*) fillets during superchilling process and storage. J. Food Eng. 114 (2), 242–248.

Kaale, Lilian Daniel, Eikevik, Trygve Magne, Rustad, Turid, Kolsaker, Kjell, 2011. Superchilling of food: a review. J. Food Eng. 107 (2), 141–146.

Kang, T., Her, J.Y., Hoptowit, R., Wall, M.M., Jun, S., 2019. Investigation of the effect of oscillating magnetic field on fresh-cut pineapple and agar gel as a model food during supercooling preservation. Trans. ASABE 62 (5), 1155–1161.

Kiani, Hossein, Sun, Da Wen, Zhang, Zhihang, 2012. The effect of ultrasound irradiation on the convective heat transfer rate during immersion cooling of a stationary sphere. Ultrason. Sonochem. 19 (6), 1238–1245.

Kim, Su Cheong, Jong Min Shin, Su-Won Lee, Cheol Hwan Kim, Yong Chol Kwon, and Ku Young Son. 2013. "Non-freezing refrigerator." US008616008B2.

Le-Bail, A., Chevalier, D., Mussa, D.M., Ghoul, M., 2002. High pressure freezing and thawing of foods: a review. Int. J. Refrigeration 25 (5), 504–513.

Le-Bail, Alain, Jha, P.K., 2019. Application—freezing of foodstuffs. In: Cachon, R., Girardon, P., Voilley, A. (Eds.), Gases in Agro-Food Processes. Academic Press, ELSEVIER, Amestradam, pp. 241–252.

LeDanois. 1920. "Nouvelle Mèthode de Frigorification Du Poisson." FR506296A.

Li, B., Sun, D.W., 2002. Effect of power ultrasound on freezing rate during immersion freezing of potatoes. J. Food Eng. 55 (3), 277–282.

Ma, Yahong, Zhong, Lisheng, Gao, Jinghui, Liu, Lin, Hu, Huiyu, Yu, Qinxue, 2013. Manipulating ice crystallization of 0.9 wt.% NaCl aqueous solution by alternating current electric field. Appl. Phys. Lett. 102 (18), 183701.

Marand, H.L., Stein, R.S., Stack, G.M., 1988. Isothermal crystallization of poly(vinylidene fluoride) in the presence of high static electric fields. I. Primary nucleation phenomenon. J. Polym. Sci. Part B: Polym. Phys. 26 (7), 1361–1383.

Martiniano, H.F.M.C., Galamba, N., 2013. Insights on hydrogen-bond lifetimes in liquid and supercooled water. J. Phys. Chem. B 117 (50), 16188–16195.

Mok, J.H., Choi, W., Park, S.H., Lee, S.H., Jun, S., 2015. Emerging pulsed electric field (PEF) and static magnetic field (SMF) combination technology for food freezing. Int. J. Refrigeration 50, 137–145.

Mok, Jin Hong, Her, Jae Young, Kang, Taiyoung, Hoptowit, Raymond, Jun, Soojin, 2017. Effects of pulsed electric field (PEF) and oscillating magnetic field (OMF) combination technology on the extension of supercooling for chicken breasts. J. Food Eng. 196, 27–35.

Moore, E.B., Molinero, Valeria, 2011. Structural transformation in supercooled water controls the crystallization rate of ice. Nature 479 (7374), 506–508.

Naito, Munekazu, Hirai, Shuichi, Mihara, Makoto, Terayama, Hayato, Hatayama, Naoyuki, Hayashi, Shogo, Matsushita, Masayuki, Itoh, Masahiro, 2012. Effect of a magnetic field on *Drosophila* under supercooled conditions. PLoS ONE 7 (12), 1–4.

Orlowska, Marta, Havet, Michel, Le-Bail, Alain, 2009. Controlled ice nucleation under high voltage DC electrostatic field conditions. Food Res. Int. 42 (7), 879–884.

Otero, Laura, Rodríguez, Antonio C., Sanz, Pedro D., 2018. Effects of static magnetic fields on supercooling and freezing kinetics of pure water and 0.9% NaCl solutions. J. Food Eng. 217, 34–42.

Owada, Norio. 2007. "Highly-Efficient Freezing Apparatus and Highly-Efficient Freezing Method." US 7,237,400 B2.

Owada, Norio and Satoru Kurita. 2001. "Super-Quick Freezing Method and Apparatus Therefor." US 6,250,087 B1.

Owada, Norio and Shobu Saito. 2010. "Quick Freezing Apparatus and Quick Freezing Method." US 7,810,340 B2.

Pang, Xiao Feng, Bo, Deng, 2008a. Investigation of changes in properties of water under the action of a magnetic field. Sci. China Ser. G: Phys. Mech. Astronom. 51 (11), 1621–1632.

Pang, Xiao Feng, Bo, Deng, 2008b. The changes of macroscopic features and microscopic structures of water under influence of magnetic field. Phys. B: Condensed Matter 403 (19), 3571–3577.

Pruppacher, H.R., Steinberger, E.H., Wang, T.L., 1968. On the electrical effects that accompany the spontaneous growth of ice in supercooled aqueous solutions. J. Geophys. Res. 73 (2), 571–584.

Sadot, Mathieu, Curet, Sébastien, Chevallier, Sylvie, Le-Bail, Alain, Rouaud, Olivier, Havet, Michel, 2020a. Microwave assisted freezing part 2: impact of microwave energy and duty cycle on ice crystal size distribution. Innov. Food Sci. Emerg. Technol. 62, 102359 October 2019.

Sadot, Mathieu, Curet, Sébastien, Le-Bail, Alain, Rouaud, Olivier, Havet, Michel, 2020b. Microwave assisted freezing part 1: experimental investigation and numerical modeling. Innov. Food Sci. Emerg. Technol. 62, 102360 October 2019.

Salt, R.W., 1961. Effect of electrostatic field on freezing of supercooled water and insects. Science 133, 458–459.

Sato, Motohiko and Kazuhiko Fujita. 2008. "Freezer, Freezing Method and Frozen Objects." US 7,418,823 B2.

Semikhina, L.P., Kiselev, V.F., 1988. Effect of weak magnetic fields on the properties of water and ice. Sov. Phys. J. 31 (5), 351–354.

Shafel, Timothy, 2015. Supercooling of Perishable Foods for Extended Shelf Life: An Investigation of Quality. University of Hawaii at Manoa, Amestradam.

Shevkunov, Sergei V., Vegiri, Alice, 2002. Electric field induced transitions in water clusters. J. Mol. Struct.: Theochem. 593 (1–3), 19–32.

Shi, X., Datta, A.K., Mukherjee, S., 1999. Thermal fracture in a biomaterial during rapid freezing. J. Therm. Stresses 22 (3), 275–292.

Stan, Claudiu A., Tang, Sindy K.Y., Bishop, Kyle J.M., Whitesides, George M., 2011. Externally applied electric fields up to 1.6*105 V/m do not affect the homogeneous nucleation of ice in supercooled water. J. Phys. Chem. B 115 (5), 1089–1097.

Sun, Da Wen, Li, Bing, 2003. Microstructural change of potato tissues frozen by ultrasound-assisted immersion freezing. J. Food Eng. 57 (4), 337–345.

Sun, Wei, Chen, Zhong, Huang, Su-yi, 2006a. Effect of an external electric field on liquid water using molecular dynamics simulation with a flexible potential. J. Shanghai Univ. (English Ed.) 10 (3), 268–273.

Sun, Wei, Li, Xiao Hui, 2010. Study on ice construction of normal saline under alternated electric field by dielectric method, Proceedings - 2010 3rd International Congress on Image and Signal Processing, CISP 2010. Yantai, China, 6. IEEE, pp. 2933–2936.

Sun, Wei, Xu, Xiaobin, Sun, W, Ying, Lu, Xu, Chuanxiang, 2006. Effect of alternated electric field on the ice formation during freezing process of 0.9% K_2MnO_4 water, Eighth International Conference on Properties and Applications of Dielectric Materials. Bali. IEEE, pp. 774–777.

Sun, Wei, Xu, Xiaobin, Zhang, Hong, Sun, W., Xu, Chuanxiang, 2006. The mechanism analysis of NaCI solution ice formation suppressed by electric field, Eighth International Conference on Properties and Applications of Dielectric Materials. Bali. IEEE, pp. 770–773.

Sutariya, Suresh G., Sunkesula, Venkateswarlu, 2020. Food Freezing: Emerging Techniques for Improving Quality and Process Efficiency a Comprehensive Review. Elsevier, Amestradam.

Suutarinen, J., Heiska, K., Moss, P., Autio, K., 2000. The effects of calcium chloride and sucrose prefreezing treatments on the structure of strawberry tissues. LWT - Food Sci. Technol. 33 (2), 89–102.

Tian, You, Chen, Zhubing, Zhu, Zhiwei, Sun, Da Wen, 2020. Effects of tissue pre-degassing followed by ultrasound-assisted freezing on freezing efficiency and quality attributes of radishes. Ultrason. Sonochem. 67 (April), 105162.

Tian, You, Zhang, Peizhi, Zhu, Zhiwei, Sun, Da Wen, 2020. Development of a single/dual-frequency orthogonal ultrasound-assisted rapid freezing technique and its effects on quality attributes of frozen potatoes. J. Food Eng. 286 (April), 110112.

Tian, You, Zhu, Zhiwei, Sun, Da Wen, 2020. Naturally sourced biosubstances for regulating freezing points in food researches: fundamentals, current applications and future trends. Trends Food Sci. Technol. 95, 131–140 May 2019.

Toledo, Evelyn J.L., Ramalho, Teodorico C., Magriotis, Zuy M., 2008. Influence of magnetic field on physical – chemical properties of the liquid water : insights from experimental and theoretical models. J. Mol. Struct. 888, 409–415.

Urrutia Benet, Gabriel, Schlüter, O., Knorr, D., 2004. High pressure-low temperature processing. Suggested definitions and terminology. Innov. Food Sci. Emerg. Technol. 5 (4), 413–427.

Van Buggenhout, S., Lille, M., Messagie, I., Von Loey, A., Autio, K., Hendrickx, M., 2006. Impact of pretreatment and freezing conditions on the microstructure of frozen carrots: quantification and relation to texture loss. Eur. Food Res. Technol. 222, 543–553.

Vegiri, Alice., 2004. Reorientational relaxation and rotational-translational coupling in water clusters in a d.c. external electric field. J. Mol. Liq. 110 (1), 155–168.

Vegiri, Alice, Schevkunov, Sergei V., 2001. A molecular dynamics study of structural transitions in small water clusters in the presence of an external electric field. J. Chem. Phys. 115 (9), 4175–4185.

Wang, Yongfu, Zhang, Bin, Gong, Zhenbin, Gao, Kaixiong, Ou, Yujing, Zhang, Junyan, 2013. The effect of a static magnetic field on the hydrogen bonding in water using frictional experiments. J. Mol. Struct. 1052, 102–104.

Watanabe, M., Kanesaka, N., Masuda, K., Suzuki, T., 2011. Effect of oscillating magnetic field on supercooling in food freezing, 23rd International Congress of Refrigeration; refrigeration for sustainable development. Prague. IIR.

Wei, Sun, Xiaobin, Xu, Hong, Zhang, Chuanxiang, Xu, 2008. Effects of dipole polarization of water molecules on ice formation under an electrostatic field. Cryobiology 56 (1), 93–99.

Woo, M.W., Mujumdar, A.S., 2010. Effects of electric and magnetic field on freezing and possible relevance in freeze drying. Drying Technol. 28 (4), 433–443.

Xanthakis, E., Havet, M., Chevallier, S., Abadie, J., Le-Bail, A., 2013. Effect of static electric field on ice crystal size reduction during freezing of pork meat. Innov. Food Sci. Emerg. Technol. 20, 115–120.

Xanthakis, E., Le-Bail, A., Ramaswamy, H., 2014. Development of an innovative microwave assisted food freezing process. Innov. Food Sci. Emerg. Technol. 26, 176–181.

Xie, J., Zhao, Y., 2003. Improvement of physicochemical and nutritional qualities of frozen marionberry by vacuum impregnation pretreatment with cryoprotectants and minerals. J. Horticult. Sci. Biotechnol. 78 (2), 248–253.

Xu, Baoguo, Azam, Roknul S.M., Wang, Bo, Zhang, Min, Bhandari, Bhesh, 2019. Effect of infused CO_2 in a model solid food on the ice nucleation during ultrasound-assisted immersion freezing. Int. J. Refrigeration 108, 53–59.

Ye, Linzheng, Zhu, Xijing, Liu, Yao, 2019. Numerical study on dual-frequency ultrasonic enhancing cavitation effect based on bubble dynamic evolution. Ultrason. Sonochem. 59 (May), 104744.

You, Youngsang, Her, Jae-young, Shafel, Timothy, Kang, Taiyoung, Soojin, Jun, 2020a. Supercooling preservation on quality of beef steak. J. Food Eng. 274, 109840.

You, Youngsang, Kang, Taiyoung, Jun, Soojin, 2020b. Control of ice nucleation for subzero food preservation. Food Eng. Rev. 13, 15–35 October 2019.

Zhang, Mingcheng, Haili, Niu, Chen, Qian, Xia, Xiufang, Kong, Baohua, 2018. Influence of ultrasound-assisted immersion freezing on the freezing rate and quality of porcine longissimus muscles. Meat Sci. 136, 1–8 October 2017.

Zhang, Peizhi, Zhu, Zhiwei, Sun, Da Wen, 2018. Using power ultrasound to accelerate food freezing processes: effects on freezing efficiency and food microstructure. Crit. Rev. Food Sci. Nutr. 58 (16), 2842–2853.

Zhang, Xiang-xiong, Li, Xin-hao, Chen, Min, 2016. Role of the electric double layer in the ice nucleation of water droplets under an electric field. Atmos. Res. 178–179, 150–154.

Zhao, Lin, Ma, Kai, Yang, Zi, 2015. Changes of water hydrogen bond network with different externalities. Int. J. Mol. Sci. 16, 8454–8489.

Zhou, K.X., Lu, G.W., Zhou, Q.C., Song, J.H., Jiang, S.T., Xia, H.R., 2000. Monte Carlo simulation of liquid water in a magnetic field. J. Appl. Phys. 88 (4), 1802–1805.

Zhou, Zipeng, Zhao, Hongxia, Han, Jitian, 2012. Supercooling and crystallization of water under DC magnetic fields. CIESC J. 63 (5), 1408–1410.

Zhu, Zhiwei, Zhang, Peizhi, Sun, Da Wen, 2020. Effects of multi-frequency ultrasound on freezing rates and quality attributes of potatoes. Ultrason. Sonochem. 60, 104733 August 2019.

Cooling of milk on dairy farms: an application of a novel ice encapsulated storage system in New Zealand

11

Refat Al-Shannaq, Amar Auckaili, Mohammed Farid

Department of Chemical and Materials Engineering, The University of Auckland, Auckland, New Zealand

11.1 Introduction

Dairy farms are an important part of the economy in New Zealand. The Milk is not only consumed here but also exported in the form of dairy products. As Milk quality depends heavily on its temperature, the milk cooling process is of utmost important. The energy used in the milk cooling process contributes a significant part of the total energy use (30% of the daily energy use at a dairy farm). The milking of cows usually takes place in the mornings. This milk needs to be cooled and stored until collected as per the regulations set up by the government. Failure to adhere to these regulations would result in the milk not being collected.

Design efficient and reliable cold energy storage system is an important requirement in many fields such as air-conditioning, food processing, high power electronic cooling and other industrial applications (Oró et al., 2012; Lu and Tassou, 2013; Farid et al., 2004). Phase change materials (PCMs) are organic or inorganic compounds, which melt and solidify with a melting range suitable for the specific applications. If the requirement is to cool close to 0 °C, then water is the best PCM due to its high energy storage capacity (330 J/g) during its phase transition from ice to water or via versa. When snap chilling and storing milk in vats, large refrigeration plants will be required that consume heavy electrical loads for rapid chilling of the milk within the vats. However, storing cold energy in the forms of ice could be a solution to reduce peak energy demand, by shifting most of the load consumed by the chiller from peak to off-peak, which could lead to significant energy and chiller equipment cost saving.

In this chapter, the application of thermal ice storage for cooling of milk is discussed. As this application enables the farm to take advantage of the low-cost night tariff, a comparison between day and night tariffs is given (Barzin et al., 2015).

11.2 Background

In this section the background to the following topics are explained:

- New Zealand (NZ) milk cooling regulations
- Electricity tariffs

Food Engineering Innovations Across the Food Supply Chain. DOI: https://doi.org/10.1016/B978-0-12-821292-9.00006-6

- Dairy farm milk cooling operations
- Thermal energy storage

11.2.1 NZ milk cooling regulations

The NZ Ministry of Primary Industries (MPI) has put together a regulation in place for the cooling of milk. Also known as the Dairy-NZCP1[1], which clearly states the standard practice each dairy farm must follow in order to produce the highest milk quality. Cooling of milk on-farm was first introduced way back in 1955. Further, the existing regulation and the proposed new regulation are summarized in the following sections.

11.2.1.1 NZCP1 Version 5 amendment 2 (old milk cooling standards)

This version of the NZCP1 regulation was signed on July 2013 (Ministry of Primary Industries, 2013). This regulation is a comprehensive detailed document that describes the design and operation of a dairy farm.

The stated code of NZCP1 regulation defines the time/temperature requirements for milk cooling on farms. The following requirements are set standards for the dairy farms in New Zealand:

- *Primary cooling*: After milking and filtering takes place, the milk must be cooled to below 18 °C. Normally, cooling water (bore water) is used with volumetric flow rate of 2.5 times higher than the milk. In case there is an extended time required for milking, a prechilling system should be installed. This will not let the milk stay at high temperatures.
- *Subsequent milking*: If subsequent milking takes place, the temperature of the milk entering the vat must not exceed 13 °C. Most of the farms today practice two milking cycles per day.
- *Post milking*: The milk in the vat should be cooled to below 7 °C within 3 h of the completion of milking.
- *Until collection*: Up to the time when the milk is collected from the farm the milk must be kept below 7 °C.
- *Robotic milking systems*: If the farms have an installed automatic milking system, the milk must be immediately chilled to and kept below 7 °C until collection.

The code also provides a graphical representation for the time/temperature requirements. The same is shown in Fig. 11.1.

11.2.1.2 NZCP1 2017 (new milk cooling standards)

The New Zealand MPI new milk cooling requirements (NZCP1 2017) shall be applicable to all dairy farms by 1st of January 2018. The new regulations have the following requirement for the cooling of milk (Ministry of Primary Industries, 2017):

- Raw milk must be cooled to less than 10 °C within 4 h from the time milking was started.
- The milk must be cooled to 6 °C or below within 6 h from the commencement of milking and within 2 h of the completion of milking.
- The milk must be stored at less than 6 °C until collection or the next milking.
- The temperature of milk in the vat must not exceed 10 °C during subsequent milking.
- In dairy farms that have automatic milking systems installed, the milk must enter the vat at 6 °C or below.

---

[1] Code of Practice for the Design and Operation of Farm Dairies to form part of a Risk Management Programme.

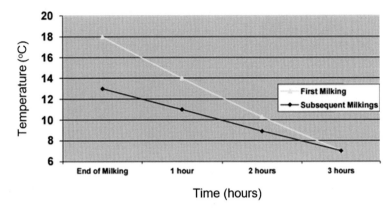

FIG. 11.1 Time/temperature targets for first and subsequent milking (Ministry of Primary Industries, 2013).

11.2.2 Electricity Tariffs

In New Zealand, electricity retailers are providing an option for choosing a time of use, anytime/ night tariffs. The price that the consumer pays is not dependent on the demand response pricing at the electricity market. The annual electricity bill of an average farm in year of 2018 is approximately $17,500–$20,000 (Miller, 2013). Utilizing the night tariff rates can prove to be economical than day-time tariff rates. The different types of tariffs that retailers provide are as follows:

- *Anytime tariff*: This tariff is normally used everywhere. There is a fixed charge per energy unit all day. The use of electricity is measured with a single meter.
- *Day/night tariff*: The use of energy here is measured with two different meters. The night charges are less than the anytime tariff by almost 2.3 times. The anytime tariff is charged from 7 a.m. to 11 p.m. and the night tariff is charged from 11 p.m. to 7 a.m.

11.2.3 Milk cooling operations

A typical on-farm milk cooling system in New Zealand is shown in Fig. 11.2. The operations for milk cooling can be broadly categorized into two stages as shown in Fig. 11.2:

- *Precooling*: milk is precooled using a plate heat exchanger prior to entering the cooling vat.
- *Cooling in storage vat*: further cooling to the desired temperature achieved at storage vat.

11.2.3.1 Precooling

The main driver for an improved precooler design is faster cooling of milk in the vat to achieve 6 °C or below within 6 h since the commencement of milking and within 2 h of the completion of milking. With precooling, it is possible to cool the milk to a temperature of about 2–3 °C above the water inlet temperature (coolant) to the plate cooler using a proper choose of plate heat exchanger size. In most dairy farms, the cooling water is obtained from a bore or dam. For example, the Graejo Trust farm in Southland New Zeeland uses cooling water pumped from a bore, while Coldstream Downs farm uses the water from a dam. Fig. 11.3 shows the water temperatures profile at different time of the year and

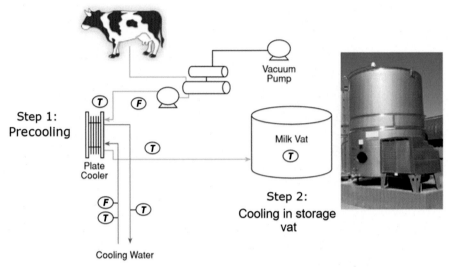

Where, *T* and *F* are the milk/coolant temperature (°C) and volumetric flowrate (L/s) respectively.

FIG. 11.2 Typical on-farm milk cooling system in New Zealand (Morison et al., 2007).

it is affected with a change in air temperature. When the milk leaves the plate cooler its temperature has been reduced to 2–3 °C above the water inlet temperature (Morison et al., 2007). This difference between the milk outlet and water inlet temperatures is normally referred to as the "temperature approach" of the heat exchanger.

11.2.3.2 Cooling in storage vat

The precooled milk in the primary stage is pumped to the milk vat refrigeration unit (secondary stage) to further decrease the temperature to the desired value (6 °C or below). The milk vat refrigeration unit is equipped with a cooling pad attached to the bottom of the milk vat and supplied with refrigerant from a refrigeration unit located nearby (Fig. 11.4). To prevent the milk from freezing, the vat has a motor-driven agitator that continuously moves the milk across the base of the vat. On very large vats, a second cooling pad is fitted to the lower wall of the vat and this is supplied from a second refrigeration unit.

11.3 Options for further cooling of milk

The milk is cooled slowly in an agitated cooling vat as there is a chance for the freezing of milk if the evaporator temperature is too low and insufficient mixing of the milk occurs. Furthermore, the refrigerants that are used in the dimple-pad cooling systems in NZ are likely to be banned for their high Global Warming Potential. Some of the current farm set-ups may not be able to satisfy the requirements of the new milk cooling regulations. Therefore, the best practice appears to be in snap chilling, probably using chilled water. This water will be circulated through a second stage plate heat exchanger to drop

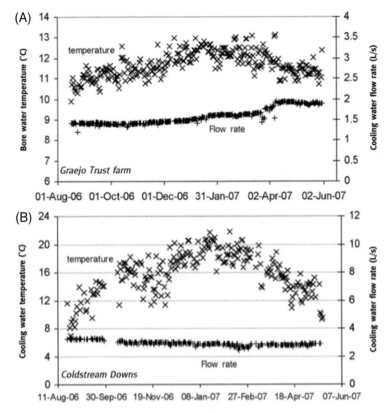

FIG. 11.3 Cooling water temperature variation during 2006/2007 at (A) Graejo Trust farm and (B) Coldstream Downs farm (Morison et al., 2007). Legend x: temperature, + flow rate.

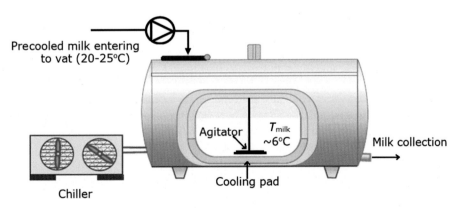

FIG. 11.4 Storage vat cooling system (direct expansion refrigeration system; DeLaval International AB, 2000).

the milk temperature to the desired value before entering to storage vat. The chilled water could be obtained from the following sources:

- Cooling towers
- Instant chilling
- Chilled water storage system
- Ice banks (ice-on-tube, packed bed ice encapsulated, and ice slabs)

11.3.1 Cooling towers

In some of the dairy farms, cooling towers are being used. It is a heat exchanger in which air and water are brought into direct contact with each other in order to reduce water temperature. As this occurs, a small volume of water is evaporated, reducing the temperature of the water being circulated through the tower. This technique is effective in areas having low air humidity. Water can be cooled to within 5 °C of the air wet-bulb temperature if the cooling towers are well designed. They operate overnight to cool a large volume of water, usually 4.5 times the volume of the daily milk yield. The high electrical energy required to drive the air fan used and the requirements for good insulation limit the widespread application of this system on farms. In addition, the efficiency of the system depends on the air humidity.

11.3.2 Instant chilling

In this case, a high capacity refrigeration system is required to cool a specific volume of food grade propylene glycol solution to -5 to -8 °C during the milking time. As shown in the schematic diagram of DeLaval Ltd. Complete milk cooling system (Fig. 11.5), milk is first cooled using tab water to around 20 °C, and then further cooled using chilled food grade propylene glycol solution to desired temperature (~6 °C) before it enters the storage vat. The chilled food grade propylene glycol solution is supplied from a chiller running on demand (during milking time). This option is not preferable in areas where energy is not available all the time (e.g., because of electricity cuts) and when the power consumption reaches the maximum (electricity peak load period).

11.3.3 Chilled water storage system

Using concrete or steel tanks to store chilled water at 4.4–5.6 °C that is produced with conventional chillers at night is a common practice in dairy farms. During milking time, cold water is pumped from the bottom of the tank to cool the milk, and after use, it returns to the top of the tank. Due to the different densities for water at different temperatures, a stable thermal stratification can be achieved.

Chilled water storage tanks are sometimes used in dairy farms to increase the efficiency of milk cooling by providing chilled water. Chilled water pumped through a second plate heat exchanger or a second stage added to an existing plate cooler enables the instant cooling of milk. Spray chilled water over the outside wall of a spherical milk vat to cool the milk could be another option. Robert Stone Stainless Steel designed and manufactured a spherical milk vat under the brand name of "Sphericool" (Morison et al., 2007). This system provides more rapid and efficient cooling than the traditional cylindrical vats fitted with direct expansion cooling pads as claimed by the manufacturer. However, this system had not been applied in New Zealand.

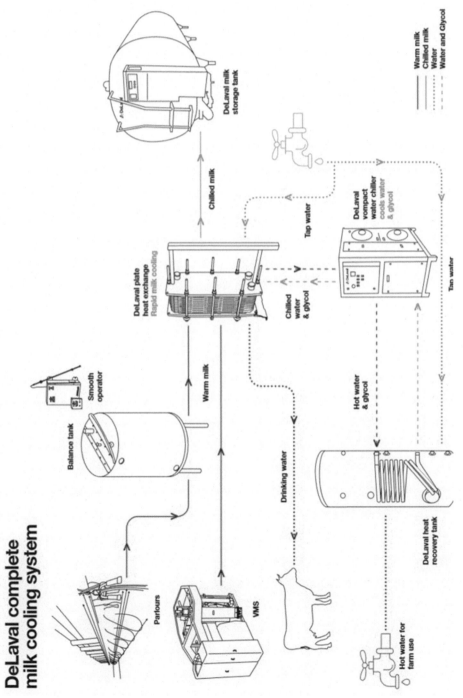

FIG. 11.5 Schematic diagram of DeLaval Ltd. complete milk cooling system (DeLaval Ltd., 2021).

Chilled water storage systems require a large storage tank volume and the volume of chilled water required is approximately three times higher than the volume of milk produced. Furthermore, they are not suitable in areas that have severe water restrictions or limited access to clean uncontaminated water. During milking time, cold water is pumped from the bottom of the tank while an equal amount of warm return water is supplied to the top of the tank and this could provide unstable chill water temperature during milking. The high cost of the storage tank and insulation due to the required large tanks limits its application.

11.3.3.1 Chilled water storage system installed in Coldstream Downs farm, New Zealand: a case study

The integrated cooling system of the chilled water storage tank, water chiller with the clip-on system, plate heat exchanger and circulating pumps was installed in Coldstream Downs farm, New Zealand (2007) (Fig. 11.6). The chilled water storage tank capacity was 25,000 L and it was made from plastic. During off-peak hours of electricity demand (11 p.m. to 7 a.m.), the chiller was starting to chill the water. During milking, chilled water was circulated through a plate heat exchanger (PHE) at a rate of 12,000 L per h. As this system was not capable of cooling the milk to the desired temperature (6 °C or below), direct expansion cooling pads in the milk vat had to be used to complete the cooling cycle (Table 11.1).

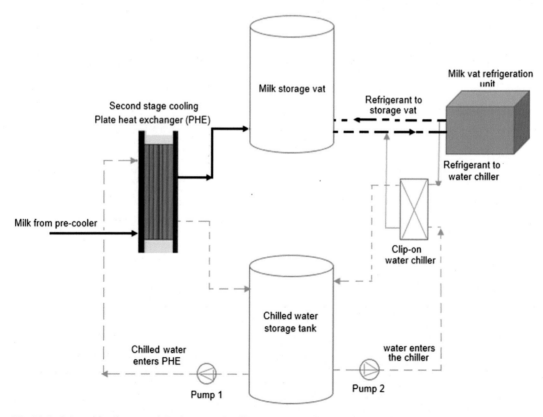

FIG. 11.6 Schematic diagram of the integrated chilled water-cooling system (Morison et al., 2007).

Table 11.1 Electricity tariff charges (Contact Energy Ltd., 2018).	
Electricity plan	**Price (NZ cents/kWh excluding GST)**
Anytime	27.284
Night	12.070

The total cost of installing this chilled water system at Coldstream Downs was approximately NZ$19,000 and the annual electricity saving is around NZ$900, which would require about 21 years to be able to pay back (Table 11.2). The main reason for installing a chilled water system was to improve the milk cooling process, and not to reduce electricity costs.

11.3.4 Ice storage systems

For about 15 years, several air conditioning systems have been built with ice storage (Lee and Jones, 1996). These storage systems benefit from the low night electricity tariffs and provide base-load support to meet the cooling demand. The ice storage systems are generally categorized by their different combinations of storage media, charging (making ice) and discharging (melting ice) mechanisms. The two main types are ice-on-tube and encapsulated ice in a spherical sphere.

11.3.4.1 Ice-on-tube storage system

This system was put into practice in the United States of America between 1960 and 1970 and in Japan in the 1980s (Okamura, 2009). In this system, the tank is fitted with an array of copper coils (known as cooling coils) filled with pressurized refrigerants. During the charging mode (off-peak hours), the chiller operates to cool a glycol solution to subfreezing temperatures, which is then circulated through the ice storage coils and forms ice around the external surfaces of these coils. During the day (on-peak electrical demand) the ice is ultimately melted and used as a cooling agent. There are two methods of ice melt, called external and internal melt.

A. External melt

The full storage charge is reached when the ice is typically 2.8–3.8 cm thick around the coils, which is approximately 65 vol.% ice of the total volume of the storage tank (Lee and Jones, 1996). Water circulates through the free space between the ice rings (35 vvol.% of total volume)

Table 11.2 Daily electricity cost for milk cooling by two alternative systems (Morison et al., 2007).			
	Average daily electricity use (kWh)	**Electricity price (cents/kWh)**	**Daily electricity cost (NZ$)**
Direct expansion cooling (anytime plan)			
	94.7	13.6	**12.88**
Chilled water cooling (day/night plan)			
Day	33.8	15.5	5.24
Night	77.9	6.5	5.06
Total	111.7		**10.30**

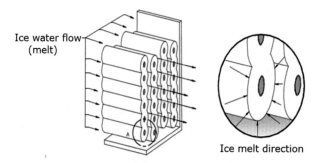

FIG. 11.7 External melt of ice (EVAPCO online website, 2017).

as shown in Fig. 11.7. The direct contact between water and ice-built rings allows the ice to melt and chill the water. Ice is melted from outside of the ring, thus it is called "external melt."

A. Internal melt

In this system, one fluid (glycol solution) works for both ice build and ice melt. The coils are placed close to each other in the storage tank and the formation of ice rings around the coils is allowed touching each other. Approximately 80 vol.% of the total volume of the storage tank is ice. During ice making cycle (charging process), subfreezing fluid is circulated through the coils and builds up ice around these coils. During ice melt cycle (discharging process), the warmer returning glycol solution is circulated through the same ice coil circuits, and ice is melted from the inside of the ring. Thus, it is called "internal melt" (Fig. 11.8). A comparison between internal and external ice melt systems is given in Table 11.3.

11.3.4.2 Packed bed ice encapsulated

A packed bed of plastic sphere filled with PCMs is used for low-temperature thermal energy storage (TES) applications (Esence et al., 2017; Anderson et al., 2015). Heat transfer fluid (HTF; typically glycol solution) circulates around the PCM spheres to freeze them at night and then melt them and release the coolness the following day (Fig. 11.9). This system has inherent advantages over other

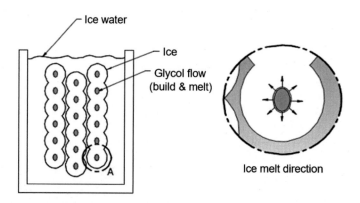

FIG. 11.8 Internal melt of ice (EVAPCO online website, 2017).

Table 11.3 Comparison between internal and external ice melt systems.

External Ice melt	Internal ice melt
• Two different fluids were used; glycol solution (ice build fluid) and water (ice melt fluid).	• One fluid (glycol solution) works for both ice build and ice melt.
• Rapid cooling (direct contact between ice storage and cooling fluid).	• Slow cooling (indirect contact between ice storage and cooling fluid).
• Ice water supplied to the system is at temperature of 1 °C or lower.	• Glycol solution supplied to the system is at temperature of 3 °C or higher.
• The ice storage cooling water is often the same as the cooling system fluid.	• Water in the storage container remains static and goes through phase changes only. This water is not circulated in the cooling system.

(A)   (B)

FIG. 11.9 (A) Packed bed phase change materials encapsulated storage system and (B) close view of packed spheres (Arkar and Medved, 2007).

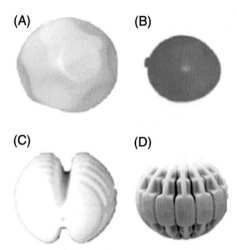

FIG. 11.10 Commercial macroencapsulated ice containers: (A) Cryogel (Cryogel, 2019), (B) Crystopia (Cool Storage Technology Guide, 2000), (C) Ice-Bon (Cool Storage Technology Guide, 2000), and (D) Macro vessel (MacroVesl online website, 2018).

configurations (e.g., ice-on-tube) such as ease of installation, simplicity, and flexibility in selecting the storage tank material and shape. Comprehensive experimental and modeling studies have been reported in the literature on a packed bed of plastic spheres filled with different types of PCM including, water (Fang et al., 2010; Wu et al., 2014; de Gracia and Cabeza, 2017).

Samples of commercially available ice encapsulated spheres are shown in Fig. 11.10 (A–C). An important aim when designing a thermal store is to achieve high rates of heat transfer between the fluid and the encapsulated PCM during the melting and freezing processes. Recently, Pure Temp Company (USA) developed a MacroVesl (Fig. 11.10D) with a high capacity of PCM and a multichambered container approximating sphere. It is designed to efficiently transfer thermal energy with fast charge and discharge even with a very small-temperature difference. No information of these systems has been provided by the manufacturer regarding their thermal performance.

11.3.5 Innovative approaches for ice encapsulation

The performance of latent heat thermal storage systems is limited by the poor thermal conductivity of PCMs used including that of water. Low thermal conductivity of PCMs results in low charging and discharging rates that cause a decrease in the overall efficiency of the latent heat thermal storage system. Nowadays, more attention has been given by different researchers to enhance heat transfer performance of PCMs such as the use of extended surfaces or fin configurations (Jegadheeswaran and Pohekar, 2009), dispersion of high conductivity particles and nanoparticles (Ho and Gao, 2009), introduction of a metal matrix (Haldera et al., 2015; Tong et al., 1996) and graphite compounded material (Mills et al., 2006), and microencapsulate the PCM with a high thermal conductivity shell materials (Cao et al., 2015; Yu et al., 2014; Al-Shannaq et al., 2016). Fig. 11.11 shows sketches of some of the common heat transfer enhancement techniques studied.

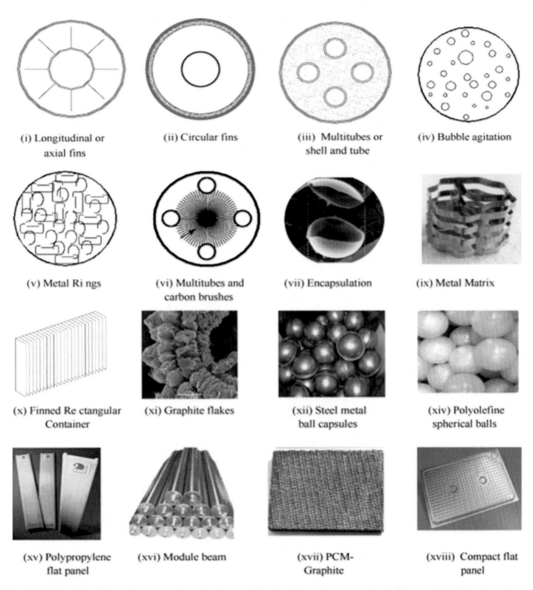

(i) Longitudinal or axial fins

(ii) Circular fins

(iii) Multitubes or shell and tube

(iv) Bubble agitation

(v) Metal Ri ngs

(vi) Multitubes and carbon brushes

(vii) Encapsulation

(ix) Metal Matrix

(x) Finned Re ctangular Container

(xi) Graphite flakes

(xii) Steel metal ball capsules

(xiv) Polyolefine spherical balls

(xv) Polypropylene flat panel

(xvi) Module beam

(xvii) PCM- Graphite

(xviii) Compact flat panel

FIG. 11.11 Heat transfer enhancement methods employed in phase change material research (Kenisarin and Mahkamov, 2007).

11.3.5.1 Packed bed of graphite sphere containing PCM (water)

Encapsulated ice storage systems limit their applications due to the low thermal conductivity of PCMs including that of water and ice. In this section, a novel solution is presented of using graphite/water composite spheres in a packed bed with high charge/discharge rates. A packed bed of spheres made of graphite as TES system was constructed in The University of Auckland laboratory. The TES system consists of: (1) An acrylic tank filled with 50 graphite composite spheres, (2) a refrigerated bath filled with a HTF, (3) a circulating magnetic drive pump and flowmeter providing flow rate up to 2.5 L/min, (4) K-type thermocouples positioned at the inlet and outlet of the storage unit, and (5) a data acquisition system (PicoLog Thermocouple Data Logger-Model TC-08) connected to a computer for continuous temperature measurements. The acrylic tank and all pipes were well insulated with polyurethane foam to minimize heat gain from the environment. Fig. 11.12 shows a schematic diagram of the packed bed of graphite composite spheres test set-up.

Fast PCM charging is essential in latent heat storage systems and should occur within the available night tariff electricity charges, which is in general between 11 p.m. and 7 a.m. (off-peak). The freezing start time is determined when the bottom sphere center temperature starts flattening at the freezing point of PCM (water), which is 0 °C for water. However, the PCM freezing time is finished when the top sphere center temperature is changing its profile toward the HTF inlet set temperature. At the end of the PCM freezing process (latent and sensible heat), the inlet and outlet temperatures of the HTF approach each other.

Fig. 11.13 shows the transient temperature variations of outlet working fluid temperature and the two graphite composite spheres center temperatures. The inlet temperature and volumetric flow rate of HTF are set at -3.5 °C and 1000 mL/min, respectively. The result shows a water freezing time of 5.3 h, which is within the time available for night tariff electricity charges. The graphite matrix has a significantly enhanced PCM freezing rate. In addition, it exhibits minimum super-cooling of 1.5 °C, indicating that the graphite matrix material plays a good role as a nucleating agent for water freezing.

For the PCM melting (discharging process), Fig. 11.14 shows the transient temperature of the HTF outlet temperature and center temperature of the graphite composite sphere positioned at the top of the storage tank. It is assumed that the PCM in the top graphite composite sphere is the last to melt, thus its center temperature determines the time of complete PCM melting. The inlet temperature and HTF

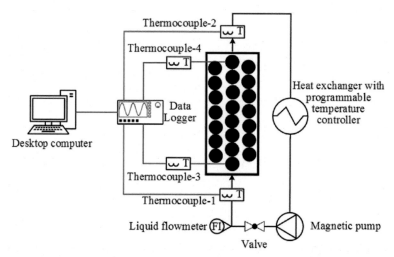

FIG. 11.12 Schematic diagram of the packed bed system made of graphite composite spheres (Al Shannaq et al., 2009).

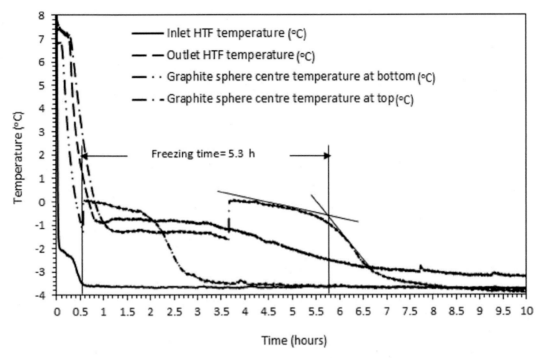

FIG. 11.13 Transient temperature variations of the HTF and graphite composite spheres (Al Shannaq et al., 2009).

volumetric flow rate were kept constant ($T_{in, HTF}$ = 10 °C, volumetric flow rate = 1500 mL/min). The result shows that the HTF outlet temperature quickly increased at the beginning of the melting process (sensible heating). However, the temperature was almost constant during the ice melting time of 1.72 h. The percentage of total energy recovered (sensible and latent), including losses during the melting process, is identified by the ratio of energy discharged to the total energy stored. The results show that for acceptable outlet temperatures of up to 7 °C (a minimum temperature difference of 3 °C), 83% of the total recoverable thermal storage capacity can be utilized.

The influence of the HTF inlet temperature and volumetric flow rate on the thermal behavior of the packed bed latent heat storage system is experimentally investigated. The results showed that both had a significant effect on freezing time. It is observed that by adjusting the HTF inlet temperature and volumetric flow rate during the discharge process, the required output temperature and average melting rate can be achieved in practical applications. The developed ice storage system can be easily adapted to a variety of cooling applications. In conclusion, the performance of the developed ice storage system is promising and can be easily adapted to a variety of cooling applications such as on farm chilling of milk.

The proof-of-concept for the innovative ice encapsulated storage system (IESS) as a TES was successfully demonstrated using a lab-scale IESS module consisting of a column filled graphite spheres. The experimental results reported and discussed above showed that the IESS TES concept

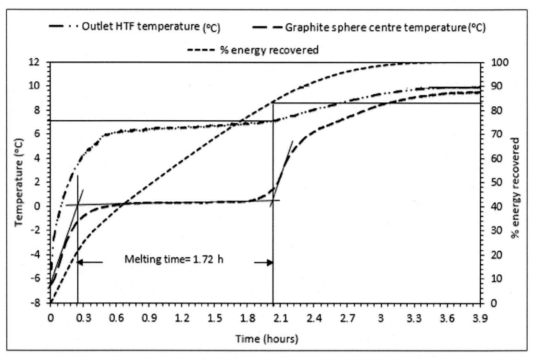

FIG. 11.14 Transient variations of HTF outlet temperature and graphite sphere center temperature as well as % of energy recovered (Al Shannaq et al., 2009).

is technically feasible and initial results were very encouraging to demonstrate the technical, manufacturing and economic feasibility with a larger system. The cost analysis and projections for high volume were estimated for three different types of comparable cold energy storage, for example, chilled water, ice-on-tube, and ice encapsulated storage systems. The estimated capital costs of this innovative IESS system are lower than other similar systems on the market; thus, is a highly competitive option for the future.

11.3.5.2 Ice slab storage system
This section focuses on explaining the performance of an experimental laboratory-scale ice slab storage system. An experimental heat storage unit consists of an assembly of $1 \times 1 \times 2$ rows of slabs containing 1 kg water plus nucleating agent each. The assembly of two rows with 14 slabs at each row was stacked in a stainless-steel basket shown in Fig. 11.15A. The assembly was placed in the stainless-steel heat storage tank where the HTF (glycol solution) is pumped and flow through the 5 mm spacing available between the slabs. The overall experimental setup is shown in Fig. 11.15B.

The flowrate was controlled manually using a rotameter, which operates in a flow rate range between 1 and 6 L/min designed for a water system. Since the HTF is denser than water (by about 5%), a calibration was required. A laboratory-scale full-featured circulating bath (20 L) system with ramp and soak capability, of the working temperature range in the range of -30 °C to 200 °C,

(A)   (B)

14 slabs stacked in a stainless
steel basket at 2 rows

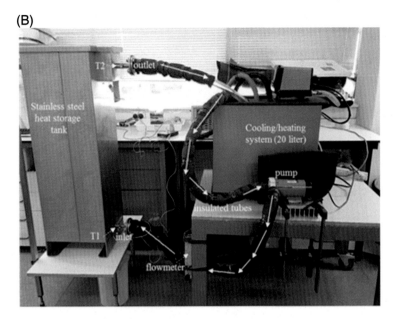

FIG. 11.15 (A) The stack of the slabs at each row. (B) The complete experimental setup.

was employed in this experimental work. As the preliminary thermal performance results on the lab scale ice slab storage system were promising, further experimental investigations will be carried out. In addition, a computational fluid dynamics model to validate the experimental results will be developed.

11.4 Pilot scale ice slab storage system

A novel ice encapsulated storage system (50 kWh) was designed, constructed and installed on a farm (794 Tauhei Rd, Hamilton 3375, New Zealand) to achieve the new milk cooling regulations (Operational code NZCP1-2017). The system can provide chilled food-grade propylene glycol solution (30 vol.% propylene glycol) on-demand, during milking time, for instant chilling of milk to 6 °C or lower before it enters the storage vat.

11.4.1 Process description

The process flow diagram of the constructed novel ice encapsulated storage system on the farm is shown in Fig. 11.16. Accordingly, two stages of milk cooling were applied as follows:

Stage 1: Precooling

The milk enters the first plate heat exchanger (PHE1) and goes through the first stage of cooling provided by a farm water source (bore or tap water). This will lower the milk temperature decrease from approximately 35 °C to 3 °C warmer than the water temperature (~20 °C). This stage of milk cooling (precooling) has already existed on the farm.

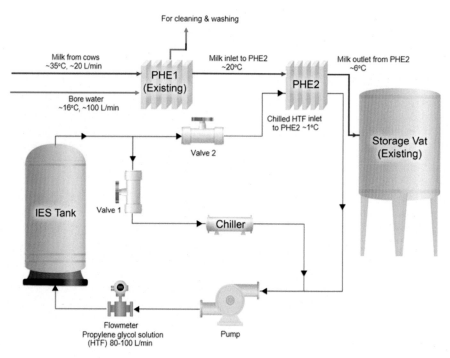

FIG. 11.16 Integrated on-farm milk cooling system with ice slabs storage unit—process flow diagram (Harris et al., 2019).

Stage 2: Final cooling

The precooled milk will then enter a second plate heat exchanger (PHE2), where the chilled food-grade propylene glycol solution (HTF) is used as the coolant, lowering the milk temperature to 6 °C or below before it enters the milk storage vat. The flow rate of chilled HTF will be at least 3 times higher than the milk flow rate. The source of chilled HTF (coolant) is provided from the novel ice encapsulated storage tank (IES). This IES tank contains plastic slabs made from high-density polyethylene and filled with water (90 vol.%) stacked on each other vertically, forming a self-assembling large heat exchanger storage system. The gap between the slabs (5 mm) provides an ideal HTF flow passage with a large heat transfer area and a high heat transfer coefficient.

11.4.2 System operation

11.4.2.1 Making ice (charging process-night)

The temperature of HTF is dropped to -5 °C using the chiller, then circulated through the IES tank causing the water to freeze inside the plastic slabs.

11.4.2.2 Melting ice (discharging process-milking period)

HTF is circulated through the IES tank where it is shock chilled down to ~1 °C and passes to the secondary plate heat exchanger (PHE2) to chill the milk and then recycled back through to the IES tank.

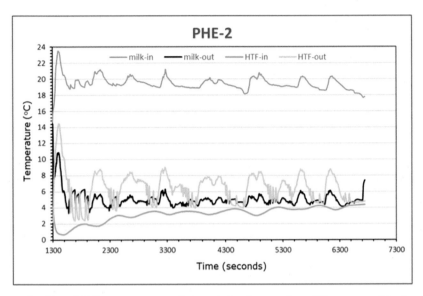

FIG. 11.17 Transit milk and HTF temperatures variations during the morning milking process (Harris et al., 2019).

11.4.3 Technical results

Fig. 11.17 shows the transit milk and HTF temperatures during milking. The milk enters PHE-2 at an average temperature of 20 °C and leaves it at an average temperature of 5 °C. The volumetric flow rates ratio of HTF to milk (4:1) is kept constant. Technically, the novel ice storage system is capable to cool the milk below the required temperature, 6 °C.

Ice slabs storage system advantages

- Easy to install, operate and integrate with the existing cooling system on farms.
- Minimal system maintenance is required.
- Indirect heat exchanger between HTF and cold storage media, HTF passes through designed free space between ice containers, so there is no risk of blocking the flow path of HTF.
- Provides almost stable HTF temperature during milking (provides near-zero discharge temperatures).
- High rate of ice making (ice thermal resistance is small).
- Compact system, providing energy storage capacity of up to 50 kWh per one cubic meter of the storage tank. Nonpressurized refrigeration system/ low risk of refrigerant leaking.
- Flexible to use in different TES applications.

Mr. Stuart Husband[2] (farm owner) commented on the milk quality after the system commissioning and said: "*The milk quality result is improved from B to A+ when the ice storage system is installed and this is because the milk is chilled down quickly so no time for bacteria growth, which is great for us.*"

---

[2] Councillor, Waihou General Constituency. Co-Chair, Integrated Catchment Management Committee. Waikato Regional Council.

11.4.4 Cost analysis

A simple capital cost analysis is shown in Table 11.4 for 50 kWh ice storage system.

Based on 50 kWh energy storage capacity, the cost of manufacturing ice slabs encapsulated system including; IES stainless steel tank, plastic slabs and pump is only NZ$10,500. The price for the IES tank (NZ$8000) is estimated based on the tank made from double layer insulated stainless steel. Alternatively, a plastic tank could be used at a much cheaper price of NZ$3000 only which will bring down the difference in the manufacturing cost by up to half.

The refrigeration unit, second stage PHE, and accessories (pipes and three-ways valve) are common components for both storage systems of ice encapsulated and ice-on-tube (commercially available), so they were excluded in the cost analysis.

In summary, the 50 kWh ice encapsulated storage system installed on-farm under the typical operating conditions ($T_{milk,inlet}$ ~20 °C, F_{milk} ~20 L/min, $T_{chilled\ HTF}$ ~1 °C, and $F_{chilled\ HTF}$ ~80 L/min) is capable of rapid cooling of the milk below the required temperature of 6 °C. Commercially, the novel ice encapsulated storage system is very competitive and lower in capital cost compared to alternative systems. Therefore, it has a high potential for commercialization.

11.5 Conclusions

Milk cooling accounts for about 30% of the total energy costs of operating a dairy (Dairying for Tomorrow).[3] Here in New Zealand, most of the power companies provide an option for night tariffs which are significantly lower than the normal tariffs. With the use of efficient cold energy storage system, it is possible to shift the energy demand from morning to night. This will allow dairy farms to save on their energy bills. Furthermore, when NZ legislation came into force in 2018 dairy farmers were required to upgrade their milk cooling systems. So, the best practice looks to lie in the direction of snap chilling, probably using chilled water cooled by ice-banks.

The key conclusions from this research were as follows:

* Application of thermal ice storage for dairy farms is commercially viable
* The thermal ice storage is a good option for dairy farms to meet with the new regulations without spending too much on completely new systems

Table 11.4 Cost analysis for 50 kWh ice storage system.

Ice encapsulated	Manufacturing cost (NZ$)	The commercially available ice bank
1. IES tank (rectangular shape)	8000	Ice bank with pump and controls
2. Plastic slabs (504 slabs)	1500	
3. Pump	1000	
Total cost	**10,500**	**15,525** + GST (selling price)

---

[3] Dairying for Tomorrow- Saving energy on dairy farms, Dairy Australia Limited (2018).

- Load shifting from day to night will enable dairy farms to take advantage of the low night-time electricity tariff
- The novel ice slabs encapsulated storage system is commercially competitive and has a high potential for commercialization

Acknowledgment

The authors gratefully acknowledge the Ministry for Primary Industries (MPI)-New Zealand for funding this project [Food Industry Enabling Technologies (FIET) Project (contract MAUX1402)].

References

Al Shannaq, R., Young, B., Farid, M., 2009. Cold energy storage in a packed bed of novel graphite/PCM composite. Energy 171, 296–305.

Al-Shannaq, R., Kurdi, J., Al-Muhtaseb, S., Farid, M., 2016. Innovative method of metal coating of microcapsules containing phase change materials. Energy 129, 54–65.

Anderson, R., Bates, L., Johnson, E., Morris, J.F., 2015. Packed bed thermal energy storage: a simplified experimentally validated model. J. Energy Storage 4, 14–23.

Arkar, C., Medved, S., 2007. Free cooling of a building using PCM heat storage integrated into the ventilation system. Solar Energy 81, 1078–1087.

Barzin, R., Chen, J., Brent, Y., Farid, M., 2015. Peak load shifting with energy storage and price-based control system. Energy 92, 505–514.

Cao, F., Kalinowski, P., Lawler, J., Lee, H., Yang, B., 2015. Synthesis and heat transfer performance of phase change microcapsule enhanced thermal fluids. J. Heat Transfer 137 (9).

Contact Energy Ltd. Retrieved from Contact Energy. http://www.contactenergy.co.nz; 2018.

Cool Storage Technology Guide. Palo Alto, CA: EPRI; 2000 (TR-111874).

Cryogel. Thermal Energy Storage. http://www.cryogel.com; 2019.

de Gracia, A., Cabeza, L.F., 2017. Numerical simulation of a PCM packed bed system: a review. Renew Sustain Energy Rev. 69, 1055–1063.

DeLaval International AB. Efficient Cooling 2000. The Netherlands, www.delavalcorporate.com/globalassets/our-products/milking-solutions/efficient-cooling.pdf.

DeLaval Ltd. Instant cooling: protect milk quality with less than half the energy requirements 2021 (www.delaval.co.nz).

Esence, T., Bruch, A., Molina, S., Stutz, B., Fourmigué, J.-F., 2017. A review on experience feedback and numerical modeling of packed-bed thermal energy storage systems. Solar Energy 153, 628–654.

EVAPCO online website. Thermal Ice Storage Application and Design Guide. http://www.evapco.eu: 2017.

Fang, G., Wu, S., Liu, X., 2010. Experimental study on cool storage air-conditioning system with spherical capsules packed bed. Energy Build. 42, 1056–1062.

Farid, M., Khudhair, A., Razack, S., AlHallaj, S., 2004. A review on phase change energy storage: materials and applications. Energy Convers. Manage. 45, 1597–1615.

Haldera, S., Singha, S., Sahab, S.K., 2015. Effect of metal matrix and foam porosity on thermal performance of latent heat thermal storage for solar thermal power plant, Third Southern African Solar Energy Conference. Kruger National Park, South Africa.

Harris, M., Archer, R., Waite, R., Al-Shannaq, R., Silver, K., 2019. Milk chilling and farm milk vats. Food New Zealand 18 (6), 17–19 December 2018/January.

Ho, C.J., Gao, J.Y., 2009. Preparation and thermophysical properties of nanoparticle-in-paraffin emulsion as phase change material. Int. Commun. Heat Mass Transfer 36, 467–470.

Jegadheeswaran, S., Pohekar, S.D., 2009. Performance enhancement in latent heat thermal storage system: a review. Renew. Sustain. Energy Rev. 13, 2225–2244.

Kenisarin, M., Mahkamov, K., 2007. Solar energy storage using phase change materials. Renew. Sustain. Energy Rev. 11, 1913–1965.

Lee, A.H.W., Jones, J.W, 1996. Modelling of an ice-on-coil thermal energy storage system. Energy Convers. Manage. 37, 1493–1507.

Lee, A.H.W., Jones, J.W, 1996. Laboratory performance of an ice-on-coil, thermal-energy storage system for residential and light commercial applications. Energy 21, 115–130.

Lu, W., Tassou, S.A., 2013. Characterization and experimental investigation of phase change materials for chilled food refrigerated cabinet applications. Appl. Energy 112, 1376–1382.

MacroVesl online website: http://www.puretemp.com/stories/products/macro-vesl; 2018.

Miller, J., 2013. Some Observations on Saving Electricity on the Farm. Millbridge Consulting Ltd, SMASH TRUST, New Zealand, CAMBRIDGE 3434.

Mills, A., Farid, M., Selman, J., Al-Hallaj, S., 2006. Thermal conductivity enhancement of phase change materials using a graphite matrix. Appl. Thermal Eng. 26, 1652–1661.

Ministry of Primary Industries, 2013. NZCP1: Code of Practice for the Design and Operation of Farm Dairies. Wellington, New Zealand.

Ministry of Primary Industries, 2017. Operational Code: NZCP1: Design and Operation of Farm Dairies. Wellington, New Zealand.

Morison, K., Gregory, W., Hooper, R., 2007. Improving Dairy Shed Energy Efficiency: Technical Report. New Zealand Centre for Advanced Engineering (CAENZ), New Zealand.

Okamura, A., 2009. Guide to Thermal Storage Technology. Ohmsha, Tokyo.

Oró, E., de Gracia, A., Castell, A., Farid, M., Cabeza, L., 2012. Review on phase change materials (PCMs) for cold thermal energy storage applications. Appl. Energy 99, 513–533.

Tong, X., Khan, J.A., RuhulAmin, M., 1996. Enhancement of heat transfer by inserting a metal matrix into a phase change material. Numer. Heat Transfer 30, 125–141.

Wu, M., Xu, C., He, Y.-L., 2014. Dynamic thermal performance analysis of a molten-salt packed-bed thermal energy storage system using PCM capsules. Appl. Energy 121, 184–195.

Yu, S., Wang, X., Wu, D., 2014. Microencapsulation of n-octadecane phase change material with calcium carbonate shell for enhancement of thermal conductivity and serving durability: synthesis, microstructure, and performance evaluation. Appl. Energy 114, 632–643.

Novel drying technologies using electric and electromagnetic fields

12

Olivier Rouaud, Michel Havet

Oniris, Université de Nantes, CNRS, GEPEA, Nantes, France

12.1 Introduction

Drying is an operation aimed at removing water impregnating a solid or a liquid. It is mainly performed at atmospheric pressure by evaporation or at a pressure below the triple point of water by sublimation (freeze-drying). In these processes, heat is the driving mechanism for water removal whereas other technologies are based on osmotic dehydration. Drying is a very old process and one of the main methods for preserving agricultural and food products, such as fruits, vegetables, and spices or other products with high water content. The process is also involved in food industries as a separate stage of processing, simply because the following processing steps and/or the final products require dried solids. Drying offers many benefits including extended shelf life, off-season availability, reduced packaging, storage, handling, and transportation costs (Moses et al., 2014). According to Mujumdar (2014), between 12% and 20% of the national industrial energy consumption of many industrialized countries is due to the thermal dehydration operations. Indeed, even today, more than 85% of the dryers found in industries are convection dryers that use hot air or gas as the drying medium (Moses et al., 2014). Due to the high latent heat of vaporization and the energy consumed in heating and dehumidifying the air blown at high flow rates, convective drying is inherently inefficient and consequently a very energy-intensive process. In a world where the population continuously increases, with shortage of food and limited energy resources, it becomes crucial to find more efficient and economical drying methods able to preserve the sanitary, nutritional and organoleptic quality of food. Food industries are therefore interested in any solution that might accelerate the process, consume less energy and optimize the quality of dried product, the safety in operation, and the controllability of the dryer.

Fig. 12.1 shows a typical drying curve for a food product and drying takes place in three/four periods. The first phase, often short or even absent if the product is already at high temperature, corresponds to the time during which sensible heat is transferred to the product. The rate of evaporation increases considerably during this period because most of the free moisture is removed. The second phase is the constant rate period, in which a film of water is always present on the product surface. During this phase, the product temperature increases gradually but very slowly due to the effects of evaporative cooling. The drying rate is high because water is still available on the surface. The third and fourth phases, or falling rate periods (first and second falling rate periods in Fig. 12.1), are phases during which the migration of moisture from the internal interstices to the external surface becomes the limiting factor, resulting in a decrease in the drying rate. When

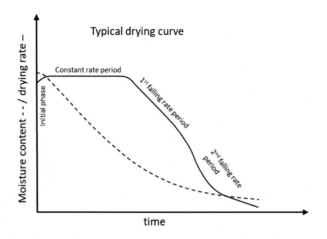

FIG. 12.1

Representation of moisture content and drying rate as a function of time for conventional drying processes.

the surface of the product is completely dry, the second falling rate period can be distinguished in most cases. Since the physical structure of the food product being dried is subject to change during drying (shrinkage, changes in porosity, etc.), the water transfer mechanisms may also change during the drying time.

In most cases, the drying time of the first falling rate period is directly dependent on the drying conditions achieved during the constant rate period. Usually, if products are dried quickly when they have high moisture content, they can continue to dry quite quickly in the low moisture range. This can be attributed to the formation of a more porous structure that promotes faster diffusion during the falling rate period and increases the exposed surface area of the material. Nevertheless, in some cases, especially for thick materials dried at high drying rates, the outer surface may dry much faster than the core and enter in a glassy state, leading to the case hardening phenomenon commonly observed during drying processes (Gulati and Datta, 2015). The resulting formation of a thin and dry shell on the outside can prevent moisture transfer and consequently reduce the drying rate. In the case of convective drying, increasing the drying medium temperature, and therefore decreasing its relative humidity, or increasing the air flow rate during the constant rate period could reduce the drying time. However, since high air temperature can be detrimental to heat sensitive products, it may be preferable to increase the heat and mass transfer coefficient. Improving the falling rate periods of drying without acting on the constant rate period is more difficult because it requires an enhancement of heat and mass transfer through the material itself. This can be done through the introduction of other technologies in dryers such as microwave or ultrasound, but when these improvements are implemented, they can lead to a change in product quality (Nazir and Azaz Ahmad Azad, 2019). Thus, the poor quality of the final dried product and the low efficiency of convective dryers have led scientists to combine several techniques and energy sources. In recent decades, attention has been focused on several hybrid drying technologies using electromagnetic and/or electric fields, such as microwave-assisted drying, radio frequency drying and more recently electrohydrodynamic (EHD) drying.

12.2 Microwave and radio frequency drying

Electromagnetic waves can be used to enhance the drying efficiency of food products. Whereas infrared radiation is limited to surface heating due to a low penetration depth, microwaves (MW) and radio frequencies (RF) have the ability to heat up volumetrically the liquid to be vaporized (i.e., water). The heat transfer resistance is virtually removed in the material and the transport of moisture out through the food can be increased due to the higher mobility of water at higher temperatures. Moreover the transport of vapor due to the internal pressure gradient toward the material surface is improved. Because dielectric heating has the unique capacity to generate heat within the food, microwave, and radio frequency drying is suitable for products with a relatively small surface-to-volume ratio.

The frequencies used for industrial MW and RF heating applications are regulated to avoid the risk of interference with radio communications and radar. Therefore, a number of frequencies have been reserved for such radiation drying and the most frequently used are 13.56 and 27.12 MHz for RF and 915 and 2450 MHz for MW. RF and MW heating can be applied to most food products because these ones are dielectric materials that are characterized by poor electrical conduction properties.

The dipole orientation and the ionic orientation are the two main mechanisms that describe the coupling between electromagnetic field and dielectric materials. Polar molecules in food products, such as water, align themselves to the electrical field because their positive and negative charges centers do not coincide. If the electrical field is alternating, the polarization leads to dipole rotation and the resulting friction with adjacent molecules generates heat. Unlike polar molecules, dissociative ions of food are free to move toward regions of opposite charges. In case of alternating electromagnetic field, the ionic migration can cause friction between molecules, resulting in heat generation. Due to the difference in frequency range, the ionic orientation is generally more important in RF drying than in MW heating where both ionic conduction and dipole relaxation can be dominant heating mechanism.

The power dissipated per unit volume of dielectric material (P_v, W/m^3) can be expressed as follows:

$$P_v = 2\pi f \varepsilon_0 \varepsilon'' |E|^2 \tag{12.1}$$

where f is the frequency (Hz), E is the electric field strength in food load (V/m), and ε_0 is the dielectric constant in vacuum (8.854×10^{-12} F/m). The parameter ε'' (loss factor) is the imaginary part of the relative complex permittivity ε^* ($\varepsilon^* = \varepsilon' - j\varepsilon''$) that governs the interaction between dielectric materials and MW or RF energy. The real part ε', termed as dielectric constant, describes the ability of dielectric products to store energy. Because dielectric properties ε' and ε'' of a food product vary with frequency, water content, temperature, porosity and composition, they are key factors in considering the relevance of dielectric drying. For some food, the dielectric properties increase with increasing temperatures but they always decrease during drying due to the decreasing water content. Control of dielectric drying has to include the competition between the increasing trend of loss factor giving rise to larger temperature differences inside the material ("thermal runaway") and the moisture content dependent-loss factor contributing to uniform drying.

According to Eq. (12.1), the heating power is proportional to the frequency. In case of loss factors of the same magnitude at any frequency, the usual way to compensate the low frequencies is to operate RF drying with higher electrical field strengths, which must, however, remain below 30,000 V/m to avoid voltage breakdown of air (arcing). Conversely, the penetration depth of the wave is inversely proportional to the frequency, as shown in Eq. (12.2).

$$d_p = \frac{c}{2\pi f \sqrt{2\varepsilon' \left[\sqrt{1 + \left(\varepsilon'' / \varepsilon'\right)^2} - 1 \right]}}$$

(12.2)

where c is the velocity of the light in free space (3×10^8 m/s). Assuming an exponential decay of the electromagnetic power, the distance d_p theoretically defines the depth where the intensity of power falls to $1/e$ of the incident power at the food surface. Thus, RF heating may be preferred if the food load is very large and MW heating can be used for food in the centimeter or millimeter range. The parameter d_p is very useful to select appropriate size of food to ensure uniform heating in RF and MW applications. Sosa-Morales et al. (2010) indicate that the penetration depth of microwaves in food with high moisture content at room temperature is between 0.3 and 7 cm, depending on the salt content and frequency. The penetration of RF energy can be more to 10 times greater.

Because dielectric heating is more powerful than conventional heating, they have often to be used in combination with conventional drying techniques to overcome some drawbacks such as excessive temperature along the corner or edges of food products, a nonuniform heating and sometimes textural damages. Several comprehensive reviews on the contribution of electromagnetic field for drying processes can be found in Chandrasekaran et al. (2013), Kumar and Karim (2019), and Zhou and Wang (2019). Common applications of dielectric heating include RF- and MW-assisted hot air drying, MW vacuum drying and MW freeze-drying. MW and RF can also be combined with spouted bed drying and osmotic drying.

Postbaking drying for crackers and cookies was the first commercial application using RF in the food industry (Piyasena et al., 2003). Usually, the production capacity is improved and factory floor space reduced because treatments are shorter and RF is not detrimental to the product quality. More recently, a RF system combined with a hot-air system was successfully used for drying in-shell nuts. In such a process hot air carries away moisture from the nut and lowers its surface temperature whereas RF energy is used to heat the nuts above the air temperature. Zhou and Wang (2019) mention an approximately 52% reduction of hot air drying time by combining RF energy. However, despite the great potential of RF-related combination drying for improving heating uniformity and energy efficiency, commercial RF drying applications are currently predominantly limited to the wood, textile and bakery industries.

MW was successfully combined with hot air for drying of high moisture fruits and vegetables, especially during the falling rate period where the diffusion is rate-limiting. Substantial reduction in the drying time has been achieved in MW drying with hot air when compared with conventional drying. The saving in drying time is often greater than 25% and the reduction can reach a factor of two for apple and up to a factor of ten for *Penne* short cut pasta (Chandrasekaran et al., 2013). Applying MW energy in a pulsed manner is more advantageous than continuous application as it can overcome the problems of overheating, uneven heating and too rapid transport of moisture which can lead to quality damage in the food texture by "puffing." Vacuum drying, usually used for highly perishable commodities and high heat sensitive products, can be combined with MW to improve the thermal efficiency and drying time. The loss of nutritional qualities of food is reduced due to the absence of air and the absence of heat. MW vacuum drying can improve the quality of dried snacks, such as the crispy texture (Zielinska et al., 2020). Freeze-drying is also considered a gentle dehydration technique for heat sensitive food but is very time consuming because of low drying rates. Experimental studies of MW freeze-drying have shown that the assistance of MW could provide a 50-75% reduction in drying time in comparison with conventional freeze-drying methods (Khaing Hnin et al., 2019).

In summary combinations of MW with other drying techniques are almost unlimited and they can reduce energy consumption and improve product qualities. Thus MW-assisted drying methods may become the future trend in the development of drying technologies. To fulfill the potential for novel combined RF- and MW-assisted drying methods, future research has to concern industrial equipment, process parameters, the combination orders between different heat sources, the properties of materials being dried, the influence of pretreatments, and the energy aspects.

12.3 Electrohydrodynamic drying

One of the obvious ways to intensify drying rates in convectional drying is to increase the convective heat and mass transfer rate. The evaporation rate can be significantly enhanced when using impinging flow configurations rather than a flow parallel to the surface to be dried. The main advantage of EHD drying is precisely to generate a secondary flow directly to the product. Generally speaking, EHD investigates the flow of electrically charged fluids. A gas flow may be generated through a corona discharge which is obtained when a high voltage is applied between two electrodes whose radii of curvature are significantly different. The EHD drying device usually consists of a vertically movable electrode (sharply needle or thin wire) suspended above a fixed grounded electrode on which the food to be dried is placed (Fig. 12.2). From a threshold voltage, usually of the order of kV up to 10 kV, the high electric field at the vicinity of the discharge electrode causes ionization of gas around the electrode. While the process is rather complex the effect of ionization is that ions of the same polarity drift to the grounded electrode. A space charge is formed and a very low electric current of the order of μA flows between both electrodes. The gas motion is generated by the ions that collide with the neutral molecules of gas. This secondary flow termed ionic wind perturbs the boundary layer created on the food surface and consequently enhances the heat and mass transfer coefficient.

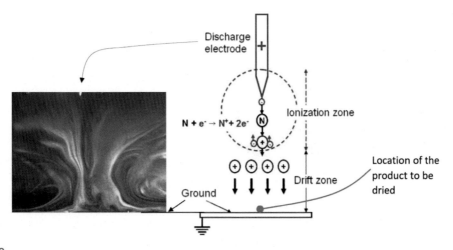

FIG. 12.2

Example of EHD drying configuration in case of positive corona discharge: flow visualization by particle image velocimetry (PIV) and physical sketch (ionization zone not on scale).

Contrary to dielectric drying for which the electromagnetic field generates heat inside the food, the electric field in EHD drying has an effect on charge transport in gas media. The concept of volume force can be introduced to describe the additional movement of air and the force F_v below is included in the Navier–Stokes equation (Martynenko and Kudra, 2016).

$$F_v = \rho_c E - \frac{\varepsilon_0}{2}|E|^2 \nabla \varepsilon_r + \frac{\varepsilon_0}{2}\nabla\left(|E|^2 \rho \frac{\partial \varepsilon_r}{\partial \rho}\right) \tag{12.3}$$

where ε_r is the relative dielectric permittivity of gas, ρ_c is the charge density (C/m^3), and ρ is the density of gas (kg/m^3). In case of air with constant temperature, the second term of the equation, representing the dielectroforetic force, and the third term called electrostriction can be neglected. Thus the Coulomb force, which is the product of the charge density and the electric field, is the main force generating ionic wind. Its mean velocity can be evaluated from Eq. (12.4) (Robinson, 1961):

$$u = k\sqrt{\frac{j}{\rho b}} \tag{12.4}$$

where j is the current density at the grounded electrode (A/m^2), k is a parameter depending on the geometry of electrodes, and b is the ion mobility in gas (m^2/(s V)). Velocities of the order of 5 m/s can be reached in EHD drying.

Because of the nongeneration of heat, EHD drying is considered as a novel nonthermal processing method particularly suitable for heat-sensitive materials. The process has received considerable attention in recent years because it ensures high product quality preservation and promises a reduction of the energy consumption. Because of low temperature drying, less shrinkage is usually obtained in EHD drying compared with conventional drying (Alemrajabi et al., 2012). Moreover, experimental investigations have shown that less color degradation of vegetables and fruits resulted from EHD drying (Martynenko et al., 2021). EHD drying can also have favorable effects on the shelf life of food materials. Several studies have reported that the respiration rates of lettuce, spinach, and some fruits such as pear, plum, banana, and cranberries were reduced following EHD treatment (Singh et al., 2012).

Drying kinetics with EHD is similar to the kinetics of convective drying, suggesting that EHD drying effects are essentially convective (Bardy et al., 2016). Cross-flow of air, parallel to the food to be dried, is often necessary, to evacuate moisture from the material. It was clearly shown that a cross-flow with too high velocity can suppress the positive effect of ionic wind (Ould Ahmedou et al., 2009). In addition to its negative effect, additional airflow implies increasing of energy consumption (due to blower and temperature and humidity control).

Numerous publications refer to energy advantages of EHD drying. Based on the energy calculated from current and applied voltage and the energy associated with primary air blower (Bardy et al., 2015) have shown, through an energy analysis, that EHD improved efficiency of forced convection especially at the beginning of the process with an eightfold greater efficiency. However, as mentioned by Kudra and Martynenko (2015), the energy calculated in most papers does not include the energy efficiency of high voltage converter which can range between 0.01 and 0.17. The real energy consumed for EHD drying remains certainly lower than the one for air drying but the savings effect might be less than the one observed in several studies. This does not, however, call into question the interest of using EHD for heat-sensitive products.

EHD drying has been basically investigated by laboratory-scale tests and working small scales prototypes are scarce (Lai, 2010). Process parameters, such as voltage settings, electrode geometries and configurations, gap between electrodes and cross air-flow velocity, have to be optimized to develop EHD industrial units. For example, the drying rate increases with the increase of electric field strength through either the decrease of the electrode gap or the increase of voltage. If some results are more or less consensual, the influence of some parameters needs to be further investigated. For example, multiple needles/wires configuration seems to be necessary in case of industrialization but the number of electrodes per dried surface area remains under debate. Because food can be dried more uniformly, Defraeye and Martynenko (Defraeye and Martynenko, 2019) have shown that the wire-to-mesh configuration shows better potential than conventional wire to plate, but more configurations have to be studied. Finally, some important parameters, such as air humidity, which clearly influences the onset of ionization, have never been extensively considered.

In conclusion, the lack of knowledge on the influence of process parameters and the incertitude concerning the real energetic gain are the main reasons why commercial dryers with EHD enhancement are not yet available. Thus, efforts remain to be done to deploy this technology with huge potential.

12.4 Conclusions and perspectives

EHD drying and dielectric drying and more generally drying with the help of electromagnetic and/or electric fields have a great potential of innovation and application, especially for preserving organoleptic quality of foods and enhancing energy efficiency of the drying process. Several studies remain to be carried out to examine the application of MW, RF, and EHD for the enhancement of evaporation processes and to fill the gap between laboratory scales and industrial scales, especially for EHD drying, which is still under development. Besides additional experimental studies, numerical simulation is a very promising way to enhance the knowledge of physical phenomena involved in such new drying processes (Defraeye and Martynenko, 2018; Malekjani and Jafari, 2018; Ould Ahmedou et al., 2009; Shi et al., 2017). The development and the optimization of new drying technologies can favorably benefit from numerical models able to predict fluid flow and heat and mass transfer. The design and the performance assessment of the innovative drying process as well as its dynamic control should also include the detailed and simultaneous analysis of food quality/energy efficiency issues. Such an integrated modeling approach requires the development of more comprehensive multiphysical models.

References

Alemrajabi, A.A., Rezaee, F., Mirhosseini, M., Esehaghbeygi, A., 2012. Comparative evaluation of the effects of electrohydrodynamic, oven, and ambient air on carrot cylindrical slices during drying process. Dry. Technol. 30 (1), 88–96. https://doi.org/10.1080/07373937.2011.608913.

Bardy, E., Hamdi, M., Havet, M., Rouaud, O., 2015. Transient exergetic efficiency and moisture loss analysis of forced convection drying with and without electrohydrodynamic enhancement. Energy 89, 519–527. https://doi.org/10.1016/j.energy.2015.06.017.

Bardy, E., Manai, S., Havet, M., Rouaud, O., 2016. Drying kinetics comparison of methylcellulose gel versus mango fruit in forced convective drying with and without electrohydrodynamic enhancement. J. Heat Transfer 138 (8) , 084504-1-084504-5. https://doi.org/10.1115/1.4033390.

Chandrasekaran, S., Ramanathan, S., Basak, T., 2013. Microwave food processing—a review. Food Res. Int. 52 (1), 243–261.

Defraeye, T., Martynenko, A., 2018. Electrohydrodynamic drying of food: new insights from conjugate modeling. J. Cleaner Prod. 198, 269–284. https://doi.org/10.1016/j.jclepro.2018.06.250.

Defraeye, T., Martynenko, A., 2019. Electrohydrodynamic drying of multiple food products: Evaluating the potential of emitter-collector electrode configurations for upscaling. J. Food Eng. 240, 38–42. https://doi.org/10.1016/j.jfoodeng.2018.07.011.

Gulati, T., Datta, A.K., 2015. Mechanistic understanding of case-hardening and texture development during drying of food materials. J. Food Eng. 166, 119–138. https://doi.org/10.1016/j.jfoodeng.2015.05.031.

Khaing Hnin, K., Zhang, M., Mujumdar, A.S., Zhu, Y., 2019. Emerging food drying technologies with energy-saving characteristics: a review. Dry. Technol. 37 (12), 1465–1480. https://doi.org/10.1080/07373937.2018.1510417.

Kudra, T., Martynenko, A., 2015. Energy aspects in electrohydrodynamic drying. Dry. Technol. 33 (13), 1534–1540. https://doi.org/10.1080/07373937.2015.1009540.

Kumar, C., Karim, M.A., 2019. Microwave-convective drying of food materials: a critical review. Crit. Rev. Food Sci. Nutr. 59 (3), 379–394. https://doi.org/10.1080/10408398.2017.1373269.

Lai, F.C., 2010. A prototype of EHD-enhanced drying system. J. Electrostat. 68 (1), 101–104. https://doi.org/10.1016/j.elstat.2009.08.002.

Malekjani, N., Jafari, S.M., 2018. Simulation of food drying processes by computational fluid dynamics (CFD); recent advances and approaches. Trends Food Sci. Technol. 78, 206–223. https://doi.org/10.1016/j.tifs.2018.06.006.

Martynenko, A., Bashkir, I., Kudra, T., 2021. Electrically enhanced drying of white champignons. Dry. Technol. 39 (2), 234–244. https://doi.org/10.1080/07373937.2019.1670672.

Martynenko, A., Kudra, T., 2016. Electrically-induced transport phenomena in EHD drying – a review. Trends Food Sci. Technol. 54, 63–73. https://doi.org/10.1016/j.tifs.2016.05.019.

Moses, J., Norton, T., Alagusundaram, K., Tiwari, B., 2014. Novel drying techniques for the food industry. Food Eng. Rev. 6, 43–55.

Mujumdar, A., 2014. Perspectives on international drying symposium series – past, present and future prospects, Proceedings of the 19th International Drying Symposium.

Nazir, S., Azaz Ahmad Azad, Z.R., 2019. Ultrasound: A Food Processing and Preservation Aid. In: Malik, A., Erginkaya, Z., Erten, H. (Eds.), Health and Safety Aspects of Food Processing Technologies. Springer, Cham, pp. 613–632. https://doi.org/10.1007/978-3-030-24903-8_22.

Ould Ahmedou, S., Rouaud, O., Havet, M., 2009. Assessment of the electrohydrodynamic drying process. Food Bioprocess Technol. 2, 240–247.

Piyasena, P., Dussault, C., Koutchma, T., Ramaswamy, H.S., Awuah, G.B., 2003. Radio frequency heating of foods: principles, applications and related properties—a review. Crit. Rev. Food Sci. Nutr. 43 (6), 587–606. https://doi.org/10.1080/10408690390251129.

Robinson, M., 1961. Movement of air in the electric wind of the corona discharge. Trans. Am. Inst. Electr. Eng. Part I: Commun. Electron. 80 (2), 143–150. https://doi.org/10.1109/TCE.1961.6373091.

Shi, C.A., Martynenko, A., Kudra, T., Wells, P., Adamiak, K., Castle, G.S.P, 2017. Electrically-induced mass transport in a multiple pin-plate electrohydrodynamic (EHD) dryer. J. Food Eng. 211, 39–49. https://doi.org/10.1016/j.jfoodeng.2017.04.035.

Singh, A., Orsat, V., Raghavan, V., 2012. A comprehensive review on electrohydrodynamic drying and high-voltage electric field in the context of food and bioprocessing. Dry. Technol. 30 (16), 1812–1820. https://doi.org/10.1080/07373937.2012.708912.

Sosa-Morales, M.E., Valerio-Junco, L., López-Malo, A., García, H.S., 2010. Dielectric properties of foods: reported data in the 21st century and their potential applications. LWT - Food Sci. Technol. 43 (8), 1169–1179. https://doi.org/10.1016/j.lwt.2010.03.017.

Zhou, X., Wang, S., 2019. Recent developments in radio frequency drying of food and agricultural products: a review. Dry. Technol. 37 (3), 271–286. https://doi.org/10.1080/07373937.2018.1452255.

Zielinska, M., Ropelewska, E., Xiao, H.-W., Mujumdar, A.S., Law, C.L., 2020. Review of recent applications and research progress in hybrid and combined microwave-assisted drying of food products: quality properties. Crit. Rev. Food Sci. Nutr. 60 (13), 2212–2264. https://doi.org/10.1080/10408398.2019.1632788.

Electrostatic spray drying of high oil load emulsions, milk and heat sensitive biomaterials

13

A.K.M. Masum, Juhi Saxena, Bogdan Zisu

Spraying Systems Co., Fluid Air, Truganina VIC, Australia

13.1 Introduction

Preservation of biological materials is often achieved by removing free water and lowering the water activity. In traditional commercial settings, this is often achieved by using established technologies such as high-heat spray drying and low-temperature freeze drying. Both are effective; however, each technique is also limited to specific applications. Spray drying, for example, operates at high temperatures and is unsuitable for drying biologically active material susceptible to thermal degradation. Living cells, microorganisms and many active ingredients often result in denaturation, product degradation and loss of quality when heated above specific temperatures (Ananta et al., 2005; Chávez and Ledeboer, 2007). The commercially viable alternative to high-heat spray drying for the preservation of microbiological samples and other biological materials is subzero freeze drying. Although the technology is well established, the low operating temperatures affect the survival rate of microorganisms (Carvalho et al., 2003). However, the underlying limitation of commercial freeze drying is generally not temperature related but rather batch processing under vacuum, thereby limiting throughput.

There is a need for continuous-operation commercial-scale drying technologies that maintain the thermal integrity of a product. The future of food and nutraceutical manufacturing is driven by innovation, and high value adding nutritional, functional, and bioactive ingredients are key to sustainability. Consequently, the drive for high quality ingredients also requires innovation in manufacturing technology necessary to support emerging markets and novel product development. This gap in process capability was recently filled by the PolarDry range of electrostatic spray dryers. By delivering an electrostatic charge during the atomization process of liquid droplets, water is evaporated at lower temperatures than possible in traditional high-heat spray drying. The technology is implemented commercially across north America, Europe, Asia, and Australia with the bulk of commercial applications dedicated to the preservation of biomaterials (e.g., viable cells) and encapsulation of volatile compounds (e.g., oils and flavors). Evaporation capacity in commercial dryers currently ranges from 4 to 200 kg/h.

Heat transfer to the atomized droplets is based on latent heat transfer allowing powders to be dried at exhaust temperatures as low as 30 °C. Successful applications include the drying of biological solutions such as colostrum and lactoferrin where there is little or no loss in biological activity. Other suitable applications include drying of microalgae and living microorganisms. Probiotic microorganisms, agricultural bacteria and various other species associated with the human microbiome have been dried successfully using a polysaccharide carrier to obtain >50% biomass to dry-mass ratio. Survival postdrying is generally high with expected viable losses of approximately half a log reduction or less.

Unlike traditional high-heat spray drying, electrostatic spray drying (ESD) takes place in an inert gas environment where oxygen is replaced by nitrogen. This expands applicability to oxygen sensitive materials and not only appeals to anaerobic microorganisms but is extremely well suited to spray drying of encapsulated oils. Operating costs are reduced by recycling the nitrogen in a closed loop system and the inert nature of the nitrogen gas eliminates the risk of powder explosion from the electrical current delivered during atomization. By electrostatic charging of the active components based on polarity, the surface charge distribution of the atomized droplets changes during the drying process and this becomes evident in resulting powders. In powders with high fat content, some of the surface fat is replaced by protein and carbohydrate. When using a carbohydrate carrier and protein stabilizer, oil retention in the powder reaches 50–80% (w/w). Interest in oil encapsulation is driven by the processing of highly volatile and unsaturated lipids, oil soluble flavor and aroma compounds, nutritional formulations, and cannabinoid oils.

ESD technology was created specifically for processing value adding materials, not only by adopting a unique drying mechanism but also based on throughput. The maximum evaporation capacity of current day electrostatic spray dryers is substantially lower than the largest traditional high heat spray dryers, however, value is created in ESD products by retaining qualities characteristics otherwise lost. In comparison to freeze drying where biological retention is high, ESD technology is at least as effective, and this was demonstrated in the drying of probiotic microorganisms and biologically active lactoferrin (commercially sensitive data not shown). A distinguishing aspect between freeze drying and ESD is, however, efficiency. Freeze drying as a batch process limits productivity, and on the contrary, the ESD process is a continuous operation increasing throughput and reducing production costs.

13.2 Principles of electrostatic spray drying

ESD systems consist of four major components: a pumping system, an electrostatic two-fluid spray nozzle, high voltage power supply, and inert drying gas (Nguyen et al., 2016). During electrostatic atomization, the conductive liquid feed is injected by the pump through a two-fluid nozzle to which the electrostatic charge is applied by a high voltage generator (usually <30 kV). Product viscosity is a limiting factor in all pumping operations, however, a peristaltic pump fitted to all ESD machines provides substantial flexibility in product handling capability. In the ESD process, the electrostatic effect stratifies the components of atomized droplets based on the polarity of the materials. In a feedstock containing polar solvent, the insoluble materials are driven to the droplets' core and the solvent to the droplets' surface. This prevents early shell formation and reduces the time spent in the falling-rate drying period. Droplet size and distribution are not only manipulated by controlling the atomizing gas pressure but these parameters are also controlled by changing the atomizing tip and air cap configuration.

In microencapsulation, an emulsion consists of three major components: a solvent (water or other solvent), a carrier (starch and other high molecular weight polysaccharides), and an active ingredient (oil, vitamin, or other). The components of each emulsion have different polarities. The solvent and carrier, being the high polar components, will have the largest electric dipole, whereas the active ingredient will have the smaller electric dipole due to a less polar nature. The solvent and carrier, being the highest electric dipole, will repel each other and migrate to the outer surface of the drying droplets. The active component, on the other hand, will remain at the core of the drying droplets (Fig. 13.1). This allows efficient drying of the materials at mild temperatures without shell formation.

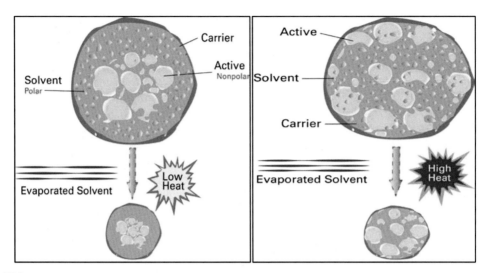

FIG. 13.1

Components of atomized droplets during low-heat electrostatic spray drying (left) and high-heat traditional spray drying (right) (Electrostatic Spray Drying, n.d.).

13.3 Applications of electrostatic spray drying

13.3.1 Whole milk, skim milk, and infant milk formulae

Milk provides a rich source of proteins, lipids, carbohydrates, minerals, and vitamins. Infant milk formula (IMF) is a substitute for human milk and is blended from various dairy and nondairy ingredients to fulfill the nutritional requirements of infants. These milk products are often converted to powder to extend shelf life and for ease of transportation and storage. Traditional high heat spray drying often contributes to the degradation of powder quality during manufacturing (i.e., protein denaturation) and storage (i.e., oxidation, browning). Alternative low heat ESD can minimize the impact of these handling steps to produce powders with unique physicochemical and functional properties.

ESD of whole milk, skim milk, and IMF wet mix was carried out at typical ESD operating conditions specified in Table 13.1. The pressure required to atomize the liquid feed ranges from 100 to 300 kPa and the droplet size is dependent on the pressure applied, with high operating pressures reducing

Table 13.1 Drying conditions for electrostatic spray drying (ESD) and high-heat spray drying (SD) of whole milk, skim milk, and infant milk formula wet mix.

Parameter	ESD	SD
Inlet temperature (°C)	90	180
Exhaust temperature (°C)	35	90
Electrostatic charge (kV)	10	NA

droplet size and vice versa. The electrostatic charge (kV) operates between 1 and 30 kV. The IMF powder was prepared to a target composition of 26% fat (w/w), 15% protein (w/w) and 59% lactose (w/w) (Masum et al., 2019, 2020a). To compare the physicochemical properties of powders, identical formulations of whole milk, skim milk, and IMF were spray dried by the traditional high heat process at conditions reported in literature (Kelly et. al., 2003; Ishwarya and Anandharamakrishnan, 2017; Montagne et al., 2009; Masum et al., 2020b).

The typical moisture content and water activity (a_w) of spray dried whole milk powder (WMP), skim milk powder (SMP), and IMF powder are presented in Table 13.2. Irrespective of the drying technique, and despite low ESD drying temperatures, the moisture content and a_w of all powders were similar and within acceptable limits. The typical moisture content of milk powders should be less than 4% (Pisecky, 2012) as higher moisture content (>4.0%) and water activity (>0.25) shortens shelf life owing to changes in enzymatic, microbial, and oxidative activity. The moisture content in commercial IMF powders usually varies within 2–4% (Hanely et al., 2011).

Surface composition of powders plays a critical role in determining their physicochemical behavior during storage. The surface composition of powders significantly influences their functional properties including solubility, wettability, and oxidative stability (Kim et al., 2003). Understanding powder surface composition can be used to improve quality. In an ideal scenario, the surface of the dairy powders is expected to be mainly composed of protein (due to its high surface activity) while fat is encapsulated at the core of the powder particle. The results for bulk composition analysis of ESD and SD WMP showed 21% fat, 25% protein, and 48% carbohydrate; however, notable differences were observed in the surface chemistry of these powders (Table 13.3). The distribution of fat, protein, and carbohydrates was fat > protein > carbohydrate. Fat was over-represented on the surface, despite constituting only ~21% of the bulk powder formulation. Such results have previously been reported by Kim et al. (2003)

Table 13.2 Typical moisture content and water activity of whole milk powder (WMP), skim milk powder (SMP), and infant milk formula (IMF) powder dried by electrostatic spray drying (ESD) and high heat spray drying (SD).

	WMP	SMP	IMF
	Moisture content (%)		
ESD	2.7–3.4	3.2–3.6	2.1–2.3
SD	1.8–2.0	2.4–2.6	1.8–2.0
	Water activity		
ESD	0.05–0.10	0.09–0.12	0.09–0.13
SD	0.06–0.09	0.17–0.18	0.19–0.21

Table 13.3 Differences in the surface composition of whole milk powder dried with electrostatic spray drying (ESD) and traditional high heat spray drying (SD).

	Surface composition		
	Fat	Protein	Carbohydrate
SD	78%	17%	6%
ESD	↓9%	↑5%	↑4%

suggesting the segregation between the components during drying which essentially results in accumulation of fat on the surface. Surface coverage of spray dried powders was approximately 78% fat and ESD powders due to electrostatic rearrangement of macromolecules had approximately 9% lower surface fat. The electrostatic effect not only lowers the surface fat but this was replaced by carbohydrate (~4% increase in lactose) and protein (~5% increase) on the surface.

Remarkable differences were observed in the morphology of ESD and SD powders. Scanning electron microscope images of WMP, SMP, and IMF are presented at 2000× magnification in Fig. 13.2. ESD dried powders showed agglomeration and the primary particles were predominantly spherical in appearance. High heat spray dried powders were mostly nonagglomerated, and the primary particles were larger than those of ESD powders.

Maillard reactions are a frequent occurrence in thermally processed foods. These are nonenzymatic, sugar–amine reactions that occur between lysine-rich proteins and reducing sugars like lactose, glucose, and fructose. Maillard reactions are accompanied by the formation of brown colored complexes in the advanced stages where the extent of browning is directly correlated with the formation of advanced Maillard reaction products such as 5-hydroxymethyl furfural (HMF). The HMF content in WMP, SMP, and IMF was reported as an indicator of Maillard browning and is shown in Table 13.4. High heat spray drying conditions (180 °C inlet and 90 °C exhaust), accelerated the Maillard reactions and HMF values ranged from approximately 103–111 µg/100 g in WMP, 81–89 µg/100 g in SMP, and 34–40 µg/100 g sample in IMF powders. HMF content was lower in ESD powders manufactured at the lower drying temperatures (90 °C inlet and 35 °C exhaust). HMF was approximately 33% lower in WMP, 57% lower in SMP, and 11% lower IMF.

Sample	WMP	SMP	IMF
ESD			
SD			

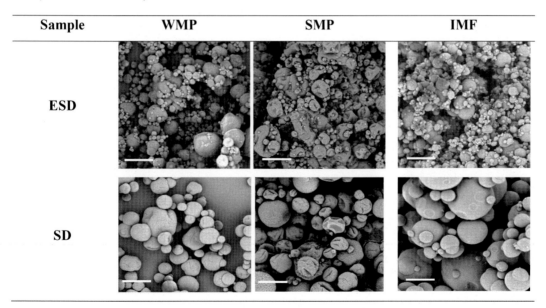

FIG. 13.2

Scanning electron microscope (SEM) images of whole milk powder (WMP), skim milk powder (SMP), and infant milk formula (IMF) dried with electrostatic spray drying (ESD) and traditional high heat spray drying (SD). Magnification = 2000×, scale bar = 30 µm.

Table 13.4 Maillard browning (HMF in µg/100 g sample) in whole milk powder (WMP), skim milk powder (SMP), and infant milk formula (IMF) powder dried with electrostatic spray drying (ESD) and traditional spray drying (TSD).

	WMP	SMP	IMF
SD	103–111	81–89	34–40
ESD	↓33%	↓57%	↓11%

13.3.2 Colostrum and lactoferrin powders

Colostrum is a complex nutrient-rich fluid produced by mammals during the first 4 days of calving (Gopal and Gill, 2000) and is loaded with immune growth (immunoglobulins (Ig)) and tissue repair factors which help develop immunity in the newborn. Lactoferrin (Lf) is a multifunctional glycoprotein, primarily derived from bovine milk, which possesses antibacterial, antioxidant, and anticarcinogenic properties and acts as an iron transfer agent in the human body (Lönnerdal and Iyer, 1995). Currently, lactoferrin and colostrum-based products are manufactured in countries like Australia, New Zealand, USA, and China (Cao et al., 2007). Commercially, these products are either spray-dried or freeze-dried (Yu et al., 2015); however, extremely high and low processing temperatures expose the product to thermal damage.

In this example, a high-drying and low-drying temperature study shows the significance of the spray dryer inlet temperature. Liquid colostrum and lactoferrin were dried with an electrostatic spray dryer (Model 001, PolarDry Electrostatic Spray Dryer, Spraying Systems Co., Naperville, IL, USA) at high inlet temperature (150 °C) and low inlet temperature (90 °C). The active IgG content was measured in colostrum powders and active lactoferrin was determined in Lf powders by the ELISA Quantitation method (ELISA Kit, Catalog No, E10-126, Bethyl Laboratories, Montgomery, TX, USA). A significantly greater retention of the immunomodulatory compounds (~16% higher bioactive yield retention) was observed in colostrum powders dried at the lower temperature (90 °C). Lactoferrin powders dried at the lower inlet temperature (90 °C) retained close to 100% of the starting biological activity and approximately 10% was lost at 150 °C.

13.3.3 Yoghurt powders

Yoghurt is made by the fermentation of lactose using lactic acid bacteria, often with co-cultures of *Streptococcus thermophilus* and *L. delbrueckii* subsp. *bulgaricus*. Yoghurt may be dried to a powder by spray-drying for direct consumption by reconstitution or as an ingredient in confectionaries, bakery foods, soup bases, and dips (Koc et al., 2010). In this example, skim milk containing 10% solids (w/w) was fermented to pH 4.8 then dried by ESD (Model 001, PolarDry Electrostatic Spray Dryer, Spraying Systems Co., Naperville, IL, USA) at inlet temperatures below 95 °C. The viable cell counts (cfu/mL) for *Streptococcus thermophilus* and *L. delbrueckii* subsp. *bulgaricus* in yoghurt were quantified and the powder had no-loss of cell viability (Fig. 13.3), highlighting the effectiveness of low temperature ESD in preservation of viable microorganisms.

13.3.4 Oil encapsulation

Encapsulation of oil in a solid matrix preserves the inherent functional characteristics of the oil, including aroma and antioxidative properties. The process involves entrapment of the oil within a solid

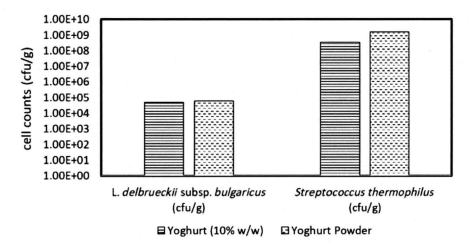

FIG. 13.3

Cell counts (cfu/g) in yogurt (10% w/w) fermented to 4.8 pH and yogurt powders dried by low temperature ESD.

matrix which acts as a barrier to oxygen, water and light (Turchiuli et al., 2005). The encapsulation of oils in powders has been common practice in the food and pharmaceutical industries for decades. Vegetable oils are successfully spray-dried to make powders with ~30–50% oil encapsulation often in a solid matrix composed of maltodextrin and acacia gum at an inlet air temperature of 220 °C (Turchiuli et al., 2005; Fuchs et al., 2006). In this example, ESD was used to emphasize the encapsulation efficiency of vegetable oil in emulsions containing maltodextrin, sodium caseinate and starch at low operating temperatures. Different ratios of the ingredients were mixed to prepare emulsions with 50–70% oil load. Typical ESD drying conditions are shown in Table 13.5. The moisture content of ESD powders was <2.5% and a_w < 0.20. Powders dried with 50–60% oil load had an encapsulation efficiency of ~98% with surface free fat < 1.5%. Approximately 80% encapsulation efficiency was achieved at 70% oil load.

The physical appearance and morphology of oil encapsulated powders is shown in Fig. 13.4. Scanning electron microscopy images of ESD oil encapsulated powders at 10,000× magnification showed a wide distribution of spherical and agglomerated primary particles. There were no cracks or fissures observed in the particle morphology implying good retention of oil within the particles.

Table 13.5 Typical operating parameters for electrostatic spray drying of oil encapsulated emulsions.

Parameter	Drying conditions
Inlet temperature (°C)	110–140
Exhaust temperature (°C)	40–60
Electrostatic charge (kV)	5–10

Oil load	Powder	10,000× Magnification
50%		

FIG. 13.4

Physical appearance of 50% (w/w) oil load powder (left) and scanning electron microimages after electrostatic spray drying. Magnification = 10,000×, scale bar = 8 μm.

13.4 Conclusions

ESD is an innovative technology designed to meet the demands and challenges associated with conventional spray- and freeze drying by providing alternative commercial solutions. ESD combines gas–liquid atomization and electrostatic charge, which allows drying of atomized liquid droplets at temperatures lower than conventional high heat spray drying. ESD is suitable for processing agricultural, food, pharmaceutical, and biomedical materials and in this chapter the potential of this technology was demonstrated by drying milk and milk products, bioactive proteins, living microorganisms, and high oil load emulsified products. The lower drying temperatures associated with ESD were shown to preserve functional and bioactive properties and the electrostatic element encouraged spontaneous agglomeration of the atomized particles. By rearranging the macromolecular composition of emulsified products based on electrostatic charge, ESD produced powders with high oil load, low surface free fat, and demonstrated high encapsulation efficiency.

References

Ananta, E., Volkert, M., Knorr, D., 2005. Cellular injuries and storage stability of spray-dried *Lactobacillus rhamnosus* GG. Int. Dairy J. *15* (4), 399–409.

Cao, J., Wang, X., Zheng, H., 2007. Comparative studies on thermoresistance of protein G-binding region and antigen determinant region of immunoglobulin G in acidic colostral whey. Food Agric. Immunol. *18* (1), 17–30.

Carvalho, A.S., Silva, J., Ho, P., Teixeira, P., Malcata, F.X., Gibbs, P., 2003. Impedimetric method for estimating the residual activity of freeze-dried *Lactobacillus delbrueckii* ssp. bulgaricus. Int. Dairy J. *13* (6), 463–468.

Chávez, B.E., Ledeboer, A.M., 2007. Drying of probiotics: optimization of formulation and process to enhance storage survival. Dry. Technol. *25* (7-8), 1193–1201.

Electrostatic Spray Drying (n.d.). Electrostatic spray drying – microencapsulation. Retrieved on December 22, 2020 from https://www.fluidairinc.com/AU/encapsulation.html.

Fuchs, M., Turchiuli, C., Bohin, M., Cuvelier, M.E., Ordonnaud, C., Peyrat-Maillard, M.N., Dumoulin, E., 2006. Encapsulation of oil in powder using spray drying and fluidised bed agglomeration. J. Food Eng. 75 (1), 27–35.

Gopal, P.K., Gill, H.S., 2000. Oligosaccharides and glycoconjugates in bovine milk and colostrum. Br. J. Nutr. 84 (S1), 69–74.

Hanley, K.J., Cronin, K., O'Sullivan, C., Fenelon, M.A., O'Mahony, J.A., Byrne, E.P., 2011. Effect of composition on the mechanical response of agglomerates of infant formulae. J. Food Eng. 107 (1), 71–79.

Ishwarya, S.P., Anandharamakrishnan, C., 2017. Spray drying. In: Anandharamakrishnan, C. (Ed.), Handbook of Drying for Dairy Products, first ed. Wiley, United States, pp. 57–94.

Kelly, A.L., O'Connell, J.E., Fox, P.F., 2003. Manufacture and properties of milk powders. In: Fox, P.F., McSweeney, P.L.H. (Eds.), Advanced Dairy Chemistry Volume 1: Proteins, third ed., Kluwer Academic Publishers, USA, pp. 1027–1061.

Kim, E.H.J., Dong Chen, X., Pearce, D, 2003. On the mechanisms of surface formation and the surface compositions of industrial milk powders. Dry. Technol. 21 (2), 265–278.

Koc, B., Yilmazer, M.S., Balkır, P., Ertekin, F.K., 2010. Spray drying of yogurt: optimization of process conditions for improving viability and other quality attributes. Dry. Technol. 28 (4), 495–507.

Lönnerdal, B., Iyer, S., 1995. Lactoferrin: molecular structure and biological function. Annu. Rev. Nutr. 15 (1), 93–110.

Masum, A.K.M., Chandrapala, J., Adhikari, B., Huppertz, T., Zisu, B, 2019. Effect of lactose-to-maltodextrin ratio on emulsion stability and physicochemical properties of spray-dried infant milk formula powders. J. Food Eng. 254, 34–41.

Masum, A.K.M., Chandrapala, J., Huppertz, T., Adhikari, B., Zisu, B, 2020b. Influence of drying temperatures and storage parameters on the physicochemical properties of spray-dried infant milk formula powders. Int. Dairy J. 105, 104696.

Masum, A.K.M., Huppertz, T., Chandrapala, J., Adhikari, B., Zisu, B, 2020a. Physicochemical properties of spray dried model infant milk formula powders: influence of whey protein-to-casein ratio. Int. Dairy J. 100, 104565.

Montagne, D.H., Van Dael, P., Skanderby, M., Hugelshofer, W., 2009. Infant formulae – powders and liquids. In: Tamime, A.Y. (Ed.), Dairy Powders and Concentrated Products first ed. Wiley-Blackwell, West Sussex, UK, pp. 294–331.

Nguyen, D.N., Clasen, C., Van den Mooter, G., 2016. Pharmaceutical applications of electrospraying. J. Pharm. Sci. 105 (9), 2601–2620.

Pisecky, J., 2012. Achieving product properties. In: Westergaard, V., Refstrup, E. (Eds.), Handbook of Milk Powder Manufacture second ed. GEA Process Engineering A/S, Copenhagen, Denmark, pp. 163–198.

Turchiuli, C., Fuchs, M., Bohin, M., Cuvelier, M.E., Ordonnaud, C., Peyrat-Maillard, M.N., Dumoulin, E., 2005. Oil encapsulation by spray drying and fluidised bed agglomeration. Innovative Food Sci. Emerg. Technol. 6 (1), 29–35.

Yu, H., Zheng, Y., Li, Y., 2015. Shelf life and storage stability of spray-dried bovine colostrum powders under different storage conditions. J. Food Sci. Technol. 52 (2), 944–951.

Dairy encapsulation systems by atomization-based technology

Yong Wang[a], Bo Wang[b], Cordelia Selomulya[a]

[a]*School of Chemical Engineering, UNSW, Kensington, NSW, Australia*
[b]*School of Behavioural and Health Sciences, Australian Catholic University, Kensington, NSW, Australia*

14.1 Introduction

Microencapsulation technology aims to entrap bioactive compounds in droplets and/or particles using a layer of coating or physicochemical matrix at the microscale (Dickinson, 1992; McClements, 1999). The entrapped bioactive compounds are usually called "active," "core," "internal phase," or "fill," while the materials used to encapsulate the bioactive compounds are referred to as "shell," "wall," "external phase," or "membrane". Generally, the "shell" material is immiscible and should not react with the "core" compounds. In the microencapsulation delivery systems for food applications, common "shell" materials are carbohydrate polymers (sugars, polysaccharides, and gums), proteins, natural fats (lipids and waxes), and synthetic polymers (Gibbs et al., 1999).

One reason to apply microencapsulation technology in food applications is to stabilize and protect susceptible ingredients against external environmental stresses such as elevated temperature, oxygen, light, and the presence of other reactive food ingredients/components (Barrow et al., 2013). The undesirable color, flavor, and taste of the entrapped compound can be significantly masked with encapsulation to prevent interference with the sensory profile of the final food products (Galindo-Cuspinera, 2011). In the case where a dehydration process is involved, ease of handling can be achieved by converting liquid microcapsules into free-flowing powder (Barrow et al., 2013). Microencapsulation can also control the release of bioactive compounds. Various release triggers such as temperature, pH, mechanical stresses, and enzymatic decomposition can be used to ensure targeted delivery of the bioactive ingredients (e.g., small or large intestines). This includes a delayed or sustained release of the encapsulated compounds over time (Madene et al., 2006).

Dairy ingredients are often used in microencapsulation as either wall material or core material, or both. The protein components of milk, namely casein and whey proteins, have good emulsifying ability and capability to encapsulate bioactive ingredients (Ramos et al., 2019). In addition, the pleasant flavor and high melting temperature of milk fat render it suitable as a wall material in spray chilling applications (Queiros et al., 2020). Lactose itself is usually not sufficient for use as a sole wall material, but is often used in combination with other ingredients (e.g., proteins and polysaccharides) to increase encapsulation efficiency and shelf-life stability (Li et al., 2017). Active ingredients derived from dairy, such as peptides and lactoferrin, can be microencapsulated to improve bioavailability by increasing the tolerance through either processing and/or in the digestive tract (Darmawan et al., 2020). Dairy products are processed using atomization-based technology to produce powders. Thus, dairy ingredients have been extensively studied, especially on spray drying, from a single droplet behavior to large-scale industrial optimization

Food Engineering Innovations Across the Food Supply Chain. DOI: https://doi.org/10.1016/B978-0-12-821292-9.00023-6

and simulation (Schuck et al., 2016). This existing knowledge also lays the foundation for exploring the role of dairy ingredients in microencapsulation. Due to their availability, excellent functional properties (emulsification and encapsulation, etc.) and compatibility in food systems, dairy ingredients will continue to be one of the key research topics in microencapsulation (Shivaram and Saini, 2019).

Microcapsules can be prepared via either physical or chemical means. The physical techniques include homogenization, spray drying, spray chilling/cooling, spinning disk, fluidized bed coating, and extrusion. The chemical techniques include phase separation/transition, solvent evaporation, single and complex coacervation, and liposomes (Gouin, 2004). The industrial characteristics of each microencapsulation technique are shown in Fig. 14.1. Among these techniques, spray drying, spray chilling/cooling, and fluidized bed coating can be grouped as atomization-based techniques since they involve spraying droplets either to convert them into powders or to form the shell layers. Contemporarily, most of the microencapsulated food ingredients are manufactured using atomization-based techniques. Around 80–90% of microcapsules food ingredients are currently produced using the spray drying technique (Mahdavi et al., 2014). Thus, this review provides an overview of recent updates on this topic, together with the application of these microcapsules in the food products.

14.2 Atomization-based technology for encapsulation
14.2.1 Spray drying

Spray drying is one of the most common processing methods for microencapsulation due to its broad applicability, ease of scale-up, and relatively low production costs (Eun et al., 2019). The conventional

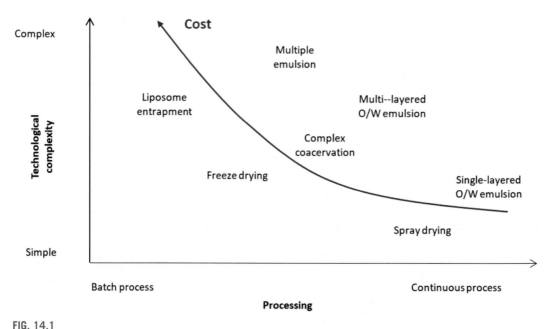

FIG. 14.1

Characteristics of different microencapsulation techniques (modified with permission from Barrow et al., 2013).

process is to mix the wall material and the core material in the pretreatment steps (e.g., homogenization of emulsion) before spray drying. Additional downstream processing, such as fluidization or granulation, may follow for some products. The main parameters that can be adjusted during the spray drying process include the inlet/outlet temperatures (usually there is a correlation between the two) of the drying air, wall: core material ratio, solid content, feed rate, etc. (Drosou et al., 2017). For microcapsules, the main challenge is usually the drying temperature, as the encapsulated ingredients can be sensitive to heat damage (Huang et al., 2017). Another challenge is nonhomogeneous product, which could cause variation of the microcapsule characteristics. A feasible solution is to produce uniform particles using microfluidic drier, where the spray nozzle is modulated by the microfluidic controller to break down the liquid flow into droplets with approximately similar size, as shown in Fig. 14.2 (Waldron et al., 2016; Liu et al., 2016; Amelia et al., 2011).

14.2.2 Spray chilling

Similar to the principle of spray drying, spray chilling uses cold air instead of hot air. Spray chilling has a unique advantage in the production of microcapsules; that is, the low temperature to maintain the functionality of raw materials to the greatest extent. However, it is only suitable for raw materials that can be solidified quickly to form particles at the low temperature in a relatively short residence time (several seconds before it reaches the bottom of the chilling chamber). Among the dairy ingredients, milk fat has the potential for this purpose (Queiros et al., 2020).

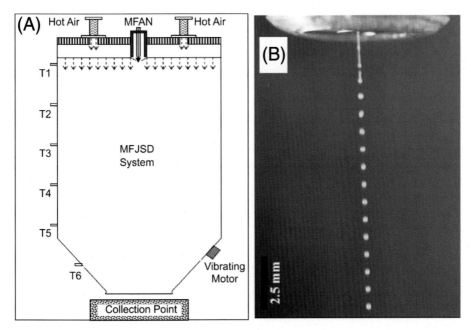

FIG. 14.2

(A) Microfluidic jet spray dryer (MFJSD) schematic diagram (not to scale), T1–T6 indicate the thermocouples; (B) Monodisperse droplet formation by the microfluidic jet spray (with permission from Amelia et al., 2011).

14.2.3 Fluidized bed coating

Fluidized bed coating systems include bottom spray (Wurster), top spray, and tangential spray (rotary) systems, among which the bottom spray system is the most commonly used equipment to create the coating for small particles (diameter <6.35 mm; KuShaari et al., 2006). The bottom spray fluidized bed coating technique was developed by Dr. Dale E. Wurster in the 1960s (Wurster, 1963, Lindlof and Wurster, 1964). In this fluidized bed coating system, the particles with lower density move upward within the drying chamber, concurrently with the spray direction (Trojanowska et al., 2017). Due to rapid drying, the potential for particle agglomeration and attrition are minimized. So far, this technique has been widely used for encapsulation for a wide range of food ingredients, pharmaceutical tablet, confectionary items, fertilizers, chemicals, and even grains (Teunou and Poncelet, 2002; Tzika et al., 2003; Liu and Lister, 1993).

14.3 Dairy ingredients as wall materials for encapsulation

14.3.1 Dairy proteins (casein/whey)

The proteins in dairy products can be divided into two categories, caseins and whey, which account for about 80% and 20% of the protein content of dairy milk, respectively (Phillips and Williams, 2011). In industrial production, casein or whey are usually divided into different quality grades according to the protein content. Since whey is a by-product of cheese processing, it is more commonly used as a food ingredient than casein (Chen et al., 2019). For microencapsulation, there are more published studies on casein because of its functionality, although whey has also been studied extensively as wall material (Augustin and Oliver, 2014).

Casein can form micelles in the milk, which carry negative charges on their surface as a result of phosphorylation and tend to bind nanoclusters of amorphous calcium phosphate (Glab and Boratynski, 2017). Due to its excellent emulsifying ability, casein has been used in many studies to encapsulate oil/fat as wall materials, including conjugated linoleic acid, fish oil, flaxseed oil, chia oil, sunflower oil, etc. (Zhuang et al., 2018; Moisio et al., 2014; Binsi et al., 2017; Shiga et al., 2017; Vaucher et al., 2019). Casein has been reported to show high encapsulation efficiency as wall material for clove oil (97%) to produce particles via spray drying (Sahlan et al., 2019). When interacting with other components, casein's encapsulation capability can be further enhanced. Casein was able to form complexes with pectin and was reported to have higher encapsulation efficiency (67%) for fish oil after spray drying compared to gum arabic (56%; Vaucher et al., 2019). When conjugated with lactose via the Maillard reaction, casein was reported to be able to achieve stable microcapsules with canola oil loading as high as 80% after spray drying (Li et al., 2017).

Casein is also widely used as a wall material for bioactive compounds (Reineccius, 2019). One reason is that casein is able to interact with the bioactive compounds via electrostatic attraction or molecular entanglement and entrap them into the inner part of the microcapsule, thus provide protection from high temperatures during spray drying (Pan et al., 2013). Casein has been used to encapsulate curcumin using the microfluidic jet spray dryer with an (inlet temperature of 180 °C) powders displaying high curcumin retention rates (>95%) and 86% of the antioxidant activity of pure curcumin by the 2,2-azino-bis (3-ethylbenzothiazoline-6-sulfonic acid) radical scavenging assay method. In this study, curcumin was found to bind with casein via hydrophobic sites without altering the micellar structure of caseins (Khanji et al., 2018). Casein can protect the bioactivity and provide desirable digestion profile when binding with emodin via hydrophobic force, while the spray drying inlet temperature was kept as low as 130 °C (Yang et al., 2020).

Whey is often used as a wall material for oil/fat microencapsulation because of its emulsifying properties. Whey protein isolate (Hilmar 9000, 93.0% protein, dry base) can encapsulate tributyrin and also reduce the surface oil to 24% by spray drying (35% surface oil for the control sample dried by a two-fluid nozzle), improving the encapsulation efficiency and rehydration properties (Shi and Lee, 2020). Similar to casein, whey proteins can also achieve higher encapsulation efficiency when it is combined with polysaccharides, compared to protein alone. The formation of whey proteins-alginate double-layer emulsions can protect flaxseed oil against oxidation during storage, with the highest encapsulation efficiencies of 84% (calculated by the percentage of encapsulated oil in total oil used) in spray-dried powders (Fioramonti et al., 2019). Whey protein isolate-agar gum and gellan gum complex can form double shells in the encapsulated tuna oil powders when spray-dried at 160 °C (inlet) with high encapsulation efficiency of 95.8% (Wang et al., 2019). Additionally, the gel-forming ability of whey is considered to contribute to the stability of the encapsulation system. When pregelled at 80 °C, whey protein isolates can form a hard shell for microcapsules of acetaminophen during spray drying and may be modulated for controlled release purpose (Tan et al., 2020). The pregelled whey protein isolate can also achieve a retention rate of encapsulated pepsin of around 84.3–89.4% with spray drying inlet temperature from 110 °C to 190 °C (Tan et al., 2019).

The capability of whey protein as a wall material for encapsulation purpose can be further enhanced via the modification of their molecular conformation. Ethanol desolvation (using up to 50% v/v ethanol) was reported to enable the exposure of embedded hydrophobic amino acids of whey protein to riboflavin to facilitate the formation of riboflavin-whey complexes (Ye et al., 2019). Moreover, crosslinking using calcium in addition to ethanol desolvation can tune the digestion release profile of the spray-dried powder, as shown in Fig. 14.3 (e.g., either rapid peptic digestion in less than 30 min, or excellent gastric resistance and intestinal release; Ye et al., 2019). Similarly, modified whey protein using ethanol desolvation treatment was successfully used to lower the unbound curcumin content to <5% (w/w) in the spray-dried powder and the powder showed better water solubility, stability, and bioaccessibility than raw curcumin (Ye et al., 2021).

For microencapsulation of probiotics, some studies have found that whey may have a unique advantage compared to other wall materials. Whey performed well as wall materials for protecting bacterial cells from high temperatures during spray drying, with the *Lactobacillus plantarum* viability of 10.68 and 10.26 log CFU per gram before and after the drying process, respectively (Eckert et al., 2017). Highly concentrated sweet whey (20–30% w/w) was reported to significantly increase bacteria survival rate up to 70% after spray drying (inlet temperature of 200 °C; Huang et al., 2016). Cheese whey also showed a good potential for the production of encapsulated lactic acid bacteria cultures via spray drying (Rama et al., 2019). Interestingly, when *Lactobacilluus reuteri* was fermented using whey as the key component in the growth media and this slurry was spray- directly, an improved survival rate of this probiotics (32% higher) was observed in the simulated digestive juice (Jantzen et al., 2013).

14.3.2 Lactose

Lactose is usually not used alone as the wall material of the microcapsules, but in combination with other wall materials. The efficiency of casein-stabilized emulsion increases in the presence of lactose,

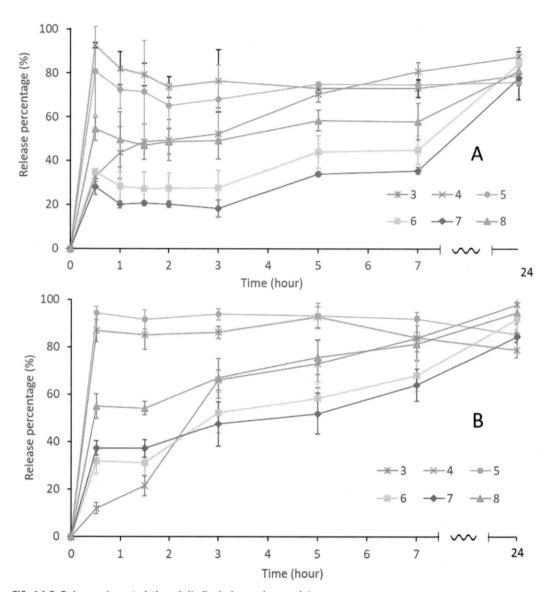

FIG. 14.3 Release characteristics of riboflavin from microparticles

(A) in the SGF with pepsin; (B) in the SIF with trypsin; [(3) microparticles prepared from pure water solvent; from 30% v/v ethanol system at (4) $[Ca^{2+}]$ = 0 mM; (5) $[Ca^{2+}]$ = 0.5 mM; (6) $[Ca^{2+}]$ = 1 mM; (7) $[Ca^{2+}]$ = 2 mM; (8) in 50% v/v ethanol system at $[Ca^{2+}]$ = 2 mM] (with permission from Ye et al., 2019).

which was due to the ability of lactose to form solid (or glassy) capsules during sudden dehydration (Vega and Roos, 2006). It has been reported that the surface fat of encapsulated powder could be reduced from 30% to less than 5% when lactose was added at equal amount of casein (Vega et al., 2007). In the encapsulation of flavorant ethyl butyrate via spray drying, the encapsulation efficiency was highest (42%) when the lactose concentration increased to 80% w/w in the wall material (Li et al., 2016). When lactose was used to replace 70% of maltodextrin during encapsulation, the carotenoid encapsulation efficiency significantly increased by 22.6% because lactose may act as a hydrophilic sealing material and reduces the diffusion of carotenoids from the particle core to the surface during spray drying (Etzbach et al., 2020).

14.3.3 Milk fat

Compared with other dairy ingredients, milk fat is not as commonly used as wall material for microcapsules, except in spray chilling. The content of saturated fat in milk fat is relatively high and results in a high melting point range (up to 40 °C depending on the milk source and separation conditions; Lopez et al., 2006; Toral et al., 2013). The solid nature at room temperature makes it feasible as a carrier for oil-soluble core materials. A recent study showed that hydrogenated anhydrous milk fat could be made into spherical shaped particles with a smooth surface by spray chilling at 2 °C, which can potentially be used to deliver active compounds (Queiros et al., 2020). It should be noted that these microcapsules are more suitable for oil-soluble active ingredients, while oil-insoluble ingredients (such as probiotics) would be limited in size to be properly encapsulated within the spray chilled particles (Procopio et al., 2018; Pedroso et al., 2013). In addition, special attention should be given to avoid aggregation during processing and storage, due to the sticky nature of the surface fat (Yin and Cadwallader, 2019).

14.3.4 Mixtures

In addition to single-component dairy products, there are many mixed dairy products used as microcapsule wall materials. Whole milk powder and skimmed milk powder are the most common materials due to their low cost and availability. Common core materials of whole milk powder are fat/oil to produce fat-filled milk powders in the food industry, the production principle of which is microencapsulation via spray drying of emulsions (Kelly and Fox, 2016). Palm oil and coconut oil are common core materials, with the oil loading of microcapsules varying from 26% to 28% (most common cases, w/w) to 50%, depending on the targeted application (Masters, 2018; Hedayatnia and Mirhosseini, 2018).

Some fats with health benefits are microencapsulated with skimmed milk powder to retain the nutritional value. For example, conjugated linoleic acid can be microencapsulated with skimmed milk powder to increase the level of high-density lipoprotein cholesterol (Rodríguez-Alcalá et al., 2013). In an attempt to embed fish oil with skim milk powder (wall material: core material = 2:1), the microcapsule powder can partially cover the odor of fish oil, but other ingredients are required to improve the encapsulation efficiency, including methylcellulose, maltodextrin, and/or lecithin (Kolanowski et al., 2007). To further improve the stability of the encapsulated fish oil, multiple atomization-based technique can be used such as the spray drying process using acacia gum-skim milk powder-grape juice system as the internal wall material, followed by a spray cooling process with hydrogenated and inter-esterified vegetable oils as the external fat coating (Fadini et al., 2019).

Curcumin can be encapsulated in skim milk powder to improve stability during storage for up to 2 months (Neves et al., 2019). In addition, skim milk powder has been reported as an effective wall material to encapsulate microbial cultures in a microcapsule matrix, with a survival rate of 64.4% when spray-dried at 75 °C (Würth et al., 2018). There are also examples of successful microencapsulation of vitamins and bioactive compounds via spray drying, such as encapsulation of vitamin D using milk powder, polyphenols from chokeberry using skim milk and maltodextrin, and the antioxidant phenolic compounds from green coffee using skim milk (Maurya et al., 2020; Ćujić-Nikolić et al., 2019; Desai et al., 2019).

Whey powder (or sweet whey) is a by-product of cheese production, which is directly dried into powder without separation, where the main components are whey protein and lactose. It can be used to microencapsulate probiotics and vitamins (Huang et al., 2016; Mahdi Jafari et al., 2019). Compared to pure dairy ingredients, mixed ingredients usually cost less, however, a larger amount is often needed, and it may be difficult to adjust the composition for encapsulation in a targeted manner.

14.4 Dairy ingredients as core materials for encapsulation

14.4.1 Lactoferrin

Lactoferrin is an iron-binding glycoprotein with a molecular weight of about 80 kDa (Bourbon et al., 2019). It has a unique nutritional value for the development of infants and the maintenance of adult's health (Wang et al., 2019). The bioactivity of lactoferrin is easily lost during processing and storage, thus microencapsulation is an effective technique for protection (Wang et al., 2019; Liu et al., 2020). β-Lactoglobulin and α-lactalbumin have been reported as effective carriers for lactoferrin via spray drying, predominantly due to the interactions among the acidic and basic amino acid residues of the proteins (Darmawan et al., 2020). Spray drying has been reported to enhance the extent of intermolecular associations between lactoferrin and oat β-glucan during encapsulation (Yang et al., 2020). In addition, the protein–polysaccharide complexes, such as the whey protein–pectin nanocomplex, have been reported to have higher encapsulation efficiency for lactoferrin than individual wall materials (Raei et al., 2017). The complexation of lactoferrin with negatively charged alginate gel matrices was able to prevent its denaturation during spray drying or in the acidic gastric environment (Schuck et al., 2016; Raei et al., 2017).

14.4.2 Peptides

Dairy proteins can be hydrolyzed by protease to release bioactive fragments during processing or storage. Milk peptides exert multifunctional properties, including antimicrobial, immunomodulatory, antioxidant, antithrombotic, and antagonistic activities against various toxic agents (Mohanty et al., 2016). Among different techniques of encapsulation, spray drying is the most economical and flexible process for reduction of hygroscopicity, masking of unpleasant flavors and increasing the stability of bioactive peptides (Sarabandi et al., 2020; Wang and Selomulya, 2020).

Starch, maltodextrin and hydrocolloids are the most common microcapsule wall materials for dairy peptides. Studies have found that using maltodextrin to microencapsulate casein peptides through spray drying can reduce the bitterness while maintaining 90–98% of antioxidant properties of peptides (Sarabandi et al., 2018). After encapsulation with a maltodextrin–gum arabic blend, casein peptides can still maintain the antioxidant activity with encapsulation efficiency higher than 80% (Rao et al., 2016). Whey peptides can be encapsulated by alginate-arabic gum, and 85% of the angiotensin-converting enzyme activity of the whey peptide can be retained in the dried powder in the in vitro digestion experiments (Alvarado et al., 2019). In the spray drying of peptides for encapsulation, the inlet/outlet temperatures are important. Some studies have found that when the inlet temperature is higher than 180 °C, the amino acid residues of whey peptides will be easily conjugated with other components, and may cause agglomeration and reduced solubility (Darmawan et al., 2020).

14.5 Summary

Technologies such as spray drying and spray chilling have been successfully used to produce a variety of microcapsules, with dairy ingredients used as wall materials for microcapsules. In addition, functional ingredients derived from dairy products such as lactoferrin and dairy peptides have also been used as core materials. The physicochemical, morphological, thermal, and sensorial properties, in addition to encapsulation efficiency and release profile, are relevant characteristics to assess the quality of the microcapsule particles. Brief examples of microencapsulation for functional food applications are provided, including ongoing research currently taking place.

There are several areas for future research in the encapsulation systems by atomization-based technology. Current dairy-based encapsulation delivery systems mostly produce homogenous structures (uniformly distributed core material), while core-shell structures are rare (possible by enriching the outer layer with specific components, or by forming a thin layer of hydrogel on the surface). The ability to produce powders that approximate the structure of a core-shell by atomization-based technology from a single nozzle would potentially improve the protection of core materials. The size distribution and morphology of particles produced by current industrial atomization-based technology are relatively varied. Although monodisperse particles can be obtained by microfluidic jet spray dryer, currently the application is limited for research or for small-scale production. In addition, there is a lack of elucidation of mechanisms of interaction between dairy proteins and polysaccharides (from plant, animal, and/or microorganism), although they are crucial for achieving higher encapsulation efficiency. Most studies are focused on a particular protein-polysaccharide system, and there is no general approach to guide the design of future delivery systems. The increasing use of plant proteins to replace animal proteins is growing. However, a complete replacement of dairy proteins with plant proteins may not be practical considering the nutritional value of dairy proteins and often technically difficult since the functionality of plant proteins in most cases is inferior to that of dairy proteins. The development of a dual protein combination may be a reasonable compromise to take into account the benefits of both plant and animal proteins, but still require further research to construct an efficient delivery system.

References

Alvarado, Y., et al., 2019. Encapsulation of antihypertensive peptides from whey proteins and their releasing in gastrointestinal conditions. Biomolecules 9 (5), 164.

Amelia, R., et al., 2011. Microfluidic spray drying as a versatile assembly route of functional particles. Chem. Eng. Sci. 66 (22), 5531–5540.

Augustin, M.A., Oliver, C.M., 2014. Use of milk proteins for encapsulation of food ingredients. Microencapsulation in the Food Industry: A Practical Implementation Guide. Academic Press, San Diego, CA, USA, pp. 211–226.

Barrow, C., et al., 2013. Spray drying and encapsulation of omega-3 oils. In: Jacobsen, C. et al (Ed.), Food Enrichment with Omega-3 Fatty Acids. Woodhead Publishing Limited., Cambridge, UK, pp. 194–219.

Binsi, P.K., et al., 2017. Structural and oxidative stabilization of spray dried fish oil microencapsulates with gum arabic and sage polyphenols: characterization and release kinetics. Food Chem. 219, 158–168.

Bourbon, A.I., et al., 2019. 6 - Nanoparticles of lactoferrin for encapsulation of food ingredients. In: Jafari, S.M. (Ed.), Biopolymer Nanostructures for Food Encapsulation Purposes. Academic Press, London EC2Y 5AS, United Kingdom, pp. 147–168.

Chen, H., et al., 2019. Nanoparticles of casein micelles for encapsulation of food ingredientsBiopolymer Nanostructures for Food Encapsulation Purposes. Elsevier, London EC2Y 5AS, United Kingdom, pp. 39–68.

Ćujić-Nikolić, N., et al., 2019. Chokeberry polyphenols preservation using spray drying: effect of encapsulation using maltodextrin and skimmed milk on their recovery following in vitro digestion. J. Microencapsul. 36 (8), 693–703.

Darmawan, K.K., et al., 2020. High temperature induced structural changes of apo-lactoferrin and interactions with β-lactoglobulin and α-lactalbumin for potential encapsulation strategies. Food Hydrocolloids 105, 105817.

Desai, N.M., et al., 2019. Microencapsulation of antioxidant phenolic compounds from green coffee. Prep. Biochem. Biotechnol. 49 (4), 400–406.

Dickinson, E., 1992. An Introduction to Food Colloids. Oxford Science Publishers, Oxford, UK.

Drosou, C.G., Krokida, M.K., Biliaderis, C.G., 2017. Encapsulation of bioactive compounds through electrospinning/electrospraying and spray drying: a comparative assessment of food-related applications. Dry. Technol. 35 (2), 139–162.

Eckert, C., et al., 2017. Microencapsulation of *Lactobacillus plantarum* ATCC 8014 through spray drying and using dairy whey as wall materials. LWT - Food Sci. Technol. 82, 176–183.

Etzbach, L., et al., 2020. Effects of carrier agents on powder properties, stability of carotenoids, and encapsulation efficiency of goldenberry (*Physalis peruviana* L.) powder produced by co-current spray drying. Curr. Res. Food Sci. 3, 73–81.

Eun, J.-B., et al., 2019. A review of encapsulation of carotenoids using spray drying and freeze drying. Crit. Rev. Food Sci. Nutr. 60 (21), 1–26.

Fadini, A.L., et al., 2019. Optimization of the production of double-shell microparticles containing fish oil. Food Sci. Technol. Int. 25, 359–369.

Fioramonti, S.A., et al., 2019. Spray dried flaxseed oil powdered microcapsules obtained using milk whey proteins-alginate double layer emulsions. Food Res. Int. 119, 931–940.

Galindo-Cuspinera, V., 2011. Taste masking: trends and technologies. Prepared Foods 180, 51–56.

Gibbs, B.F., et al., 1999. Encapsulation in the food industry: a review. Int. J. Food Sci. Nutr. 50 (3), 213–224.

Glab, T.K. and J. Boratynski, Potential of casein as a carrier for biologically active agents.Top Curr. Chem. (Cham), 2017. 375(4): p. 71.

Gouin, S., 2004. Microencapsulation: industrial appraisal of existing technologies and trends. Trends Food Sci. Technol. 15 (7-8), 330–347.

Hedayatnia, S., Mirhosseini, H., 2018. Quality of reduced-fat dairy coffee creamer: affected by different fat replacer and drying methods. Descriptive Food Sci. IntechOpen, London, SW7 2QJ, United Kingdom, p. 115.

Huang, S., et al., 2016. Double use of highly concentrated sweet whey to improve the biomass production and viability of spray-dried probiotic bacteria. J. Funct. Foods 23, 453–463.

Huang, S., et al., 2017. Spray drying of probiotics and other food-grade bacteria: a review. Trends Food Sci. Technol. 63, 1–17.

Jantzen, M., Gopel, A., Beermann, C., 2013. Direct spray drying and microencapsulation of probiotic Lactobacillus reuteri from slurry fermentation with whey. J. Appl. Microbiol. 115, 1029–1036.

Kelly, A.L., Fox, P.F., 2016. Manufacture and properties of dairy powdersAdvanced Dairy Chemistry. Springer. New York, USA, pp. 1–33.

Khanji, A.N., et al., 2018. Structure, gelation, and antioxidant properties of curcumin-doped casein micelle powder produced by spray-drying. Food Funct. 9 (2), 971–981.

Kolanowski, W., et al., 2007. Sensory assessment of microencapsulated fish oil powder. J. Am. Oil Chem. Soc. 84 (1), 37–45.

KuShaari, K., et al., 2006. Monte Carlo simulations to determine coating uniformity in a Wurster fluidized bed coating process. Powder Technol. 166 (2), 81–90.

Li, K., et al., 2017. Enhancing the stability of protein-polysaccharides emulsions via Maillard reaction for better oil encapsulation in spray-dried powders by pH adjustment. Food Hydrocolloids. 69, 121–131.

Li, R., Roos, Y.H., Miao, S., 2016. Flavor release from spray-dried amorphous matrix: effect of lactose content and water plasticization. Food Res. Int. 86, 147–155.

Li, R., Roos, Y.H., Miao, S., 2017. Characterization of mechanical and encapsulation properties of lactose/maltodextrin/WPI matrix. Food Hydrocolloids 63, 149–159.

Lindlof, J.A., Wurster, D.E., 1964. U.S. Patent No. 3,117,027. Washington, DC: U.S. Patent and Trademark Office.

Liu, H., et al., 2020. Kinetic modelling of the heat stability of bovine lactoferrin in raw whole milk. J. Food Eng. 280, 109977.

Liu, L.X., Lister, J.D., 1993. Spouted bed seed coating: the effect of process variables on maximum coating rate and elutriation. Powder Technol. 74, 215–230.

Liu, W., et al., 2016. On enhancing the solubility of curcumin by microencapsulation in whey protein isolate via spray drying. J. Food Eng. 169, 189–195.

Lopez, C., et al., 2006. Milk fat thermal properties and solid fat content in emmental cheese: a differential scanning calorimetry study. J. Dairy Sci. 89 (8), 2894–2910.

Madene, A., et al., 2006. Flavour encapsulation and controlled release – a review. Int. J. Food Sci. Technol. 41 (1), 1–21.

Mahdavi, S.A., et al., 2014. Spray-drying microencapsulation of anthocyanins by natural biopolymers: a review. Dry. Technol. 32, 509–518.

Mahdi Jafari, S., Masoudi, S., Bahrami, A., 2019. A Taguchi approach production of spray-dried whey powder enriched with nanoencapsulated vitamin D3. Dry. Technol. 37 (16), 2059–2071.

Masters, K., *Spray processing of fat-containing foodstuffs.* Lipid Technologies and Applications, 2018.

Maurya, V.K., Bashir, K., Aggarwal, M., 2020. Vitamin D microencapsulation and fortification: trends and technologies. J. Steroid Biochem. Mol. Biol. 196, 105489.

McClements, D.J., 1999. Food Emulsions: Principles, Practice and Techniques. CRC Press, Boca Raton, FL.

Mohanty, D.P., et al., 2016. Milk derived bioactive peptides and their impact on human health – a review. Saudi J. Biol. Sci. 23 (5), 577–583.

Moisio, T., et al., 2014. Interfacial protein engineering for spray-dried emulsions - part I: effects on protein distribution and physical properties. Food Chem. 144, 50–56.

Neves, M.I.L., et al., 2019. Encapsulation of curcumin in milk powders by spray-drying: physicochemistry, rehydration properties, and stability during storage. Powder Technol. 345, 601–607.

Pan, K., Zhong, Q., Baek, S.J., 2013. Enhanced dispersibility and bioactivity of curcumin by encapsulation in casein nanocapsules. J. Agric. Food Chem. 61 (25), 6036–6043.

Pedroso, D.L., et al., 2013. Microencapsulation of Bifidobacterium animalis subsp. lactis and Lactobacillus acidophilus in cocoa butter using spray chilling technology. Braz. J. Microbiol. 44 (3), 777–783.

Phillips, G.O., Williams, P.A., 2011. Handbook of food proteins Woodhead Publishing Series in Food Science, Technology and Nutrition. Woodhead Pub., Cambridge, p. 457 1 online resource.

Procopio, F.R., et al., 2018. Solid lipid microparticles loaded with cinnamon oleoresin: characterization, stability and antimicrobial activity. Food Res. Int. 113, 351–361.

Queiros, M.S., et al., 2020. Dairy-based solid lipid microparticles: a novel approach. Food Res. Int. 131, 109009.

Raei, M., et al., 2017. Application of whey protein-pectin nano-complex carriers for loading of lactoferrin. Int. J. Biol. Macromol. 105, 281–291.

Rama, G.R., et al., 2019. Potential applications of dairy whey for the production of lactic acid bacteria cultures. Int. Dairy J. 98, 25–37.

Ramos, O.L., et al., 2019. 3 Nanostructures of whey proteins for encapsulation of food. Biopolymer Nanostructures for Food Encapsulation Purposes: Volume 1. the Nanoencapsulation in the Food Industry series. Academic Press, London EC2Y 5AS, United Kingdom, pp. 69.

Rao, P.S., et al., 2016. Encapsulation of antioxidant peptide enriched casein hydrolysate using maltodextrin–gum arabic blend. J. Food Sci. Technol. 53 (10), 3834–3843.

Reineccius, G., 2019. Use of proteins for the delivery of flavours and other bioactive compounds. Food Hydrocolloids 86, 62–69.

Rodríguez-Alcalá, L.M., et al., 2013. CLA-enriched milk powder reverses hypercholesterolemic risk factors in hamsters. Food Res. Int. 51 (1), 244–249.

Sahlan, M., et al., 2019. Microencapsulation of clove oil using spray dry with casein encapsulator and activity test towards Streptococcus mutans. AIP Conf. Proc. 2193 (1), 030006.

Sarabandi, K., et al., 2018. Microencapsulation of casein hydrolysates: physicochemical, antioxidant and microstructure properties. J. Food Eng. 237, 86–95.

Sarabandi, K., Gharehbeglou, P., Jafari, S.M., 2020. Spray-drying encapsulation of protein hydrolysates and bioactive peptides: opportunities and challenges. Dry. Technol. 38 (5-6), 577–595.

Schuck, P., et al., 2016. Recent advances in spray drying relevant to the dairy industry: a comprehensive critical review. Dry. Technol. 34 (15), 1773–1790.

Shi, X., Lee, Y., 2020. Encapsulation of tributyrin with whey protein isolate (WPI) by spray-drying with a three-fluid nozzle. J. Food Eng. 281, 109992.

Shiga, H., et al., 2017. Effect of oil droplet size on the oxidative stability of spray-dried flaxseed oil powders. Biosci. Biotechnol. Biochem. 81 (4), 698–704.

Shivaram, S.H., Saini, R., 2019. Spray drying-assisted fabrication of passive nanostructures: from milk protein, Nanotechnology Applications in Dairy Science: Packaging, Processing, and Preservation. Apple Academic Press, Oakville, ON L6L 0A2, Canada, p. 45.

Tan, S., Zhong, C., Langrish, T., 2019. Microencapsulation of pepsin in the spray-dried WPI (whey protein isolates) matrices for controlled release. J. Food Eng. 263, 147–154.

Tan, S., Zhong, C., Langrish, T., 2020. Pre-gelation assisted spray drying of whey protein isolates (WPI) for microencapsulation and controlled release. LWT 117, 108625.

Teunou, E., Poncelet, D., 2002. Batch and continuous fluid bed coating—review and state of the art. J. Food Eng. 53, 325–340.

Toral, P.G., et al., 2013. Short communication: diet-induced variations in milk fatty acid composition have minor effects on the estimated melting point of milk fat in cows, goats, and ewes: insights from a meta-analysis. J. Dairy Sci. 96 (2), 1232–1236.

Trojanowska, A.N., et al., 2017. Technological solutions for encapsulation. Phys. Rev. Res. 9, 1–20.

Tzika, M., Alexandridou, S., Kiparissides, C., 2003. Evaluation of the morphological and release characteristics of coated fertilizer granules produced in a Wurster fluidized bed. Powder Technol. 132, 16–24.

Vaucher, A., et al., 2019. Microencapsulation of fish oil by casein-pectin complexes and gum arabic microparticles: oxidative stabilisation. J. Microencapsul. 36 (5), 459–473.

Vaucher, A.C.d.S., et al., 2019. Microencapsulation of fish oil by casein-pectin complexes and gum arabic microparticles: oxidative stabilisation. J. Microencapsul. 36 (5), 459–473.

Vega, C., Goff, H.D., Roos, Y.H., 2007. Casein molecular assembly affects the properties of milk fat emulsions encapsulated in lactose or trehalose matrices. Int. Dairy J. 17 (6), 683–695.

Vega, C., Roos, Y.H., 2006. Invited review: spray-dried dairy and dairy-like emulsions–compositional considerations. J. Dairy Sci. 89 (2), 383–401.

Waldron, K., et al., 2016. On spray drying of uniform mesoporous silica microparticles. Mater. Today: Proc. 3 (2), 646–651.

Wang, B., Adhikari, B., Barrow, C.J., 2019. Highly stable spray dried tuna oil powders encapsulated in double shells of whey protein isolate-agar gum and gellan gum complex coacervates. Powder Technol. 358, 79–86.

Wang, J., et al., 2019. An advanced near real dynamic in vitro human stomach system to study gastric digestion and emptying of beef stew and cooked rice. Food Funct. 10 (5), 2914–2925.

Wang, Y., Selomulya, C., 2020. Spray drying strategy for encapsulation of bioactive peptide powders for food applications. Adv. Powder Technol. 31 (1), 409–415.

Wurster, D.E., 1963. U.S. Patent No. 3,089,824. Granulating and coating process for uniform granules. Washington, DC: U.S. Patent and Trademark Office.

Würth, R., Foerst, P., Kulozik, U., 2018. Effects of skim milk concentrate dry matter and spray drying air temperature on formation of capsules with varying particle size and the survival microbial cultures in a microcapsule matrix. Dry. Technol. 36 (1), 93–99.

Yang, M., et al., 2020. Effect of ultrasound on binding interaction between emodin and micellar casein and its microencapsulation at various temperatures. Ultrason. Sonochem. 62, 104861.

Yang, W., et al., 2020. Structures, fabrication mechanisms, and emulsifying properties of self-assembled and spray-dried ternary complexes based on lactoferrin, oat β-glucan and curcumin: a comparison study. Food Res. Int. 131, 109048.

Ye, Q., et al., 2021. On improving bioaccessibility and targeted release of curcumin-whey protein complex microparticles in food. Food Chem. 346, 128900.

Ye, Q., Woo, M.W., Selomulya, C., 2019. Modification of molecular conformation of spray-dried whey protein microparticles improving digestibility and release characteristics. Food Chem. 280, 255–261.

Yin, Y., Cadwallader, K.R., 2019. Spray-chilling encapsulation of 2-acetyl-1-pyrroline zinc chloride using hydrophobic materials: storage stability and flavor application in food. Food Chem. 278, 738–743.

Zhuang, F., et al., 2018. Effects of casein micellar structure on the stability of milk protein-based conjugated linoleic acid microcapsules. Food Chem. 269, 327–334.

Three-dimensional (3D) food printing—an overview

Peter Watkins[a], Amy Logan[a], Bhesh Bhandari[b]

[a]*CSIRO Agriculture and Food, Werribee, VIC, Australia*
[b]*School of Agriculture and Food Sciences, University of Queensland, Brisbane, QLD, Australia*

15.1 Introduction

The process of three-dimensional (3D) printing of food materials is a form of additive manufacturing, and an emerging field over the last decade or so. Additive manufacturing is the industrial definition to describe a computer-controlled process that creates 3D objects by depositing materials, usually in layers. For food, a structured product formed from food "ink" ingredients, using a computer-aided design (CAD) mass production while having economic and environmental impacts (Le-Bail et al., 2020). The food products can be fabricated with customized color, shape, flavor, texture, nutritional loading, and delivery systems, with the concept of 4D printing, where the product changes with time, starting to be realized (Le-Bail et al., 2020). It is an area which has been covered by a number of reviews (Dankar et al., 2018; Dick et al., 2019; Godoi et al., 2015, 2016; Guo et al., 2019a; Hemsley et al., 2019; Le-Bail et al., 2020; Lipton et al., 2015; Pitatachaval et al., 2018; Piyush et al., 2020; Portanguen et al., 2019; Rhee, 2018; Sayem et al., 2020; Sun et al., 2015a, 2015b; Sun et al., 2018a; Topuz et al., 2018; Webb et al., 2016; Wegrzyn et al., 2012; Fan Yang et al., 2017; Voon et al., 2019) and a book (Godoi et al., 2019). In this work, we aim to provide an overview of the technique, as well as provide description of some recent innovations entering the 3D food printing market, which has a global value of around $485.5M in 2020 and is estimated to grow to around $1B by 2025, primarily in confectionery and bakery products (Anonymous, 2020a).

15.2 Overview

Fig. 15.1 provides an overview of the 4D printing process. In this case, the process follows that for 3D printing with an initial step that consists of designing a suitable 3D model using CAD software (Guo et al., 2019a). The model and design are transferred into an intermediate stereolithography file format which can be used for 3D manufacturing. The stereolithography file is then sliced to get the outline of every layer, which are sequentially deposited or formed using the appropriate materials by the printer to form the desired model and consequently the food product (Guo et al., 2019a). The product can also undergo postprocessing activities if needed, for example, cooking a 3D printed pizza. Different examples of associated software used for developing the 3D models and the intermediate steps are described elsewhere (Guo et al., 2019a), with a good example which illustrates the complete process

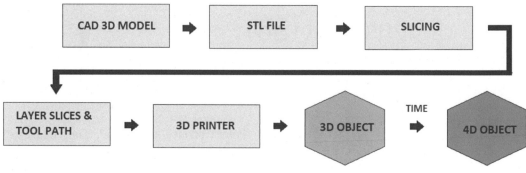

FIG. 15.1

Workflow of 4D printing process (after Piyush et al., 2020).

for the manufacture of a tuna analogue provided by Kouzani et al. (2017). The aspect of time can be introduced into the printing process as well to create 4D printing. This is demonstrated in Fig. 15.1 which shows the color change of a 3D printed product with time; see, for example, Ghazal et al. (2019).

15.3 Hardware

For food applications, there are four different types of hardware ("printers") that are based on different working principles (Topuz et al., 2018; Table 15.1). The first approach is based on extrusion, where the food ink materials are often hydrated and forced to flow through a shaped hole or die under varied temperatures and pressures at a convenient rate to form the desired product (Alam and Aslam, 2020; Menis-Henrique et al., 2020). The extrusion process is sometimes either "hot" (Zipeng Liu et al., 2020) or "cold" (Gholamipour-Shirazi et al., 2019). Contemporary practice often sees hot extrusion used in 3D printing for foods, due to its high printing accuracy and flexibility to handle a range of food ingredients (Zipeng Liu et al., 2020; Sun et al., 2015). For cold extrusion, the ink is made of a self-supporting material, and the process is generally performed at room temperature (Gholamipour-Shirazi et al., 2019; Sun et al., 2018b). Exemplar food inks for extrusion include egg and rice flour blends (Anukiruthika et al., 2020), potato starch (Zipeng Liu et al., 2020), tomato paste as a model system

Table 15.1 Three-dimensional food printer types with related inks and food type[a].

Printer type	Printer material ("ink")	Food type
Extrusion	Polymers, hydrogels, bio-gels	Soft foods (e.g., meat puree, chocolate, cheese)
Inkjet printing	Liquid/solid phase (e.g., fruit concentrate)	Low viscosity materials (e.g., fruit puree, pizza suice)
Binder jets	Sugar/starch mixtures	Powdered materials (e.g., sugar/starch)
Selective laser sintering	Powdered materials (nonsticky, not likely to agglomerate)	Powdered materials (e.g., fat chocolate)

[a]Değerli and El (2017), Zhenbin Liu and Zhang (2019), and Topuz et al. (2018).

for water-based systems (Zhu et al., 2019) and hydrocolloids such as xanthan and other plant-based gums (Gholamipour-Shirazi et al., 2019). The second approach is based on inkjet printing, which uses an array of pneumatic nozzle-jets that produces a layer of the printing material that is deposited onto a moving object (Le-Bail et al., 2020). One commercially available food inkjet printer is from Foodjet (https://www.foodjet.com/), and generally uses low viscous materials to print drawings on flat products, rather than construct complex, structured food materials (Le-Bail et al., 2020). Suitably viscous inks currently used for 3D food applications include cheese, chocolate, liquid dough, sugar icing, meat paste, jams, and gels (Godoi et al., 2016; Pitatachaval et al., 2018). The third approach, binder jetting, also uses inkjets, however in this case, a liquid binder is ink-jetted onto a powder base (Holland et al., 2018a). After one layer is printed, a fresh layer of powder is deposited using a roller or a blade prior to the application of the next layer. This process is performed sequentially until the final structure is produced (encased in unbound powder) and remove from the powder bed (Holland et al., 2018a). In this case, suitable inks that have been described result from the blending of cellulose, xanthan gum, and glucomannan (Holland et al., 2018b). The fourth approach for 3D printing is laser sintering. In this case, a laser is used as the sintering source and fuses the base material to form a solid layer without a liquefaction step (Le-Bail et al., 2020). Once done, a new layer is distributed onto the previous layer using a roller and completely sequentially until the final product is formed. One advantage of this approach is that complex food systems can be developed quickly but the process is complicated due to the many variables involved (Le-Bail et al., 2020). Suitable printing materials include sugar, chocolate and Nesquick (Le-Bail et al., 2020). A range of 3D printers are commercially available with a summary provided in Table 15.2. Most of the printers shown in Table 15.2 can be regarded as bench-top models suitable for small niche applications while the BeeHex Deco-pod has semi-industrial use; in this case, as an autonomous cake decorating system.

15.4 Inks

Along with the hardware used for 3D food printing, the feedstock is an important consideration for the construction of food materials. As we have noted above, a synergy will exist between the hardware and the ink for the food product to be constructed. For example, the use of a highly viscous material is not recommended with the use of a screw-based extruder for 3D printing (Guo et al., 2019b). A key requirement of any raw material to be used as an ink is that it can flow smoothly from the print cartridge to the printing platform (Topuz et al., 2018). At present, most published work stems from research as ongoing investigations are required to understand the role of material properties in relation to food components in the 3D printing process (Liu et al., 2017).

Proteins, along with other food ingredients, offer potential to provide nutrients and functional ingredients in 3D printed foods (Lille et al., 2018). A range of proteins have been investigated for their suitability as feedstocks; for example, soy protein isolate was mixed with sodium alginate and gelatin, and found to be suitable materials for 3D printing (Chen et al., 2019) where the soy protein isolate mixture with gelatin produced excellent geometries and alginate improved the related hardness and chewiness of the formed products. Milk protein concentrate (MPC) has been used with sodium caseinate as a potential feedstock for extrusion-based 3D printing (Zipeng Liu et al., 2020). It was found that increasing total protein content improved the 3D structure, along with other related properties (e.g., apparent viscosity and yield stress). Whey protein isolate (WPI) has also been mixed with MPC with

Table 15.2 Summary of companies producing 3D printers suitable for food production[a].

Company	Model	Price*	Country	Build volume	Web page
Natural Machines (Foodini)		$4k	Spain	250 × 165 × 120 mm	https://www.naturalma-chines.com/
WASP (World's Advanced Saving Project) Deltawasp					https://www.3dwasp.com/en/
3DSystems – Chef Jet Pro		$5 to 10 K		203.2 × 203.2 × 203.2 mm	https://www.3dsystems.com/culinary
nūfood 3D printing robot			UK		http://www.nufood.io/
byFlow	Focus	€3.3K	Netherlands	208 × 228 × 150 mm	https://www.3dbyflow.com/
Mmuse	Choc. 3D	$4.5K	China	160 × 120 × 150 mm	https://www.3dprintersonlinestore.com/mmuse-desktop-food-3d-printer
Pancake Bot		$300	Norway	445 × 210 × 15 mm	http://www.pancakebot.com/
BeeHex B2B	Nutripod/Decopod				https://www.beehex.com/
Structur3d Printing (paste extrusion)					https://www.structur3d.io/
Procusini					https://www.procusini.com/
Createbot		$2.5K	China	150 × 150 × 100 mm	https://www.creatbot.com/en/
Choc Edge	Choc Creator V2.0 Plus	£2.4K	UK	180 × 180 × 40 mm	http://chocedge.com/
Micromake Food	3D printer	$1K	China	100 × 100 × 15 mm	https://www.aniwaa.com/product/3d-printers/micro-make-food-3d-printer/
ORD Solutions	RoVaPaste	$1K	Canada	100 × 100 × 15 mm	https://www.ordsolutions.com/rovapaste-filament-3d-printer/
ZBOT	Commercial Art Pancakes Printer F5	$0.5 - $5K	China	180 × 180 × 15 mm	http://www.zbot.cc/index.php?m=wenbon&a=a&t=product&id=34&l=en
ZMorph	VX	$3K	Poland	250 × 235 × 165 mm	https://zmorph3d.com/product/zmorph-vx

[a]Lansard (2020).

an optimal MPC:WPI ratio of 5:2 which gave the most desirable material for use with extrusion-based 3D printing (Liu et al., 2018a). Skim milk powder, when used as a paste base and which is also a good protein source, was reported to provide good printing precision and shape stability for extrusion-based printing (Lille et al., 2018).

Plant-based proteins have also been evaluated using concentrates prepared from rye bran, faba bean, and oat (Lille et al., 2018). Pureed tuna has been used as a protein source for the formation of food products intended for people with swallowing difficulties (dysphagia; Kouzani et al., 2017). Fish surimi gel has been evaluated as a printer ink with promising results indicated with its use for printing complex food constructs (Wang et al., 2018). Egg yolk and egg white have been incorporated with rice flour as a potential ink for extrusion-based printing (Anukiruthika et al., 2020). The flour as a filler agent had a significant effect on the stability of the printed egg material, and the egg yolk could be 3D printed with a high degree of precision.

Animal muscle tissue (red meat) is another protein source where interest exists in using a 3D printing approach to create meat-based products on-demand (Dick et al., 2019; Godoi et al., 2015; Lupton and Turner, 2018; Portanguen et al., 2019; Webb et al., 2016). However, there has been little published work describing the printability of fibrous meat materials (e.g., pork, turkey, chicken, fish), with none for beef (Dick et al., 2019). There is much which needs to be done in this area to provide cost effective and suitable delivery systems. For example, Webb et al. (2016) found the flow properties of a meat concentrate could not be controlled upon extrusion. Similarly, costs associated with powdered meat production as an ink ingredient may prove inhibiting for some applications. However, it is feasible that low value meat (e.g., emulsified meat trimmings) can be used to produce high value meat products with 3D printing. Recent work sponsored by Meat and Livestock Australia has evaluated this approach with promising outcomes (Dahm, 2018; Rubinsky, 2018a, 2018b). Some "proof-of-concept" designs of recombined meat products for use in aged care facilities are shown in Fig. 15.2.

Additives are also often needed as feedstocks for 3D printing to assist in the formation of stand-alone structured foods (Portanguen et al., 2019). Such additives could include hydrocolloids from different sources (e.g., animal, dairy, plant, and algal) which can be combined to a meat protein mixture (Godoi et al., 2015). A recent study used a modified large deformation approach to map the rheological flow properties of meat-alginate pastes, in combination with other factors such as the particle size of the dried meat powder (Lindström, 2019). This work provides insights into the boundaries that govern ingredient suitability for 3D printing applications and demonstrates the influence of different drying techniques (e.g., ultrasound assisted) on the ingredient appearance and physicochemical function.

Food inks based on the use of xanthan gums have also been described (García-Segovia et al., 2020; Holland et al., 2018a; Kim et al., 2017; Lee et al., 2019); for example, the use of xanthan and konjac gums with syrup to form gels for 3D printing (García-Segovia et al., 2020). Similarly, xanthan gum has been formulated for use with an ink jet printer, using amorphous cellulose as a substrate (Holland et al., 2018b). Moreover, xanthan gum was deployed in a food ink system comprised of spinach powder of different particle size (Lee et al., 2019), where increasing particle size had a corresponding increase on the mechanical strength. Other materials include callus formed from carrot tissues (Park et al., 2020), potato starch (Liu et al., 2020b), by-products from potato and yam processing (Feng et al., 2020), fruit and vegetables (Derossi et al., 2018; Liu et al., 2018b; Ricci et al., 2019; Zhu et al., 2019), mixtures of gelatin and kappa-carrageenan (Warner et al., 2019), gels formed by lemon juice gels with starch (Yang et al., 2018), pectin (Vancauwenberghe et al., 2019), and dough for cereal-based products (Caporizzi et al., 2019; Liu et al., 2019; Pulatsu et al., 2020; Severini et al., 2016; Zhang et al., 2018).

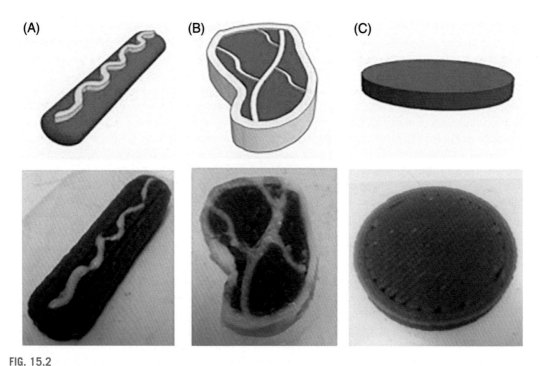

FIG. 15.2

Hypothetical meat constructs for aged care facilities: (A) sausage, (B) steak "recombined meat," and (C) patty (Dick et al., 2019).

Consumer acceptance needs to be considered in the development of 3D printed foods, especially when novel ink ingredient or approaches are considered. For example, a survey of Australian consumers conducted in 2016 suggested little support for cultured meat and insect-based foods (Lupton and Turner, 2018). However, recent evidence in the media and growth in the number of emerging start-up companies involving cellular protein and insect farming indicates there might be a step-change in consumer acceptance, providing greater social license for advancements in this space. A recent study has examined the potential to form self-standing food structures upon 3D printing, from dried tomato powder enriched with insect protein and rehydrated to varying levels (van Rijssel, 2019). Fig. 15.3 demonstrates the influence of the powder to water ratio on the product structure and appearance. Results indicate the best stand-up was achieved for a 150 g serving size from a blend of 33 g tomato powder and 10 g of cricket (insect) protein, without significant differences in appearance compared to the rehydration of 100% tomato powder.

15.5 Example applications

3D printing of food allows for the manufacture of food products with desired shape and structure, which can be rapidly produced, customized, personalized with precision control (Yang et al., 2017; Derossi et al., 2018; Hua et al., 2018). A wide range of food products have been fabricated with 3D

FIG. 15.3 Tomato and cricket (insect) powdered blends hydrated to varying levels:
1:2 (A), 1:3 (B), 1:4 (C), and 1:45 (D) (powder to water ratio) to form a paste and printed using a layer-by-layer extrusion-based 3D printer (van Rijssel, 2019).

printing, including chocolate, cheese, some cereal products, gels, doughs, and candies (Caporizzi et al., 2019; Lanaro et al., 2017, 2019; Ricci et al., 2019; Topuz et al., 2018). At present, most products have been developed and reported to demonstrate the technique's potential for food manufacturing, rather than disclose what would be regarded as trade secrets. Traditional foods such as pizza have been produced using 3D printers (Lipton et al., 2015). A demonstration was performed for the New York Times where the dinner was entirely 3D printed, and parameters such as the thickness of the dough, sauce, and cheese were controlled in the layering process (Lipton et al., 2015). A smoothie has been constructed using 3D printing using a range of selected fruit and vegetables (Severini et al., 2018), along with a fruit-based snack for children (Derossi et al., 2018) while the use of yams for 3D printed snack foods has described in Feng et al. (2020). Chocolate has also been the subject of investigation by several authors (Ferreira and Alves, 2017; Hao et al., 2019; Lanaro et al., 2019; Mantihal et al., 2017; Zhao et al., 2018), along with processed cheese (Le Tohic et al., 2018).

Food printing is particularly suited to niche food applications which have a strong emphasis on individualized food design or customized manufacturing (Wegrzyn et al., 2012). A good example is the development of texture modified foods for people with swallowing difficulties (dysphagia, Hemsley et al., 2019; Hua et al., 2018; Kouzani et al., 2017). Dysphagia impacts on people with a range of health conditions (e.g., stroke, cerebral palsy, etc.) which creates difficulty with their swallow actions (Kouzani et al., 2017). Usually, foods need to be texture modified (e.g., soft, mashed, puree) or presented as thickened liquids to meet the specific guidelines for safe foods to be consumed, but can become unacceptable in terms of appearance, taste and texture to those on long-term modified diets (Kouzani et al., 2017). The International Dysphagia Diet Standardisation Initiative provides a framework to describe food textures and drink thickness (Anonymous, 2020c) which has been endorsed as best practice by relevant national groups (e.g., speech pathologists in Australia; Anonymous, 2020b). Food products can be constructed using 3D printing that are appropriate and acceptable for people on texture-modified methods as well as the same or better than those foods produced by conventional means (Hemsley et al., 2019). The printed construct looks the same as the "real" food but is much more to easier to swallow and digest (Hua et al., 2018); for example, a construct that resembles a fish using tuna puree as the feedstock has been described by Kouzani et al. (2017). Fig. 15.2 shows example constructs of meat products for use in aged care facilities. Additionally, 3D printing has been deployed to produce portrait images (Zhao et al., 2018). Sugar painting is a traditional Chinese folk art, requiring the skill of an experienced artist to create detailed figures. However, 3D printing has been deployed to produce images, in some cases, on food materials (see, e.g., Fig. 15.4; Zhao et al., 2018).

Some food materials are yet to fully realize their potential with 3D printing, due to several reasons. One relates to the need to understand the fundamental principles which 3D printing has on food microstructure which, in some areas, is lacking (e.g., dairy, Ross et al., 2019). Additionally, the wide variation that exists in the physico-chemical properties of complex food materials also creates issues for 3D printing (Godoi et al., 2016). Other challenges include lack of specific rules and regulations, ingredient limitations and post processing (Dankar et al., 2018). Some of these could be overcome by developing specific guidelines for food safety, shelf life, facilities, etc., as well as the development of simulation models for printing behavior which can be used to optimize the process for certain products (Dankar et al., 2018).

Another potential area for growth and opportunity is personalized nutrition. The concept of personalized nutrition resulted after the sequencing of the human genome in 2000 and involves tailoring dietary advice specifically to an individual's characteristics (Anonymous, 2015). It was anticipated that, with the identification of gene-nutrient interactions, an individual's response and susceptibility to diets could be better understood and, through individualized dietary advice, appropriate dietary advice/modifications could be used to optimize health and lower disease risk (Anonymous, 2015). It has been

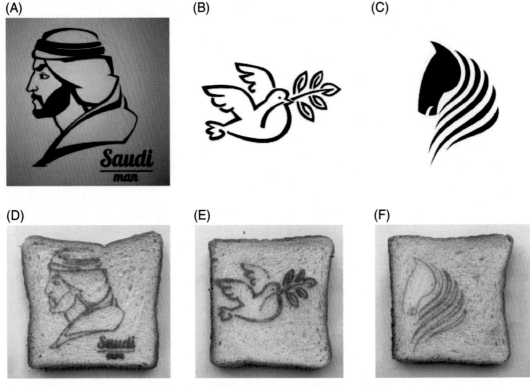

FIG. 15.4

3D printed images on sliced bread (Zhao et al., 2018).

suggested that technologies such as 3D printing, along with robotic systems, could be expected to advance opportunities in personalized nutrition through start-ups and small players (Spitzer, 2019). An example of this is Nourished (https://get-nourished.com/) who manufacture customized vitamin products ("Nutrition Stacks") using 3D printing, that are home delivered. This disruptive business relies on product customization by the customer, effectively bringing the consumer closer to the manufacturer. It also allows, and promotes, a different shopping experience that introduces a new business model allow further uptake of the technology. There are also indications that significant growth is possible in this area with projected annual revenues as high as $64 billion by 2040, according to one market analysis (Fitzgerald, 2020).

15.6 Commercial activity

While it still can be regarded as nascent, there has been recent applied commercial activity in the deployment of 3D printing for the manufacture of food products with focus on faux meats. Two start-up companies have been developing analogue meat systems that are plant based and use 3D printing to form the final products. Based in Israel, Redefine Meat (www.redefinemeat.com) is producing their own printers which use plant-based materials to produce product which aims to have the same form, function and taste as animal tissue (Askew, 2020; Vinoski, 2019). The company claims to have a 3D printing process which enables the "precise composition of meat layer by layer to bring every fiber of real meat to fruition in a way that conventional alt meat processes cannot" (Askew, 2020). The other company, NovaMeat based in Spain (www.novameat.com/) aims to mimic meat with plant-based analogues of fibrous meats and seafoods (Southey, 2020). Using 3D printing to produce a plant-based beef steak in 2018, the company has also claimed to have formed version 2.0, a product which exhibits both the texture and appearance of a whole beef cut (Southey, 2020). Additionally, this company has also announced a plant-based analogue for pork meat as well (Lamb, 2020). Seafood applications are also starting to emerge. Legendary Vish (https://www.legendaryvish.com/) is using 3D printing to produce salmon fillets made from plant-based materials.

Food waste is being utilized as a source material for the creation of new food products. Known as "upscaling," Upprinting Foods (www.upprintingfood.com) based in the Netherlands is utilizing ingredients from "residual food flows" that are used to create purees for use with 3D printing. The resulting prints are baked and dehydrated for long-term storage and stability. Recovered foods used include discarded bread, fruit, and vegetable as well as rice. The company also aims to work with chefs to minimize their food waste, in order to create new and innovative food designs. Such responsiveness is a characteristic often associated with start-ups. The ability to pivot according to a customer's needs can lead to the development of successful business plans (Rasmussen and Tanev, 2016), that also be disruptive as well (Voinea et al., 2019).

Other companies are endeavoring to provide consumer experiences with 3D printing as part of that experience. Food Ink (www.foodink.io), based in the United Kingdom, is incorporating 3D food printing as part of a larger experience for consumers where all of the food, utensils and furniture are completely produced through 3D printing as a pop-up event (Cecchini, 2018). In Japan, Open Meals plans to create an experience where personalized nutrition almost coming to fruition. They are planning to open a restaurant, called "Sushi Singularity," which will use biological samples supplied by the customer (saliva, feces, and urine) to create 3D-printed sushi that is tailored to their nutritional needs (Yalcinkaya, 2019). Open Meals (http://www.open-meals.com/sushisingularity/index_e.html) has three unique concepts as part of its business plan; the first, "Food Fabrication Machine," relates to the development of future kitchens were 3D food printers will appear in every home while the second, "Food Operating System," is a software suite that can be deployed to digitally design food and the third, "Health Identification," is aimed at personalized nutrition where DNA, urine and intestinal testing will be used to form the optimal diet for the individual.

Considerable effort is being undertaken in the area of 3D printing of food, with many patents related to this area. As of March 2020, there were over 250 patents identified with "3D printing food" used as a query string from a search from the World Intellectual Property Organization (WIPO). This indicates the interest which exists in this area which undoubtedly will grow in the years ahead as the potential of this technique is fully realized.

15.7 Conclusion

3D printing of food materials has been an emerging field of research over the last decade. With the appropriate use of "ink" ingredients, CAD can be used to form a food product of either complex shape or pattern, enabling mass production while having economic and environmental impact. The food products can be fabricated with customized color, shape, flavor, texture, and nutritional loading and delivery systems. In this chapter, we have provided an overview of the technique as well as describe some recent innovations relating to the 3D food printing market. Opportunity still exists within the area for the development of new materials and applications. An example of this is the development of 4D printing where characteristics of the food change with time. Further fundamental research is still required though, for example, understanding the fundamental principles relating to food structure, but this represents the opportunity which can be developed with the technique's growth. Applied commercial activity remains nascent but start-ups have commenced activity which exploits 3D printing in the fabrication of new food products, for example, "faux" meats. These start-ups have the chance to pivot in response to customer needs, thus being disruptive as well. No doubt, the development of different food products will continue in future, along with the creation of newer consumer markets and experiences.

Acknowledgments

The authors would like to acknowledge the input and supervision of Henry Sabarez and Thu McCann, as well as Filip Janakievski and Sofia Oiseth (CSIRO Agriculture and Food, Werribee, Australia) for the novel work of Viktor Lindström and Arjan van Rijssel, respectively, introduced in this chapter.

References

Alam, M.S., Aslam, R., 2020. Extrusion for the production of functional foods and ingredientsReference Module in Food Science. Elsevier, Amsterdam, Netherlands. https://doi.org/10.1016/B978-0-08-100596-5.23041-2.

Anonymous. (2015, March 31). Final Report Summary - FOOD4ME (Personalised nutrition: An integrated analysis of opportunities and challenges). https://cordis.europa.eu/project/id/265494/reporting.

Anonymous. (2020a). *3D Food Printing* (No. FOD093A). BCC Research LLC.

Anonymous. (2020b). *Modified Foods and Fluids Terminology*. https://www.speechpathologyaustralia. org.au/SPAweb/Resources_for_the_Public/Modified_Foods_and_Fluids_Terminology/SPAweb/ Resources_for_Speech_Pathologists/Professional_Resources/Modified_Foods_and_Fluids_Terminology. aspx?hkey=822fd30c-b7d4-45c3-9071-a20a7b38bb52.

Anonymous. (2020c). What is the IDDSI Framework? https://iddsi.org/framework/.

Anukiruthika, T., Moses, J.A., Anandharamakrishnan, C., 2020. 3D printing of egg yolk and white with rice flour blends. J. Food Eng. *265*, 109691. https://doi.org/10.1016/j.jfoodeng.2019.109691.

Askew, K. (2020, June 24). Redefine meat targets "meat-lovers" with 3D printed solutions: "We're closer than you think." Food Navigator. https://www.foodnavigator.com/Article/2020/06/04/Redefine-Meat-targets-meat-lovers-with-3-D-printed-plant-based-solutions-We-re-closer-than-you-think?utm_source=copyright&utm_medium=OnSite&utm_campaign=copyright.

Caporizzi, R., Derossi, A., Severini, C., 2019. Chapter 4 - cereal-based and insect-enriched printable food: from formulation to postprocessing treatments. Status and perspectives. In: Godoi, F.C., Bhandari, B.R., Prakash, S., Zhang, M. (Eds.), Fundamentals of 3D Food Printing and Applications. Academic Press, Boston, Massachusetts, pp. 93–116. https://doi.org/10.1016/B978-0-12-814564-7.00004-3.

Cecchini, C. (2018, October 21). Edible Carving: The World's First 3D Printing Restaurant. https://thespoon.tech/edible-carving-the-worlds-first-3d-printing-restaurant/.

Chen, J., Mu, T., Goffin, D., Blecker, C., Richard, G., Richel, A., Haubruge, E., 2019. Application of soy protein isolate and hydrocolloids based mixtures as promising food material in 3D food printing. J. Food Eng. *261*, 76–86. https://doi.org/10.1016/j.jfoodeng.2019.03.016.

Dankar, I., Haddarah, A., Omar, F.E.L., Sepulcre, F., Pujolà, M, 2018. 3D printing technology: the new era for food customization and elaboration. Trends Food Sci. Technol. 75, 231–242. https://doi.org/10.1016/j.tifs.2018.03.018.

Değerli, C., El, S.N., 2017. A review on food production with 3 dimensional (3D) printing technology. Turk. J. Agricul.Food Sci. Technol. *5* (6), 593–599.

Derossi, A., Caporizzi, R., Azzollini, D., Severini, C., 2018. Application of 3D printing for customized food. A case on the development of a fruit-based snack for children. 3D Printed Food Des. Technol. *220*, 65–75. https://doi.org/10.1016/j.jfoodeng.2017.05.015.

Dahm, C. (2018). *3D Printing Demonstration* (V.RMH.0075). Meat and Livestock Australia.

Dick, A., Bhandari, B., Prakash, S., 2019. 3D printing of meat. Meat Sci. 153, 35–44. https://doi.org/10.1016/j.meatsci.2019.03.005.

Feng, C., Zhang, M., Bhandari, B., Ye, Y., 2020. Use of potato processing by-product: Effects on the 3D printing characteristics of the yam and the texture of air-fried yam snacks. LWT 125, 109265. https://doi.org/10.1016/j.lwt.2020.109265.

Ferreira, I.A., Alves, J.L., 2017. Low-cost 3D food printing. Materiais 29 (1), e265–e269. 2015. https://doi.org/10.1016/j.ctmat.2016.04.007.

Fitzgerald, M. (2020, January 19). Personalized nutrition could be the next plant-based meat, worth $64 billion by 2040, says UBS [Cnbc.com]. https://www.cnbc.com/2020/01/19/personalized-nutrition-could-be-the-next-plant-based-meat-worth-64-billion-by-2040-says-ubs.html.

García-Segovia, P., García-Alcaraz, V., Balasch-Parisi, S., Martínez-Monzó, J., 2020. 3D printing of gels based on xanthan/konjac gums. Innovative Food Sci. Emerg. Technol., 102343. https://doi.org/10.1016/j.ifset.2020.102343.

Ghazal, A.F., Zhang, M., Liu, Z., 2019. Spontaneous color change of 3D printed healthy food product over time after printing as a novel application for 4D food printing. Food Bioprocess Technol. 12 (10), 1627–1645. https://doi.org/10.1007/s11947-019-02327-6.

Gholamipour-Shirazi, A., Norton, I.T., Mills, T., 2019. Designing hydrocolloid based food-ink formulations for extrusion 3D printing. Food Hydrocolloids. *95*, 161–167. https://doi.org/10.1016/j.foodhyd.2019.04.011.

Godoi, F.C., Bhandari, B.R., Prakash, S., Zhang, M., 2019. Front MatterFundamentals of 3D Food Printing and Applications. Academic Press, Boston, Massachusetts, p. iii. https://doi.org/10.1016/B978-0-12-814564-7.01001-4.

Godoi, F.C., Prakash, S., Bhandari, B.R., 2015. Review of 3D Printing and Potential Red Meat Applications (V.RMH,0034). Meat and Livestock, Australia. https://www.mla.com.au/download/finalreports?itemId=3310.

Godoi, F.C., Prakash, S., Bhandari, B.R., 2016. 3D printing technologies applied for food design: status and prospects. J. Food Eng. 179, 44–54. https://doi.org/10.1016/j.jfoodeng.2016.01.025.

Guo, C., Zhang, M., Bhandari, B., 2019a. Model building and slicing in food 3D printing processes: a review. Compr. Rev. Food Sci. Food Saf. 18 (4), 1052–1069. https://doi.org/10.1111/1541-4337.12443.

Guo, C.-F., Zhang, M., Bhandari, B., 2019b. A comparative study between syringe-based and screw-based 3D food printers by computational simulation. Comput. Electron. Agric. 162, 397–404. https://doi.org/10.1016/j.compag.2019.04.032.

Hao, L., Li, Y., Gong, P., Xiong, W., 2019. Chapter 8 -- Material, process and business development for 3D chocolate printing. In: Godoi, F.C., Bhandari, B.R., Prakash, S., Zhang, M. (Eds.), Fundamentals of 3D Food Printing and Applications. Academic Press, Boston, Massachusetts, pp. 207–255. https://doi.org/10.1016/B978-0-12-814564-7.00008-0.

Hemsley, B., Palmer, S., Kouzani, A., Adams, S., Balandin, S., 2019. Review informing the design of 3D food printing for people with swallowing disorders: constructive, conceptual, and empirical problems, HICSS 52 : Proceedings of the 52nd Annual Hawaii International Conference on System Sciences, 5735–5744.

Holland, S., Foster, T., MacNaughtan, W., Tuck, C., 2018a. Design and characterisation of food grade powders and inks for microstructure control using 3D printing. 3D Printed Food Des. Technol. 220, 12–19. https://doi.org/10.1016/j.jfoodeng.2017.06.008.

Holland, S., Tuck, C., Foster, T., 2018b. Selective recrystallization of cellulose composite powders and microstructure creation through 3D binder jetting. Carbohydr. Polym. 200, 229–238. https://doi.org/10.1016/j.carbpol.2018.07.064.

Hua, W.S., Na, L., Dong, Z.Y., 2018. GW29-e0057 3D food printing can help elder to digest and swallow foods, The 29th Great Wall International Congress of Cardiology China Heart Society Beijing Society of Cardiology, 72, C210. Supplement. https://doi.org/10.1016/j.jacc.2018.08.911.

Kim, H.W., Bae, H., Park, H.J., 2017. Classification of the printability of selected food for 3D printing: development of an assessment method using hydrocolloids as reference material. J. Food Eng. 215, 23–32. https://doi.org/10.1016/j.jfoodeng.2017.07.017.

Kouzani, A., Adams, S., Whyte, D.J., Oliver, R., Hemsley, B., Palmer, S., 2017. 3D printing of food for people with swallowing difficulties. KnE Eng. 2 (1), 23–29. https://doi.org/10.18502/keg.v2i2.591.

Lamb, C. (2020, May 1). Novameat Develops 3D-printed Pork Alternative to Feed Plant-based Meat Demand. https://thespoon.tech/novameat-develops-3d-printed-pork-alternative-to-feed-plant-based-meat-demand/.

Lanaro, M., Desselle, M.R., Woodruff, M.A., 2019. Chapter 6 -- 3D printing chocolate: properties of formulations for extrusion, sintering, binding and ink jetting. In: Godoi, F.C., Bhandari, B.R., Prakash, S., Zhang, M. (Eds.), Fundamentals of 3D Food Printing and Applications. Academic Press, Boston, Massachusetts, pp. 151–173. https://doi.org/10.1016/B978-0-12-814564-7.00006-7.

Lanaro, M., Forrestal, D.P., Scheurer, S., Slinger, D.J., Liao, S., Powell, S.K., Woodruff, M.A., 2017. 3D printing complex chocolate objects: platform design, optimization and evaluation. J. Food Eng. 215, 13–22. https://doi.org/10.1016/j.jfoodeng.2017.06.029.

Lansard, M. (2020, February 6). Food 3D printing: 10 food 3D printers available in 2020. https://www.aniwaa.com/buyers-guide/3d-printers/food-3d-printers/.

Le-Bail, A., Maniglia, B.C., Le-Bail, P., 2020. Recent advances and future perspective in additive manufacturing of foods based on 3D printing. Curr. Opin. Food Sci. 35, 54–64. https://doi.org/10.1016/j.cofs.2020.01.009.

Le Tohic, C., O'Sullivan, J.J., Drapala, K.P., Chartrin, V., Chan, T., Morrison, A.P., Kerry, J.P., Kelly, A.L., 2018. Effect of 3d printing on the structure and textural properties of processed cheese. 3D Printed Food Des. Technol. 220, 56–64. https://doi.org/10.1016/j.jfoodeng.2017.02.003.

Lee, J.H., Won, D.J., Kim, H.W., Park, H.J., 2019. Effect of particle size on 3D printing performance of the food-ink system with cellular food materials. J. Food Eng. 256, 1–8. https://doi.org/10.1016/j.jfoodeng.2019.03.014.

Lille, M., Nurmela, A., Nordlund, E., Metsä-Kortelainen, S., Sozer, N., 2018. Applicability of protein and fiber-rich food materials in extrusion-based 3D printing. 3D Printed Food Des. Technol. 220, 20–27. https://doi.org/10.1016/j.jfoodeng.2017.04.034.

Lindström, V., 2019. Ultrasonic assisted drying and its effect on 3D printability of minced beef and other foods. M. Sc. thesis. Lund University, Sweden.

Lipton, J.I., Cutler, M., Nigl, F., Cohen, D., Lipson, H., 2015. Additive manufacturing for the food industry. Trends Food Sci. Technol. 43 (1), 114–123. https://doi.org/10.1016/j.tifs.2015.02.004.

Liu, Y., Liu, D., Wei, G., Ma, Y., Bhandari, B., Zhou, P., 2018a. 3D printed milk protein food simulant: improving the printing performance of milk protein concentration by incorporating whey protein isolate. Innovative Food Sci. Emerg. Technol. 49, 116–126. https://doi.org/10.1016/j.ifset.2018.07.018.

Liu, Y., Liang, X., Saeed, A., Lan, W., Qin, W., 2019. Properties of 3D printed dough and optimization of printing parameters. Innovative Food Sci. Emerg. Technol. 54, 9–18. https://doi.org/10.1016/j.ifset.2019.03.008.

Liu, Z., Bhandari, B., Prakash, S., Zhang, M., 2018b. Creation of internal structure of mashed potato construct by 3D printing and its textural properties. Food Res. Int. 111, 534–543. https://doi.org/10.1016/j.foodres.2018.05.075.

Liu, Z., Chen, H., Zheng, B., Xie, F., Chen, L., 2020. Understanding the structure and rheological properties of potato starch induced by hot-extrusion 3D printing. Food Hydrocolloids. 105, 105812. https://doi.org/10.1016/j.foodhyd.2020.105812.

Liu, Z., Zhang, M., Bhandari, B., Wang, Y., 2017. 3D printing: printing precision and application in food sector. Trends Food Sci. Technol. 69, 83–94. https://doi.org/10.1016/j.tifs.2017.08.018.

Liu, Z., Zhang, M, 2019. Chapter 2 -- 3D food printing technologies and factors affecting printing precision. In: Godoi, F.C., Bhandari, B.R., Prakash, S., Zhang, M. (Eds.), Fundamentals of 3D Food Printing and Applications. Academic Press, Boston, Massachusetts, pp. 19–40. https://doi.org/10.1016/B978-0-12-814564-7.00002-X.

Lupton, D., Turner, B., 2018. Food of the future? Consumer responses to the idea of 3D-printed meat and insect-based foods. Food Foodways 26 (4), 269–289.

Mantihal, S., Prakash, S., Godoi, F.C., Bhandari, B., 2017. Optimization of chocolate 3D printing by correlating thermal and flow properties with 3D structure modeling. Emerg. Technol. Reference IUFoST 44, 21–29. https://doi.org/10.1016/j.ifset.2017.09.012.

Menis-Henrique, M.E.C., Scarton, M., Piran, M.V.F., Clerici, M.T.P.S, 2020. Cereal fiber: extrusion modifications for food industry. Curr. Opin. Food Sci. 33, 141–148. https://doi.org/10.1016/j.cofs.2020.05.001.

Park, S.M., Kim, H.W., Park, H.J., 2020. Callus-based 3D printing for food exemplified with carrot tissues and its potential for innovative food production. J. Food Eng. 271, 109781. https://doi.org/10.1016/j.jfoodeng.2019.109781.

Pitatachaval, P., Sanklong, N., & Thongrak, A. (2018). A review of 3D food printing technology. 213. https://www.matec-conferences.org/articles/matecconf/abs/2018/72/matecconf_acmme2018_01012/matecconf_acmme2018_01012.html.

Piyush, R.K., Kumar, R, 2020. 3D printing of food materials: a state of art review and future applications. Mater. Today: Proc.. https://doi.org/10.1016/j.matpr.2020.02.005.

Portanguen, S., Tournayre, P., Sicard, J., Astruc, T., Mirade, P.-S., 2019. Toward the design of functional foods and biobased products by 3D printing: a review. Trends Food Sci. Technol. 86, 188–198. https://doi.org/10.1016/j.tifs.2019.02.023.

Pulatsu, E., Su, J.-W., Lin, J., Lin, M., 2020. Factors affecting 3D printing and post-processing capacity of cookie dough. Innovative Food Sci. Emerg. Technol. 61, 102316. https://doi.org/10.1016/j.ifset.2020.102316.

Rasmussen, E.S., Tanev, S., 2016. Lean start-up: making the start-up more successful. Start-Up Creation: The Smart Eco-Efficient Built Environment, 39–56. https://doi.org/10.1016/B978-0-08-100546-0.00003-0.

Rhee, J.-K., 2018. Cryogenic grinding and 3D printing techniques for establishing "disperse and absorb" brick-type constructs of food materials. FASEB J. *32* (1_supplement), 801.9. https://doi.org/10.1096/fasebj.2018.32.1_supplement.801.9.

Ricci, I., Derossi, A., Severini, C., 2019. Chapter 5 – 3D printed food from fruits and vegetables. In: Godoi, F.C., Bhandari, B.R., Prakash, S., Zhang, M. (Eds.), Fundamentals of 3D Food Printing and Applications. Academic Press, Boston, Massachusetts, pp. 117–149. https://doi.org/10.1016/B978-0-12-814564-7.00005-5.

Ross, M.M., Kelly, A.L., Crowley, S.V., 2019. Chapter 7 – Potential applications of dairy products, ingredients and formulations in 3D printing. In: Godoi, F.C., Bhandari, B.R., Prakash, S., Zhang, M. (Eds.), Fundamentals of 3D Food Printing and Applications. Academic Press, Boston, Massachusetts, pp. 175–206. https://doi.org/10.1016/B978-0-12-814564-7.00007-9.

Rubinsky, D., 2018. *Stage 2 Upscaling 3D Printed Meat* (V.RMH.0087). Meat and Livestock, Australia.

Rubinsky, D., 2018. *3d Printing Of Meat - Cryolithography - (Rs3d) - Stage 1 Development* (V.Rmh.0001). Meat and Livestock, Australia.

Sayem, A.S.M., Shahariar, H., Haider, J, 2020. An overview on the opportunities for 3D printing with biobased materials. In: Hashmi, S., Choudhury, I.A. (Eds.), Encyclopedia of Renewable and Sustainable Materials. Elsevier, Amsterdam, Netherlands, pp. 839–847. https://doi.org/10.1016/B978-0-12-803581-8.10942-7.

Severini, C., Derossi, A., Azzollini, D., 2016. Variables affecting the printability of foods: preliminary tests on cereal-based products. Innovative Food Sci. Emerg. Technol. *38*, 281–291. https://doi.org/10.1016/j.ifset.2016.10.001.

Severini, C., Derossi, A., Ricci, I., Caporizzi, R., Fiore, A., 2018. Printing a blend of fruit and vegetables. New advances on critical variables and shelf life of 3D edible objects. 3D Printed Food Des. Technol. *220*, 89–100. https://doi.org/10.1016/j.jfoodeng.2017.08.025.

Southey, F., 2020. NovaMeat develops 'world's first' meat analogue with look and feel of whole beef muscle cut. Food Navigator https://www.foodnavigator.com/Article/2020/01/07/NovaMeat-develops-meat-analogue-with-look-and-feel-of-whole-beef-muscle-cut?utm_source=copyright&utm_medium=OnSite&utm_campaign=copyright.

Spitzer, V., 2019. Digital health -- personalised nutrition of the future. Vitafoods Insights *1*, 10–15.

Sun, J., Peng, Z., Yan, L., Fuh, J.Y.H., Hong, G.S, 2015a. 3D food printing an innovative way of mass customization in food fabrication. Int. J. Bioprint. *1* (*1*). *2015*. https://doi.org/10.18063/IJB.2015.01.006.

Sun, J., Peng, Z., Zhou, W., Fuh, J.Y.H., Hong, G.S., Chiu, A, 2015. A review on 3D printing for customized food fabrication, 43rd North American Manufacturing Research Conference, NAMRC 43, 8-12 June 2015. UNC Charlotte, North Carolina, United States, *1*, 308–319. https://doi.org/10.1016/j.promfg.2015.09.057.

Sun, J., Zhou, W., Huang, D., 2018. 3D printing of foodReference Module in Food Science. Elsevier, Amsterdam, Netherlands. https://doi.org/10.1016/B978-0-08-100596-5.21893-3.

Sun, J., Zhou, W., Huang, D., Fuh, J.Y.H., Hong, G.S., 2015. An overview of 3D printing technologies for food fabrication. Food Bioprocess Technol. *8* (8), 1605–1615. https://doi.org/10.1007/s11947-015-1528-6.

Sun, J., Zhou, W., Yan, L., Huang, D., Lin, L., 2018b. Extrusion-based food printing for digitalized food design and nutrition control. 3D Printed Food – Des. Technol. *220*, 1–11. https://doi.org/10.1016/j.jfoodeng.2017.02.028.

Topuz, F.C., Bakkalbaşı, E., Cavidoğlu, İ., 2018. The current status, development and future aspects of 3d printer technology in food industry. Int. J. 3D Print. Technol. Digital Industry *2* (3), 66–73.

Vancauwenberghe, V., Baiye Mfortaw Mbong, V., Vanstreels, E., Verboven, P., Lammertyn, J., Nicolai, B., 2019. 3D printing of plant tissue for innovative food manufacturing: encapsulation of alive plant cells into pectin based bio-ink. J. Food Eng. *263*, 454–464. https://doi.org/10.1016/j.jfoodeng.2017.12.003.

Van Rijssel, A., 2019. Development of 3D Printed Food Structures M.Sc. Thesis. Wageningen University, Netherlands.

Vinoski, J. (2019, September 25). With food-grade 3-D printing, redefine meat is out to… well, redefine meat. Forbes. https://www.forbes.com/sites/jimvinoski/2019/09/25/with-food-grade-3-d-printing-redefine-meat-is-out-to-well-redefine-meat/#7ab4abcf75c7.

Voinea, C.L., Logger, M., Rauf, F., Roijakkers, N., 2019. Drivers for sustainable business models in start-ups: multiple case studies. Sustainability (Switzerland) *11* (24). https://doi.org/10.3390/su11246884.

Voon, S.L., An, J., Wong, G., Zhang, Y., Chua, C.K., 2019. 3D food printing: a categorised review of inks and their development. Virtual Phys. Prototyping *14* (3), 203–218. https://doi.org/10.1080/17452759.2019.1603508.

Wang, L., Zhang, M., Bhandari, B., Yang, C., 2018. Investigation on fish surimi gel as promising food material for 3D printing. 3D Printed Food Des. Technol. *220*, 101–108. https://doi.org/10.1016/j.jfoodeng.2017.02.029.

Warner, E.L., Norton, I.T., Mills, T.B., 2019. Comparing the viscoelastic properties of gelatin and different concentrations of kappa-carrageenan mixtures for additive manufacturing applications. J. Food Eng. *246*, 58–66. https://doi.org/10.1016/j.jfoodeng.2018.10.033.

Webb, L., Swanepoel, T., Green, P., 2016. *Final Report – Review of Market Acceptance and Value Proposition for 3D Printed Meat* (Final V.RMH.0039). Meat and Livestock, Australia https://www.mla.com.au/download/finalreports?itemId=3279.

Wegrzyn, T.F., Golding, M., Archer, R.H., 2012. Food layered manufacture: a new process for constructing solid foods. Trends Food Sci. Technol. *27* (2), 66–72. https://doi.org/10.1016/j.tifs.2012.04.006.

Yalcinkaya, G. (2019), https://www.dezeen.com/2019/04/02/sushi-singularity-3d-printed-sushi-customised/.

Yang, F., Zhang, M., Bhandari, B., 2017. Recent development in 3D food printing.. Crit. Rev. Food Sci. Nutr. *57* (14), 3145–3153. https://doi.org/10.1080/10408398.2015.1094732.

Yang, F., Zhang, M., Bhandari, B., Liu, Y., 2018. Investigation on lemon juice gel as food material for 3D printing and optimization of printing parameters. LWT *87*, 67–76. https://doi.org/10.1016/j.lwt.2017.08.054.

Zhang, L., Lou, Y., Schutyser, M.A.I, 2018. 3D printing of cereal-based food structures containing probiotics. Food Struct. *18*, 14–22. https://doi.org/10.1016/j.foostr.2018.10.002.

Zhao, H., Wang, J., Ren, X., Li, J., Yang, Y.-L., Jin, X., 2018. Personalized food printing for portrait images. CAD/Graphics 2017 (70), 188–197. https://doi.org/10.1016/j.cag.2017.07.012.

Zhu, S., Stieger, M.A., van der Goot, A.J., Schutyser, M.A.I., 2019. Extrusion-based 3D printing of food pastes: correlating rheological properties with printing behaviour. Innovative Food Sci. Emerg. Technol. *58*, 102214. https://doi.org/10.1016/j.ifset.2019.102214.

Mathematical modeling— Computer-aided food engineering

16

Ferruh Erdogdu[a], Ashim Datta[b], Olivier Vitrac[c], Francesco Marra[d], Pieter Verboven[e], Fabrizio Sarghini[f], Bart Nicolai[e]

[a]*Department of Food Engineering, Ankara University, Golbasi-Ankara, Turkey*
[b]*Department of Biological and Environmental Engineering, Cornell University, Ithaca, NY, United States*
[c]*UMR 0782 SayFood Paris-Saclay Food and Bioproduct Engineering Research Unit, INRAE, AgroParisTech, Université Paris-Saclay, Massy, France*
[d]*Dipartimento di Ingegneria Industriale, Università degli studi di Salerno, Fisciano, SA, Italy*
[e]*Division of Mechatronics, Biostatistics and Sensors (MeBioS), Biosystems Department, KU Leuven, Leuven, Belgium*
[f]*Department of Agricultural Sciences, University of Naples Federico II, Naples, Italy*

16.1 Introduction

New product development drives innovation in the food industry rather than process technology. Food processing is usually based on traditional systems and technological applications adapted from other industries (Jousse, 2008; Erdogdu et al., 2017b). These choices are inherited from culinary practices and the unique characteristic of food products and their transformation. As a result, food engineering is a rather new discipline compared to its parents: chemical and mechanical engineering. However, the complexity of the phenomena (physical, chemical, and biological) involved has nothing to envy the other disciplines. Food safety and quality are essential to food, and they are controlled by complex temperature-dependent biochemical and microbiological reactions during processing and storage. Several couplings mediated by water transfer, including shrinkage, swelling, dissolution, structural changes, etc. complicate the mathematical description of the food transformation and its evolution dramatically. For example, the detailed simulation of food drying may intricate up to three levels of description: (i) the anisothermal flow of the drying medium (air, steam, oil,...), (ii) coupled heat and mass transfer inside the food product, and (iii) the mechanical behavior of the solid matrix with its hygrothermal history. Finally, the evolution of the properties (thermal, transport, rheological) of the food product as a function of composition and temperature, and possible hysteresis closes the food drying description.

Modern food process modeling was conceptualized and introduced by Loncin and Merson (1979) in their pioneering book and derived from similar principles of chemical engineering. Conservation laws, thermodynamics, and kinetic descriptions are combined in coupled partial differential equations for the concept of the food process modeling. Mathematical modeling associated with current computational power helps to virtualize food product development and may eventually accelerate innovation in the food industry. In recent years, simulations in complex geometries and configurations moved from supercomputers to standard ones. Physical modeling of problems, which can be treated separately in a sequential manner, has become accessible to almost any scientist and professional. In other words, the inactivation of microorganisms or the Maillard reaction evolution can be derived from a first heat transfer simulation because they do not affect heat transfer but depend upon the temperature change with respect to the process time. Besides, possibly simplified calculations finalized in a few minutes can be integrated into optimization loops for both product and process design. However, expectations vary greatly depending on the physical processes and the number of scales involved. Multiscale phenomena, such as phase separation, mixing, grinding, wettability and impregnability problems, microstructure creation, or mechanical failure, are difficult to simulate numerically. The term "multiphysics" refers to a general resolution methodology where several physical fields (e.g., heat and Maxwell's equations in microwave heating) need to be resolved simultaneously.

Solutions of the equations defining the physical problem can be exact or numerically estimated. Exact solutions are less general, as they may exist only for simplified geometries and specific initial and boundary conditions. Numerical solutions are more general as they can be inferred at the mesh nodes approximating the real geometry in 1D, 2D, or 3D. The crudest approach with a staircase mesh leads to the finite difference method. It was the first accessible method. It has, however, little interest beyond 1D or regular geometries. Finite volume and finite element techniques accommodate unstructured meshes and are more general. They have been implemented in various computational fluid dynamics (CFD) or "multiphysics" packages. Even though the name implies a fluid related problem, CFD numerical resolution tools cover a broad range of physical problems involving transports, conservation laws, and source terms. The most advanced implementations can handle complex phenomena in foods such as volume and structural changes. CFD and multiphysics codes popularize physics-based modeling without having to write a program. They usually include a mesh generator facilitating the tailoring for various geometries. Typical steps include:

- geometry definition and mesh generation,
- setting governing equations with proper boundary and initial conditions for the considered physics,
- setting property data and their variation with calculated intensive quantities (temperature, water content, concentration, pressure, chemical potential, etc.)
- choosing a numerical solver.

Once the solution is discretized in space and integrated in time, postprocessing is the last step to retrieve the results of interest or comparable with the experience. Validation against experimental results from literature or against already validated (or exact) solutions is mandatory. According to the end-user experience, additional validation such as sensitivity analysis and uncertainty estimations may be carried out.

Physics-based modeling tends to replace progressively data-driven modeling techniques such as the response surface methodology. These iterative techniques relying on specific experimental results are time-consuming and far less cost-efficient than direct simulation. Alternatively, curve fitting based on an existing analytical solution has been the common practice to determine a constant mass and heat diffusion coefficients. Though the practice is highly commendable when the solution reflects the proper physics, it is strongly questionable in food process modeling, when effective properties replace a set of coupled equations. Combining the right physics and the proper numerical technique can be highly beneficial for the experimentalist by reducing unnecessary tests and trials. The first example was reported in the food engineering literature by Teixeira et al. (1969). The authors solved the heat conduction equation in a cylinder by finite differences to better estimate bacterial destruction kinetics. This model was further applied to reduce the loss of thermolabile nutrients in thermal processing. As a follow-up, Datta and Teixeira (1988) resolved numerically coupled Navier-Stokes and heat transfer equations in a canned liquid to reconstruct the temperature distribution in cans during retorting. Several numerical studies followed to determine temperature changes during food processing. Although they were mostly verified in lab-scale systems, predictions were extended for some of them at industrial scale, leading eventually to optimized operation conditions and new designs.

Based on the personal experience of the authors of this manuscript, it is now an accepted fact that mechanistic modeling may offer insights beyond experimental capabilities (Datta, 2008). Computer-aided food and process engineering have gradually earned their scientific reputation. Because their assumptions and predictions are refutable, they can be applied with confidence in various domains ranging from process and formulation optimization to food safety evaluation. This manuscript aims to present the implementation challenges and future directions of food and process computer-engineering.

16.2 Engines of computer-aided food engineering: mechanistic modeling frameworks

16.2.1 Frameworks: general discussion

As indicated, food processes cover many physics phenomena. Thus, computer-aided engineering of food processes cannot progress much with ad-hoc models of individual processes where a solution is signified for a problem. Therefore, frameworks are required for various classes of processes where there is significant commonality in physics within a class. Individualization of a product/process is addressed using product properties and process parameters within that class.

Fortunately, several frameworks have been developing within the last 25 years and have come to a level of maturity. If cooking and drying are considered as food processes where more than one physics is involved (e.g., simultaneous heat and mass transfer with phase change), three particular frameworks are common (Datta, 2016):

1. A sharp boundary phase change formulation,
2. A distributed phase change with porous media formulation, and
3. A multiscale porous medium model.

The second one of this list was summarized in more detail by Datta (2016). This modeling framework considered a solid or semisolid food product as a homogenized macroscale porous medium that was also deformable/swellable and, within it, multiphase transport of water, vapor and other components, along with distributed evaporation were present. This framework has been successful in modeling a number of important processes, including drying that incorporated the case hardening (Gulati and Datta, 2015), rehydration (Weerts et al., 2003), baking (Zhang and Datta, 2006; Nicolas et al., 2014), frying (Yamsaengsung and Moreira, 2002; Halder et al., 2007), meat cooking (Dhall and Datta, 2011), microwave heating (Ni et al.,1999), microwave puffing (Rakesh and Datta, 2013), and freeze drying (Warning et al., 2015).

16.2.2 Details of one framework: distributed phase change with multiphase transport in a porous medium

In this framework, food materials are treated as hygroscopic and capillary-porous. A porous media formulation homogenizes the real porous material and treats it as a continuum where the pore scale information is no longer available. The results, therefore, are valid with a resolution of the representative elementary volume. Once homogenized, standard formulations of mass and energy balances for the individual phases (e.g., water, water vapor, and air) are developed, which include the various modes of transport such as capillarity, pressure-driven flow, and molecular diffusion. The more commonly used equations such as the simple diffusion equation sometimes used to model drying are adaptations of (i.e., can be derived from) the above general equation, by making appropriate simplifications. To include shrinkage and swelling as structural changes, the porous medium is treated as deformable, using the well-known poromechanics framework (Dhall and Datta, 2011). When a food is being processed without any externally applied stress on it, the effective stress on the solid skeleton of the food matrix is related to the average fluid pressure in the matrix, using the standard equation for force balance under static equilibrium. This fluid pressure in the matrix is a volumetric mean of the liquid and gas pressures, computed from the transport equations. The effective stress on the solid skeleton is related to the strain and thus the deformation in the material using material models that are typically elastic, hyper-elastic, viscoelastic or plastic. These material models come from measurement, although very limited data are available.

Auxiliary relationships required in the formation have now been fairly well developed. Evaporation inside the food product during a process can be estimated assuming the food as a porous medium and the evaporation occurs from surfaces of internal pores into the pore space. Liquid pressure that drives the liquid convective transport is made up of gas pressure, capillary pressure, swelling pressure, and gravity. Capillary pressure driven transport would be converted into capillary diffusivity that comes from measured data. Swelling pressure can be estimated from measured water holding capacity data using relationships that have been proven using fundamental physics. These results in a set of transport equations, coupled partial differential equations, that can be readily implemented in a computational software package. A survey of most of the existing modeling approaches in the food literature shows that of the fundamental physics-based approaches, this framework is broad, flexible in accommodating many different processes, and easier (than other frameworks) to understand and implement in a commercial software (Dhall and Datta, 2011).

16.2.3 Extending the above framework to quality and safety

The fundamental concept in extending the above framework to quality is that quality such as texture, color, and flavor are first functions of histories of temperature history. Besides, moisture, structure, composition, and other variables affect the quality, and all these variables can be implemented in a framework model. For example, in a French-fry process, if Young's modulus can be considered a measure of texture, this modulus can be computed at any point during frying by relating the local moisture levels throughout the food obtained from the framework, to local modulus using experimental moisture–modulus relation. A homogenization process would provide the overall modulus of the French-fry at any time (Thussu and Datta, 2012). In an analogous manner, color can be predicted by combining the kinetics of reactions (e.g., Maillard) that depend on temperature and moisture, with these variables already available from the framework, as was mentioned. The main idea in quality prediction, combining the rate of kinetics of reactions with local temperature, moisture, and other calculated variables from the process framework, can be extended to safety. This idea is not new and is the basis of Ball's formula in canning from almost 100 years ago (Simpson et al., 2006). However, the latest quantitative understanding on bacterial kinetics can be used easily in the computational framework. More details of the bacterial behavior such as transport (motility, Brownian motion, chemotaxis), attachment and communication can be combined with the computed process variables to obtain the most realistic computation of safety (Ranjbaran et al., 2020). Following the same analogy with microbiological kinetics, chemical kinetics can be included to estimate chemical safety like the formation of carcinogens (Halder et al., 2011).

16.3 Properties for the mechanistic models—prediction and integration

While an available framework model is rather useful tool for further predictions of a process, simulating the transformation of food may be highly frustrating as it involves a large number of thermodynamic, transport, and mechanical properties as summarized above. However, it is a poor tabulation in the literature is also a fact (Gulati and Datta, 2013). Where point values are available, they may not reflect biological variability, compositional and structural changes during processing and storage. The physical transformation of food constituents (gelation, gelatinization, glass transition, swelling) complicates the extrapolation of properties from one condition to another. Until recently, the only viable solution was to fit semiempirical models to experimental data (Gulati and Datta, 2013). Thinking about foods in terms of their purpose (i.e., their final sensory or nutritional properties with the required safety), rather than describing their evolution in a given process requires, coupling one or more organization scale with the macroscopic one in the modeling framework. At the lowest scale, molecular modeling at the atomistic scale offers a competitive alternative to experiments in amorphous materials (Nguyen et al., 2017a), homo-polymers (Gillet et al., 2009, 2010; Vitrac and Gillet, 2010), and block-polymers (Nguyen et al., 2017b). Similar approaches have been utilized to derive diffusivities in glassy and rubber polymers (Fang et al., 2013; Fang and Vitrac, 2017; Zhu et al., 2019b). Molecular modeling provides a direct approach to encode the chemical information in models. Temperature and pressure effects, mutual diffusion, viscosity, phase diagram, mechanical rupture can be studied without approximation. However, size and time scale are the main limitations. Millions (a cube of side a few tens of nanometers) to billions (a cube of side a few of hundreds of nanometers) of atoms can be simulated today for dynamic time up to hundreds of nanoseconds. At the cost of a large amount of memory and massively parallel computation, the maximum achievable size approaches that of a small bacterium, but it is still much smaller than a food product. Besides, the span of the time frame is limited by the smallest time step used in the implicit integration scheme, which is determined by the vibration frequency of the hydrogen atom (about femtosecond or 10^{-15} s). Dropping hydrogen atoms and merging heavy atoms or motion groups into blobs (like replacing small beads by large ones in a necklace) provides extra integration time from two to four orders of magnitude (Durand et al., 2010). Compared to the hard-core potentials used to represent atoms, the soft-core potentials describing blobs accept overlap and have fewer excluded volumes. They are not defined uniquely and require prior parameterization based on radial distributions (see Boltzmann-inversion procedure in Eriksson et al., 2009). For food and bioproducts, the force field (López et al., 2009; Marrink et al., 2007) is the most generic and natively available in the popular and open-source molecular dynamics package GROMACS (Berendsen et al., 1995). Coarse-graining leads directly, through integration (bottom-up approach) or by intuition (analogy reasoning), to a broad range of meshless simulation methods, which can be used to infer macroscopic quantities from microscopic descriptions. At the supramolecular scale (tens and hundreds of nanometers), DPD or dissipative particle dynamics (Español and Warren, 1995, 2017; Hoogerbrugge and Koelman, 1992) is the best alternative to describe flow and wetting properties in complex mixtures as well as mechanical fracture in solids. The method has been illustrated recently on hydrogels (Palkar et al., 2020). At the microscopic scale (from micrometers to centimeters), SPH or smoothed particle hydrodynamics (Cundall and Strack, 1979; Gingold and Monaghan, 1977) has been proposed to describe a wide range of phenomena, including morphological changes during drying (Karunasena et al., 2014), oral processing (Harrison et al., 2014), aroma extraction from coffee (Ellero and Navarini, 2019), and 3D food printing (Makino et al., 2017). The DPD and SPH methods cover the full range of length scales in food and are implemented in the generic open source molecular dynamics program LAMMPS (Plimpton, 1995). For transport phenomena, the kinetic Monte-Carlo (KMC) coupled with first passage algorithms (a time-continuous hominization technique replacing high-frequency evolutions by a low frequency one), which has been originally proposed to describe hopping diffusion in alloys and glassy polymers (Vitrac and Hayert, 2007), offers gains in time integration of several magnitude orders. It is very effective to describe oil percolation in fried products at cellular scale in the presence of an incondensable phase (air) and damaged tissues (Patsioura et al., 2015; Vauvre et al., 2015). For pure diffusional transports in inhomogeneous foods, a generic solver based on Langevin dynamics (LD) and an explicit definition of the local thermodynamic equilibrium has been devised to estimate effective diffusion coefficients regardless of the fractal character of the structure and its evolution with time (Vitrac and Hayert, 2020). The KMC ad LD methods take advantage of widely available 3D microscopic observations from laser scanning confocal microscopy and microcomputed X-ray tomography to estimate the effective properties of a real food. In specific cases, DPD, SPH, KMC, and LD techniques can integrate across the scales without the need for an effective medium approximation (EMA), that is, without replacing an inhomogeneous medium at the microscopic scale by a medium without microscopic fluctuations. Simulation workflows for designing food and bioproducts with controlled properties from molecular properties have been reviewed by Kontogeorgis et al. (2019) and Zhang et al. (2016). EMA remains necessary when mass transfer and transport are combined with heat transfer, phase changes, or when they involve time scales longer than hours (Matouš et al., 2017).

16.4 Multiphysics and multiscale

Erdogdu et al. (2017a) discussed the general complexity in modeling and virtualization of food processing, which resides in both complexity of food system structures, their properties and complexity due to simultaneous phenomena (such as heat, mass, momentum, structural mechanics, and so on) characterizing a given food process, often in more than one phase. As explained above, there is no doubt that food processes are multiphysics by nature (meaning that very rarely it is possible to reduce the analysis of a food process to the analysis of just one physics phenomenon). Is it possible to analyze heat transfer in a food product without taking into account the mass transfer or the loss of quality attributes or the transformation at structure level? No, it is not possible in general or it is possible with rather extensive assumptions and simplifications.

A query put on scopus.com using the following keyword string "multiphysics AND modeling AND food" provided 98 sources (overall cited 983 times, with h-index 19), 64% being original research articles, the first being Edmondson et al. (2005) where the term "Multiphysics" associated to modeling and food was introduced. A very interesting introduction to multiscale modeling can be found in Ho et al. (2013). However, back in the 90s, multiphysics approach was already presented (Datta, 1991) with another fundamental contribution in understanding processes driven by electromagnetic fields like in the case of microwave processing (Rakesh and Datta, 2013). In a milestone work published by Datta (2008), status of physics-based models in the design of food products, processes, and equipment was presented. In these studies, coupling of different types of physics with heat transfer promoted by microwaves in food processing was exemplified. While the core of the modeling was constituted by heat transfer, it was not the only physics phenomenon to be considered in these processes considering that the heat transfer was promoted by the interaction of the food system within an electromagnetic field (which requires its own mathematical description). Heat transfer also causes the evaporation and moisture changes with structural deformations. This, in turn, forces the model to take into account the required equations describing the displacement of the electromagnetic field in the food system, heat balance, the latent heat, and eventually the structural modification of the food as well. This multiphysics nature of such a process would also require the knowledge of thermophysical properties: density, specific heat, thermal conductivity, and possibly viscosity while dielectric properties and electrical conductivity with structure-based physical properties are additional. It should also be noted that all these properties are to be functions of temperature, applied frequency, moisture, etc. In addition to considering the changes in the food product, the geometry of the process system may be an additional variable for the multiphysics framework model. The multiphysics modeling approach has found successful application in the optimal design of complex systems for food processing, like in the case of continuous radio frequency assisted thawing as discussed by Erdogdu et al. (2017a) in a computational study to design process conditions. This study described the significant impact of multiphysics modeling for design and optimization of industrial scale radio frequency heating and thawing processes, where a coupling between heat transfer, displacement of the electromagnetic field, and the dynamics of instantaneous movement of the electrodes within the cavity with simultaneous movement of the food product along the conveyor belt while facing a phase change.

With the term of multiscale models, the authors of this manuscript refer to a hierarchy of a model which operates at different spatial scale, allowing the prediction of the macroscale behavior of a food from microscale factors (i.e., what is the effect at a macroscale process on microscale processes in foods?). Fundamental contributions to the advancement of multiscale modeling in food processing, proposing models for the prediction of bulk properties was presented by Van Der Sman (2017). A detailed discussion for the macro- to microscale models was also given in Section 16.3 of this chapter.

While the bottle neck of multiscale approaches resides in the difficulty of providing a detailed representation of the 3-D multiscale geometry of the food product, multiscale modeling can still be adopted as a tool for structural engineering of foods, design of food products, and further design and optimization of food processing and related process systems.

Recently, Welsh et al. (2021) discussed the capability of multiscale modeling approach to capture the heterogeneity of food materials. The multiscale modeling approach allows to overcome the approximation of the food material's structure as a fictitious continuum, which limits the ability of the models to capture the heterogeneity of a food material. The multiscale modeling also leads to advance the understanding of transport processes at microscale level.

16.5 Process design and optimization

As explained, food process engineering relies on mathematical modeling based process innovations for further design and optimization. This objective can be to determine the optimum conditions of a process or a whole processing plant while there is also the potential to develop processing systems. For example, the design and optimization may focus on the process conditions of a radio frequency thawing process, while the radio frequency system's geometry and design may be another challenge that can be solved as long as an experimentally validated mechanistic physics-based model is present.

Optimization is simply to choose the best alternative to suit the given objective function under the applied restrictions (constraints) among a specified set of alternatives. Hence, describing the set of alternatives (decision variables) and the decision mechanism to decide upon the best case scenarios are required (Norback, 1980; Evans, 1982). Based on this, an optimization problem should have:

- set of variables to specify the alternatives,
- set of requirements–restrictions–constraints to achieve or satisfy (e.g., required sterility for process safety), and
- the objective function to compare the alternatives.

Based on the number of objective functions for the performance criteria describing the performance of the system, the optimization may be a single- or multiobjective problem. Such problems are also known as multiple criterion decision making problems (Deb, 2002). Optimization studies in the food engineering literature were mostly trial and error approaches in lab-scale processes based on experimental data. Response surface methodology was the preferred solution for these problems. The study by Saguy and Karel (1979) was the first to apply a mathematically challenging optimization algorithm (Pontryagin's maximum principle) to determine the optimal process temperature to maximize the nutrient retention for a canned food while the earliest study for computer-aided optimization in food engineering dates back to 1969 where computer-based optimization of nutrient retention in the thermal processing of conduction-heated foods was carried out (Teixeira et al., 1969).

In addition to the process based optimization studies, system geometry design for process improvement are more recent challenges. One of such cases was reported by Sarghini et al. (2018). In this geometry optimization, based on a CFD solution of a viscous flow, a constrained optimization approach for designing a pasta extrusion bell and ice-cream flow splitter was applied. The use of a complex CDF model combined with an optimization algorithm demonstrated the significance of this approach for rather complex design and optimization cases (Fig. 16.1).

The above figure by Erdogdu (2009) demonstrates the design of a computer-aided-based optimization problem for this purpose. Banga at al. (2003) presented the computer-aided optimization as an ultimate tool to improve food processing, and the requirement of a mathematical model to define the

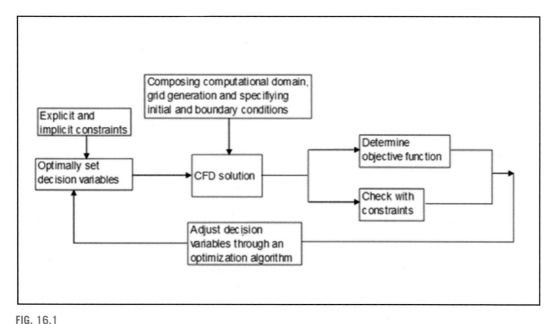

FIG. 16.1

Algorithm for a computer-aided-based optimization problem (reprinted from Erdogdu, 2009).

process was also indicated to combine with the optimization algorithm. The optimization approaches combined with a multiphysics model may also be used for packaging design purposes.

16.6 Food packaging design

With the progressive banning of epoxy-phenolic coatings in cans and of single-use plastics worldwide, it is imperative to reinvent the concept of food shelf-life and its packaging, but without taking any risks. Changing the materials, the shape and thinning the walls may affect mass transfer in a complex fashion: higher permeation of water and oxygen, aroma scalping, loss of the modified atmosphere, higher release of chemicals (e.g., with recycled materials), risk of collapse are issues that may be encountered. The number of possible configurations (food × process × design × supply chain and consumer practices) is so large and affect potentially all manufactured food product leaving the modeling approaches as the only feasible solution. An integrated framework has recently been developed for a single multicriteria optimization problem: multicomponent mass transfer, mechanical resistance, waste reduction, maximization of recycled material, 3D geometry optimization, risk minimization, shelf-life maximization (Zhu et al., 2019a). The framework reuses several concepts previously presented: nested multiscale simulations, an efficient finite volume solver FMECAengine (Vitrac and Nguyen, 2014-2019) and an inference language, so-called key2key, to quickly implement complex simulation scenarios at the scale of the supply chain (Nguyen et al., 2013). The same framework has been used recently to demonstrate a generalized cross-mass transfer between all layers of the packaging in or out of contact (Nguyen et al., 2017c, 2019).

16.7 Challenges in implementation

While the time has certainly come for the presented multiphysics approaches to be incorporated in practical applications, and the frameworks have made this significantly simpler, there are still certain challenges in the application and implementation. Since the goals of a food process model, especially in a computer-aided engineering context, are to understand the physics governing the process and observe the effects of parameter variation, none of these challenges are show-stoppers, but important to be noted. Some of the challenges can be summarized as follows:

1. The models need properties as function of temperature and composition (with additional requirements of frequency effects for electromagnetic processes or viscosity as a function of both temperature and shear rate especially for non-Newtonian fluids) since all these would be changing during a process. Such detailed properties data are currently not available, but prediction equations are available that can estimate properties to within 10% accuracy.
2. Acquiring exact geometry at multiple scales to make the solutions more realistic. Significant work is being done worldwide to obtain such geometries.
3. Computation time and resource requirement is also a challenge in realistic multiphysics calculations. Compromise then has to be made following the paradigm "As simple as possible, but no simpler." This quote is attributed to Albert Einstein, commonly given as "Everything should be made as simple as possible, but no simpler."
4. Finding human resources that have strong computational skills, training in the relevant physics and an appreciation/interest in food as a complex material is also a challenge for which new university courses, or short courses presented by academics and industry professionals, and learning material will play a significant role.

16.8 Conclusions and future directions

Considering the above summarized challenges in implementation, rather unique opportunities have become available. Considering the availability and increase power of computing have brought a new era for multiphysics modeling and eventually for computer-aided food processing. Within the last decade, several developments in the current digitalization era serve well for a more widespread uptake of food modeling in applications. Among them is the concept of digital twins. Digital twins are virtual equivalents of the actual process (Tao and Qi, 2019), taking real time measured data as input and returning outputs that are directly useful; in- or at-line, for process control or mixture of the two. It has its origin in other engineering areas, and with the increasing use of sensor and communication technologies applications are becoming more successful in the manufacturing industry (Lu and Xu, 2019; Tao et al., 2018, 2019), health care (Bruynseels et al., 2018), and recently also the food industry (Verboven et al., 2020; Defraeye et al., 2019). Verboven et al. (2020) explained the concept of the digital twin with respect to the food processing operations and food process models in detail where the digital twin was defined to be the virtual replicate of an actual process, connected to the realty worked by sensor data and advanced big data analytical tools. Via the availability of a comprehensive mathematical model defining the process and the required on-line data, it would be possible to improve and revise the process if required. The digital twin concept will also benefit from ICT infrastructure such as the internet of things (IoT), providing enhanced connectivity between objects such as machines and sensors using standard internet protocols and cloud based computing and database centers. IoT makes it possible to implement precision technology, automation, and robotics more efficiently and effectively in the agrifood industry (Kodan et al., 2019), including supply-chain wide applications related to tracking and tracing, helping to ensure consumer trust, and public health (Astill et al., 2019). Furthermore, through user-friendly apps that run on mobile devices, both operator and manager have, at anytime and anywhere access to rich decision support data. For running the ever more computationally demanding mathematical models, cloud computation services are appearing, such as ANSYS Cloud running on Microsoft Azure, Siemens Cloud Solutions, or the completely cloud based platform SimScale. Apart from running the conventional numerical solvers, the digital twin could also incorporate machine and deep learning algorithms such that an intelligent modeling engine is created that continuously gets better at its tasks, as more data of the actual process becomes available with time.

References

Astill, J., Rozita, A.D, Campbell, M., Farber, J.M., Fraser, E.D.G., Shayan, S., Yada, R.Y., 2019. Transparency in food supply chains: a review of enabling technology solutions. Trends Food Sci. Technol. 91, 240–247.

Banga, J.R., Balsa-Canto, E., Moles, C.G., Alonso, A.A., 2003. Improving food processing using modern optimization methods. Trends Food Sci. Technol. 14, 131–144.

Berendsen, H.J.C., Van Der Spoel, D., Van Drunen, R, 1995. GROMACS: a message-passing parallel molecular dynamics implementation. Comput. Phys. Commun. 91, 43–56.

Bruynseels, K., Santoni de Sio, F., van den Hoven, J., 2018. Digital twins in health care: ethical implications of an emerging engineering paradigm. Front. Genet. 9, 1–11.

Cundall, P.A., Strack, O.D.L, 1979. A discrete numerical model for granular assemblies. Geotechnique 29, 47–65.

Datta, A.K., 1991. Mathematical modeling of biochemical changes during processing of liquid foods and solutions. Biotechnol. Prog. 7, 397–402.

Datta, A.K., 2008. Status of physics-based models in the design of food products, processes, and equipment. Comprehens. Rev. Food Sci. Food Safety 7, 121–129.

Datta, A.K., 2016. Toward computer-aided food engineering: mechanistic frameworks for evolution of product, quality and safety during processing. J. Food Eng. 176, 9–27.

Datta, A.K., Teixeira, A.A., 1988. Numerically predicted transient temperature and velocity profiles during natural convection heating of canned liquid foods. J. Food Sci. 53, 191–195.

Deb, K., 2002. Multi-Objective Optimization Using Evolutionary Algorithms. Wiley, New York, NY.

Defraeye, T., Tagliavini, G., Wu, W., Prawiranto, K., Schudel, S., Kerisima, M.A., Verboven, P., Bühlmann, A., 2019. Digital twins probe into food cooling and biochemical quality changes for reducing losses in refrigerated supply chains. Resour. Conserv. Recycl. 149, 778–794.

Dhall, A., Datta, A.K., 2011. Transport in deformable food materials: a poromechanics approach. Chem. Eng. Sci. 66, 6482–6497.

Durand, M., Meyer, H., Benzerara, O., Baschnagel, J., Vitrac, O., 2010. Molecular dynamics simulations of the chain dynamics in monodisperse oligomer melts and of the oligomer tracer diffusion in an entangled polymer matrix. J. Chem. Phys. 132, 194902.

Edmondson, P.T., Grammatika, M., Fryer, P.J., Handy, B., 2005. Modelling of heat transfer, mass transfer and flavour development in chocolate crumb. Food Bioprod. Process. 83, 89–98.

Ellero, M., Navarini, L., 2019. Mesoscopic modelling and simulation of espresso coffee extraction. J. Food Eng. 263, 181–194.

Erdogdu, F., 2009. Computational fluid dynamics for optimization in food processing. In: Erdogdu, F. (Ed.), Optimization in Food Engineering. CRC Press, Boca Raton, FL, USA chapter 11.

Erdogdu, F., Altin, O., Marra, F., Bedane, T.F., 2017a. A computational study to design process conditions in industrial radio-frequency tempering/thawing process. J. Food Eng. 213, 99–112.

Erdogdu, F., Sarghini, F., Marra, F., 2017b. Mathematical modeling for virtualization in food processing. Food Eng. Rev. 9, 295–313.

Eriksson, A., Jacobi, M.N., Nyström, J., Tunstrøm, K., 2009. A method for estimating the interactions in dissipative particle dynamics from particle trajectories. J. Phys. Condens. Matter 21, 095401.

Español, P., Warren, P., 1995. Statistical mechanics of dissipative particle dynamics. Europhys. Lett. (EPL) 30, 191–196.

Español, P., Warren, P.B., 2017. Perspective: dissipative particle dynamics. J. Chem. Phys. 146, 150901.

Evans, L.B., 1982. Optimization theory and its application in food processing. Food Tech. 36 (7), 88.

Fang, X., Domenek, S., Ducruet, V., Refregiers, M., Vitrac, O., 2013. Diffusion of aromatic solutes in aliphatic polymers above glass transition temperature. Macromolecules 4, 874–888.

Fang, X., Vitrac, O., 2017. Predicting diffusion coefficients of chemicals in and through packaging materials. Crit. Rev. Food Sci. Nutr. 57, 275–312.

Gillet, G., Vitrac, O., Desobry, S., 2009. Prediction of solute partition coefficients between polyolefins and alcohols using a generalized Flory-Huggins approach. Ind. Eng. Chem. Res. 48, 5285–5301.

Gillet, G., Vitrac, O., Desobry, S., 2010. Prediction of partition coefficients of plastic additives between packaging materials and food simulants. Ind. Eng. Chem. Res. 49, 7263–7280.

Gingold, R.A., Monaghan, J.J., 1977. Smoothed particle hydrodynamics: theory and application to non-spherical stars. Mon. Not. R. Astron. Soc. 181, 375–389.

Gulati, T., Datta, A.K., 2013. Enabling computer-aided food process engineering: property estimation equations for transport phenomena-based models. J. Food Eng. 116, 483–504.

Gulati, T., Datta, A.K., 2015. Mechanistic undertanding of case-hardening and texture development during drying of food materials. J. Food Eng. 166, 119–138.

Halder, A., Dhall, A., Datta, A.K., 2007. An improved, easily implementable, porous media based model for deep-fat frying: part I: model development and input parameters. Food Bioprod. Process. 85, 209–219.

Halder, A, Dhall, A., Datta, A.K., Black, D.G., Davidson, P.M., Li, J., Zivanovic, S., 2011. A user-friendly general-purpose predictive software package for food safety. J. Food Eng. 104, 173–185.

Harrison, S.M., Eyres, G., Cleary, P.W., Sinnott, M.D., Delahunty, C., Lundin, L., 2014. Computational modeling of food oral breakdown using smoothed particle hydrodynamics. J. Texture Stud. 45, 97–109.

Ho, Q.T., Carmeliet, J., Datta, A.K., Defraeye, Delele, M.A., Herremans, E., Opara, L., Ramon, H., Tijskens, E., Van Der Sman, R., Van Liedekerke, P., Verboven, P., Nicolaï, B.M, 2013. Multiscale modeling in food engineering. J. Food Eng. 114, 279–291.

Hoogerbrugge, P.J., Koelman, J.M.V.A, 1992. Simulating microscopic hydrodynamic phenomena with dissipative particle dynamics. Europhys. Lett. (EPL) 19, 155–160.

Jousse, F., 2008. Modeling to improve the efficiency of product and process development. Comprehens. Rev. Food Sci. Food Safety 7, 175–181.

Karunasena, H.C.P., Senadeera, W., Brown, R.J., Gu, Y.T, 2014. A particle based model to simulate microscale morphological changes of plant tissues during drying. Soft Matter 10, 5249–5268.

Kodan, R., Parmar, P., Pathania, S., 2019. Internet of things for food sector: status quo and projected potential. Food Rev. Int. 36, 584–600.

Kontogeorgis, G.M., Mattei, M., Ng, K.M., Gani, R., 2019. An integrated approach for the design of emulsified products. AIChE J. 65, 75–86.

Loncin, M., Merson, R.L., 1979. Food Engineering – Principles and Selected Applications. Academic Press, New York, NY.

López, C.A., Rzepiela, A.J., de Vries, A.H., Dijkhuizen, L., Hünenberger, P.H., Marrink, S.J., 2009. Martini coarse-grained force field: extension to carbohydrates. J. Chem. Theory Comput. 5, 3195–3210.

Lu, Yuqian, Xu, Xun, 2019. Cloud-based manufacturing equipment and big data analytics to enable on-demand manufacturing services. Rob. Comput. Integr. Manuf. 57, 92–102.

Makino, M., Fukuzawa, D., Murashima, T., Furukawa, H., 2017. Simulation of 3D food printing extrusion and deposition. SPIE Smart Structures and Materials + Nondestructive Evaluation and Health Monitoring Nanosensors, Biosensors, Info-Tech Sensors and 3D Systems, 10167. https://doi.org/10.1117/12.2261409.

Marrink, S.J., Risselada, H.J., Yefimov, S., Tieleman, D.P, de Vries, A.H., 2007. The MARTINI force field: coarse grained model for biomolecular simulations. J. Phys. Chem. B 111, 7812–7824.

Matouš, K., Geers, M.G.D., Kouznetsova, V.G., Gillman, A, 2017. A review of predictive nonlinear theories for multiscale modeling of heterogeneous materials. J. Comput. Phys. 330, 192–220.

Nguyen, P.-M., Dorey, S., Vitrac, O., 2019. The ubiquitous issue of cross-mass transfer: applications to single-use systems. Molecules 24, 3467.

Nguyen, P.-M., Goujon, A., Sauvegrain, P., Vitrac, O., 2013. A computer-aided methodology to design safe food packaging and related systems. AIChE J. 59, 1183–1212.

Nguyen, P.-M., Guiga, W., Dkhissi, A., Vitrac, O., 2017b. Off-lattice Flory-Huggins approximations for the tailored calculation of activity coefficients of organic solutes in random and block copolymers. Ind. Eng. Chem. Res. 56, 774–787.

Nguyen, P.-M., Guiga, W., Vitrac, O., 2017a. Molecular thermodynamics for food science and engineering. Food Res. Int. 88 (Part A), 91–104.

Nguyen, P.-M., Julien, J.-M., Breysse, C., Lyathaud, C., Thébault, J., Vitrac, O., 2017c. Project safe food pack design: case study on indirect migration from paper and boards. Food Addit. Contam. 34, 1703–1720.

Ni, H., Datta, A.K., Parmeswar, R., 1999. Moisture loss as related to heating uniformity in microwave processing of solid foods. J. Food Process Eng. 22, 367–382.

Nicolas, V., Salagnac, P., Glouannec, P., Ploteaum, J.-P., Jury, V., Boillereaux, L., 2014. Modeling of heat and mass transfer in deformable porous media: application to bread baking. J. Food Eng. 130, 23–35.

Norback, J.B., 1980. Techniques for optimization of food processes. Food Technol. 34 (2), 86.

Palkar, V., Choudhury, C.K., Kuksenok, O., 2020. Development of dissipative particle dynamics framework for modeling hydrogels with degradable bonds. MRS Adv. 5, 927–934.

Patsioura, A., Vauvre, J.-M., Kesteloot, R., Jamme, F., Hume, P., Vitrac, O., 2015. Microscopic imaging of biphasic oil-air flow in French-fries using synchrotron radiation. AIChE J. 61, 1427–1446.

Plimpton, S., 1995. Fast parallel algorithms for short-range molecular dynamics. J. Comput. Phys. 117, 1–19.

Ranjbaran, M., Solhtalab, M., Datta, A.K., 2020. Mechanistic modeling of light-induced chemotactic infiltration of bacteria into leaf stomata. PLoS Comput. Biol. 16 (5), e1007841.

Rakesh, V., Datta, A.K., 2013. Transport in deformable hygroscopic porous media during microwave puffing. AIChE J. 59, 33–45.

Saguy, I., Karel, M., 1979. Optimal retort temperature profile in optimizing thiamin retention in conduction-type heating of canned foods. J. Food Sci. 44, 1985–1990.

Sarghini, F., 2018. Application of constrained optimization techniques in optimal shape design of food equipment, Presented in Conference of Food Engineering – CoFE, September 9-12. Minneapolis, MN.

Simpson, R., Almonacid, S., Teixeira, A.A., 2006. Bigelow's general method revisited: development of a new calculation technique. J. Food Sci. 68, 1324–1333.

Tao, F., Cheng, J., Qi, Q., Zhang, M., Zhang, H., Sui, F., 2018. Digital twin-driven product design, manufacturing and service with big data. Int. J. Adv. Manufact. Technol. 94, 3563–3576.

Tao, F., Qi, Q., 2019. Make more digital twins. Nature 573 (7775), 490–491.

Tao, F., Zhang, H., Liu, A., Nee, A.Y.C, 2019. Digital twin in industry: state-of-the-art. IEEE Trans. Ind. Inf. 15, 2405–2415.

Teixeira, A.A., Dixon., J.R., Zahradnik, J.W., Zinmeister, G.E, 1969. Computer optimization of nutrient retention in the thermal processing of conduction-heated foods. Food Technol. 23 (6), 138–142.

Thussu, S., Datta, A.K., 2012. Texture prediction during deep frying: a mechanistic approach. J. Food Eng. 108, 111–121.

Van der Sman, R.G.M., 2017. Model for electrical conductivity of muscle meat during Ohmic heating. J. Food Eng. 208, 37–47.

Vauvre, J.-M., Patsioura, A., Kesteloot, R., Vitrac, O., 2015. Multiscale modeling of oil uptake in fried products. AIChE J. 61, 2329–2353.

Verboven, P., Defraeye, T., Datta, A.K., Nicolai, B., 2020. Digital twins of food process operations: the next step for food process models? Curr. Opin. Food Sci. 35, 79–87.

Vitrac, O., Gillet, G., 2010. An off-lattice Flory-Huggins approach of the partitioning of bulky solutes between polymers and interacting liquids. Int. J. Chem. Reactor Eng. 8 (1), Article A6.

Vitrac, O., Hayert, M., 2007. Effect of the distribution of sorption sites on transport diffusivities: a contribution to the transport of medium-weight-molecules in polymeric materials. Chem. Eng. Sci. 62, 2503–2521.

Vitrac, O., Hayert, M., 2020. Modeling in food across the scales: towards a universal mass transfer simulator of small molecules in food. SN Appl. Sci. 2, 1509. accepted for publication.

Vitrac, O. and Nguyen, M. 2014-2019. FMECAengine 0.63: a failure mode effects and criticality analysis applied to mass transfer in packaging applications (GPL2 compatible license), https://github.com/ovitrac/FMECAengine.

Warning, A.D., Arquiza, J.M.R., Datta, A.K., 2015. A multiphase porous medium transport model with distributed sublimation front to simulate vacuum freeze drying. Food Bioprod. Process. 84, 637–648.

Weerts, A.H., Lian, G., Martin, D., 2003. Modeling rehydration of porous biomaterials: anisotropy effects. J. Food Sci. 68, 937–942.

Welsh, Z.G., Khan, M.I.H., Karim, M.A, 2021. Multiscale modeling for food drying: a homogenized diffusion approach. J. Food Eng. 292, 110252.

Yamsaengsung, R., Moreira, R.G., 2002. Modeling the transport phenomena and structural changes during deep fat frying: part I: model development. J. Food Eng. 53, 1–10.

Zhang, J., Datta, A.K., 2006. Mathematical modeling of bread baking process. J. Food Eng. 75, 78–89.

Zhang, L., Babi, D.K., Gani, R., 2016. New vistas in chemical product and process design. Annu. Rev. Chem. Biomol. Eng. 7, 557–582.

Zhu, Y., Guillemat, B., Vitrac, O., 2019a. Rational design of packaging: toward safer and ecodesigned food packaging systems. Front. Chem. 7 (349).

Zhu, Y., Welle, F., Vitrac, O., 2019b. A blob model to parameterize polymer hole free volumes and solute diffusion. Soft-Matter 15, 8912–8932.

Chlorine dioxide technologies for active food packaging and other microbial decontamination applications

Christopher J. Doona[a,b,c], F.E. Feeherry[c], K. Kustin[d], C. Charette[c], E. Forster[e], A. Shen[f]

[a]Massachusetts Institute of Technology, Cambridge, MA, United States
[b]John A. Paulson School of Engineering and Applied Sciences, Harvard University, Cambridge, MA, United States
[c]Combat Capabilities Development Command Soldier Center (CCDC SC), Natick, MA, United States
[d]Department of Chemistry, Brandeis University, Waltham, MA, United States
[e]Graduate School of Biomedical Sciences, Tufts University, Boston, MA, United States
[f]School of Medicine, Molecular Biology and Microbiology, Tufts University, Boston, MA, United States

17.1 Introduction

Research scientists at the US Army-Natick Soldier RD&E Center (now named the US Army-CCDC Soldier Center) have invented a diverse ensemble of antimicrobial technologies based on the action of the biocide chlorine dioxide. These antimicrobial technologies have different attributes, since they were developed for different uses and were intended for different applications with different goals and different constraints, but all with the common purpose of inactivating target microorganisms using chlorine dioxide.

Two next-generation chlorine dioxide technologies were developed for use as concepts for active food packaging systems with antimicrobial activity with potential applications in the food industry to ensure microbial safety, extend shelf-life, and reduce waste and economic losses associated with perishable foods such as fresh fruits, vegetables, and berries. One system produces chlorine dioxide from precursors adsorbed onto films of the biopolymer polylactic acid (PLA) that can be used in food packaging. The second is a smart food packaging system designed with chemical precursors incorporated in an insertable sachet made of a superabsorbent hydrogel polymer called the Compartment of Defense (CoD). The CoD packaging system has the characteristic of being humidity-activated for the time-released, controlled production of ClO_2 inside the container. The CoD technology inactivates pathogens and prevents mold growth on perishable foods (fresh berries, fruits, and vegetables) that respire and create humidity in-packaging.

As an adaptable antimicrobial technology that works well on the species and strains of vegetative pathogens and bacterial spores that are causes of concern in the food industry for food safety and spoilage, these chlorine dioxide technologies work equally well in textile-related applications (parachutes and clothing) and in biodecontamination. For example, chlorine dioxide can be used to prevent mold growth that can occur in humid or warm wet environments and degrade military textiles during storage, such as parachute sleeves, clothing (uniforms, undergarments, socks, boots), tents and rigid-walled shelters, vehicle mats, and even metal surfaces coated with polyurethane containing antimicrobial compounds. Additionally, chlorine dioxide can inactivate microorganisms on clothing can cause cutaneous rashes or irritations, especially in instances when clothing is worn for prolonged periods without access to laundering and/or adequate hygiene facilities. Further, chlorine dioxide can inactivate bacterial spores of *Bacillus anthracis* (the causative agent of "Anthrax") that have been used in bioterrorism attacks.

In addition to the development of the active antimicrobial packaging made with films of PLA biopolymer and the humidity-activated CoD food packaging system, additional chlorine dioxide technologies were invented and developed for other applications. Presented herein is the ensemble consisting of next-generation chlorine dioxide decontamination technologies that can be adapted for use in myriad antimicrobial applications, either in the food industry, for the military, or to meet other needs. Specific examples of these novel technologies and their validation are demonstrated, which include technologies for sanitizing hard surfaces, disinfecting graywater, decontaminating porous materials, preventing mold growth on parachute sleeves, inactivating spores of *Clostridiodes difficile*, the pathogen notorious in hospitals and healthcare settings, and self-decontaminating textiles incorporated with a stimuli-responsive polymer for hospital gowns and other personal protective equipment. These demonstrations should encourage new uses of these antimicrobial technologies in the food industry, such as treating fresh produce before shipping to consumers, sanitizing hard surfaces in food processing facilities and food handling environments, and cleaning-in-place methods for production lines, and in other applications where controlling microorganisms is a concern.

17.2 Current uses of chlorine dioxide

The earliest set of chlorine dioxide decontamination technologies that were developed were designed to be portable, electricity-free, power-free, and environmentally friendly ("green"), and they were intended for use in austere or otherwise resource-constrained circumstances of military field deployments, crisis responses, natural disaster relief, or humanitarian aid (Doona et al., 2014, 2015; Setlow et al., 2009). These disinfectant technologies include a Novel Chemical Combination (NCC); a Portable Chemical Sterilizer (PCS) to sterilize surgical instruments (20 lbs, 2.1 ft3, 10 oz of water, 0 kW of electricity); a collapsible handheld trigger-sprayer (Disinfectant-sprayer For ENvironmentally-friendly Sanitization, D-FENS), and Field Decontamination Kits (FDKs).

The FDK technology uses gaseous chlorine dioxide and was invented at US Army CCDC SC (formerly Natick Soldier RD&E Center). This technology was deployed for use by WHO, Doctors Without Borders (MSF), NIH/NIAID, Public Health Canada, USAMRIID, and other global public health organizations during the Ebola crisis in West Africa in 2014–2015. Because of its suitability for use in remote or austere environments with constrained availability of power and its lightweight portability for rapid mobility, FDKs were used to sterilize Ebola-contaminated medical instruments and electronics and protect healthcare workers (Doona et al., 2015). Commercial industry has licensed these technologies and marketed salable products to civilian consumers for additional uses in laboratories, offices, and homes.

This early set of disinfectant/sterilization technologies were based primarily on the biocide chlorine dioxide, which inactivates a broad spectrum of microorganisms (e.g., viruses, bacterial cells and spores, yeasts, molds and mildews, and fungal spores) without acquiring increased resistance. Chlorine dioxide

(ClO_2) is a well-known and versatile disinfectant, whose biocidal efficacy in either the gaseous state or in aqueous solution is well-established (Schaufler, 1933; Young and Setlow, 2003; Doona et al., 2014, 2015). Regulatory agencies have approved chlorine dioxide for different uses. The US Department of Agriculture (USDA) allows its uses in crop production and livestock production and for processed products (USDA, 2000); the US Food and Drug Administration (US FDA) allows its uses in poultry processing water and water used to wash fruits and vegetables (US FDA, 1998); and the US Environmental Protection Agency (US EPA) Emerging Pathogen Program lists chlorine dioxide as effective against coronavirus on hard surfaces (U.S. EPA, 2017).

Descriptions of the industrial-scale methods for generating ClO_2 can be found in chemical technology handbooks and encyclopedias (Vogt et al., 2003). Chlorine dioxide is employed safely in commercial uses across multiple industries and cross-cutting applications, with its largest consumers (ca. 4.5 million pounds used per day worldwide) being the pulp and paper industry and the approximately 700–900 municipal water systems in the United States that disinfect the water supply and render potability (Hoehn, 1992). Small-scale applications of chlorine dioxide include its use in personal hygiene consumer products such as mouthwashes and toothpastes (The ClO_2 Fact Sheet, 2013). Sterilizing medical electronics devices in austere environments in remote areas during the Ebola crisis in 2014–2015 (Doona et al., 2015) requires the use of dry chemical precursors that mix safely and controllably in water or aqueous solution to generate ClO_2 on-site, at-will, and at point-of-use for their intended application.

17.2.1 Chemical methods of generating ClO_2

The uses of ClO_2 and the methods for its generation rely on the unique chemical attributes of the element chlorine (chemical symbol Cl). As a member of Group VIIA (halogens) of the periodic table, Cl is found in diverse chemical species with different reactivities and the Cl atom in different oxidation states and (Table 17.1). Many of the chemical methods that generate ClO_2 mentioned above are categorized as occurring by processes involving the: (i) reduction of chlorate, (ii) acidification of chlorite, (iii) oxidation of chlorite ion, and (iv) reduction of chlorite ion.

Chemical systems in categories (i), (ii), and (iii) all have substantial practical drawbacks that make them unsuited for small-scale, electricity-free, point-of-use applications. In the method of category (i) reduction of chlorate (ClO_3^- with Cl in the +5 oxidation state), the chemical reaction involves the reaction of the small molecule chlorate in strong acid solution with a reductant such as sulfur dioxide, methanol, hydrogen peroxide, or hydrochloric acid. The reduction of chlorate accounts for more than 95% of the ClO_2 used in the world today, primarily in the pulp and paper industry, through the large-scale production of ClO_2 for the manufacture of high-quality white paper products. The chemical

Table 17.1 Chemical species and names, symbols, and oxidation states of chlorine-containing molecules.

Name	Chemical symbol	Oxidation state
Chloride	Cl^-	−1
Chlorine gas (dichlorine)	Cl_2	0
Hypochlorite (bleach) hypochlorous acid	OCl^- HOCl	+1
Chlorite chlorous acid	ClO_2^- $HClO_2$	+3
Chlorine dioxide	ClO_2	+4
Chlorate	ClO_3^-	+5
Perchlorate	ClO_4^-	+7

reaction takes place in large-scale equipment and with the use of strong acid solutions. In the method of category (ii) acidification of chlorite (ClO_2^-, Cl in the +3 oxidation state), the addition of acid to chlorite solution forms unstable chlorous acid ($HClO_2$) that subsequently disproportionates to produce ClO_2. In the method of category (iii) oxidation of chlorite, chlorite (ClO_2^-) is oxidized by chemical reaction with dichlorine (Cl_2) gas. In all three of these cases, the uses of hazardous materials (acids in categories i and ii and dichlorine gas in category iii) require special considerations for transportation, storage, and disposal. Acids are also incompatible with many surface materials, and transporting large-scale equipment or cylinders of gas for the on-site, small-scale production of ClO_2 in high-intensity, rapid-mobility environments is impractical.

In contradistinction, chemical systems that produce ClO_2 by the method of category (iv) reduction of chlorite ion (ClO_2^-) do so handily by adding safe, dry reagents (chlorite and reductants) to water to effectuate the controlled generation of chlorine dioxide on-site with distinct advantages in terms of their ease-of-use, convenience, and safety to users and the environment. In the method of category (iv) reduction of chlorite ion (ClO_2^-), chlorite reacts with a reductant to produce chlorine dioxide (ClO_2, with the Cl in the +4 oxidation state).

17.2.2 Chemical mechanism proposed for the reduction of chlorite

The astute chemist might notice the paradox that the process of category (iii) is the *oxidation* of chlorite (ClO_2^-, Cl in the +3 oxidation state) to chlorine dioxide (ClO_2, Cl in the +4 oxidation state), and that the process of category (iv) is the *reduction* of ClO_2^- to produce ClO_2. While the former situation for the loss of an electron seems fairly straightforward ($ClO_2^- \rightarrow ClO_2 + e^-$), the latter situation might seem counterintuitive, unless one understands the role(s) of transient intermediates formed in the chemical reaction system.

Consider the reaction of ClO_2^- with the reductant sulfite (SO_3^{2-}). This chemical reaction is thermodynamically favorable (molar enthalpy of reaction $\Delta H = -648.3$ kJ/mol) but kinetically inert—it does not proceed to produce ClO_2 and/or exothermic heat on any type of practical timescale. However, adding substoichiometric amounts of an electron-transfer effector that reacts readily with ClO_2^- is postulated to establish new reaction pathways through the formation of oxidizing transient intermediates ($ClO^\bullet$), which even react with the reductant sulfite (SO_3^{2-}). Table 17.2 depicts a reaction scheme for this chemical reaction with the $ClO^\bullet$ intermediate.

In Table 17.2, proposed reaction step i denotes *formation* of the transient oxidizing intermediate chlorine monoxide radical (designated as $ClO^\bullet$, Cl in the +2 oxidation state) by the ClO_2^--effector reaction; steps ii–iii denote the *propagation* of $ClO^\bullet$ by reactions involving the reductant; and step iv

---

Table 17.2 Mechanism with transient intermediates.

Overall reaction

$ClO_2^- + R^{2-} \rightarrow$ no reaction
Overall effector-driven reaction
$ClO_2^- + R^{2-} + Eff^- \rightarrow ClO_2, SO_4^{2-}, C_2O_4^{2-}, CO_2$ (fast reaction, exothermic)
Effector-driven reaction mechanism
i $ClO_2^- + Eff^- \rightarrow ClO^\bullet + Eff^{\bullet-}$
ii $ClO^\bullet + R^{2-} \rightarrow ClO^- + R^{\bullet-}$
iii $ClO_2^- + R^{\bullet-} \rightarrow ClO^\bullet + salt$
iv $ClO_2^- + ClO^\bullet \rightarrow ClO^- + ClO_2$

denotes the *termination* of ClO$^•$ and concomitant formation of ClO$_2$ by the reaction with ClO$_2^-$. Other mechanisms and intermediates (e.g., Cl$_2$O$_2$) could plausibly account for this reaction, but the overall result of an electron-transfer chemical effector accelerating an otherwise inert chemical reaction by initiating additional chemical reaction pathways involving transient intermediates capable of oxidizing ClO$_2^-$ would be the same.

17.2.3 Microbiological validation of the PCS and D-FENS

The PCS and D-FENS systems use a three-component reduction of ClO$_2^-$ to rapidly produce concentrated ClO$_2$ solutions with the evolution of copious exothermic heat. As a suitcase sterilizer, the PCS is a modern field autoclave designed to off-gas ClO$_2$ that acts in concert with heat and steam to inactivate bacterial spores and sterilize surgical instruments. The D-FENS, a collapsible handheld sprayer for surface decontamination applications in laboratories, offices, and homes, uses the same reaction chemistry as the PCS, but D-FENS invokes the principles of kinetics control and postreaction dilution to produce relatively dilute ClO$_2$ solutions without concomitant heat production.

Achieving sterility with the PCS was demonstrated using live cultures of *Bacillus stearothermophilus* spores and indicators of *B. stearothermophilus* or *Bacillus atrophaeus*. The efficacy of the D-FENS system was validated by spraying the solution onto a representative porous surface consisting of Petri dishes containing agar and Baird-Parker nutritive media supplemented with egg yolk tellurite and inoculated with a three-strain cocktail of *Staphylococcus aureus*. Plates sprayed with the D-FENS solution showed no subsequent colony growth, whereas untreated plates showed the growth of dark colonies characteristic of *S. aureus*. In-house cross-laboratory testing of *S. aureus* dried on metal coupons showed that the D-FENS solution (100 ppm ClO$_2$) inactivated > 7-logs of *S. aureus*, whereas hydrogen peroxide or ozone inactivated <3.4-logs and <4.8-logs, respectively (Doona et al., 2014, 2015).

17.3 Next-generation ClO$_2$ technologies

While the PCS and D-FENS accomplished their goals of portable power-free sterilization and surface decontamination, certain attributes of these systems (e.g., producing concentrated ClO$_2$ and copious heat for the PCS; requiring three-components and two-step mixing for D-FENS) might not be desirable for other microbial decontamination applications, and so other custom-made systems and configurations were needed for those instances. Accordingly, three new chemical reductions of chlorite were invented and developed to produce ClO$_2$ for antimicrobial food packaging concepts, disinfecting graywater, sanitizing hard surfaces, decontaminating porous materials (i.e., textiles), self-decontaminating protective garments, and other innovative applications.

All of these next-generation chemical systems that produce chlorine dioxide could be used in the food industry in applications such as treating fresh produce before shipping to consumers, sanitizing surfaces in food processing facilities, and cleaning-in-place methods for production lines. An active food packaging system with antimicrobial properties was developed using one of the next generation chemical systems in conjunction with a film made of PLA biopolymer as a biodegradable food packaging material, because of the interest in recent years of reducing food packaging waste and increasing environmental sustainability (Ray et al., 2013). Additionally, an active food packaging system was also developed that inserted a sachet consisting of superabsorbent hydrogel polymer incorporated with chemical precursors into a clamshell plastic container and called the CoD. The CoD packaging system is activated by high humidity for the time-released, controlled, sustained production ClO_2 in-container. Both of these antimicrobial food packaging concepts could be used by the food industry to inactivate pathogens, prevent mold growth, and extend the safe shelf-life of perishable foods that respire in-packaging (fresh berries, fruits, and vegetables) and thereby reduce economic losses.

These next-generation technologies can also be adapted for other applications. For example, the active food packaging system (called CoD) can be used in other military applications in which mold growth is also a concern, such as clothing (uniforms, undergarments, socks, boots), tents and rigid-walled shelters, and vehicle mats, especially when stored in warm and humid environments. Additionally, the next-generation chlorine dioxide technologies could be used in conventional cleaning and disinfectant applications, such as disinfecting surgical or laboratory instruments, sanitizing surfaces in hospitals, nursing care facilities, kitchens or galleys, or other janitorial-type applications (showers, laundries, latrines, public restrooms, and corridors), or cleaning surfaces in vehicles. Descriptions of these next-generation chlorine dioxide technologies, their chemical bases for generating chlorine dioxide, demonstrations of their operation, and tests validating their microbiological efficacy in: (*i*) preventing mold growth on parachute sleeves, (*ii*) inactivating the notorious pathogen in hospitals and healthcare settings *Clostridiodes difficile*, and (*iii*) creating self-decontaminating textiles for protective garments that inactivate bacterial spores are presented below.

17.3.1 Disinfectant for environmentally friendly decontamination, all-purpose (D-FEND ALL)

D-FEND ALL was developed to be especially convenient for field use in applications on a slightly larger scale than a typical handheld spray-bottle (approx. 1 L) could accommodate. For example, the dry precursors of the D-FEND ALL system were added to a beaker of water (6 < pH < 8) and stirred, to produce the pale yellow color of ClO_2 almost immediately. With variations in the stating conditions, ClO_2 concentrations in the range of 5–5900 ppm were produced quickly in the 3–15 min timeframe and showed no noticeable generation of exothermic heat. The reactivity of the reductant is such that it produces dilute concentrations of ClO_2 rapidly and without generating substantial exothermic heat. This chemical system was an integral and essential component of an environmentally friendly graywater recycling system to render potable water in expeditionary base camps.

The D-FEND ALL system was used to generate ClO_2 solutions in the concentration range of 20–200 ppm, as determined using a ClO_2 indicator strip. For microbiological validation, individual strips (1" × 2") of military fabric samples using the Advanced Combat Uniform (ACU, made of 50%:50% nylon/cotton) or an experimental weatherproof fabric printed with a camouflage pattern were spot-inoculated with spores of *Bacillus amyloliquefaciens* and *Bacillus anthracis* Sterne. Inoculated fabric strips were immersed in the 20–200 ppm ClO_2 solutions in pouches and paddle-mixed with a Stomacher for 2–4 min. After 10 min, excess ClO_2 was quenched with reductant, and the solution was serially diluted and spread-plated on agar with ST-1 and nutrient agar, respectively. The ClO_2 solutions at concentrations of 20, 40, 80, and 100 ppm inactivated 0, 1.3, 2.0, and 7.3-logs of *B. amyloliquefaciens* spores, respectively. All of the ClO_2 concentrations (20–100 ppm) inactivated 7.5-logs of *B. anthracis* Sterne spores. Further, the ACU samples were unaffected visually by exposure to ClO_2 at all of the concentrations tested, whereas the experimental weatherproof fabric samples were unaffected at concentrations at or below 80 ppm, but became bleached at the 200 ppm concentration.

17.3.2 Active food packaging concept using PLA

The potential for PLA to be used as antimicrobial active packaging systems has been reported previously (Ray, 2013). PLA can be synthesized from renewable bioderived monomers and is an alternative to conventional petroleum-based polymers for commercial use in food packaging applications with a focus on films and coatings that are suitable for short shelf life and ready-to-eat food products. The use of PLA as a chlorine dioxide-releasing film for the microbial decontamination of fresh produce used acidification chemistry with the chemical precursors incorporated into the PLA films. Moisture in the PLA package activated the release of gaseous ClO$_2$.

The chemical precursors used in the D-FEND ALL (see Section 17.3.1) system and in the CoD system (see Section 17.3.3) were tested with PLA as a candidate active food packaging material. Samples of PLA film (16 in. wide, 0.2 mm thickness) were cut into 2-in. square coupons and spotted with drops of concentrated solutions of either the oxidant and reductant on two separate PLA squares, or of both reagents onto spatially discrete regions of a single PLA coupon. After drying over night at $T = 25$ °C in a covered glass dish, the spotted coupons were placed in a beaker of water at the same time and the yellow color of ClO$_2$ appeared over the next 3–15 min. These results indicate that these chemical systems could be used with the biopolymer PLA as active food packaging that produced the antimicrobial chlorine dioxide. The oxidation–reduction chemical precursors could also be incorporated into a plastic spray bottle or container that, when filled with water, would generate a sanitizing solution for spraying onto or receiving microbially contaminated objects, respectively.

17.3.3 The Compartment of Defense active food packaging concept

A third chemical system was invented that used another alternative reductant. Tests were carried out using a 100 ppm ClO$_2$ solution made by mixing chlorite and this reductant in water to inactivate spores of the foodborne pathogen *Bacillus cereus*. The 100 ppm ClO$_2$ solution inactivated 5.5×10^4 CFU/mL of the *B. cereus* spores directly in solution or by immersing fabric samples spot-inoculated with the *B. cereus* spores into the disinfectant solution, paddle-mixing the sample, quenching the remaining ClO$_2$ with reductant, and plating on nutrient agar medium.

The third chemical system was also used in conjunction with a polymeric material (i.e., superabsorbent hydrogel) to absorb humidity and controllably generate gaseous ClO$_2$ in-container as an active food packaging system called the CoD. To demonstrate the operation and microbial validation with the CoD system, a superabsorbent hydrogel pad was impregnated with dry ClO$_2$-producing precursors and placed inside a plastic clamshell container. The container was made of the biopolymer PLA or polyethylene terephthalate (PET) and had vent holes to allow the free-flow exchange of moisture with its surroundings. Additionally, bioindicator strips of *G. stearothermophilus* or *B. atrophaeus* spores and a ClO$_2$ chemical indicator strip were placed inside the plastic container. The entire container was placed inside a desiccator containing a saturated solution of KCl to created a humid (87–90 %RH) environment. The entire desiccator with all its contents were placed in a warm ($T = 25$–35 °C) incubator. After 24–48 h of incubation, the ClO$_2$ indicator showed approx. 25 ppm level of ClO$_2$ had been reached and the spore bioindicators were inactivated. This example demonstrated the proof-of-concept for the CoD active packaging concept. Specifically, the test confirmed that the oxidation–reduction reaction had taken place through the absorption of moisture from the humid environment water into the absorbent polymeric pad substrate to produce ClO$_2$. The low-level concentration of ClO$_2$ was produced over time in a time-released manner. This packaging concept can be used for fresh produce, vegetables, berries, plants, and other tissues that respire in-packaging to create a humid environment and trigger the controlled production of ClO_2.

17.3.4 The Biospray technology and the inactivation of *Clostridiodes difficile* spores

The fourth next-generation system, called Biospray, uses a biological reagent to generate ClO_2 solution in a sprayer to disinfect equipment and facilities in microbiology and molecular biology laboratories, particularly those that routinely handle pathogens or infected biological tissues. These reagents could be incorporated into PLA films used as food packaging or in collapsible plastic spray bottles to sanitize food processing and handling equipment, food contact surfaces, or surfaces in kitchens, galleys, hospitals, or nursing care settings.

Biospray was validated against spores of *Clostridiodes difficile*, a notorious opportunistic pathogen in hospitals and nursing care settings. Infections with *C. difficile* constitute a public health threat, with over half a million cases and 29,000 deaths estimated to have occurred in the United States in 2011 (Lessa et al., 2015). *C. difficile* spores germinate, grow out, and produce toxins that cause severe digestive infections and enteric lesions. Eliminating this environmental contaminant from surfaces, objects, foods, etc. would likely reduce the incidence and expenses of illnesses from *C. difficile* infection.

To demonstrate that ClO_2 inactivates spores of *C. difficile* irrespective of the chemical precursors or methods used, the Biospray and CoD (see Section 17.3.3) chemistries were tested as aqueous ClO_2 solutions on spores of *C. difficile*. Specifically, spores of *C. difficile* wild-type 630Δ*erm* strain were treated with 0–25 ppm aqueous solutions of ClO_2 produced using CoD or Biospray. At 15 min, excess ClO_2 was quenched, and the spore suspension was plated on brain heart infusion-supplemented agar with 0.1% taurocholic acid (TCA). The plates were incubated at $T = 35\ °C$ for ~24 h, and germinated spores were counted as CFU (Fig. 17.1). Samples were done in two biological replicates with three technical replicates of each. While each concentration of ClO_2 inactivated *C. difficile* spores to some

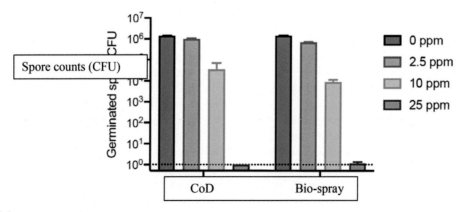

FIG. 17.1

Inactivation of *C. difficile* spores by ClO_2 from two different sources.

extent (1–2 logs at 10 ppm ClO$_2$), the 25 ppm concentration of ClO$_2$ from both solutions sterilized (inactivated 6-logs) of the *C. difficile* spores, thereby demonstrating that it is the action of ClO$_2$ rather than its method of production that is responsible for decontaminating the microorganisms.

Phase contrast microscopy provided further interesting observations of the ClO$_2$-inactivated *C. difficile* spores. Specifically, phase contrast images of untreated control *C. difficile* wild-type 630Δ*erm* spores showed typical phase-bright character (Fig. 17.2A). However, phase contrast images taken after treatments of the *C. difficile* spores with 25 ppm aqueous ClO$_2$ produced using the CoD or Biospray systems also showed phase-bright character (Fig. 17.2B, C). Similar results were found in the inactivation of *Bacillus subtilis* spores by ClO$_2$ (Young and Setlow, 2003), in which the ClO$_2$-inactivated *B. subtilis* retained their phase-bright character. And although the precise mechanism of ClO$_2$ killing *B. subtilis* spores was not identified, ClO$_2$ was thought to target a location on the spores' inner membrane (IM).

The interactions of *Bacillus* and *Clostridia* spores with high hydrostatic pressures offer some insight into future research needs for inactivating bacterial spores with chlorine dioxide. First, *Bacillus* and *Clostridia* spores have some key morphological and physiological differences that account for very different responses of *B. subtilis*, *C. difficile*, and *C. perfringens* spore to germination by high hydrostatic pressures (Doona et al., 2016). High hydrostatic pressure treatments of *Bacillus* spores at conditions of $P = 150$ MPa and $T = 37$ °C stimulate GRs on the IM and conditions of $P = 550$ MPa and $T = 50$ °C stimulate IM SpoVA channels, so that in both instances, DPA subsequently releases from the spore core, the cortex lytic enzymes (CLEs) become activated, and hydrolysis of the spore cortex peptidoglycan effectuates the conversion to phase-dark spores.

In contradistinction, *C. difficile* spores did *not* respond to high hydrostatic pressure conditions of $P = 150$ MP and $T = 37$ °C, because *C. difficile* does not have IM GRs (GRs are located in the outer layers (OL) in a few *C. difficile* spores). HPP at $P = 550$ MPa and $T = 50$ °C causes DPA release with cortex hydrolysis in *Bacillus* spores, but DPA release *without* cortex hydrolysis for *C. difficile* spores. *Bacillus* spores have two CLEs that are located in spores' OLs and that degrade peptidoglycan spore cortex, whereas *C. difficile* has one CLE and *C. difficile* spores have a specific Csp protease(s) located in the OLs that activates a CLE zymogen (pro-SleC) leading to subsequent cortex hydrolysis, whereas *Bacillus* spores do not have this Csp protease. Accordingly, with conditions of $P = 550$ MPa and $T = 50$ °C, *C. difficile* spores rapidly release approx. 90% of the DPA from the spore core *without*

(A)   (B)   (C)

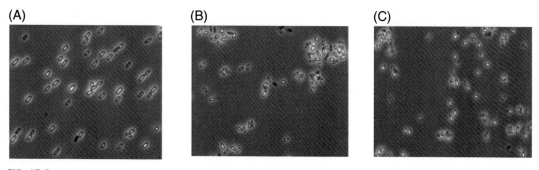

FIG. 17.2

(A–C) Phase contrast images of *C. difficile* spores as phase-bright untreated controls (A) and retaining phase-bright character after treatment with 25 ppm ClO$_2$ from the CoD (B) or Biospray (C) systems.

conversion to phase-dark spore (<2%), apparently because the Csp protease and SleC (the CLE) were not activated and the cortex was not hydrolyzed. Apparently, ClO_2 inactivates both *Bacillus* and *C. difficile* spores without casing cortex hydrolysis in either. These results suggest that there might be similar loci for spore inactivation by ClO_2 for both *Bacillus* and *Clostridial* spores. This finding warrants further research into the mechanism of bacterial spore inactivation by ClO_2, and including the effects of ClO_2 on *C. perfringens* spores.

17.3.5 Chlorine dioxide to control mold (fungal spores)

As discussed above, ClO_2 is a potent biocide that functions at low concentrations to inactivate a broad spectrum of microorganisms on surfaces, even the sensitive surfaces of whole tomatoes, fresh berries, sliced apples, and other fresh produce commodities, without damaging the tissue during treatment (Setlow et al., 2009). The CoD system (Section 17.3.3) was developed to prevent mold growth that causes spoilage of fresh produce and berries. Mold growth causes a similar problem with cotton parachute sleeves. Cotton, if stored wet or improperly dried, can support the growth of mold that degrades tensile strength. Therefore, tests were carried out to determine the ability of ClO_2 produced with the CoD system to prevent mold growth on cotton parachute sleeves. The versatile biocide ClO_2 from the CoD system will be shown to prevent mold growth on the cotton without compromising the mechanical strength of the cotton fabric.

The purpose of the cotton deployment sleeve is to control the opening of the parachute during deployment for increased safety and reliability of the system. The jumper's entire parachute is packed into this narrow, almost 20-foot long sleeve, which is then packed into a bag and container that is worn on the jumper's back. After the jumper exits the aircraft, the parachute deployment sequence initiates, the container and bag open to release the sleeve, then the parachute inflates as it exits from the sleeve and creates intense heat due to friction. The rapid inflation process occurs in less than 1 s and simultaneously pushes the sleeve off the parachute.

With the paramount goal of protecting the safety of the jumpers, maintaining the tensile strength of the textiles used in parachutes and sleeves for multiple uses over a large number of deployments is critical in ensuring the reliability and durability of the parachute systems over many years of parachute service life. Cotton fabric is used as the construction material of the sleeve, because cotton's thermo-resistance allows it to withstand the intense frictional heat associated with deployment without compromising the tensile strength of either the sleeve or the parachute. In contrast, high-tech, synthetic fibers have been investigated for use as the construction material for deployment sleeves and found not suitable for these purposes.

A cotton parachute sleeve previously deployed in a wet landing environment and exhibiting spots of a black mold was tested with three (3) different disinfectants. Mold from the contaminated cotton sleeve was viewed with phase contrast microscopy, to confirm the presence of refractile (phase-bright) spores. Examining the moldy sleeve with light microscopy showed black spores (presumably *Aspergillus niger* spp.) with hyphae on the cotton. Samples (1-in^2) were cut from the cotton sleeve, placed in individual Petri dishes containing sterile potato dextrose agar (PDA), wetted with a few drops of sterile water to ensure contact with the agar, and incubated at 25 °C or 30 °C. At both temperatures, the PDA plates exhibited luxuriant mold growth within 3–4 days. One of the moldy cotton samples was transferred to a fresh, sterile Petri dish of PDA with sterile water. A second moldy cotton sample was placed in a plastic pouch containing 10 mL of 100 ppm ClO_2 solution, mixed in a paddle-Stomacher

Table 17.3 Tensile strength and elongation of cotton cloth treated with aqueous decontaminants.

Treatment	Breaking strength (lbs)
Requirement	80–96
ClO_2	92.5
ClO_2	94.5
ClO_2	83.5
Avg	90.2 ± 5.9
H_2O_2	37
H_2O_2	29
H_2O_2	38
Avg	34.7 ± 4.9
OCl^-	N/A

Note: All tests in warp direction.

for 2-min intervals over a 20-min period, then transferred to a fresh Petri dish with PDA. Both samples were stored at 30 °C for 5 days. After incubation, the untreated moldy sample exhibited luxuriant mold growth. The ClO_2-treated sample showed no discernible indications of mold growth or damage to the cotton, thereby demonstrating the efficacy of ClO_2 in inactivating fungal spores (*Aspergillus niger* spp.).

To test the effects of ClO_2 treatment on the tensile strength of the textile substrate, the cotton fabric samples were immersed for 30 min in disinfectant solutions of either ClO_2, hydrogen peroxide (H_2O_2), or household bleach (OCl^-), then rinsed in deionized water and dried. Samples were tested for tensile strength, also called ultimate breaking strength, as measured by applying a controlled force to strips of the 4 oz cotton fabric used in the deployment sleeve according to standard method ASTM D5035-11 (ASTM, 2019).

The break strength requirement for the cotton fabric is 80 lbs. Results in Table 17.3 show that ClO_2 is the only disinfectant solution that does not degrade the break strength of cotton (average = 90.15 lbs) below minimum requirements. Conversely, results showed that hydrogen peroxide solutions weakened the break strength of cotton by more than 50% (average = 34.53 lbs), and treatment with typical household chlorine bleach (OCl^-) degraded the integrity of cotton so severely that the sample could not be prepared and tested with the standard procedure (Table 17.3). Future R&D of mold prevention technologies for parachute textiles will focus on gaseous ClO_2 applications using the CoD system to simultaneously prevent mold growth and retain cotton's mechanical properties.

17.4 Nonthermal processing for inactivating *B. anthracis* spores

High pressure processing (HPP) is perhaps the most well-recognized nonthermal processing technology that is growing worldwide in its use for food pasteurization. The inactivation of bacterial spores for food sterilization with HPP is possible, but the food processing industry has not implemented this application commercially. Hydrostatic pressure has also been used to decontaminate *B. anthracis* spores in buffer solutions (Cléry-Barraud et al., 2004). ClO_2 is a recognized nonthermal technology as a chemical sanitizer for rinsing fresh produce, and ClO_2 inactivates bacterial spores (Doona et al., 2015).

Accordingly, ClO$_2$ in solution (D-FEND ALL), gaseous ClO$_2$ (PCS), and HPP were used to inactivate *B. anthracis* spores on novel fabrics intended for use in self-decontaminating protective garments.

17.4.1 Decontaminating bacterial spores on protective garment fabrics

A series of novel fabrics were tested for their abilities to self-decontaminate when challenged with bacterial spore surrogates of *B. anthracis*. In general, the fabrics consisted of a repellant omniphobic coating on the outer surface of a fabric; a cloth layer adsorbed with the antimicrobial compounds (8-hydroxyquinoline, abbreviated HQ, and 1,2-benzisothiazol-3(2H)-one, abbreviated BIT); and an interior liner impregnated with activated carbon. The premised function of these test fabrics was that when challenged with dry aerosols of bacterial spores on their outer surface, humidity would trigger the release of the antimicrobial compounds from the cloth layer to inactivate the bacterial spores. Spores of *B. anthracis* Delta Sterne, a surrogate of the bioweapon *B. anthracis*, were inoculated on the exterior surfaces of fabric samples as dry aerosols using a unique biodispersal chamber (Fig. 17.3) in accordance with the standard method ASTM E2894-12 (Harnish et al., 2014).

17.4.2 Dry aerosol inoculation of fabrics

Spore crops were prepared using standard media concentrations to yield smaller crop densities and prevent agglomeration of the spores. Mild heating inactivated vegetative cells prior to cleaning by centrifugation and scraping off debris. The clean spores were lyophilized using a Labconco Freeze Dry System and stored frozen. For testing, preweighed amounts of lyophilized spores were deposited into the loading region (a) of the chamber located in the square black section in the upper, right-side region of the chamber (Fig. 17.3), and the system was turned "on." With air flowing, spores transferred out of the loading area as a dry aerosol, conveyed through the stainless steel cylinder and tubing on top of the chamber (b), released into glass-enclosed biodispersal chamber (c, the entire left-hand side of

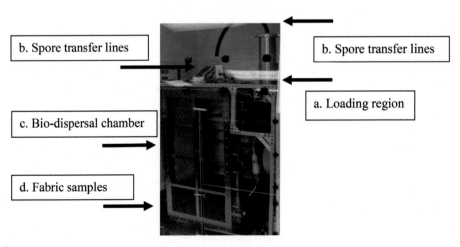

FIG. 17.3

Biodispersal chamber for inoculating samples with dry aerosols of spores.

Table 17.4 Results of challenging fabrics with spores and prolonged high-humidity storage.

	Dry aerosol deposition time	Humidity incubation (%RH, T, t)	Recovered (CFU/cm^2)	% Kill
Control fabric	24 h	0	2.39×10^6	0
Control fabric	24 h	18%, 37 °C, >24 h	$\sim 10^6$	0
Fabric-1	24 h	0	1.14×10^6	0
Fabric-1	24 h	89%, 37 °C, 15 d	2.21×10^6	0
Fabric-1	24 h	89%, 37 °C, 28 d	2.76×10^6	0
Fabric-2	24 h	0	1.80×10^6	0
Fabric-2	24 h	89%, 37 °C, 15 d	1.67×10^6	0
Fabric-2	24 h	89%, 37 °C, 28 d	1.71×10^6	0

Fig. 17.3), and deposited onto the multiple 1-in^2 fabric samples located at the bottom of the biodispersal chamber on a rotating surface (d).

After inoculation, the test samples were placed in humidity-controlled glass containers (covered desiccators containing saturated KCl solutions producing conditions of 89%RH and $T = 37$ °C) and the entire desiccator was stored isothermally in incubators (Feeherry et al., 2003). Inoculated control samples were covered and stored at $T = 35$ °C and 18 %RH. The inoculated fabric test samples were withdrawn at intervals over a 28-day period (over a 60-day period for control samples) for enumeration.

To recover and enumerate the spores from the inoculated fabrics, samples were placed in sterile pouches with sterile water and mixed with a paddle stomacher (Feeherry et al., 2003), and the water was subsequently serially diluted and plated on Nutrient Agar. In all cases, 100% of the spore load was recovered (Table 17.4). The fabrics did not inactivate contaminating bacterial spores and the fabrics are not inherently self-decontaminating.

17.4.3 Alternative methods of decontamination

To decontaminate the bacterial spores on the inoculated fabrics, alternative methods of applying external agents were tested. Initial experiments used aqueous solutions of ClO_2 or bleach (OCl^-) to treat the fabrics, according to the following methods to determine the quantity of spores removed from the fabrics by the mechanical agitation of the rinsing process versus the quantity of spores inactivated by the chemical decontaminating agent.

In the first method, the inoculated fabric samples were immersed in 20 mL of sterile water in a Stomacher bag labeled RINSED and agitated mildly by gently moving the fabric sample through the water. Next, the fabric sample was moved to a second Stomacher bag containing 20 mL of sterile water and labeled RECOVERED. The count of spores removed by the rinsing process were determined by serially diluting the solution in the RINSED stomacher bag, spread-plating on NA plates, incubating, then enumerating the plates (Table 17.5).

The counts of bacterial spores remaining on the fabric samples *after* the water rinse were determined by vigorously agitating the sample in the RECOVERD bag with a Stomacher, then serially diluting the RECOVERED water, spread-plating it on NA plates, incubating the plates, and enumerating survivors (Table 17.5). Carrying out the same process with immersing the fabric samples in 20 mL of water (RINSE), then moving the fabric samples to 20 mL of household bleach (5–6% hypochlorite, OCl^-) and adding a small quantity of solid reductant at 10 min of exposure to quench the OCl^- yielded

Table 17.5 Results of physical decontamination by aqueous rinsing.

	Inoculum (CFU/cm^2)	RINSED (CFU/mL)	RECOVERED (CFU/mL)	Log-kill (removed)
Fabric-1	1.61×10^6			
Fabric-1		1.4×10^4	2.7×10^4	0.46
Fabric-1		8.6×10^3	9.0×10^3	0.31
Fabric-1		3.2×10^3	5.9×10^3	0.45
		$\mathbf{8.6 \pm 5.4 \times 10^3}$	$\mathbf{1.4 \pm 1.1 \times 10^4}$	$\mathbf{0.41 \pm 0.08}$
Fabric-2	2.51×10^6			
Fabric-2		9.6×10^2	1.01×10^3	0.31
Fabric-2		3.2×10^3	1.81×10^3	0.19
Fabric-2		7.2×10^3	5.9×10^3	0.26
		$\mathbf{3.8 \pm 3.2 \times 10^3}$	$\mathbf{2.9 \pm 2.6 \times 10^3}$	$\mathbf{0.25 \pm 0.06}$
Fabric-1 Bleach rinse—10 min	$\mathbf{1.61 \times 10^6}$	1.1×10^2	0	>4
Fabric-2 Bleach rinse—10 min	$\mathbf{2.51 \times 10^6}$	2.0×10^0	0	>6

different results from the water-only rinse. Table 17.5 shows that rinsing the inoculated fabrics with copious water effectuated physical decontamination to only a minor extent (less than 0.5-log reduction), and that exposure to OCl$^-$ for both types of fabrics completely inactivated all of the spores remaining after the rinsing process.

In the second approach, the inoculated 1-in^2 fabric samples were cut in half with sterile scissors. The control half was enumerated without further treatment, while the second half of the sample was subjected to a lethal treatment (aqueous bleach OCl$^-$, aqueous or gaseous ClO$_2$, or HPP) before enumerating survivors. Treatment with a bleach rinse (5–6% hypochlorite, OCl$^-$) for 30 min inactivated all contaminating spores (>5-logs, see Table 17.6). ClO$_2$ sterilized contaminating bacterial spores on all of the inoculated fabric samples, irrespective of whether the inoculated fabric samples were immersed in aqueous solutions of ClO$_2$ made using the D-FEND ALL technology (data not shown) or whether the inoculated fabric samples were treated with gaseous ClO$_2$ in the PCS for 30 min (>4.60-log inactivation, see Table 17.6). Treating the inoculated fabric samples with HPP at conditions of $P = 550$ MPa, $T = 65$ °C, time $t = 100$ min effectuated >6-log inactivation of all contaminating *B. anthracis* Sterne spores (Table 17.6). To prevent cross-contamination of the pressure-transmitting fluid, the inoculated fabric samples were sealed in sterile pouches for the HPP experiments. It is important to note that 1 mL of sterile water was added to the pouch with the inoculated fabric sample to achieve the >6-log inactivation. HPP treatments of the same inoculated fabrics at the same HPP conditions *without* adding water to the pouch did *not* inactivate the bacterial spore dry powder.

Table 17.6 Alternative treatments to decontaminate spores on fabrics.

Fabric sample	Control (CFU/mL)	Bleach—30 min (CFU/mL)	Log (kill)	Control (CFU/mL)	Gaseous (ClO$_2$)—30 min	Log (kill)	Control (CFU/mL)	HPP	Log (kill)
Fabric-2	1.10×10^5	0	5.04	4.00×10^4	0	4.60	1.34×10^6	0	6.13
Fabric-1	3.90×10^6	0	6.59	5.40×10^5	0	5.73	1.28×10^6	0	6.11

Fabric samples inoculated with the surrogate *B. anthracis* Delta Sterne spores (10^6 CFU/cm^2) using the ASTM E2894-12 dry aerosol method and incubated for 28 days at elevated humidity-temperature conditions (89 %RH and 37 °C) showed no discernible inactivation of spores—the spores were not even activated by this prolonged humidity incubation. The presence of the antimicrobial compounds HQ and BIT in the fabrics did not interact with and inactivate the bacterial spores, presumably because the repellant coating acts as a physical barrier preventing any such interaction. The HQ and BIT were not even leached from the fabrics to the surface to effectuate inactivation of the contaminating spores. The repellant coating, however, does not protect spores from externally applied sterilants, whether aqueous or gaseous ClO_2 or chemical-free high hydrostatic pressures, and all three methods sterilized the fabrics.

17.5 Conclusions

In 2014–2015, a gaseous chlorine dioxide technology invented at US Army CCDC SC (Natick Soldier RD&E Center) was adapted for use in the Ebola-stricken areas of West Africa by USAMRIID, NIH (NIAID), Doctors without Borders, Public Health Canada, and other global public health organizations for use in remote environments with limited infrastructure for power and clean water. This technology was also used to decontaminate the biodispersal chamber prior to its use in inoculating protective fabrics with *B. anthracis* Sterne. As the inventory of CCDC SC's innovative chlorine dioxide technologies continues to grow with D-FEND ALL, the CoD, and Bio-spray, the uses of these technologies for food packaging, sanitizing hard surfaces, decontaminating porous materials, preventing mold growth, or disinfecting pathogens such as *B. cereus, C. difficile*, Ebola virus, and perhaps Coronavirus to prevent the spread of COVID-19 and meet the needs in broad-based or niched dual-use applications for military and civilian consumers will likely continue to increase.

References

ASTM ASTM International, 2019. ASTM D5035-11, Standard Test Method for Breaking Force and Elongation of Textile Fabrics (Strip Method). Available at www.astm.org.

Cléry-Barraud, C., Gaubert, A., Masson, P., Vidal, D., 2004. Combined effects of high hydrostatic pressure and temperature for inactivation of *Bacillus anthracis* spores. Appl. Environ. Microbiol. 70 (1), 635–637.

Doona, C.J., Feeherry, F.E., Setlow, P., Malkin, A.J., Leighton, T.J., 2014. The portable chemical sterilizer (PCS), D-FENS, and D-FEND ALL: novel chlorine dioxide decontamination technologies for the military. J. Vis. Exp. 88, e4354. doi:10.3791/4354.

Doona, C.J., Feeherry, F.E., Kustin, K., Olinger, G.G., Setlow, P., Malkin, A.J., Leighton, T., 2015. Fighting Ebola with novel spore decontamination technologies for the military. Front. Microbiol. 6, 663. https://www.frontiersin.org/articles/10.3389/fmicb.2015.00663/full.

Doona, C.J., Feeherry, F.E., Set low, B., Wang, S., William, Li, Nichols, F.C., Talukdar, P.K., Sarker, M.R., Li, Y-Q, Shen, A., Setlow, P., 2016. Effects of high-pressure treatment on spores of *Clostridium* species. Appl. Environ. Microbiol. 82 (17), 5287–5297.

Feeherry, F.E., Doona, C.J., Taub, I.A., 2003. Effect of water activity on the growth kinetics of *Staphylococcus aureus* in ground bread crumb. J. Food Sci. 68 (3), 982–987.

Harnish, D., Heimbuch, B., McDonald, M., Kinney, K., Dion, M., Stote, R., Rastogi, V., Smith, L., Wallace, L., Lumley, A., Schreuder-Gibson, H., Wander, J., 2014. Standard method for deposition of dry aerosolized *Bacillus* spores on inanimate surfaces. J. Appl. Microbiol. 117 (1), 40–49.

Hoehn, R.C., 1992. Chlorine dioxide use in water treatment: key issues, Conference Proceedings, Chlorine Dioxide: Drinking Water Issues: Second International Symposium. Houston, TX.

Lessa, F.C., Mu, Y., Bamberg, W.M., Beldavs, Z.G., Dumyati, G.K., Dunn, J.R., Farley, M.M., Holzbauer, S.M., Meek, J.I., Phipps, E.C., Wilson, L.E., Wnston, L.G., Cohen, J.A., Limbago, B.M., Fridkin, S.K., Gerding, D.N., McDonals, C., 2015. Burden of *Clostridium difficile* in the United States. N. Engl. J. Med. 372 (9), 825–834.

Ray, S., Jin, T., Fan, X., Liu, L., Yam, K.L., 2013. Development of chlorine dioxide releasing film and its application in decontaminating fresh produce. J. Food Sci. 78 (2), M276–M284.

Schaufler, C., 1933. Antiseptic effect of chlorine solutions from interactions of potassium chlorate and hydrochloric acid. Zentralhl Chir 60, 2497–2500.

Setlow, P., Doona, C.J., Feeherry, F.E., Kustin, K., Sisson, D., Chandra, S., 2009. Enhanced safety and extended shelf life of fresh produce for the military. In: Fan, X., Niemira, B.A., Doona, C.J., Feeherry, F.E., Gravani, R.B. (Eds.), Microbial Safety of Fresh Produce. IFT Press Wiley Blackwell, Ames, IA, pp. 263–288.

The ClO2 Fact Sheet, 2013The ClO2 Fact Sheet. 2013. Available at http://www.lenntech.com/faqclo2.htm (accessed May 24, 2013).

USDA U. S. Department of Agriculture - Agricultural Marketing Service, 2000. The National List of Allowed and Prohibited Substances. 7 CFR Part 205 Subpart G.

U.S. EPA U. S. Environmental Protection Agency (EPA), 2017. Emerging Viral Pathogen Guidance for Antimicrobial Pesticides. Available at https://www.epa.gov/pesticide-registration/emerging-viral-pathogen-guidance-antimicrobial-pesticides accessed August 14, 2020.

U.S. FDA Food and Drug Administration, U.S. Department of Health and Human Services, 1998. Secondary Direct Food Additives Permitted in Food for Human Consumption. 21 CFR. Part 173.300 Chlorine Dioxide.

Vogt, H., 2003, sixth ed.Ullmann's Encyclopedia of Industrial ChemistryVol. 8 Wiley-VCH, Weinheim, pp. 281–327.

Young, S.B., Setlow, P., 2003. Mechanisms of killing of *Bacillus subtilis* spores by hypochlorite and chlorine dioxide. J. Appl. Microbiol. 95 (1), 54.

Polymer packaging for in-pack thermal pasteurization technologies

Chandrashekhar R. Sonar, Juming Tang, Shyam S. Sablani

Department of Biological Systems Engineering, Washington State University, Pullman, WA, United States

18.1 Introduction

Thermal pasteurization is a milder heat treatment compared to sterilization and is normally carried out at temperatures in the range of 70–100 °C (Holdsworth and Simpson, 2007). Pasteurized products have a limited shelf life of a few days to a few weeks at refrigerated conditions depending on the process design (Peng et al., 2017). Pasteurization of low-acid (pH > 4.6) food products inactivates vegetative cells and some spores (*Clostridium botulinum* type E) of pathogens of public health concern known to cause foodborne illnesses. In contrast, pasteurization of high-acid (pH < 4.6) food products inactivate pathogens as well as a number of spoilage microorganisms (Silva and Gibbs, 2010). Pasteurization cannot inactivate highly thermally resistant vegetative pathogen cells or spore-forming microorganisms (*C. botulinum* type A and *Bacillus* species), but it can reduce spoilage microorganisms that may grow during storage and distribution, making an effective cold chain an essential criterion to ensure food safety (Peng et al., 2017). A pasteurization process can be designed to achieve 6-log inactivation of *Listeria monocytogenes* ($P_{70\,°C} = 2$ min), or hepatitis A virus ($P_{72\,°C} = 10$ min or $P_{80\,°C} = 3$ min), or vegetative cells and spores of psychrotrophic nonproteolytic *C. botulinum* ($P_{90\,°C} = 10$ min; Peng et al., 2017; Tang et al., 2018). According to the European Chilled Food Federation, a process equivalent to $P_{70\,°C} = 2$ min and $P_{90\,°C} = 10$ min provides a shelf life of ≤10 days and 6–12 weeks at 5 °C, respectively (ECFF, 2006).

Commercially, food products are thermally pasteurized using hot-filled (cook-chill) or in-package process technologies (Peck, 1997). In a hot-filled process, the product is first cooked and pasteurized (70–90 °C) in a steam jacketed kettle, filled into containers while still hot, and then sealed, followed by rapid chilling. Polymer-based semirigid and flexible packaging are commonly used for pasteurized products. There are possibilities of cross-contamination during hot-fill pasteurization and packaging operations. In-package pasteurization involves vacuum packing of the product into containers followed by thermal treatment in a hot water bath (70–90 °C) or a microwave system, and then rapid chilling after desired microbial inactivation is achieved. In-package processing reduces the chance for postprocessing contamination. However, during in-package processing, the packaging is also subjected to the heating conditions, unlike with a hot-fill process. Therefore, polymer package material must be able to withstand process temperature and water exposure during pasteurization while maintaining its structural integrity, visual appearance, and barrier properties after processing.

Oxygen and water vapor barrier properties of packaging materials may influence the shelf life of the pasteurized products. Specifically, the oxygen transmission rate (OTR) of the packaging material determines quality losses due to oxidation, while the water vapor transmission rate (WVTR) controls moisture losses during storage. For sterilized products, high barrier packaging, that is, low OTR and WVTR, is required to achieve a shelf life of more than 1 year at ambient temperature. However, pasteurized products have a limited shelf life under refrigerated storage; therefore, low/medium barrier packaging (high/medium OTR and WVTR) can be used. Hence, the selection of the optimal packaging OTR & WVTR for pasteurized products is required, which normally depends on product composition and desired shelf-life.

18.2 Packaging material options

Many different polymers with a range of physical properties are commercially available, which can be used to fabricate flexible films and then converted into pouches to be used for in-package thermal processing. Table 18.1 summarizes the oxygen and water vapor barrier, thermal, and mechanical properties of commonly used polymers for making packages for in-package thermal pasteurization technologies. Film/pouches for in-package pasteurization must be heat sealable, mechanically and thermally robust to withstand processing conditions, be able to provide enough barrier properties to maintain stipulated shelf life, and meet regulatory requirements. In addition to these requirements, the cost of the packages also influences the final selection.

Monolayer films/pouches are not commonly used for in-package processing of products intended for long shelf life due to limitations for each polymer. For example, nonpolar polymers like PE and PP have high oxygen permeabilities, whereas polar materials like EVOH and PA are hydrophilic and tend to absorb moisture, reducing their oxygen barrier properties. Hence, multilayer structures are formed to attain desirable properties at an optimum cost. Generally, multilayer laminates vary from 3-ply to 9-ply with an inner food contact/heat sealable layer (PP/PE), a barrier layer (EVOH, nylon, and/or metal oxide-coated PET), and an outer heat stable, water-resistant and printable layer (PET/PE). These structures are either co-extruded or laminated using a tie layer of adhesives.

Table 18.1 Properties of commonly used polymers for in-package thermal processing.

Polymer	Barrier properties		Thermal properties		Mechanical properties		
	OTR (cm³/m²/day)	WVTR (g/m²/day)	T_g (°C)	T_m (°C)	Tensile strength (MPa)	Elastic modulus (MPa)	Elongation at break (%)
LDPE	7400	12.5	−110	115	8–20	300–500	500–1000
HDPE	1600	3.7	−90	137	19–31	600–1400	20–50
EVOH[a]	0.2	80	72	191	55–65	2000–2300	100–225
PA	18–35	90–280	40–57	185–277	49–69	700–980	200–300
PET	55	20	69	265	48–72	2800–4100	30–300
PP	1200–3040	4.6–8.2	−18	176	30–40	1000–2000	400–900

Sources: Bhunia et al. (2013) and Robertson (2013).
EVOH, ethylene vinyl alcohol; HDPE, high-density polyethylene, LDPE, low-density polyethylene; PA, polyamide; PET, polyethylene terephthalate; PP, polypropylene.
OTR and WVTR were measured at 23 °C, 0% RH and 38 °C, 90% RH, respectively and based on 25 μm film thickness.
[a] 32% ethylene content.

Table 18.2 Barrier properties of some commercially available multilayer packaging films suitable for in-package thermal pasteurization (adapted from Sonar, 2020).

Multilayer films	Thickness (μm)	OTR (cm^3/m^2/day)	WVTR (g/m^2/day)
PA/HDPE/PE/Tie/EVOH/PE/Tie/ HDPE/PA	108 ± 4.07	0.22 ± 0.02	1.76 ± 0.05
PET/adhesive/EVOH-PE blend	81.2 ± 0.75	0.86 ± 0.03	4.72 ± 0.07
PET/LLDPE/LDPE/Tie/Nylon66/Tie/ LLDPE/LDPE	102 ± 2.73	0.99 ± 0.05	3.94 ± 0.03
PET/barrier PET/adhesive/PP	84.4 ± 1.36	1.05 ± 0.06	5.11 ± 0.06
Nylon/Nylon/Tie/LLDPE	194 ± 3.37	10.3 ± 0.43	3.92 ± 0.13
LDPE/Tie/Nylon/Tie/LDPE	86.4 ± 1.36	29.8 ± 1.38	4.18 ± 0.06
PET-PE-based	70.0 ± 1.26	80.9 ± 2.15	6.59 ± 0.03
LDPE-based	50.2 ± 2.71	104 ± 2.47	9.69 ± 0.03

OTR and WVTR were measured at 23 °C, 55% RH and 38 °C, 100% RH, respectively.

Multilayer laminate structures vary considerably depending on the application and can provide a range of barrier properties. Table 18.2 provides oxygen and water vapor barrier properties of some of the commercially available multilayer structures that are suitable for in-package thermal pasteurization. High barrier structures are critical for shelf-stable food products to maintain extended shelf life, whereas these high barrier films are not economical for pasteurized food products, which have limited shelf life under refrigeration. Usually, the higher the barrier of a film the higher is the cost of the package. Commercially pasteurized products are packaged in films/pouches with a wide array of oxygen barrier properties, but generally, products prone to oxidation are packaged in higher oxygen barrier pouches (Table 18.3). Within pasteurized food products, the appropriate selection of packages may depend on the nature of the product, type of pasteurization process (hot-filled or in-package), and desired shelf life.

18.3 Packaging selection criteria

During in-package pasteurization processes, packaging materials may be subjected to high temperatures, high humidity or hot water, high pressure, and intense electromagnetic waves. Food packages must withstand these conditions and retain visual integrity and barrier properties during and after processing. Hence, package thermal properties, gas barrier properties, mechanical properties, and food-package interactions need to be investigated when selecting appropriate packaging materials.

Table 18.3 Barrier properties of packaging films used for some thermally pasteurized commercial food products (adapted from Sonar, 2020).

Product	Pasteurization	Thickness (μm)	OTR (cm^3/m^2/day)	WVTR (g/m^2/day)
Imitation crab	In-package	100	39.8 ± 1.15	5.18 ± 0.02
Chicken	In-package	85	3.18 ± 0.02	12.4 ± 0.04
Pulled pork	In-package	100	4.00 ± 0.04	14.0 ± 0.05
Mashed potato	Hot-filled	45	85.5 ± 0.20	13.4 ± 0.11
Mac and cheese	Hot-filled	50	87.1 ± 0.13	12.5 ± 0.13

OTR and WVTR were measured at 23 °C, 55% RH and 38 °C, 100% RH, respectively.

18.3.1 Visual integrity

The package must maintain its physical shape and seal integrity during and after thermal processing. Any kind of damage such as pinholes, broken seal, shrinkage, or delamination will compromise the safety and quality of the product. Thermal properties (melting temperature, T_m and glass transition temperature, T_g) of the packaging polymer can determine whether the package can withstand thermal processing conditions. T_m of packaging polymer should be higher than the pasteurization temperature to prevent the melting of the package, causing any shrinkage or seal damage, while T_g should be low enough to provide flexibility to film.

18.3.2 Gas barrier properties

Food products are a multicomponent system containing lipids, proteins, nutrients, color pigments, among others, and these components are prone to oxidative degradation. Pasteurized products are also high in water content, and a small amount of water change could influence the sensory quality of the products. Hence, gas barrier properties, OTR and WVTR, of packages are crucial for maintaining the desired shelf life of products. The barrier properties of packaging films may change following in-package thermal pasteurization (Bhunia et al., 2016; Halim et al., 2009; Ramalingam et al., 2015; Sonar et al., 2020a, 2020b) which may influence product shelf stability (Sonar et al., 2020a, 2020b). The extent of barrier property deterioration after processing depends on the type of film structure and the severity of the thermal process. Therefore, to maintain product shelf life, the deterioration of the film barrier should be minimal.

18.3.3 Migration

In packaging, migration is termed as the transfer of compounds (usually unwanted) from packaging material to the packed food. The transfer of intentional compounds such as antioxidants, antimicrobials, etc. in the case of active packaging is not considered under migration. Migration could be *overall migration* (OM), which refers to the sum of all substances transferred from the packaging or *specific migration*, which refers to the individual or identifiable compound only (Robertson, 2013). As per EU Directive 10/2011, the limit of OM is 60 mg/kg of simulant or food or 10 mg/dm² of the contact area. Compounds like unreacted monomers, oligomers, plasticizers, antioxidants, light stabilizers, lubricants, antistatic agents, slip compounds, and thermal stabilizers may migrate from packaging into food (Bhunia et al., 2013). Hence, the migration of any compound from packaging material should either be none or within the regulatory requirements to ensure food safety. At the same time, packaging must not absorb any aroma, flavor, or other compounds from the product during storage, resulting in quality losses.

18.4 Process–packaging interaction

Thermal processing may deteriorate packaging film barrier properties, despite maintaining physical appearance and visual integrity by package after treatment. In many cases, the package deteriorations are not easily detected by visual inspection, and most of those deteriorations occur at the microscopic level in forms of changes in polymer morphology, thermal, and dielectric properties of the packaging materials.

18.4.1 Gas barrier properties

The OTR and WVTR are important for maintaining product quality and storage stability of pasteurized food. Table 18.4 summarizes the influence of pasteurization on gas barrier properties of synthetic as well as biodegradable and/or compostable packaging films. Both conventional and microwave-assisted

Table 18.4 Changes in barrier properties after in-package thermal pasteurization.

Film	Pasteurization	OTR (cm³/m²/day)		WVTR (g/m²/day)		References
		Before	After	Before	After	
PET/barrierPET/ tie/PE	Conventional (36 min at 93 °C)	0.7	1.5			Bhunia et al., 2016
		0.1	1.1			
PET/tie/Nylon-6/PP		2.3	6.0			
PET/LLDPE/LDPE/ Tie/Nylon66/Tie/ LLDPE/LDPE						
PET/barrierPET/ tie/PE	Microwave-assisted (30 min holding at 61 °C, 2 min microwave heating at 93 °C, 20 min holding at 93 °C	0.7	1.1			
		0.1	1.7			
PET/tie/Nylon-6/PP		2.3	7.9			
PET/LLDPE/LDPE/ Tie/Nylon66/Tie/ LLDPE/LDPE						
PET/LLDPE/LDPE/ Tie/Nylon66/Tie/ LLDPE/LDPE	Conventional (60 min at 92 ± 1 °C)	0.99 ± 0.05	1.77 ± 0.03	3.93 ± 0.04	4.99 ± 0.08	Sonar et al., 2020a
		10.3 ± 0.43	8.97 ± 0.11	3.92 ± 0.13	5.16 ± 0.07	
Nylon/Nylon/Tie/ LLDPE		29.9 ± 1.38	28.4 ± 0.73	4.11 ± 0.15	2.84 ± 0.09	
		80.9 ± 2.09	119 ± 5.23	6.60 ± 0.03	11.7 ± 0.16	
LDPE/Tie/Nylon/ Tie/LDPE						
PET-PE based						
PET/LLDPE/LDPE/ Tie/Nylon66/Tie/ LLDPE/LDPE	Microwave assisted (30 min holding at 51 °C, 3.2 min microwave heating at 91 °C, 10 min hold-ing at 91 °C	0.99 ± 0.05	1.76 ± 0.10	3.93 ± 0.04	5.12 ± 0.20	
		10.3 ± 0.43	9.10 ± 0.29	3.92 ± 0.13	2.08 ± 0.13	
Nylon/Nylon/Tie/ LLDPE		29.9 ± 1.38	27.5 ± 0.67	4.11 ± 0.15	2.73 ± 0.04	
		80.9 ± 2.09	111 ± 1.06	6.60 ± 0.03	5.97 ± 0.06	
LDPE/Tie/Nylon/ Tie/LDPE						
PET-PE based						
Chemically modified PLA-PBAT blend	Conventional (15 min at 72 ± 0.5 °C)	330 ± 1.63	358 ± 3.44	37.7 ± 1.58	217 ± 2.31	Sonar et al., 2020b
		541 ± 8.20	>1000	47.6 ± 0.43	254 ± 3.61	
Heat sealable PLA-layer/PLA core/Heat sealable PLA-layer		619 ± 24.0	>1000	48.8 ± 1.46	288 ± 3.38	
		80.9 ± 2.15	110 ± 0.63	6.59 ± 0.03	9.53 ± 0.07	
Heat sealable PLA-layer/PLA core/Heat sealable PLA-layer						
PET-PE based						

PBAT, polybutylene adipate terephthalate; PLA, polylactic acid.
OTR and WVTR were measured at 23 °C, 55% RH and 38 °C, 100% RH, respectively.

pasteurization (MAPS) significantly decreased barrier properties of PET and nylon-based films, while MAPS showing higher changes in OTR than conventional pasteurization (Bhunia et al., 2016). The larger barrier changes in MAPS were due to the longer preheating time in the first generation of MAPS (52 min) than in conventional pasteurization (36 min). Later, improvement in MAPS design showed higher retention of oxygen barrier in multilayer films (Sonar et al., 2020a). Although biodegradable and/or compostable films have limited thermal stability, PLA and PBAT-based composite films have shown suitability for in-package pasteurization at $P_{70\,°C}$ processing conditions (Sonar et al., 2020b).

Significant changes in barrier properties were observed in co-extruded PE/Nylon6/PE (N6), PE/NanoNylon6/PE (N6/Nano), and PE/Nylon6/EVOH/Nylon6/PE films after pasteurization (75 °C, 30 min) process (Halim et al., 2009). OTR increased in N6 and N6/Nano films by 13% and 76%, respectively, whereas WVTR of all films increased by 41–97%. Similarly, a significant increase was observed in OTR (14–28%) and WVTR of PP/Nylon6/PP, PET/Nylon6/CPP, SiO_x-PET/Nylon6/CPP following a pasteurization (75 °C, 30 min) process (Ramalingam et al., 2015). Thus, the deterioration of barrier properties depends upon the nature of the material and the process conditions (time and temperature).

18.4.2 Polymer morphology and thermal properties

Polymers are semicrystalline, having both crystalline and amorphous regions in their structure. The crystalline region has an orderly arrangement of molecules making a closely packed structure, whereas the molecular chains are disordered and without regular structure in the amorphous region (Robertson, 2013). Gas barrier properties of polymers along with other properties like thermal properties (T_g and T_m), transparency, density, and stiffness depend on the degree of crystallinity of polymers. The higher the degree of crystallinity the better is the oxygen barrier properties of polymers (Mokwena and Tang, 2012). High temperature and moist conditions during in-package pasteurization may influence the orientation of crystal structure, affecting the morphology, gas barrier, and thermal properties. The morphological changes in polymer induced by the absorbed moisture and thermal processing correlate well with the changes in gas barrier properties. Several analytical techniques, including X-ray diffraction, positron annihilation lifetime spectroscopy, Fourier transform infrared spectroscopy, scanning electron microscopy, and differential scanning calorimetry are used to characterize thermal and structural properties (Bhunia et al., 2016; Dhawan et al., 2014; Halim et al., 2009; Ramalingam et al., 2015).

Hydrophilic polymers like EVOH and nylon and hydrophobic polymers like PET, PLA, and PBAT (when heated above T_g) tend to absorb moisture during processing, and this causes the plasticization of the molecular chain by weakening the interchain hydrogen bonding, increasing chain mobility, and fractional free volume (Lopez-Rubio et al., 2003; Dhawan et al., 2014; Mokwena et al., 2011). This can be observed in the lowering of T_g in PLA and PBAT-based films after 72 °C, 15 min pasteurization process (Sonar et al., 2020b). An increase in the degree of overall crystallinity does not always correspond to improved barrier properties. The gas barrier properties of PET and nylon-based films were significantly decreased following pasteurization due to the fragmented crystal structure and reduction in crystal size, although overall crystallinity was increased after processing (Bhunia et al., 2016; Sonar et al., 2020a). Nylon and EVOH-based films showed similar observations after pasteurization, where OTR increased due to plasticization despite an increase in overall crystallinity (Halim et al., 2009). Similarly, crystal fragmentation and polymeric chain degradation were observed, causing a reduction in barrier properties of nylon-based films after pasteurization (Ramalingam et al., 2015).

18.4.3 Dielectric properties

Packaging polymers have generally much lower dielectric properties than food; hence most polymer-based packaging is suitable for microwave-based thermal processes. The dielectric parameters, dielectric loss (ε") and loss tangent (tan δ) are sensitive to the moisture content of the polymers; hence a small amount of water absorption by the polymer after thermal processing can be measured easily using split postdielectric resonance technique (Bhunia et al., 2016; Sonar et al., 2020a, 2020b). Hydrophilic polymers like polyamides and EVOH showed 2–6 times increase in ε" and tan δ after thermal processing. However, the absorbed moisture does not always lead to greater changes in dielectric properties, as observed in the case of PET, PLA, and PBAT-based films (Bhunia et al., 2016; Sonar et al., 2020a, 2020b). The type of dipole moments of the functional group, mole percentage of polar groups, hydrogen bonding, degree of crystallinity, and physical state of polymers determines the changes in dielectric properties (Bhunia et al., 2016). The functional ester group in PET, PLA, and PBAT has a lower dipole moment than the functional amide group in nylon, resulting in lower loss changes in the former compared to the latter (Bhunia et al., 2016). During the storage of packaged food in a low relative humidity (RH) environment, the packaging film usually loses the absorbed moisture, which results in the reduction of dielectric properties and recovery of oxygen barrier of the film (Sonar et al., 2020b; Zhang et al., 2017).

18.4.4 Migration

Food-package interactions are important in terms of regulatory requirements as well as product safety and quality. Limited studies have investigated the OM from packaging into food simulants following thermal pasteurization and during storage. OM from five different film samples was <1 mg/dm^2 in food simulants following thermal treatment at 40 °C for 10 days, 80 °C for 30 min, and 121 °C for 30 min (Galotto and Guarda, 1999). Similarly, OM from PA/PE film was between 0.60 and 7.35 mg/dm^2 in food simulants following storage and various thermal treatments at 20–121 °C for different intervals of time (Galotto and Guarda, 2004). Mauricio-Iglesias et al. (2010) analyzed the migration of antioxidants (Irganox 1076) and ultraviolet light absorbers (Uvitex OB) from LLDPE films into food simulants after thermal pasteurization at 63 °C for 30 min, and after storage at 40 °C for 10 and 26 days. They observed the highest migration in olive oil as compared to other simulants. Lactic acid, lactide, and oligomers migration were studied for PLA sheets at different time and temperature intervals using food simulants such as water, 4% acetic acid, and 20% ethanol (Mutsuga et al., 2008). The total migration was 0.28–15 μg/dm^2 at 40 °C after 6 months, 0.73–2840 μg/dm^2 at 60 °C after 10 days, and 2.04–49.63 μg/dm^2 at 95 °C for 30, 60, and 120 min. There is a need to investigate the migration of other compounds from packaging into food following conventional and microwave-assisted pasteurization.

18.5 Storage studies of in-package pasteurized food products

Packaging plays an essential role in maintaining the storage stability of processed food. Food products are rich in oxygen-sensitive compounds like lipids, proteins, vitamins, and pigments, which are prone to oxidation affecting sensory and nutritive value of products and thus shelf life. Therefore, oxygen present/entering a package is an important factor in keeping the chemical and nutritive quality of food. Usually, the amount of oxygen present in the headspace is a function of residual oxygen after vacuuming, oxygen permeated into the pouch based on film OTR, oxygen entering through the seal, and oxygen being consumed by the food components during chemical reactions.

Studies have investigated the effect of packaging materials with varied gas barrier properties on the quality of thermally pasteurized food products such as seafood, vegetable products, and mashed potatoes (Table 18.5). Many different food quality parameters such as lipid oxidation, vitamins degradation,

Table 18.5 Storage studies of in-package thermally pasteurized food products packaged in packaging films of varied gas barrier properties.

Product	Package	OTR	WVTR	Storage conditions	Quality indices	Refs.
Spinach soup	**a.** PET/Aluminum/PP **b.** Nylon/PE/Nylon/PE/ Nylon/LLDPE **c.** Multilayer nylon	*0 *0.098±0.002[a] *0.270±0.043[a]		18 days at 10 °C	TPC, Color, ascorbic acid, chlorophyll	Kim et al., 2003
Fresh blue mussels	**a.** mLLDPE/LLDPE/ LLDPE/tie/Nylon/tie/ LLDPE/LLDPE/mLLDPE **b.** LDPE/tie/Nylon/tie/LDPE **c.** PET/tie/EVOH-PP	62[b] 40[b] 3[b]	3.88 4.34 1.14	60 days at 3.5 ± 0.5 °C	E_h potential, TBARS, hardness	Bhunia et al., 2017
Shredded carrot	**a.** mLLDPE/LLDPE/ LLDPE/tie/Nylon/tie/ LLDPE/LLDPE/mLLDPE **b.** LDPE/tie/Nylon/tie/LDPE **c.** PET/tie/EVOH-PP	62[b] 40[b] 3[b]	3.88 4.34 1.14	60 days at 3.5 ± 0.5 °C	Color, total carotenoids, total polyphenols, firmness	Bhunia, 2016
Cooked ham	**a.** Nylon6/LLDPE/LDPE **b.** Nanoclay modified nylon6/LLDPE/LDPE **c.** Multilayer commercial	44±1[b] 96±3[b] 20[b]		27 days at 5 ± 2°C	Headspace gas composition, a_w, pH, color, aerobic count, sensory (color)	Lloret et al., 2016
Carrot puree	**a.** PET/LLDPE/LDPE/Tie/ Nylon66/Tie/LLDPE/ LDPE **b.** LDPE/Tie/Nylon/Tie/ LDPE **c.** PET-PE based	0.99 ± 0.05 29.8 ± 1.38 80.9 ± 2.15	3.94 ± 0.03 4.18 ± 0.06 6.59 ± 0.03	100, 80, and 45 days at 4 °C, 8 °C, and 13 °C, respectively	Weight loss, pH, color, β-carotene, ascorbic acid	Sonar et al., 2019a
Red cabbage, beet, and pea purees	**a.** PET/LLDPE/LDPE/Tie/ Nylon66/Tie/LLDPE/ LDPE **b.** LDPE/Tie/Nylon/Tie/ LDPE **c.** PET-PE based	0.99 ± 0.05 29.8 ± 1.38 80.9 ± 2.15	3.94 ± 0.03 4.18 ± 0.06 6.59 ± 0.03	80 days at 7 °C	Weight loss, anthocyanins, betalains, chlorophylls, color, pH, TPC	Sonar et al., 2019b

(continued)

Table 18.5 Storage studies of in-package thermally pasteurized food products packaged in packaging films of varied gas barrier properties. *Continued*

Product	Package	OTR	WVTR	Storage conditions	Quality indices	Refs.
Cream cheese mashed potato	**a.** PET/LLDPE/LDPE/Tie/ Nylon66/Tie/LLDPE/ LDPE **b.** Nylon/Nylon/Tie/LLDPE **c.** LDPE/Tie/Nylon/Tie/ LDPE **d.** PET-PE based	0.99 ± 0.05 10.3 ± 0.43 29.8 ± 1.38 80.9 ± 2.15	3.94 ± 0.03 3.92 ± 0.13 4.18 ± 0.06 6.59 ± 0.03	90 days at 5 °C	Vitamin A, E, and C, color, pH, TPC	Sonar et al., 2020a
Beet mixed mashed potato Salmon in sauce	**a.** Chemically modified PLA-PBAT blend **b.** Heat sealable PLA-layer/ PLA core/Heat sealable PLA-layer **c.** Heat sealable PLA-layer/ PLA core/Heat sealable PLA-layer **d.** PET-PE based	330 ± 1.63 541 ± 8.20 619 ± 24.0 80.9 ± 2.15	37.7 ± 1.58 47.6 ± 0.43 48.8 ± 1.46 6.59 ± 0.03	10 days at 4 °C	Weight loss, headspace oxygen concentra- tion, color, betalains, TBARS, vitamin C, pH, TPC	Sonar et al., 2020b

OTR and WVTR were measured at 23 °C, 55% RH and 38 °C, 100% RH, respectively, otherwise as mentioned in superscript.
[a]Oxygen permeability in mmol/m/h/ atm at 10 °C.
[b]23 °C, 0% RH.

sensory attributes, textural and color changes, pigment content, weight loss, oxidation-reduction potential, and microbiology have been studied.

18.5.1 Weight loss

Weight (moisture) loss during storage in food products may result in economic as well as quality losses, which depend upon the WVTR of the packaging, storage temperature, and storage time. Vegetable purees packaged in 3.94–6.59 g/m^2.day WVTR pouches have shown less than 0.5% weight losses when stored under refrigeration for up to 100 days (Sonar et al., 2019a, 2019b). Similarly, beet mixed mashed potato and salmon in sauce packaged in biobased pouches (WVTR: 217–288 g/m^2 day after processing) demonstrated up to 1.7% weight loss after 10 days storage at 4 °C (Sonar et al., 2020b). The weight loss during storage may or may not affect the food quality, as demonstrated from the studies on sterilized food products showing losses between 2% and 16% (Patel et al., 2019; Zhang et al., 2019). Overall, the weight/moisture loss in the pasteurized products is insignificant and may not affect sensory quality during the shorter storage period and at refrigerated conditions.

18.5.2 Color

The color of processed products is one of the primary quality attributes which appeals to consumers. The change in color of processed products during storage depends on the pigment (chlorophylls, carotenoids, anthocyanins, betalains, and myoglobin) losses, lipid oxidation, browning reactions, light conditions, moisture loss, and product composition in addition to the headspace gas composition which may be governed by the gas permeation of the packaging film. Studies have shown the effect of OTR on color values in different food products, as shown in Table 18.5. Several food products have shown color degradation (L^*, a^*, b^*, chroma, total color difference, sensory score) during storage; however, the rate of color change was dependent on package OTR.

The loss of chlorophyll in spinach soup was correlated with the loss of color values during storage and packaging type (Kim et al., 2003). Cooked ham packed in high barrier film had the lowest oxygen concentration in the headspace and retained the best color after storage (Lloret et al., 2016). Carotenoids or β-carotene in pasteurized carrot puree have shown very good stability during storage at refrigeration temperature (Sonar et al., 2019a). Shredded carrots and carrot puree packaged in high barrier film showed significantly lower color difference compared to two medium/low barrier films at the end of refrigerated storage (Bhunia, 2016; Sonar et al., 2019a). Red cabbage, beet, and pea puree demonstrated a similar trend with the OTR of the package, and it correlated with the anthocyanins, betalains, and chlorophylls contents, respectively (Sonar et al., 2019b). Betalains exhibited the highest sensitivity (4–49% losses) toward film OTR (Fig. 18.1), whereas anthocyanins were relatively stable with <4% losses irrespective of package OTR. On the other hand, chlorophylls continued to degrade

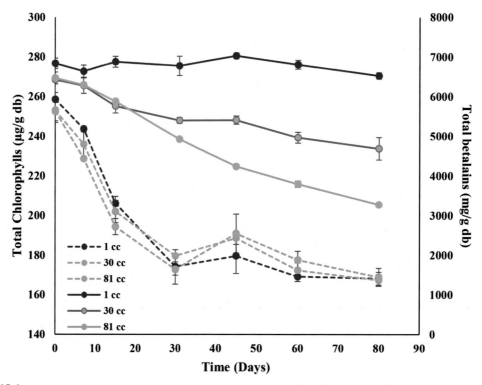

FIG. 18.1

Total chlorophylls (dotted line) in pea puree and total betalains (solid line) in beet puree packed in different packaging films during storage at 7 °C (adapted from Sonar et al., 2019b).

(33–35%) during storage irrespective of package film OTR (Fig. 18.1). Though betalains are sensitive to oxygen (Sonar et al., 2019b), during short storage (10-day) period, they have shown very good stability in the presence of vitamin C (Sonar et al., 2020b).

18.5.3 Lipid oxidation

The lipid oxidation of fat-rich products is a very good quality indicator, which often correlates well with the oxygen level present in the package. Studies on salmon and mussels have demonstrated the effect of OTR on the lipid oxidation (Bhunia et al., 2017; Sonar et al., 2020b). The film with higher OTR (62 cm^3/m^2 day has shown to increase lipid oxidation in pasteurized mussels in red sauce as compared to lower OTR (3 cm^3/m^2 day) film during storage at 4 °C for 60 days (Bhunia et al., 2017). They also observed higher headspace oxygen concentration in higher OTR film, which correlated well with the lipid oxidation. Salmon in sauce packaged in PLA and PBAT-based pouches along with PE-based control showed similar results stored at 4 °C for 10 days, as shown in Fig. 18.2 (Sonar et al., 2020b). Lipid oxidation was significantly higher for all types of films at the end of storage with a positive correlation (*R*: 0.878–0.986) with the headspace oxygen concentration and product packed in

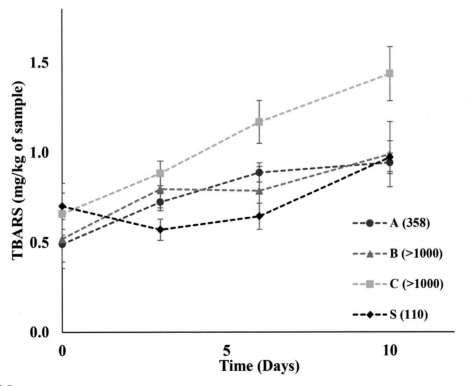

FIG. 18.2

Lipid oxidation in salmon in sauce packed in different packaging films during storage at 4 °C (values in parentheses denotes OTR (cm^3/m^2/day) of films after processing; adapted from Sonar et al., 2020b).

>1000 cm^3/m^2 day OTR film showed the maximum increase in the TBARS (thiobarbituric acid reactive substances) value.

18.5.4 Vitamins

Vitamins are essentials for normal health and growth, and when present in the food provide nutritive value to it. Vitamin C is susceptible to oxidative degradation, and usually, food matrix composition affects the storage stability of vitamin C (Gregory, 2008), while vitamin A and vitamin E are relatively stable. The limited studies have focused on oxygen sensitivity of vitamin A, vitamin E, and vitamin C in pasteurized foods during refrigerated storage. Vitamin A and vitamin E were stable during 90-day storage in cream cheese mashed potatoes regardless of package barrier properties (Sonar et al., 2020a). Vitamin C fortified in carrot puree was very sensitive toward package OTR and storage temperature (Sonar et al., 2019a). Vitamin C losses were in the range of 11–100% during storage at three different temperatures (4, 8, and 13 °C) with retention of 85–89% in the film with OTR 1 cm^3/m^2 day and 54–68% in 30 cm^3/m^2 day at the end of storage. Similarly, for a short shelf-life product, losses in the range of 14–40% were observed in beet mixed mashed potato packaged in biobased films, including synthetic control film (Sonar et al., 2020b). Encapsulation helped in the improved retention of vitamin C by more than 50% in low barrier film compared to the nonencapsulated vitamin C in cream cheese mashed potatoes (Sonar et al., 2020a). These studies suggested that food composition plays an important role in determining the storage stability of vitamin C in pasteurized food products.

18.5.5 Microbiology

Microbiology is an important consideration in determining the safety of pasteurized food products as pasteurization does not completely eliminate food pathogens and spoilage microorganisms. Hence, microbial analyses during storage studies involving different packaging materials can be useful in confirming product safety. Usually, the microbial growth during storage in the pasteurized products depends on the native microflora, initial count, food composition, and thermal treatment level. As per Fig. 18.3, total plate count in pasteurized red cabbage and beet puree packaged in three different OTR pouches was 1–3 log CFU/g at the end of 80-day storage against 5–6 log CFU/g for pea puree (Sonar et al., 2019b). *Bacillus* spores are mainly responsible for spoilage in pasteurized food products stored under refrigeration as they are more heat resistant and subpopulations can survive the process designed for 6-log reduction of nonproteolytic *Clostridium botulinum* (Carlin et al., 2000; Peng et al., 2017).

18.6 Summary and future development

In-package thermal pasteurization technologies have the potential to produce safe, healthy, and high sensory quality food products with shelf life ranging from a few days to weeks at refrigerated temperature. Medium and low gas barrier polymeric packaging are generally suitable for providing the desired shelf life to pasteurized food products. Higher barrier packaging may be required for food containing oxygen-sensitive components such as vitamin C and betalains. Thermal pasteurization may affect gas barrier properties of flexible packaging; hence, the selection of films with appropriate oxygen and water vapor barrier is important to achieve the desired shelf life of pasteurized products. PBAT and

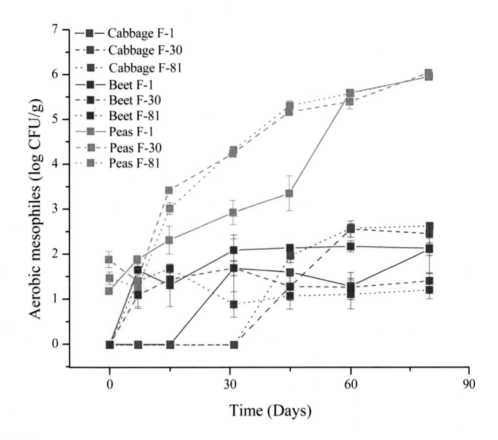

FIG. 18.3

Aerobic mesophilic count of three types of puree packed in different packaging films during storage at 7 °C (F-1, F-30, and F-81 are films with 1, 30, and 81 cm^3/m^2 day of OTR, respectively; adapted from Sonar et al., 2019b).

PLA composite films have shown the potential of extending their application to in-package thermal pasteurization of short shelf life products. Research is needed to develop biodegradable/compostable polymer packaging for in-package pasteurization technologies. Advanced encapsulation strategies should be investigated to improve the storage stability of oxygen-sensitive components in food.

References

Bhunia, K., 2016. Polymeric Packaging Films for Thermal Pasteurization Processes. Washington State University Dissertation.

Bhunia, K., Ovissipour, M., Rasco, B., Tang, J., Sablani, S.S., 2017. Oxidation–reduction potential and lipid oxidation in ready-to-eat blue mussels in red sauce: criteria for package design. J. Sci. Food Agric. 97 (1), 324–332.

Bhunia, K., Sablani, S.S., Tang, J., Rasco, B., 2013. Migration of chemical compounds from packaging polymers during microwave, conventional heat treatment, and storage. Comprehens. Rev. Food Sci. Food Safety 12 (5), 523–545.

Bhunia, K., Zhang, H., Liu, F., Rasco, B., Tang, J., Sablani, S.S., 2016. Morphological changes in multilayer polymeric films induced after microwave-assisted pasteurization. Innovative Food Sci. Emerg. Technol. 38, 124–130.

Carlin, F., Guinebretiere, M.H., Choma, C., Pasqualini, R., Braconnier, A., Nguyen-the, C., 2000. Spore-forming bacteria in commercial cooked, pasteurised and chilled vegetable purees. Food Microbiol. 17 (2), 153–165.

Dhawan, S., Varney, C., Barbosa-Cánovas, G., Tang, J., Selim, F., Sablani, S., 2014. The impact of microwave-assisted thermal sterilization on the morphology, free volume, and gas barrier properties of multilayer polymeric films. J. Appl. Polym. Sci. 131, 40376.

ECFF (2006). Recommendations for the production of pre-packaged chilled food.

Galotto, M.J., Guarda, A., 1999. Comparison between thermal and microwave treatment on the overall migration of plastic materials intended to be in contact with foods. Packaging Technol. Sci. 12 (6), 277–281.

Galotto, M.J., Guarda, A., 2004. Suitability of alternative fatty food simulants to study the effect of thermal and microwave heating on overall migration of plastic packaging. Packaging Technol. Sci. 17 (4), 219–223.

Gregory III, J.F., 2008. Vitamins. In Damodaran, S., Parkin, K.L., Fennema, O.R. (Eds.), Fennema's Food Chemistry. CRC Press, Boca Raton, FL, pp. 439–521.

Halim, L., Pascall, M.A., Lee, J., Finnigan, B., 2009. Effect of pasteurization, high-pressure processing, and retorting on the barrier properties of nylon 6, nylon 6/ethylene vinyl alcohol, and nylon 6/nanocomposites films. J. Food Sci. 74 (1), N9–N15.

Holdsworth, S.D., Simpson, R., 2007. Thermal Processing of Packaged Foods. Springer, New York, NY.

Kim, G.T., Paik, H.D., Lee, D.S., 2003. Effect of different oxygen permeability packaging films on the quality of *sous-vide* processed seasoned spinach soup. Food Sci. Biotechnol. 12 (3), 312–315.

Lloret, E., Picouet, P.A., Trbojevich, R., Fernández, A., 2016. Colour stability of cooked ham packed under modified atmospheres in polyamide nanocomposite blends. LWT-Food Sci. Technol. 66, 582–589.

Lopez-Rubio, A., Lagaron, J.M., Gimenez, E., Cava, D., Hernandez-Muñoz, P., Yamamoto, T., Gavara, R., 2003. Morphological alterations induced by temperature and humidity in ethylene–vinyl alcohol copolymers. Macromolecules 36 (25), 9467–9476.

Mauricio-Iglesias, M., Jansana, S., Peyron, S., Gontard, N., Guillard, V., 2010. Effect of high-pressure/temperature (HP/T) treatments of in-package food on additive migration from conventional and bio-sourced materials. Food Addit. Contam. 27 (1), 118–127.

Mokwena, K.K., Tang, J., 2012. Ethylene vinyl alcohol: a review of barrier properties for packaging shelf stable foods. Crit. Rev. Food Sci. Nutr. 52 (7), 640–650.

Mokwena, K.K., Tang, J., Laborie, M.P., 2011. Water absorption and oxygen barrier characteristics of ethylene vinyl alcohol films. J. Food Eng. 105 (3), 436–443.

Mutsuga, M., Kawamura, Y., Tanamoto, K., 2008. Migration of lactic acid, lactide and oligomers from polylactide food-contact materials. Food Addit. Contam. 25 (10), 1283–1290.

Patel, J., Al-Ghamdi, S., Zhang, H., Queiroz, R., Tang, J., Yang, T., Sablani, S.S., 2019. Determining shelf life of ready-to-eat macaroni and cheese in high barrier and oxygen scavenger packaging sterilized via microwave-assisted thermal sterilization. Food Bioprocess Technol. 12 (9), 1516–1526.

Peck, M.W., 1997. *Clostridium botulinum* and the safety of refrigerated processed foods of extended durability. Trends Food Sci. Technol. 8 (6), 186–192.

Peng, J., Tang, J., Barrett, D.M., Sablani, S.S., Anderson, N., Powers, J.R., 2017. Thermal pasteurization of ready-to-eat foods and vegetables: Critical factors for process design and effects on quality. Crit. Rev. Food Sci. Nutr. 57 (14), 2970–2995.

Ramalingam, R., George, J., 2015. Effect of pasteurization and retort processing on spectral characteristics, morphological, thermal, physico-mechanical, barrier and optical properties of nylon-based food packaging materials. Packaging Technol. Sci. 28 (5), 425–436.

Robertson, G.L., 2013. Food Packaging: Principles and Practice. CRC Press, Boca Raton, FL.

Silva, F.V., Gibbs, P.A., 2010. Non-proteolytic *Clostridium botulinum* spores in low-acid cold-distributed foods and design of pasteurization processes. Trends Food Sci. Technol. 21 (2), 95–105.

Sonar, C.R., 2020. In-Package Thermal Pasteurization: Evaluating Performance of Flexible Packaging and Oxygen Sensitivity of Food Components. Washington State University Dissertation Approved.

Sonar, C.R., Al-Ghamdi, S., Marti, F., Tang, J., Sablani, S.S., 2020b. Performance evaluation of biobased/biodegradable films for in-package thermal pasteurization. Innov. Food Sci. Emerg. Technol. 66, 102485.

Sonar, C.R., Paccola, C., Al-Ghamdi, S., Rasco, B., Tang, J., Sablani, S.S., 2019a. Stability of color, β-carotene, and ascorbic acid in thermally pasteurized carrot puree to the storage temperature and gas barrier properties of selected packaging films. J. Food Process Eng. 42 (4), e13074.

Sonar, C.R., Parhi, A., Liu, F., Patel, J., Rasco, B., Tang, J., Sablani, S.S., 2020a. Investigating thermal and storage stability of vitamins in pasteurized mashed potatoes packed in barrier packaging films. Food Packaging Shelf Life 24, 100486.

Sonar, C.R., Rasco, B., Tang, J., Sablani, S.S., 2019b. Natural color pigments: oxidative stability and degradation kinetics during storage in thermally pasteurized vegetable purees. J. Sci. Food Agric. 99 (13), 5934–5945.

Tang, J., Hong, Y.K., Inanoglu, S., Liu, F., 2018. Microwave pasteurization for ready-to-eat meals. Curr. Opin. Food Sci. 23, 133–141.

Zhang, H., Bhunia, K., Munoz, N., Li, L., Dolgovskij, M., Rasco, B., Sablani, S.S., 2017. Linking morphology changes to barrier properties of polymeric packaging for microwave-assisted thermal sterilized food. J. Appl. Polym. Sci. 134 (44), 45481.

Zhang, H., Patel, J., Bhunia, K., Al-Ghamdi, S., Sonar, C.R., Ross, C.F., Sablani, S.S., 2019. Color, vitamin C, β-carotene and sensory quality retention in microwave-assisted thermally sterilized sweet potato puree: effects of polymeric package gas barrier during storage. Food Packaging Shelf Life 21, 100324.

Innovations in Australia—A historical perspective

19

Janet L. Paterson
The University of New South Wales, Sydney, Australia

19.1 Introduction

Although operating on completely different principles from European agriculture, the Australian aboriginal nations established sustainable food supply chains that supplied their needs for a nourishing diet for large and small gatherings. Using stone, clay, wood, fiber, bone, and shell they constructed shelters, planted and harvested crops, processed and stored food, used fire to manage the environment in such a way that fences for animals were unnecessary and constructed engineering works: mines, dams, wells, channels, and traps.

Postcolonial food supply followed British practices. Many early food engineering principles were devised to overcome the distance from European markets.

Critical engineering developments in 1880 to 1900 were:

- Roller milling of cereals, specifically wheat
- Pasteurization of milk and cream
- Refrigeration of meat and dairy products
- Cool storage of fruit
- Introduction of mechanical dehydration

Into the 20th century scientific research institutes investigated principles of food preservation, production and control, and the use of instruments and machinery.

This chapter gives a historical perspective of progress in food engineering in Australia and outlines how some of these technologies have had global impact. There have been significant other developments in food engineering in Australia during the last 50 years and although some of these are briefly mentioned, these are not covered in depth in this chapter.

19.2 Aboriginal food engineering

19.2.1 The food supply

The most recent estimate for the minimum duration of aboriginal people in Australia is 65,000 years (Clarkson et al., 2017).

"Aboriginal people did build houses, did cultivate and irrigate crops, did sew clothes and were not hapless wanderers across the soil, mere hunter-gatherers" (Pascoe, 2014a).

There are repeated references in European journals to people building dams and wells; planting, irrigating, and harvesting seed; preserving the surplus and storing it in houses, sheds or secure vessels; and manipulating the landscape.

By bringing a food production perspective over millennia, we can catch a glimpse of Australia as aboriginals saw it (Pascoe, 2014b, 2014c; Pascoe, 2018).

19.2.2 Large-scale engineering works and traditional fish preservation

Aboriginal people dammed rivers and swamps; they cut channels through watersheds; they used fire to replace one plant community with another (Gammage, 2011a; Anon b).

For example, an understanding of water flow; fish spawning cycles; weather patterns; and food supply and distribution are all explicitly illustrated in extensive fish trap arrays. Two sites, Brewarrina in NSW and Budj Bim in Victoria, involve stone construction of water channels and dams for fish farming, but have different construction methods. Both sites are still in operation.

Brewarrina is the oldest village on earth (Pascoe, 2018) and the *Brewarrina fish traps* are over 40,000 years old and one of the oldest man-made structures on earth. This elaborate network of rock weirs and pools stretches for around half a kilometer along the riverbed and was built by ancient tribes to catch fishes as they swam upstream (Tan, 2015). A simulation of the structure and operation of the traps can be seen (Tan, 2015). The weirs are constructed by a drystone wall method and the traps depend on variable flow to enhance and retard the movement of the fish. The rocks surround 12 teardrop-shaped pools across half a kilometer (Fig. 19.1; Tan, 2015). One family's weir is 200 feet

FIG. 19.1

Brewarrina fish traps 1880–1923. Reproduced with permission from Powerhouse Museum Sydney (Tan, 2015).

long and 5 feet high (Gammage, 2011b). In the past, fishes were herded through small openings that the locals would quickly close shut with a few rocks. The pen walls are at different heights, allowing them to be used at different water levels, and have proved resistant to damage in the face of high and fast water flows (Tan, 2015).

When these fish traps were added to Australia's national heritage list in 2005, they were described as "the largest traps recorded," showing a thorough understanding of "dry stone wall construction techniques, river hydrology, and fish ecology." Owing to their size, design, and complexity, they were considered "exceptionally rare" (Tan, 2015). The site *Brewarrina Aboriginal Fish Traps* was listed on the New South Wales State Heritage Register in 2000 (Anon a). It was designated a World Heritage Site in 2000 (Engineers Australia, 2019).

Bill Gammage quotes William Mayne who wrote in in 1848 of the Darling River at Brewarrina:

"In a broad but shallow part . . . . where there are numerous rocks, the Aborigines have formed several enclosures or Pens . . . . into which the fish are carried. To form these must have been the work of no trifling labour, and no slight degree of ingenuity and skill must have been exercised in their construction, as I was informed by men who had passed several years in the vicinity, that not even the heaviest floods displace the stones forming these enclosures" (Brown, 2019; Gammage, 2011c).

The *Budj Bim Cultural Landscape* in South-Eastern Victoria consists of three serial components (Fig. 19.2) containing one of the world's most extensive and oldest aquaculture systems (UNESCO, 2019).

It is a rugged lava flow terrain of basalt rises, swampy depressions, and waterways formed as a result of the eruption of Mt Eccles (Budj Bim) at least 30,000 years ago (McNiven, 2019). The lava flows provide the basis for the complex system of channels, weirs, and dams developed by the Gundidjmara to catch kooyang or short-finned eel—*Anguilla australis* (Fig. 19.3). Young eels grew in wetlands for 10–20 years (live storage) (McNiven, 2019; McNiven and Bell, 2010). Mature eels have high calorific value and protein; returning to the sea to spawn they were caught in conical woven traps (Bulith, 2002; McNiven and Bell, 2010). The extensive Aboriginal fish-trapping systems contain hundreds of meters of excavated channels and basalt block dam walls (McNiven et al., 2015). The basalt blocks moved weighed many hundreds of tones (McNiven, 2019).

The highly productive aquaculture system (now Lake Condah) provided an economic and social base for Gunditjmara society for six millennia. This interrelationship of Gunditjmara cultural and environmental systems is documented through present-day Gunditjmara cultural knowledge, practices, material culture, scientific research, and historical documents. It is seen in the aquaculture system itself and in the interrelated geological, hydrological, and ecological systems (UNESCO, 2019).

The Gunditjmara harnessed the wetlands on the Budj Bim lava flow by creating, modifying, and maintaining an extensive hydrological engineering system that manipulated water flow in order to trap, store, and harvest kooyang that migrate seasonally through the system. The key elements of this system are the interconnected clusters of constructed and modified water channels, weirs, dams, ponds, and sinkholes in combination with the lava flow, water flow, and ecology and life-cycle of kooyang. The eels were harvested with conical nets designed so that larger eels were caught and smaller moved to another pond. Large eels were smoked in hollows of the Manna Gum tree (*Eucalyptus viminalis*) and traded (UNESCO, 2019).

With attention to the principles of mechanics they cut 300 m canals into bedrock, 50 m aqueducts a meter high, dug kilometers of channels to join and extend eel ranges and abundance. Even in drought

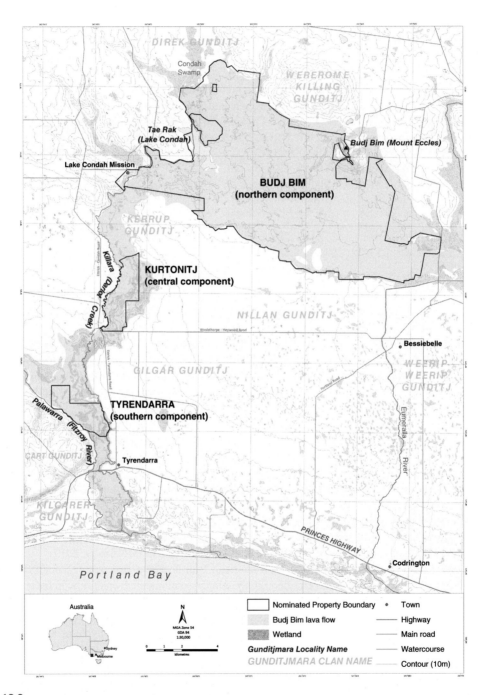

FIG. 19.2

Budj Bim Cultural Landscape – Lake Condah, Victoria (UNESCO, 2019).

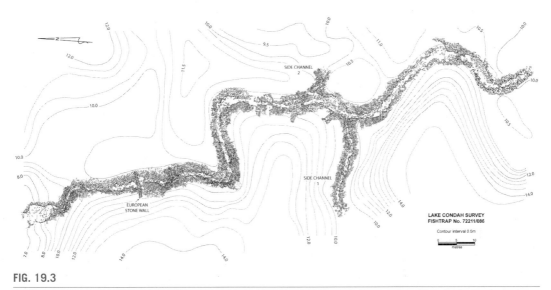

FIG. 19.3

A 200-metre long fish trap channel mapped by Peter Coutts's team at Lake Condah. Victoria Archeological Survey. Image supplied by Professor Ian McNevin, Monash University (McNiven, 2019).

or flood the systems regulated flows so that pot traps worked in water coming or going (Gammage, 2011d). This, the world's oldest aquaculture, is still in use. UNESCO designated Budj Bim a World Heritage Site in 2019 (UNESCO, 2019) and Engineers Australia (EA) recognized it as one of the country's top engineering achievements (Brown, 2019). EA CEO Peter McIntyre said the UNESCO World Heritage listing is welcome news, and highlights the important contributions of Australia's first engineers. He said, "Budj Bim is an extraordinary feat of engineering by the Gunditjmara people. For thousands of years, engineers have been using the tools available to them to improve lives and build communities" (Brown, 2019).

For maps, site views, a video of fish movement and an account of the UNESCO bid see Engineers Australia in (Brown, 2019) and Google 'Brewarrina Aboriginal Fish Traps designated as World Heritage Site'.

19.3 Colonial and postcolonial food engineering in Australia

19.3.1 The beginning

The British colonized Australia from 1788, bringing food habits and embryonic food technology of the Britain of that time. The fertility the English pastoralists encountered on first entering the country was the result of careful management (Pascoe, 2014d; Gammage, 2011e). Although starvation threatened the colonists they ignored the traditional knowledge of aboriginal people (Farrer, 2005a).

The colonists used European practices: keeping herds and farming; storing dry cereal crops; drying and smoking meat; salting fish and vegetables; drying fruit; milk stored as cheese; sugar and jams; alcoholic fermentation of juices; distillation; milling using water-powered stone mills; screening milled grain; bread-making. Advances in food processing and preservation in Europe were applied in Australia. Steam milling, introduced in 1815, became common in the late 19th century and canning in the mid-19th century (Farrer, 2005b).

19.3.2 Australian innovation in food engineering and technology

19.3.2.1 Overview

"Throughout the 19th century food science and technology everywhere developed slowly in a series of unequal commodity-oriented steps isolated from each other. In Australia this happened first in the self-contained village technologies and, from the middle of the 19th century, with the newer ones. Heat processing began later than it did in the United States but expanded rapidly with some incremental innovation. In refrigeration, however, Australia and New Zealand were world leaders. Despite the constraints of distance, travel time, and very limited educational opportunities, overseas advances in the sugar, milling and brewing industries in the latter part of the 19th century were adopted quite quickly. Australia was innovative in canning and dehydration; science [and engineering] was applied in the control of the sugar and brewing industries, and in the introduction of compositional standards for foods; and Australian research contributed to the scientific knowledge of yeasts and, especially, cereals. In general, the advances made, many of them in engineering and equipment, were isolated within their commodity groups. The milling, sugar and dairy industries illustrate this isolation, although the use of the vacuum pan did become common to sugar and dairy production. Yeast had, of course, long been essential for producing fermented beverages and bread, and refrigeration was rapidly applied to more than meat. But there is no study of the food systems themselves" (Farrer, 2005c).

The principal forces driving innovation were the distance from British markets, and the necessity of shipping foods through the tropics.

19.3.2.2 Meat canning for the British market

In 1843 the Sydney Salting Company introduced meat salting by immersion in brine. Gases were removed by vacuum (Farrer, 2005d). There were advances in filling and sealing. In 1845–1846 whale oil instead of calcium chloride brine was introduced as a heating medium for sealed cans. Town gas was used for heating in factory scale production. Elevated temperature rooms were used for accelerated testing of canned meat stability (Farrer, 2005e). Can manufacture was improved in 1864 by cutting and bending tinplate and devising new soldering tools. By the 1870s processed cans were water-cooled to reduce over-processing, there were innovations in production lines and waste treatment, and the outside of tinplate cans were lacquered with a methylated spirits solution of the gum of the grass tree (genus *Xanthorrhoea*; Farrer, 2005f).

19.3.2.3 Refrigeration

Development of meat freezing, storage, and transport. Markets preferred unprocessed whole meat to canned meat. Technical challenges caused by distance of Australia and New Zealand from Europe and shipping through the tropics drove research into mechanical refrigeration.

Milk was chilled to improve the quality for domestic consumption.

In 1851 James Harrison in Geelong, Victoria had an ethyl ether compressor patent from 1855 and produced the first successful ice-making plant in the world (Farrer, 2005g; Thévenot, 1979). E.D. Nicolle's first ice-making patent was registered in 1861 and, with the financial support of Thomas

Sutcliffe Mort, in 1862 the Sydney Ice Company (later the New South Wales Fresh Food & Ice Co.) achieved heat exchange by the liquification of ammonia. Mort continued his support over the next 10 years. In that time they designed, patented, and produced a variety of refrigerating devices, including one for domestic use and another for making powdered milk, equipped a large cold-store at Darling Harbour, Sydney, in 1872, and produced refrigerated railway vans for meat and milk, but failed to develop a machine ideally suited to ships. The seven main apparatuses Nicolle constructed relied, successively, on systems based on ammonia absorption, air expansion, low pressure ammonia absorption, and ammonia reabsorption. He was a pioneer in developing these heat exchange systems and the mechanical contrivances by which they were made effective (Barnard, 1974; Thévenot, 1979).

Principles of refrigeration by air expansion were presented to the Royal Society of Victoria in 1868 (Farrer, 2005h).

The first frozen storage facilities in the world (1875), although built for meat in Darling Harbour, Sydney, were used for milk, and cooled milk depots were established. During these years, the main impediment to a successful meat and milk supply chain was overland transport over long distances in a hot climate to freezing works and to ships. Iced railway vans for milk and meat began operating in the 1880s (Farrer, 2005h).

Research became concentrated on effective international transport. The Northern hemisphere had natural cold sources and short transport times in temperate zones. Canadian meat was frozen in winter with natural ice and was insulated for shipping to Europe (Farrer, 2005i).

In Sydney (1877) the installation of mechanical refrigeration by Nicolle and Mort in the ship *Northam* was partially successful but the first successful international shipment of mutton frozen on land was from South America to France, using ammonia absorption refrigeration in the *Paraguay* (Farrer, 2005j).

In 1879, meat was frozen on board the *SS Strathleven*. The refrigeration system was based on air compression and expansion, a controlled closed system, insulated with charcoal. The voyage was 59 days from Melbourne to London. Despite variable cold-room temperatures during the voyage, the meat arrived in excellent condition (Farrer, 2005k). Australian freezing works were built in Melbourne (1879) for frozen meat export.

In 1882 the sailing ship, *Dunedin* used air-cycle refrigeration (Bell-Coleman). Frozen lamb and mutton carcasses arrived in London in perfect condition after a 98-day voyage from New Zealand (Farrer, 2005k).

There was successful frozen meat export from Australia from the 1890s (Farrer, 2005k).

Development of meat chilling, freezing, storage and transport. Despite the success of Australia in transporting frozen meat to distant markets, there were economic and technical difficulties. For example, although the Australian Frozen Meat Company sent carcasses direct to terrestrial freezing works, transport problems still dominated the supply chain. Frozen carcasses were often delayed between freezing and ships. Although freezing plants were established in country towns near the ports to minimize transport of fresh carcasses, refrigerated rail transport, and frozen storage in port towns were still inadequate (1886) (Farrer, 2005k).

Drought, high cargo rates and lower prices in England, fire, and cyclones affected some Australian businesses. Because of high costs and low prices for meat, freezing works were converted to butter export in 1886 (Farrer, 2005l).

Frozen meat prices fell as chilled meat became available. Freezer burn during transport and drip on thawing were a disadvantage for frozen shipping. Chilled carcasses from South America gave better products but mould developed during the longer voyages from Australia and New Zealand.

In Australia the Council for Scientific and Industrial Research (CSIR) meat group prolonged the life of chilled beef by holding it in an atmosphere of carbon dioxide (1933–1936). With reduction of microbial load on the meat and of surface water activity by drying during chilling, and transport at low temperatures, the necessary 60 days of storage required to reach Britain in good condition was achieved and Australia's meat trade was transformed (Farrer, 2005m).

After 1954 the export of chilled beef quarters from Australia ceased. In 1971 chilled cuts sealed in polyethylene liners, relying on the build-up of carbon dioxide derived from the meat, for transport at −1 °C during a 9-week voyage, were successfully transported to Britain (Vickery, 1990a).

Insulated refrigerated rail cars were cooled with ice or ice and salt until about 1950. E.W. Hicks of the Commonwealth Scientific and Industrial Research Organisation (CSIRO), modeled temperature variability and thermal capacity during the 1950s. Mechanical refrigeration was more expensive than ice. Liquid nitrogen (1965) gave good temperature control but was even more expensive. Road and rail transport relied on the load being as cold as possible before loading. Temperature distribution, air movement, and controlled atmosphere in containers were investigated at CSIRO until the 2010s (Vickery, 1990b).

Improved hygiene was essential. CSIRO research into rapid freezing, postmortem pH, and controlled atmosphere (10% CO_2) benefited the industry by mid-20th century (Farrer, 2005n).

Development of cool storage of fruit. There was research into optimal storage temperatures of fruit. Successful shipment of fruit to Britain from Australia occurred from 1885. Cold stores for fruit, ice, and other produce were developed in the 1920s (Farrer, 2005n).

Domestic refrigeration. Australia developed the world's first domestic ice-making machines in the 1870s. Although Australian domestic mechanical refrigerators were developed in 1884 they were not widespread until the mid-20th century (Farrer, 2005o).

The *Coolgardie safe*, developed in Western Australia in the late 1890s, is a low-tech food storage unit for cooling and prolonging the life of food by water evaporation from a hessian cover. It was a common household item in Australia until the mid-20th century (Anon a).

19.3.2.4 The dairy industry
Initially the quality of Australian milk, cream, and butter was very low and the shelf-life very short. The understanding of hygienic practices, the availability of refrigeration, centrifugal separation, and pasteurization improved the originally poor quality of Australian dairy products. These technologies were adapted from overseas practices. Improvement of refrigeration in the dairy industry permitted better microbial quality of dairy products and export of butter and cheese. Dairy product quality improved with the introduction of centrifugal cream separation (Farrer, 2005p).

Control of butter making was improved by use of the lactometer to measure specific gravity and the lactocrite to determine fat (separating fat centrifugally from acidified milk in graduated tubes). This test, devised in Australia, was a precursor of the empirical Babcock test (USA, 1980) which was used for dairy factory control and was the basis for the payment of farmers for nearly a century (Farrer, 2005q).

Cottage cheese whey, which has the disadvantages of high salt and acidity, was successfully concentrated by nanofiltration to produce a product that, up to 50% milk solids nonfat (MSNF), was a satisfactory substitute of skim milk powder in ice cream; University of Western Sydney 1997 (Nguyen et al., 1997).

The CSIRO has improved the production process for whey protein isolate (Filip Janakievski, personal communication).

19.3.2.5 Sugar milling and refining

The first use of mass balance calculations for the control of a manufacturing operation was in Australian sugar processing in 1884. In 1899 the improvement of chemical analysis enabled payment to farmers for the sugar content of their cane (Farrer, 2005r).

The Sugar Research Institute (Mackay, Queensland) and others studied the physical chemistry of sugar boiling and phase changes in crystallization, enabling process control using electrical conductivity measurements that were implemented in the Kalamia mill in 1937 (Farrer, 2005s).

The *Donnelly chute* transports shredded cane, separating bagasse (sent as fuel to the boiler), and juice (sent to clarifier). The amount of cane fiber carried by cane carrier in a sugar mill varies because of nonuniformity of cane supply. The continuous variation of cane in the Donnelly chute is controlled by capacitive type sensor and solid-state digital indication system especially designed for measurement and display of Donnelly chute level (Energy Ventures Pty Ltd, 2017; Misra and Kamath, 2014).

Research continues in this field. *The Sugar Research Institute (SRI) Chute Height Sensor* developed in 2017, accurately measures cane height in chutes to control sugar mill feed rate, maximize sugar extraction, lower mill stress, and energy use. The sensor is an essential factory instrument used for the control of sugar mill throughput via mill speed, carrier speed, or Donnelly chute settings. The height is measured via stainless steel electrodes installed in the chute wall that determine the presence of material at each electrode. The *SRI Chute Height Sensor* determines the number of electrodes covered and alerts the mill control system (Sugar Research Institute Chute Height, 2017).

Pressure feeder chutes are used in a sugar cane milling unit to transfer bagasse from one set of crushing rolls to a second set to increase the amount of cane that can milled. The continuous pressure feeder was developed to provide a constant feed of bagasse under pressure to the mouth of the crushing mills. It was introduced in 1940s. Research continues to optimize performance (Sethuraman, 2011).

Mechanical harvesting of sugar cane. Sugar cane was traditionally burnt to remove leaves and trash, and harvested by hand using a machete, originally using indentured Pacific Islander laborers (Figs. 19.4 and 19.5; Levy, 2010; Wilkins, 2015).

In 1979, Australia achieved 100% conversion to mechanical cane harvesting.

Mechanical cane harvesters were developed in a number of countries over the years but Austoft in Australia was one of the first companies to market a viable harvester and went on to build an export market. Zelmer shows a sectional diagram of the mechanical harvester's operation (Zelmer, 2006).

The harvester in Fig. 19.6 gathers and cuts a row of cane, removing the leaves and chopping the cane into segments that are delivered into a bin moving alongside. The cane is transported to the crushing mills quickly to minimize enzymic degradation of the sugar.

The NSW Sugar Milling Co-operative, in partnership with Delta Electricity, saves 400,000 tons of greenhouse gas each year by cogeneration of electricity from waste cane fiber (Vallejo, 2009).

19.3.2.6 Cereals: grain milling and production

Pneumatic malting and drying of barley for brewing in 1874 permitted control of moisture, temperature and airflow, saved space and labor, reduced spoilage, and made malting possible over the whole year instead of only in cold months (Farrer, 2005t).

F.B. Guthrie, appointed chemist to NSW Department of Agriculture in 1892, developed a laboratory flour mill and techniques which, with the measurement of chemical properties, enabled him to produce 100 g samples of flour and predict from them how the wheat would behave commercially. Experimental

FIG. 19.4

Burning sugar cane in Australia (Levy, 2010).

plots of wheat plants could be smaller and screened for the milling and baking properties of new varieties before wheat for full milling trials was grown (Farrer, 2005u).

Production in 1925 of vital (undenatured) gluten by flash drying of starch and gluten changed bread-making (Farrer, 2005v; Energy Ventures Pty Ltd, 2017).

The Hastings machine for automatic and continuous production of leavened products baked from below was patented in 1947, followed by the De Jersey machine for continuous production of double-sided baking (Farrer, 2005w).

Dough rheology. Hibberd and his colleagues at the Bread Research Institute, NSW, devised a parallel plate rheometer in 1966 for the difficult study of the flow properties of wheat flour doughs. A series of mathematical models were developed and used to assess the influence of the nature of the starch granule surface, the granule size distribution, and the starch content on the rheological properties of doughs. Other effects of the composition of doughs (yeast or no yeast; oxidizing and reducing agents; cross-linking agents and commercial improvers) were investigated using this rheometer (Vickery, 1990c; Hibberd et al., 1996). Similar studies were later extended into the 1990s by Dr. Roger Tanner and the research group in the School of Mechanical Engineering, University of Sydney.

19.3.2.7 Dehydration

Freeze drying was practiced by indigenous people in the Andes for centuries.

Systematic research was done in the Cambridge Low Temperature Research Station (UK). Australian research in CSIRO Division of Food Preservation worked to improve the basic concept by reducing the rapid fall in drying rate that occurred soon after the start of the drying process. Mellor (1969)

FIG. 19.5

Cutting burnt sugarcane by hand, cropped (Wilkins, 2015).

patented equipment that reduced the drying time by 25–45% by applying pulsed pressure that increased both sublimation and diffusion rates (Vickery, 1990d; Mellor, 1969).

Bell and Mellor (1986) described a novel isothermal adsorption freeze dryer for removing water from biological material at ultra-low temperatures (Bell and Mellor, 1986).

Rice drying: a case study of a successful collaboration [Source: Australian Centre for International Research (ACIAR)/University of New South Wales (UNSW) Drying. 1980s to 2010s].

Background. Rice is the dominant staple food for South-East Asia. A high-yield and faster-maturation hybrid rice was developed in the 1970s. Two harvests a year meant that although the first harvest could be sun-dried during the dry season (the traditional method) the second harvest that matured during the wet season could not.

In response to national postharvest institutes in Malaysia, Thailand, and the Philippines, ACIAR established a project with UNSW, the Australian Ricegrowers Cooperative Ltd, and the Asian partners. ACIAR's philosophy was that Australian scientists did not simply supply solutions but worked with national organizations in the partner countries to build research capacity.

The ACIAR strategy was to construct a scientific basis for development of solutions. Each research team was first equipped with suitable equipment for determining the mechanical and thermal properties of paddy rice. In particular, they determined the rice drying rates under a wide range of different air conditions, and the isothermal moisture properties of rice. Several designs of dryers were investigated, two of which were chosen: a mobile high-moisture dryer to reduce moisture in the field; and deep bed dryers, suitable for economical slow drying over longer time periods. The Thai team also developed a solar grain dryer.

Computer modeling. Programs for a few of the main types of dryers were developed in Dr. Robert Driscoll's research program at UNSW. The simulations used heat and mass transport equations and

FIG. 19.6

Mechanical harvesting of sugar cane. Queensland, 2000. Source: Janet Paterson.

the thermal and physical properties of the rice as measured by the research teams. Rates of product deterioration were modeled, based on measurements of dry matter loss, a measure of respiration rates. The models used real weather data: it was the unfavorable weather conditions that were responsible for the problem in the first place. Simulation results were checked against pilot tests done in each of the three countries. This allowed the teams to run tests over a wide range of conditions very quickly, and so develop design solutions.

Outcomes. The first major outcome, a mobile fast dryer that used grain recirculation, was based on simulation predictions of drying rate for a falling curtain of grain in a fast air flow, designed in the Philippines. By the end of the project, around 1400 of these dryers had been constructed by local dryer manufacturers and distributed to areas using double cropping. For high moisture grain, fluidized bed dryers were chosen. They had a similar economy to world standard column dryers, but with faster drying rates and smaller space requirements. Energy efficiencies were optimized by air recirculation using a cyclone separator. The challenges to make the fluidized bed dryer a successful commercial solution were many. A temperature region that prevented rice fissuring had to be found, and careful pressure calculations to allow recirculation of high-speed air. Bed depth and mass flow rates had to be optimized. The solution was a commercial success, with substantial adoption in Thailand and extension to other South-East Asian countries.

The second major recommendation was to use in-store dryers at larger collection centers for slow efficient drying of rice under 19% moisture. These dryers are used extensively in Australia for rice. They achieve efficiencies typically well over 100% by using the natural drying capacity of available air. For South-East Asia, careful limits were placed on maximum bed depth, best air speed, amount of air heat required and allowable grain height in order to prevent grain deterioration under humid conditions. They were tested extensively in each country before commercialization in the Philippines and Thailand.

The major outcomes were the development of in-country resources and training which came from the unique philosophy of working not for countries, but with them, so that research capacity was enhanced and the solutions were owned by the people who needed them (Dr. R.H. Driscoll, UNSW, Personal communication).

19.3.2.8 Packaging

CSIRO teams worked on flexible packaging for vegetable products from 1949. They studied permeability of films to various gases, particularly oxygen and sulfur dioxide, and moisture. In 1977 they developed an improved design of cell for measuring permeability that was modified over the following decades as new instruments became available and new chemical reactions were understood.

During the 1990s and 2000s, CSIRO researchers developed new films for use as active flexible packaging that extended shelf-life by scavenging oxygen and permitting some atmospheric exchange (Vickery, 1990e; Rooney, 2005).

19.3.2.9 Separation

The spinning *cone column* was developed by a CSIRO group led by Don Casimir, in collaboration with industry. An Australian patent was granted in 1980.

Mass transfer between liquid and gas phases depends on turbulence induced not only in the liquid but also in the gaseous phase. Efficiency is increased. Modified spinning cone columns are used for the desulfuring of fruit juices, the stripping of essences at low temperature, and the removal of alcohol from alcoholic fluids (Vickery, 1990e).

This technology was commercialized by Flavourtech Pty Ltd who successfully installed this equipment in many countries. Further information on the design and operation, including explanatory figures and videos, are available from Flavourtech and ConeTech (Flavourtech Product Guide, 2019; ConeTech's Spinning Cone, 2014).

An efficient Counter-current extractor (CCE). The CSIRO-Howden CCE equipment, designed by Don Casimir in 1980, uses an inclined trough containing a rotating screw which is intermittently reversed. The extraction fluid enters at the higher end and the extract leaves at the lower end. The raw material enters from the lower end and flows in the opposite direction to the extraction fluid. The screw movement ensures multiple contacts between solid and extracting fluid, which may be hot or cold. Many materials can be extracted: fruit pomace and peels; residues from poultry, fish, and crustacea to recover juice, stock, fat, and soluble proteins (Vickery, 1990e). A description and diagram of operation can be found in (Casimir and Lang, 1985).

Australia has been markedly successful in the development of *membrane* technologies for food applications (Table 19.1).

Continuous chromatographic separation (CSEP) in protein purification. Australia's largest dairy company commercialized a novel simulated moving bed (SMB) ion exchange technology based on CSEP for the commercial manufacture of whey protein-derived food ingredients for application in a variety of functional and sports nutrition products. CSIRO's pilot-scale SMB process separated lactoferrin and bovine blood plasma proteins; and glycomacropeptides from cheese whey (Vickery, 1990a; Janakievski et al., 2016).

19.3.2.10 High-temperature short-time (HTST) canning

Flame sterilization. Filled cans are spun while held stationary over flames to improve the rate of heating of the can's contents. Aust. Pat. 469-637, 1971.

In association with Tarax Pty Ltd and Lohning Bros Pty Ltd, Don Casimir (CSIRO) developed a reversing, spin flow sterilizer in which cans, preheated in steam, are heated in a gas flame while spinning

Table 19.1 Australian innovation in industrial membrane processes. Modified from (Nguyen, 1996).

Research, implementation	Food applications that have achieved commercialization
CSIRO Dairy Research Laboratories	Ultrafiltration in cheese base production
	Cheddar Cheese commercial plant: a continuous process
AVP—Sirocurd, 1987	with 8% increased yield
	Ultrafiltration to preconcentrate gelatin prior to evaporation and recovery of gelatin from waste streams
Davis—UNSW, 1994	Concentrate and refine liquid pectin
Pectin Aust—UWSH, 1987	Process water treatment
	Wastewater treatment
UNSW—Memtech crossflow hollow fiber microfiltration system, 1987	Fruit juices clarification
Syrinx Research. Hollow fiber modules—Nguyen et al., 1989	Wine, vinegar clarification
	Desalting whey and molasses
	Liquid food concentration
Syrinx Research. Osmotic distillation, 1991	Grape juice concentration
Balstone Technology	Cane juice clarification
Aquapore	Water treatment

and finally spin-cooled in a water spray. The very high temperature of the flame combined with thin boundary layers in the turbulent contents increase the rates of heating and cooling, producing much better quality products (Vickery, 1990e) and is described in detail in (Casimir, 1975).

Induction heating. The same temperatures and turbulence were achieved in 1999 by Peter Rutledge, consultant food engineer (ex CSIRO) in association with Pacific Power (Australia) and very high temperatures were produced directly in spinning steel cans heated by induction. This Australian invention produces a novel HTST process and was patented worldwide in 2001 (Rutledge, 2001). Induction heating was an improvement on flame heating because there were no hot surfaces in the equipment, nor any flames. Induction therefore increased worker safety and minimized heat wastage to atmosphere. The models of heat transfer were similar.

In 2002–2003, a prototype of this induction heating system was evaluated at UNSW. This included homogeneous (fluid) and heterogenous (fluid plus solid vegetable pieces) to determine the optimum operating conditions that included headspace, proportion of solids, viscosity, optimum power input, power number, heat transfer, rotational speed, time, temperature, and cooling processes (Isjwara, 2003; Fig. 19.7).

19.3.2.11 Recent research

CSIRO has promoted and assisted in the introduction of the first high pressure processing (HPP) systems in Australia for cold pasteurization of juices and meat products and there are currently at least four plants operating in Australia. CSIRO also developed a processing canister enabling the use of existing cold HPP systems for high pressure thermal processing (HPTP) and now following steps closer to commercialization. This drop-in innovation is transforming HPP from a technology with limited applications to one unlocking new and exciting market opportunities for the preservation and shelf-life extension of foods and beverages in shelf- and chill-stable, low-acid food/beverage categories and for modulating the texture of foods, such as the tenderization of low-value meat cuts and the toughening of seafood products. The canister has been patented (Container for use in Food Processing, 2016).

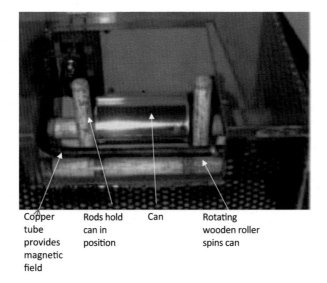

Copper tube provides magnetic field | Rods hold can in position | Can | Rotating wooden roller spins can

FIG. 19.7

Experimental rig to test model operating conditions. The supplied electricity was controlled for current, voltage, waveform and time. The rotational speed was controlled by rotation of a wooden roller. The copper tube supplied the current that induces high temperature in the spinning steel can. Source: Janet Paterson.

Recent CSIRO research that has had commercial impact includes CSEP (Janakievski et al., 2016; Vickery, 1990a), encapsulation of fish oils (Augustin et al., 2006), and the use of megasonics for the extraction of palm and olive oil (Juliano et al., 2017).

19.4 Conclusion

Australia has made an important contribution in food engineering through the whole supply chain (from ancient agricultural and food preservation practices within indigenous communities through to innovative process and packaging technologies) that has had global impact. The agrifood economy will continue to be important to Australia's prosperity and Australian food engineers in industry, universities, and research institutes will continue to innovate to sustain Australia's prosperity into the future.

Acknowledgments

Recollections of researchers who provided information and contacts: Robert Driscoll, UNSW; Minh Nguyen, WSU; Jay Sellahewa, Kai Knoerzer, and Filip Janakievski, CSIRO; Anthony Mann, QUT; Ian J. McNiven, Monash University; Don Cleland and Richard Archer, Massey University, New Zealand; Mohammed Farid, University of Auckland, New Zealand.

References

Anon a, No date. Coolgardie Safe. Wikipedia. https://en.wikipedia.org/wiki/Coolgardie_safe (accessed February 19, 2020).

Anon b, No date. Brewarrina Aboriginal Fish Traps. https://en.wikipedia.org/wiki/Brewarrina_Aboriginal_Fish_Traps (accessed 18 February 2020).

Augustin, M.A., Sanguansri, L., Bode, O., 2006. Maillard reaction products as encapsulants for fish oil powders. J. Food Sci. 71 (2), E25–E32.

Barnard A., 1974. Australian Dictionary of Biography Vol 5 MUP Mort, Thomas Sutcliffe (1816–1878). https://adb.anu.edu.au/biography/mort-thomas-sutcliffe-4258 (accessed February 19, 2020). Nicolle, Eugène Domenique (1823–1909). http://adb.anu.edu.au/biography/nicolle-eugene-dominique-4304 (accessed February 19, 2020).

Bell, G.A., Mellor, J.D., 1986. Development of the adsorption freeze-dryer. CSIRO Food Res. Quart. 46 (3), 56–58.

Brown, R., 2019. On line journal Create, 6000-year-old Aboriginal engineering feat named to World Heritage list. https://www.createdigital.org.au/6000-year-old-aboriginal-engineering-feat-named-world-heritage-list/ (accessed February 19, 2020).

Bulith, H., 2002. Gunditjmara environmental management. The Development of a Fisher-Gatherer-Hunter Society in temperate Australia. In: Crier, C., Kim, J., Uchiyama, J. (Eds.), Beyond Affluent Foragers, Oxbow Books, Oxford, 10–11 4-23.

Casimir, D.J., Lang, T.R., 1985. Counter-Current Extraction of Foods. ASEAN Food J. 1 (4). https://csiropedia.csiro.au/wp-content/uploads/2015/01/6226863.pdf (accessed February 14, 2020).

Casimir, D.J., 1975. CSIRO Food Res. Q. 35:34-9, 63-7 36:25-8. Reference details in J.R. Vickery. Food Science and Technology in Australia. A review of research since 1900, CSIRO Australia, 1990, pp. 138, 142.

Clarkson, C., Jacobs, Z., Pardoe, C., 2017. Human occupation of northern Australia by 65,000 years ago. Nature 547, 306–310.

ConeTech's Spinning Cone Column, 2014. https://www.youtube.com/watch?v=5gVVu2-2AmY (accessed February 2, 2020).

Container for use in Food Processing, Australian patent: AU2016310416; European patent: EP3341298A4 or world patent application number: WO2017031552A1. Publication year 2016.

Engineers Australia, 2019. Wonders Never Cease. Book of 100 Australian Engineering Achievements. Publisher EA Books, Sydney, Australia. ISBN978-1-925627-45-9.

Farrer, K., 2005a. To Feed a Nation: A History of Australian Food Science and Technology. CSIRO Publishing, Australia, pp. 9.

Farrer, K., 2005b. To Feed a Nation: A History of Australian Food Science and Technology. CSIRO Publishing, Australia Chapter 2.

Farrer, K., 2005c. To Feed a Nation: A History of Australian Food Science and Technology. CSIRO Publishing, Australia, pp. 201 edited.

Farrer, K., 2005d. To Feed a Nation: A History of Australian Food Science and Technology. CSIRO Publishing, Australia, pp. 29.

Farrer, K., 2005e. To Feed a Nation: A History of Australian Food Science and Technology. CSIRO Publishing, Australia, pp. 32.

Farrer, K., 2005f. To Feed a Nation: A History of Australian Food Science and Technology. CSIRO Publishing, Australia, pp. 41.

Farrer, K., 2005g. To Feed a Nation: A History of Australian Food Science and Technology. CSIRO Publishing, Australia, pp. 51.

Farrer, K., 2005h. To Feed a Nation: A History of Australian Food Science and Technology. CSIRO Publishing, Australia, pp. 52.

Farrer, K., 2005i. To Feed a Nation: A History of Australian Food Science and Technology. CSIRO Publishing, Australia, pp. 53.

Farrer, K., 2005j. To Feed a Nation: A History of Australian Food Science and Technology. CSIRO Publishing, Australia, pp. 52, 53.

Farrer, K., 2005k. To Feed a Nation: A History of Australian Food Science and Technology. CSIRO Publishing, Australia, pp. 54.

Farrer, K., 2005l. To Feed a Nation: A History of Australian Food Science and Technology. CSIRO Publishing, Australia, pp. 55.

Farrer, K., 2005m. To Feed a Nation: A History of Australian Food Science and Technology. CSIRO Publishing, Australia, pp. 124.

Farrer, K., 2005n. To Feed a Nation: A History of Australian Food Science and Technology. CSIRO Publishing, Australia, pp. 57.

Farrer, K., 2005o. To Feed a Nation: A History of Australian Food Science and Technology. CSIRO Publishing, Australia, pp. 58.

Farrer, K., 2005p. To Feed a Nation: A History of Australian Food Science and Technology. CSIRO Publishing, Australia, pp. 57.

Farrer, K., 2005q. To Feed a Nation: A History of Australian Food Science and Technology. CSIRO Publishing, Australia, pp. 102.

Farrer, K., 2005r. To Feed a Nation: A History of Australian Food Science and Technology. CSIRO Publishing, Australia, pp. 63.

Farrer, K., 2005s. To Feed a Nation: A History of Australian Food Science and Technology. CSIRO Publishing, Australia, pp. 135.

Farrer, K., 2005t. To Feed a Nation: A History of Australian Food Science and Technology. CSIRO Publishing, Australia, pp. 86.

Farrer, K., 2005u. To Feed a Nation: A History of Australian Food Science and Technology. CSIRO Publishing, Australia, pp. 111.

Farrer, K., 2005v. To Feed a Nation: A History of Australian Food Science and Technology. CSIRO Publishing, Australia, pp. 83.

Farrer, K., 2005w. To Feed a Nation: A History of Australian Food Science and Technology. CSIRO Publishing, Australia, pp. 82.

Flavourtech Product Guide, 2019. Spinning Cone Column. Flavourtech. https://flavourtech.com/products/spinning-cone-column/ (accessed February 2, 2020).

Gammage, B., 2011a. The Biggest Estate on Earth: How Aborigines Made Australia. Allen and Unwin, Sydney, pp. 3.

Gammage, B., 2011b. The Biggest Estate on Earth: How Aborigines made Australia. Allen and Unwin, Sydney, pp. 283.

Gammage, B., 2011c. The Biggest Estate on Earth: How Aborigines Made Australia. Allen and Unwin, Sydney, pp. 282 quotes. William Mayne, who wrote in in 1848 (1 Jun 1848. HRA 1, 26, 635).

Gammage, B., 2011d. The Biggest Estate on Earth: How Aborigines Made Australia. Allen and Unwin, Sydney, pp. 283 quotes Robinson , R. 1842 in P&G Ford, 241.

Gammage, B., 2011e. The Biggest Estate on Earth: How Aborigines Made Australia. Allen and Unwin, Sydney (Introduction).

Hibberd, G.E., Wallace, W.J., Wyatt, K.A., 1996. A rheometer for measuring the dynamic mechanical properties of soft solids. J. Sci. Instrum. 43 (2), 43–84.

Isjwara, Aditya, 2003. MSc Research thesis. Study of Heat Transfer Rate to the Can in Induction Heating System. UNSW, Sydney.

Janakievski, F., Glagovskaia, O., De Silva, K., 2016. Simulated moving bed chromatogrraphy in food processing. In: Knoerzer, K., Juliano, P., Smithers, G. (Eds.), Innovative Food Processing Technologies. Woodhead Publishing, Cambridge, England, pp. 133–149.

Juliano, P., Bainczyk, F., Swiergon, P., et al., 2017. Extraction of olive oil assisted by high frequency ultrasound standing waves. Ultrason. Sonochem. 38, 104–114.

Levy, R.S., 2010. Wikimedia Commons. Cane fire in Australia. https://upload.wikimedia.org/wikipedia/commons/4/49/Cane_fire_in_Australia.jpg (accessed 26.02.2020).

McNiven, I.J., Bell, D., 2010. Fishes and farmers: historicizing the Gunditjmara freshwater fishery, Western Victoria. The Latrobe Journal. The State Library of Victoria Foundation 85, 83–105.

McNiven, I., 2019. Detective work behind the Budj Bim eel traps World Heritage bid. The Conversation. http://theconversation.com/the-detective-work-behind-the-budj-bim-eel-traps-world-heritage-bid-71800. (accessed 04.02.2020).

McNiven I.J., Crouch J., Richards T., Sniderman K., Dolby N., Mirring G., Traditional Owners Aboriginal Corporation, 2015. Phased Redevelopment of an Ancient Gunditjmara Fish Trap Over the Past 800 Years. Muldoon's Trap Complex. Lake Condah, southwestern Victoria. Australian Archaeology December 2015, 81 pp. 44–58.

Mellor J.D., 1969. Freeze-drying process. Aust Pat 290-845.

Misra Y., Kamath H.R., J. Autom. Control, 2014, Vol. 2, No. 3, 62-78. Available online at http://pubs.sciepub.com/automation/2/3/2 © Science and Education Publishing DOI:10.12691/automation-2-3 -2 (12) (PDF). Analysis and Design of a Three Inputs Fuzzy System for Maintaining the Cane Level during Sugar Manufacturing. Available at: https://www.researchgate.net/publication/282601987_Analysis_and_Design_of_a_Three_Inputs_Fuzzy_System_for_Maintaining_the_Cane_Level_during_Sugar_Manufacturing (accessed February 23, 2020).

Nguyen, M., 1996. Membrane processing of liquid foods. Food Aust. 48 (5), 232–233.

Nguyen, M.H., Khan, M.M.A., Kailasapathy, K., Hourigan, J.A., 1997. Use of membrane concentrated cottage cheese whey in ice-creams. Austr. J. Dairy Technol. 52, 75–78.

Pascoe, B., 2014a. Chapter 8, Accepting History and Creating the Future. Quoted From Kindle Edition Loc2442. Dark Emu. Magabala Books, WA, Australia, pp. Loc2442.

Pascoe, B., 2014b. (Introduction) Quoted from Kindle Edition Loc54. Dark Emu. Magabala Books, WA, Australia, pp. Loc54.

Pascoe, B., 2014c. Dark Emu, Kindle Magabala Books, WA, Australia, pp. Loc63.

Pascoe, B., 2014d. Dark Emu. Magabala Books, WA, Australia, pp. Loc154 (Introduction) Quoted from Kindle edition Loc154.

Pascoe, B., 2018. A Real History of Aboriginal Australians, the First Agriculturalists, TEDx Sydney. https://www.youtube.com/watch?v=fqgrSSz7Htw (accessed February 18, 2020).

Rooney, M.L., 2005. Chapter 5 Introduction to active food packaging technologies. In: Jung, H.H., (Ed.), Innovations in Food Packaging. Academic Press, Elsevier, California, USA, pp. 63–79.

Rutledge, P.J., 2001. Inductive Heating Method and Apparatus, US patent US 6,177,662 B1.

Sethuraman, P., 2011. Master's thesis QUT.

Sugar Research Institute Chute Height Sensors, 2017. https://www.sri.org.au/chute-height-sensors/ (accessed February 23, 2020). Published 2017.

S.I. Energy Ventures Pty Ltd. 2017. https://www.energyventures.in/capacitive-type-donnelly-chute-level-sensor-for-sugar-mill.php (accessed February 20, 2020).

Tan, M., 2015. The Guardian, Australian Edition. The fish traps at Brewarrina are extraordinary and ancient structures. Why aren't they better protected? https://www.theguardian.com/australia-news/2015/jul/10/fish-traps-brewarrina-extraordinary-ancient-structures-protection (accessed 18.02.2020).

Thévenot, R., 1979. A History of Refrigeration Throughout the World. International Institute of Refrigeration, Paris, France, pp. 46 62, 130, 135.

UNESCO, 2019. Budj Bim Cultural Landscape. https://whc.unesco.org/en/list/1577 (accessed 04.02.2020).

UNESCO, 2021. https://whc.unesco.org/en/list/1577 (accessed February 4, 2020).

Vallejo J., DailyTelegraph, Sydney, September 2, 2009. http://www.dailytelegraph.com.au/property/sugar-cane-makes-power-in-world-first/story-e6frezt0-1225768416177 (accessed April 15, 2020).

Vickery, J.R., 1990a. Food Science and Technology in Australia. A Review of Research Since 1900. CSIRO, Australia, pp. 145.

Vickery, J.R., 1990b. Food Science and Technology in Australia. A Review of Research Since 1900. CSIRO, Australia, pp. 152.

Vickery, J.R., 1990c. Food Science and Technology in Australia. A Review of Research Since 1900. CSIRO, Australia, pp. 36.

Vickery, J.R., 1990d. Food Science and Technology in Australia. A Review of Research Since 1900. CSIRO, Australia, pp. 137.

Vickery, J.R., 1990e. Food Science and Technology in Australia. A Review of Research Since 1900. CSIRO, Australia, pp. 138.

Wilkins, J., 2015. Wikimedia Commons. Cutting burnt sugarcane by hand. https://commons.wikimedia.org/wiki/File:Cutting_burnt_sugarcane_by_hand.jpg (accessed 26.02.2020).

Zelmer, A.C.L., (Ed.), 2006. Cane Tram Notes. https://www.zelmeroz.com/album_rail/ctn/ctn_07.pdf (accessed February 19, 2020).

Industry 4.0 and the impact on the agrifood industry

Ingrid Appelqvist[a], Hester De Wet[b], Slaven Marusic[c], Filip Janakievski[a]

[a]*CSIRO Agriculture and Food, Werribee, VIC, Australia*
[b]*Aurecon, Energy, Resources and Water, Neutral Bay, VIC, Australia*
[c]*Aurecon, Data and Analytics, Docklands, VIC, Australia*

20.1 Introduction

It has been estimated that by 2050 there will be more than 9.5 billion people in the world with most of the population growth in the Asia-Pacific Region (United Nations, Department of Economic and Social Affairs, Population Division, 2019). The requirement for food production to meet the growing demand will need to more than double current world production according to the Food and Agricultural Organisation (FAO (Food and Agriculture Organization of the United Nations), 2011; Tilman et al., 2011) on less land and with increased competition for resources such as water, energy, and land use (Serraj et al., 2018). Added to this, there is increased pressure to use less pesticides, herbicides, and other chemicals (Hillocks, 2012) and to reduce the impact of agricultural production on greenhouse gas emissions in relation to changing climate patterns (Smith et al., 2014; IPCC, 2014). Cumulatively, this results in an unprecedented convergence of pressures that the global agrifood system will face—the so-called "perfect storm" (Serraj et al., 2018).

The complex links between environmental, societal, and economic factors and food production are increasingly present (Russell and Hedberg, 2016; Fagioli et al., 2017; Johnson and Karlberg, 2017) and leading to a total food system approach, rather than just a food supply chain view (Bullock et al., 2017), to solve the global sustainable food and nutrition security challenge. This more "holistic systems" approach has also made use of foresight and future studies and explored global trends and scenarios to define potential new and alternative pathways for a sustainable agrifood ecosystem (Bourgeois and Sette, 2017).

20.1.1 Global megatrends impacting the agrifood system

The megatrends impacting the agrifood system can be summarized by the following schematic in Fig. 20.1 adapted from CSIRO Futures (2016) and Hajkowicz and Eady (2015).

These megatrends cover social, technological, economic, environmental, and political forces of change and in general there is growing consensus and good agreement across many of the foresight studies conducted on food supply chains worldwide (Webster et al., 2014). The megatrends depicted in Fig. 20.1 represent long-term shifts in the sector that are creating new business models, social structures, and cultural paradigms. Understanding the agrifood landscape over the coming decades will be

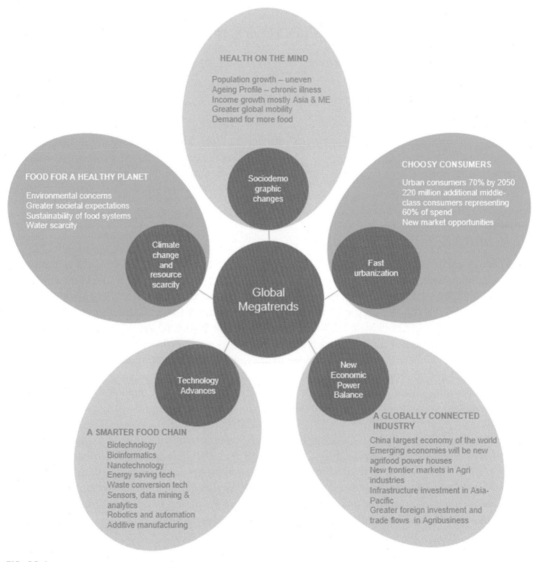

FIG. 20.1

Global megatrends impacting the agrifood system.

vital for the current stakeholders (across the agrifood ecosystem) to make strategic decisions and provide guidance to successfully position the agrifood system for success. However, there is an urgency to find solutions as the global food chain is already under pressure from global population growth, climate change, political impacts of migration, population shift from rural to urban regions, the recent COVID-19 pandemic and the demographics of an aging global population in advanced economies including parts of Asia (Hajkowicz and Eady, 2015; Maggio et al., 2019; Fedunik-Hofman, 2020). One critical area of focus has been the impact of technological advances and disruptive digital technologies (digitization) on the food production and manufacturing industry (CSIRO Futures, 2016) supporting a strong trend for sustainable circular bioeconomies. Global trends have resulted in an increasing demand for premium, trusted, and sustainable products and consumers who are demanding information about the products they are consuming, including their provenance, whether they have been produced ethically and sustainably in addition to their safety, quality, and health credentials.

20.1.2 Circular bioeconomy through the agrifood sector

The circular bioeconomy focuses on the sustainable, resource-efficient valorization of biomass in integrated, multioutput production chains (e.g., biorefineries) while also making use of residues and wastes and optimizing the value of biomass over time via cascading (Stegmann et al., 2020).

Digitization (applying digital technologies) of the agrifood ecosystem is central to developing a sustainable circular bioeconomy for which the agrifood sector can take a leading role (Bauer et al., 2017).

An example of a circular bioeconomy based on the sugar industry in Australia is shown in Fig. 20.2 (adapted from Aurecon infograph 2019. Internal publication to Aurecon).

Circular bioeconomies deliver benefits and support integrating new bioindustries with domestic and export market opportunities. These benefits include: Resilience for local economies and workforces; Productive use of "waste" products; Opportunities for efficiency and shared infrastructure; Knowledge sharing and increase in skilled workforces; Opportunities to reduce local environmental impact (Swinda et al., 2014). The adoption of digital technologies and innovation however, must respond to the grand challenges and current issues to succeed and become intricately part of the agrifood ecosystem (Scheerder et al., 2014). If successful, building a digital agrifood-based bioeconomy will deliver opportunities to reduce environmental impact, open new markets, increase productivity and profitability, and drive a sustainable agrifood sector that will meet the "perfect storm" challenges (Keogh and Henry, 2016).

20.2 Industry 4.0 applied to revolutionize the agrifood system

The application of Industry 4.0 (Yang, 2017) or Industrial Internet of Things is projected to generate $1.2–$3.7 trillion of value globally by 2025 (Ezell, 2018) through operational efficiency, predictive and preventative maintenance, supply chain management, and logistics (Manyika, 2015) in food and other manufacturing industries. We are in a time of transition where future thinking is not only changing how we manufacture food and beverages but also how consumer led solutions are influencing and disrupting food supply chains.

Industry 4.0 is the term used to describe the era of connectedness we currently find ourselves in; however, we would like to argue that this definition does not fully describe what is meant by Industry 4.0. Advances in mobile phone technology, for example, have enabled most people to carry what is a very powerful computer in their pockets. This combined with the downward trend of the cost associated with low power mobile sensing devices, we now have the capability to connect, analyze, and apply insights to our operations to gain new efficiencies and optimize processes we were unable to do easily or cost effectively before—this is the revolution of Industry 4.0.

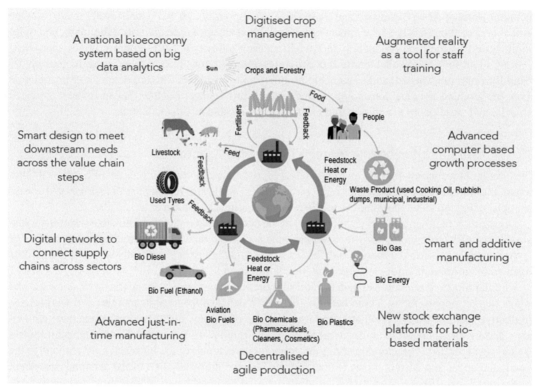

FIG. 20.2

Digitizing the bioeconomy system.

20.2.1 Developing a digitally connected agrifood sector

One of the important drivers to the Industry 4.0 discussions is the emergence of the internet of things (IoT). IoT essentially describes the ecosystem of instrumentation for measurement of physical systems/processes/events, followed by the aggregation and efficient communication of this data from distributed sensor systems via a given communication network (Atzori et al., 2010). Principally this leverages novel sensor and communications technologies for efficient transfer of essential information (Gubbi et al., 2013). Applications of this technology reach across the entire agrifood supply chain, as ag-tech receives growing attention and support, together with maturing technology (Agriculture Victoria, 2018; Maher, 2019; Johnston, 2019).

Several advanced digital-based technologies have been developed to build sustainability in the agrifood ecosystem through early value chain interventions. The CSIRO's Data61 Robotics and Autonomous Systems Group developed digital technologies that included the Ceres Tag technology a next-generation smart ear tag that helps farmers track herds; eGrazor a solar-powered device to determine pasture intake by cattle based on behavior; virtual fencing technology using GPS to prevent wondering and harm to cattle, and wireless technologies and sensors to control livestock location. Other technologies for crops include The Phenomobile Lite that measures crop characteristics such as canopy height and biomass; and WaterWise (to significantly reduce water usage), is a world-first, cloud-connected, plant-based sensor monitoring platform with advanced data analytics. Novel approaches to data analysis and access are leveraging aerial and satellite monitoring of crops and pasture for advance condition forecasting, together with 3D modeling of environments, augmented reality systems for effective visualization and interaction with this data, together with easy to use apps (such as Graincast, 1622, and Yield Prophet) placing important information in the hands of farmers (CSIRO Digital Agriculture 2019). Further developments continue for the utilization of satellite data together with locally deployed drones and robotics for more detailed monitoring and analysis of agricultural practices and potential. Such systems enable the delivery of precision agriculture, taking to another level the already widely adopted use of tools such as GPS enabled (and even autonomous) farming equipment. The end goal is better informed decision making enabled by access to useable data throughout the production process (Evans et al., 2017).

However, the level of adoption of digital technology across the value chain by agrifood sectors such as meat, dairy, horticulture, grains, etc. can be different and considered in the context of the industrial plant setting, together with the interpretation of "Industry 4.0" and associated digital technologies for data collection, analysis, and system operation, it could be said that cropping industries (grains/horticulture/rice/cotton) and dairy are more advanced than the meat/livestock sectors of the industry. This is by virtue of prevalence of processing/manufacturing operations taking associated farming outputs. This also extends to include planting and harvesting processes with associated machinery and instrumentation, where there is a strong record of precision agriculture adoption. Another indicator is the level of focused R&D investment, which may also suggest that horticultural practices are greater adopters of digital technologies and the adoption of farm management software is generally more prevalent in the cropping sector (Perrett et al., 2017).

However, when digital technology is associated with advancements in biotechnology and veterinary/breeding practice, digital data collection rates in livestock industries is higher.

In the majority of industrial plants, we will find highly automated systems, real time control of machines (including robotics), and other edge computing devices, controlled via a Supervisory Control and Data Acquisition (SCADA) system connected to a MES (Manufacturing Execution System) which consumes data from the ERP (Enterprise Resources Planning) and other control systems to manage and measure production. The ERP system contains information on sales orders, supply chain, inventory, finance, etc. and is often connected to the cloud. For a manufacturing plant to operate effectively these systems must work together, however, it is found in most process plants that these systems are owned by different stakeholders such as engineering and maintenance groups, IT, and finance. The current linear connection between these systems combined with the diverse ownership acts as a barrier for companies to optimize and utilize the data flowing through the system. Other issues encountered on site when trying to optimize the information flow include legacy hardware, limited or no network connectivity on site, and communication across legacy systems. IoT may support solving some of these issues and is further discussed in the next section (Section 20.2.2).

In the industrial setting, the manufacturing process is undergoing a similar evolution (Cui et al., 2020). Here, IoT fills a space in the technology landscape that can target low latency and low volume communications. IoT is therefore often associated with dedicated cutting-edge computer, logistic, and sensing networks fulfilling these criteria. To accommodate computer processing workloads, edge computing enables local processing and only transmission of essential information, rather than high volume data transfer and centralized processing and analysis. However, high volume data applications are now being enabled by emerging 5G telecommunications and becoming more viable. While conceptually related to SCADA systems that provide such functionality, as described below, those systems are often dedicated to specific operational tasks.

20.2.2 IoT in food manufacturing and retail: digital technologies, data insights, visualization, and interpretation

The measurement of system functions, processes, inputs, and outputs, as well as the fine tuning of these to generate the desired outcome has always been intrinsic to this sector. IoT solutions can offer greater deployment and application flexibility, together with the integration of different systems data providing business intelligence or process understanding through visibility of system/process interconnections, dependencies, and influences. System security is a key requirement, particularly with respect to deployment in or among mission/production critical systems (Industrial Internet Consortium, 2016). Deploying sensors across the ecosystem to provide information on the weather, transport and logistics, energy, security, and linking it via a secure data sharing protocol will revolutionize the food manufacturing business. Applying blockchain technology (a decentralized, distributed ledger that records the provenance of a digital asset) to data streams may provide the necessary framework for trusted data and accountability throughout the digital value chain of production (Khan and Salah, 2018; Casado-Vara et al., 2018).

The advent of accessible advanced data analytics, encompassing tools such as machine learning (ML) and artificial intelligence (AI), together with the greater ease and storage of operational data, presents an opportunity to, for example, maximize asset lifetime through:

- more accurate condition monitoring, providing early indicators of plant degradation
- the reduction of equipment failures by early fault detection or adaptive maintenance that reduces the likelihood of issues occurring between fixed maintenance schedules

The emergence of new technologies, data driven solutions and greater accessibility of granular data, will allow identification and quantification of the lost value of inefficient processes and systems and thus can be corrected. A production process that meets its operation output requirements may seem satisfactory and not warrant further analysis, but much can remain hidden by applying inadequate metrics. Examples include:

- Excess energy consumption: for example, a cooling plant not sufficiently adapted to climate variability.
- Capacity constraints: caused by inadequate monitoring of underperforming equipment.
- Nonoptimized processes: for example, alternative equipment scheduling to produce greater system yield or equipment longevity (e.g., running multiple elements at reduced capacity rather than switching single elements running at maximum capacity).
- Poor product quality and inconsistency: adapting processes and equipment for operating conditions that deliver the highest quality product.

The answers to such queries are revealed in the analysis of the corresponding operational data. These may have always been available, in part via traditional SCADA or MES systems described earlier, yet such opportunities are still missed. Deeper adoption of relevant digital technologies presents value across the entire food supply chain (Ramundo et al., 2016).

The essence of the big data concept is ease of access to larger and disparate information, enabled by more efficient storage and interrogation of data. The IoT paradigm extends this to include more flexible, distributed, and granular data collection, linking sensing, aggregation, management, communication of data that represents observable events. The capacity to retrofit sensor and communication infrastructure over existing environments is perhaps its greatest attribute (Panda et al., 2019).

Food supply chain operations are an ideal fit for future practice in the IoT, which leverage smart sensors and actuators to connect humans and machines (inspection systems, food processing equipment, or other data capturing devices along the food chain) by using the internet.

IoT for process monitoring, optimization, and control. The latest technologies in sorting/inspection could be combined with X-ray, NIR spectroscopy, cameras, and machine learning algorithms to analyze different aspects of food quality. The TensorFlow machine (TOMRA) for example has several applications including monitoring productions plants and people for safety, activity classification, or even to automate their reaction to food samples. In-line sensors could also be used in closed-loop supply chain (CLSC) control systems, so that wastes recovery and control in manufacturing contributes to improvement of the quality, process efficiencies as well as sustainable production (Herrera et al., 2018) and profitability.

For successful uptake, sensors, robotics, and IoT must be able to operate in real time and be non-invasive. Visible and near infrared optical methods as well as hyperspectral imaging can be used to improve food quality, food safety and can be utilized within inspection systems for cleaning and validation of processing lines. Incorporating an ultrasonic sensor for monitoring food texture and emulsion stability can further improve quality measurement standards currently applied across the food industry. Bowler et al. (2020) used ultrasonic sensors to monitor blending processes for honey and water and batter mixing with flour and water. Classification machine learning models were applied to the data in order to predict when mixing was complete. Sensors able to characterize extent of mixing can have a positive benefit by reducing off-specification of product, being able to detect early any process anomalies and reduce resources by not overmixing. Other benefits of ultrasonic sensors are that they are low-cost, measure in real time and can be in-line with no impact on the process and they work for opaque systems. A practical example of a real time optical sensor (Liquid Vision Innovation Ltd) has been used to assess extra virgin olive oil focusing on polyphenols. This method allowed on site monitoring of key olive oil parameters and provided oil quality and traceability through the supply chain.

IoT for food safety, fraud and provenance systems. Another opportunity to improve food safety systems involves incorporating optical fluorescence imaging and ultrasound sensing technologies to deliver data to machine learning algorithms for the monitoring of microbial debris and food particles in food processing equipment. This provides a possibility to develop and deploy new platforms for supply chain traceability and provenance for trusted supply chains, utilizing sensors, IoT, and cloud computing for food safety compliance by providing transparency for consumers, paperless records for manufacturers, improving their logistics and reducing large-scale product recalls.

Digitization of the agrifood system interestingly has been viewed to be more to do with new business models and major changes in consumer behavior (White, 2018) than digital technology, big data, and analytics per se (Klitkou et al., 2017). As consumers become more technologically savvy and with increasing affluence consumer demand for healthier functional foods, enhanced eating experience, and personalized nutrition will be greater than ever. Consumers are demanding more information about the products they are using, including their provenance, whether they have been produced ethically and sustainably in addition to their safety, quality, and health credentials. As a result, food provenance research is becoming more important and especially where the advantage is the "clean, green, and safe" brand, producers and manufacturers are seeking technologies that ensure the authenticity, transparency, and traceability of their products. Food fraud-deception for economic gain using food (Spink et al., 2019) is a world-wide problem and costs the global food industry an estimated $10–15 billion annually (Cattini, 2016). Australian food manufacturing companies are also not immune and often have to contend with less scrupulous companies around the world taking advantage of Australia's clean and green brand image to mis-label produce and charge the higher price that Australian products can attract. Australia is highly dependent upon exports (70% of agricultural produce is exported; Howden and Zammit, 2019) and as such food fraud has a huge impact on the economy. Examples of food fraud have included:

- Sales of fraudulent "Australian" branded beef are estimated at $2 billion a year (Farmonline, 2018).
- Australian horticulture particularly premium cherry and citrus exports are an attractive market for food fraudsters and cost the horticulture industry $250 million annually (The Gate, 2019) such that Tasmanian cherries have been sold in China and Vietnam in counterfeit "Tasmanian-grown" packaging.
- Counterfeiting operations that specialized in producing fake Australian wine labeled as "Penfolds" were identified in Cambodia and China (The Guardian, 2019).

Knowing the provenance of food products builds a trusted relationship with consumers and players across the entire supply chain. Insights derived from data points along the supply chain can be integrated so that producers and processors can improve their operations efficiency.

Consumers are already changing their eating habits looking to eat more sustainable plant-based diets and are keen to know where their food comes from, this includes an eagerness to track the journey of their product from paddock to plate. Consumer sentiment is also increasing on ethical raising of animals and their welfare as they more strongly advocate for sustainable production and mitigation of negative environmental impacts.

IoT for machine learning applications. The vastness of this data space both requires and allows for more advanced analysis of these events, through the application of mathematical tools such as those found within the domain of machine learning, a subset of the AI domain. Merely the implementation of computational processes (algorithms) designed to mimic brain function and human decision making (Jaquith and The Futurism Team, 2019). At the core of it is the ability to represent or model a system mathematically and derive an outcome according to some level of confidence or probability (James et al., 2013). This includes the exploratory objective in unsupervised learning, that seeks to reveal new information about the event being observed, such as relationships between different system functions (Bishop, 2006). When functions of the system (e.g., inputs/outputs) are observable, this historical information that represents its operation forms the basis for supervised learning (Haykin, 2009). That is, where the machine learning model is built from this data and the known or desired outputs are used to measure and adjust its effectiveness. This lends itself to predictive modeling and classification tasks. The application of these methods provides a data driven approach to system modeling when explicit formulation is complex or prohibitive.

For example, a novel AI/machine learning system (IBM and McCormick & Company) was developed to predict the ratio of ingredients, alternative flavor substitutes and fine tune flavor pairing, so that product developers can speed up the flavor creation process using machine learning algorithms to sort through hundreds of thousands of raw material combinations. Another example involves an AI software platform (Water Plant Inc.) to monitor, control, and optimize membrane performance in filtration systems such as Membrane Filtration, Ultra Filtration, Nano Filtration, and Reverse Osmosis. Combining product development and ingredient handling can make factories even more efficient.

Data visualization. Effective visualization of the data streams and information generated from machine learning applications enables interpretation and formation of knowledge by the end user. This may be in the form of a system alert of a threshold being exceeded, a trend line representing performance over time, a dashboard view of multiple parameters, a multidimensional representation of complex data, or an augmented reality (AR) overlay on a system view (Xi et al., 2018; Hietanen et al., 2020). This can further extend to a virtual reality (VR) display containing a 3D model of single component, entire system or complete plant environment that can be navigated by the user to whatever level of detail is provided. Additional layers of information can then be added to the visualized data, perhaps combining design specifications, operational data, and fusion of information to indicate things otherwise hidden or imperceptible, such as:

- the level of vibration or heat distribution of an internal component in machinery
- an overview of land viability accounting for soil condition, water availability, infrastructure access and long-term weather conditions
- supply chain logistics and scheduling of goods transportation

The subsequent decision-making can then be applied expertly in a manual intervention or management function (Accorsi et al., 2017) or integrated directly into system operations through process automation.

20.3 Current hurdles that are reducing uptake of digital technologies

Solutions exist for system observation, data collection, analysis, interpretation, and action, yet the broader adoption of such practices is still growing and not yet the norm (Madushanki et al., 2019). Although, IoT deployments across farming and agriculture are widespread, often focusing on water management, crop management, smart farming, livestock management, and irrigation management, applications in agrifood are fragmented, experimental and/or lack seamless integration, warranting an interoperability framework (Verdouw et al., 2019).

Some of the challenges that are slowing progress on adoption of digital technologies by the farming and food manufacturing industries include:

- Tendency for companies to be preoccupied with "business as usual" leading to complacency or becoming entrenched in established processes/equipment.
- Unfamiliarity with digital domains (digital expertise) leading to concern about data security and portability and/or mistrust of the technology.
- Not being satisfied with technology reliability with commercial derisking being difficult to overcome for some companies.
- Concern that scale of investment of resources required to implement digital technologies such as cost of software, sensors, robotics, and other hardware will be very high and will not get return on investment.
- Time pressure to deliver urgent activities rather than prioritize digital outcome savings.
- Disconnect between operational level and c-suite for implementation of IoT and digital technology in many businesses.

- Business imperative such as just cost cutting not aligned to digital technology benefits.
- Moving away from exploration and development, so need to consolidate and build growth/profitability on current assets.
- Not all the supply chain has the same level of adoption, so benefits do not always run across the value chain to maximize benefit, for example, lack of connectivity and benefits transfer. This can be related to the labor force not having appropriate skills who can understand and maintain new capital investment and the need for training.

These concerns are valid, but not insurmountable. As the reach and capability of such technologies expands, particularly with respect to the development and deployment of AI, the ethical considerations become perhaps more important than these seemingly logistical barriers. Fortunately, ethics in technology is now receiving due attention, though mostly in the form of guidelines or principles (Castelvecchi, 2016; IEEE, 2017; Dawson et al., 2019; Walsh et al., 2019; Standards Australia, 2019). This challenges stakeholders to take (or at least test) a person-centered view. For all the benefits that robotics and automation may provide throughout the production process, many challenges remain and even demand acknowledgement of the key role of people in the process, for direct as well as secondary benefit (Shamshiri et al., 2018). Retaining expertise and motivation and taking account of the social aspect of the agrifood supply chain will be vital and continuity of participation is not incidental. Lack of trust, transparency and shared benefit in what is fast becoming an information supply chain, presents related challenges (Jakku et al., 2019). These aspects require as much attention so that the gains being proposed can continue to be delivered in a fashion that strengthens and grows the agrifood sector rather than arbitrarily disrupting it.

20.4 Conclusion

There is an imperative need for agrifood to transform to meet the challenge of feeding everyone in the world. Digital technology and other advanced technologies will be central to enabling the change and meeting the challenges and providing ultimately a positive impact. Application of digital technology is very promising for realizing new levels of control and integration across the agrifood ecosystem (Sundmaeker et al., 2010; Porter and Heppelmann, 2014; Sarni et al., 2016). It will be a powerful driver that can transform farming and food manufacturing to become highly interconnected and build networked operations that are secure and can be identified, sensed, and controlled remotely (Verdouw et al., 2015). Benefits will include improved sensing and monitoring of production and farming systems such as processes and logistics, control of environmental conditions such as weather, soil quality, animal welfare, pest, and disease control.

The digitalization of agrifood manufacturing will change how products are designed, fabricated, used, operated, and serviced postsale, as well as driving transformation of the operations, processes, and energy footprint of factories and the management of production supply chains (Ezell, 2018). Benefits will be improving food traceability, provenance, and food quality as well as increasing consumer awareness of sustainable resources, health, and delivering personalized diets.

One of the main reasons why the hype around the IoT has been over inflated is mainly due to system owners not taking the time to understand what they should measure and why. Just because you can measure, it does not mean you should.

Technology debt is a term referred to when a company has over invested in a technology without gaining any new insights or efficiency, but is so far invested, that they are unable to pivot or pursue a new option including adopting new digital technologies.

However, this and many of the barriers identified could be overcome by actors across the agrifood system taking the time to understand how operational decisions are made: who makes the decisions and what data is required. There is already a huge opportunity today to take advantage of the data streams, sensors, and technology platforms available and conduct analysis to validate the data informatics and optimize decision making to make a positive significant difference. The opportunity for future agriculture and food manufacturing systems to contribute significantly to environmental sustainability, population nutrition security, and economic prosperity is great, and no doubt will have significant impact in the coming years.

References

Accorsi, R., Bortolini, M., Baruffaldi, G., Pilati, F., Ferrari, E., 2017. Internet-of-things paradigm in food supply chains control and management, 27th International Conference on Flexible Automation and Intelligent Manufacturing, FAIM2017, Procedia Manufacturing, 11, 889–895.

Agriculture Victoria, 2018. Digital Agriculture Strategy. Department of Economic Development, Jobs, Transport and Resources Agriculture.vic.gov.au/digitalag.

Atzori, L., Iera, A., Morabito, G., 2010. The internet of things: a survey. Comput. Netw. 54 (15), 2787–2805.

Bauer, F., Coenen, L., Hansen, T., McCormick, K., Palgan, Y.V., 2017. Technological innovation systems for biorefineries: a review of the literature. Biofuels Bioprod. Biorefining-Biofpr 11, 534–548.

Bishop, C.M., 2006. Pattern Recognition and Machine Learning. Springer, Cambridge.

Bourgeois, R., Sette, C., 2017. The state of foresight in food and agriculture: challenges for impact and participation. Futures 93, 115–131.

Bowler, A., Bakalis, S., Watson, N., 2020. Monitoring mixing processes using ultrasonic sensors and machine learning. Sensors 20 (1813), 1–24.

Bullock, J.M., Dhanjal-Adams, K.L., Milne, A., Oliver, T.H., Todman, L.C., Whitmore, A.P., Pywell, R.F., 2017. Resilience and food security: rethinking an ecological concept. J. Ecol. 105 (4), 880–884.

Casado-Vara, R., Prieto, J., De la Prieta, F., Corchado, J.M., 2018. How blockchain improves the supply chain: case study alimentary supply chain. Proc. Comput. Sci. 134, 393–398.

Castelvecchi, D. (2016). The Black Box of AI, Macmillan Publishers Limited, part of Springer Nature, 20, Nature, Vol. 538.

Cattini, C. (2016). https://www.ifis.org/blog/global-food-fraud#:~:text=Food%20Fraud%20Costs%20the%20 Global%20Food%20Industry%20%2410-15%20Billion%20Annually (Accessed July 2020).

CSIRO Digital Agriculture, 2019. www.csiro.au/agtechhttps://www.csiro.au/en/Research/AF/Areas/Digital-agriculture.

CSIRO Futures, 2016. Australia 2030: Navigating Our Uncertain Future. CSIRO, Canberra.

Cui, Y., Kara, S., Chan, K.C., 2020. Manufacturing big data ecosystem: a systematic literature review. Rob. Comput. Integr. Manuf. 62, 1–20.

Dawson, D., Schleiger, E., Horton, J., McLaughlin, J., Robinson, C., Quezada, G., Scowcroft, J., Hajkowicz, S., 2019. Artificial Intelligence: Australia's Ethics Framework. Data61. CSIRO, Australia.

Evans, K., Terhorst, A., Kang, B., 2017. From data to decisions: helping crop producers build their actionable knowledge. Crit. Rev. Plant Sci. 36 (2), 71–88.

Ezell, S., 2018. Why manufacturing digitization matters and how countries are supporting it. Inf. Technol. Innovation Found., 1–54.

Fagioli, F.F., Rocchi, L., Paolotti, L., Słowiński, R., Boggia, A., 2017. From the farm to the agri-food system: a multiple criteria framework to evaluate extended multi-functional value. Ecol. Indic. 79, 91–102.

FAO (Food and Agriculture Organization of the United Nations), 2011. The State of the World's Land and Water Resources for Food and Agriculture (SOLAW): Managing Systems at Risk. Earthscan, Rome: FAO, and London.

Farmonline (2018). https://www.farmonline.com.au/story/5403286/food-fraud-bites-aussie-ag-exports/.

Fedunik-Hofman, L., 2020. How Does a Global Pandemic Affect Our Food Supply Chain? https://www.science.org.au/curious/people-medicine/how-does-global-pandemic-affect-our-food-supply-chain (accessed August 2020) https://www.science.org.au/curious/people-medicine/how-does-global-pandemic-affect-our-food-supply-chain.

Gubbi, J., Buyya, R., Marusic, S., Palaniswami, M., 2013. Internet of things (IoT): a vision, architectural elements, and future directions. Fut. Generation Comput. Syst. 29 (7), 1645–1660.

Hajkowicz, S., Eady, S., 2015. Rural Industry Futures – Megatrends Impacting Australian Agriculture Over the Coming Twenty Years. CSIRO and RIRDC.

Haykin, S., 2009. Neural Networks and Learning Machines, third ed. Pearson, Ontario Canada.

Herrera, M.M., Vargas, L., Contento, D., 2018. In Applied Computer Sciences in Engineering. Figueroa Garcia, J.C., Lopez Santana, E.R., Rodriguez Molano, J.I. (Eds.), Book Series: Communications in Computer and Information Science, 915, pp. 328–339.

Hietanen, A., Pieters, R., Lanz, M., Latokartano, J., Kämäräinen, J.-K., 2020. AR-based interaction for human-robot collaborative manufacturing. Rob. Comput. Integr. Manuf. 63.

Hillocks, R.J., 2012. Farming with fewer pesticides: EU pesticide review and resulting challenges for UK agriculture. Crop Prot. 31 (1), 85–93.

Howden, M., Zammit, K., 2019. United States and Australian agriculture—A Comparison. Australian Bureau of Agricultural and Resource Economics and Sciences, Canberra. CC BY 4.0. https://doi.org/10.25814/5d6c85 84ea0f5.

IEEE, 2017. The IEEE Global Initiative on Ethics of Autonomous and Intelligent Systems. Ethically Aligned Design: A Vision for Prioritizing Human Well-being with Autonomous and Intelligent Systems, Version 2. http://standards.ieee.org/develop/indconn/ec/autonomous_systems.html.

Industrial Internet Consortium, 2016. Industrial Internet of Things Volume G4: Security Framework. IIC:PUB:G4: V1.0:PB:20160926 www.iiconsortium.org.

IPCC, 2014. Summary for policymakers. In: Edenhofer, O., Pichs-Madruga, R., Sokona, Y., Farahani, E., Kadner, S., Seyboth, K., Adler, A. et al (Eds.), Climate Change 2014: Mitigation of Climate Change. Contribution of Working Group III to the Fifth Assessment Report of the Intergovernmental Panel on Climate Change. Cambridge University Press, Cambridge, UK.

Jakku, E., Taylor, B., Fleming, A., Mason, C., Fielke, S., Sounness, C., Thorburn, P., 2019. "If they don't tell us what they do with it, why would we trust them?" Trust, transparency and benefit-sharing in Smart Farming. NJAS - Wageningen J. Life Sci. 2019, 90–91.

James, G., Witten, D., Hastie, T., Tibshirani, R., 2013. An Introduction to Statistical Learning with Applications in R. Springer, New York.

Jaquith and The Futurism Team, 2019. Understanding Machine Learning. Futurism. https://futurism.com/images/understanding-machine-learning-infographic.

Johnson, O.W., Karlberg, L., 2017. Co-exploring the water-energy-food nexus: facilitating dialogue through participatory scenario building. Front. Environ. Sci. 5, 24 art.

Johnston, M., 2019. Monash Uni using drones to reduce water wastage on farms. https://www.iothub.com.au/news/monash-uni-using-drones-to-reduce-water-wastage-on-farms-519548.

Keogh, M, Henry, M, 2016. The Implications of Digital Agriculture and Big Data for Australian Agriculture. Australian Farm Institute.

Khan, M.A., Salah, K., 2018. IoT security: review, blockchain solutions, and open challenges. Fut. Gen. Comput. Syst. 82, 395–411.

Klitkou, A., Bozell, J., Panoutsou, C., Kuhndt, M., Kuusisaari, J, Beckmann, J.P., 2017. Background Paper: Bioeconomy and Digitalisation. MISTRA the Swedish Foundation for Strategic Environmental Research.

Madushanki, A., Halgamuge, M., Wirasagoda, H., Syed, A., 2019. Adoption of the internet of things (IoT) in agriculture and smart farming towards urban greening: a review. Int. J. Adv. Comput. Sci. Applic. 10, 11–28.

Maggio, A., Scapolo, F., Crekinge, T., Serraj, R., 2019. Global drivers and megatreds in agrifood-systems. In Agriculture and food systems to 2050. World scientific series in grand public policy challenges, p. 46.

Maher, W., 2019. NSW and SA Governments tout new agtech research farms. https://www.iothub.com.au/news/nsw-and-sa-governments-tout-new-agtech-research-farms-533772 accessed March 2020.

Manyika, J., et al., 2015. The Internet of Things: Mapping the Value Beyond the Hype. McKinsey Global Institute, pp. 66.

Panda, S.K., Blome, A., Wisniewski, L., Meyer, A., 2019. IoT retrofitting approach for the food industry, 24th IEEE International Conference on Emerging Technologies and Factory Automation (ETFA). Zaragoza, Spain, 2019, 1639–1642.

Perrett, E., Heath, R., Laurie, A., Darragh, L., 2017. Accelerating Precision Agriculture to Decision Agriculture—Analysis of the Economic Benefit and Strategies for Delivery of Digital Agriculture in Australia. Australian Farm Institute and Cotton Research and Development Corporation.

Porter, M.E., Heppelmann, J.E., 2014. How smart connected objects are transforming competition. Harv. Bus. Rev.

Ramundo, L., Taisch, M., Terzi, S., 2016. State of the art of technology in the food sector value chain towards the IoT, 2016 IEEE Second International Forum on Research and Technologies for Society and Industry Leveraging a better tomorrow (RTSI), 1–6.

Russell, C., Hedberg, I., 2016. The ecology of alternative food landscapes: a framework for assessing the ecology of alternative food networks and its implications for sustainability. Landscape Res. 41 (7), 795–807.

Sarni, W., Mariani, J., Kaji, J., 2016. From dirt to data, the second green revolution and the internet of things. Deloitte Rev. (18).

Scheerder, J., Hoogerwerf, R., de Wilde, S., 2014. Horizon scan 2050: A different view of the future. Netherlands study centre for technology trends.

Serraj, R., Krishnan, L., Pingali, P., 2018. Agriculture and food systems to 2015: a synthesis. In: Serraj, R., Pengali, P. (Eds.), Agriculture and Food Systems to 2050—Global Trends, Challenges and Opportunities. World Scientific Publishing Co., Singapore.

Shamshiri, R., Weltzien, C., Hameed, I., Yule, I., Grift, T., Balasundram, S., Pitonakova, L., Ahmad, Desa, Chowdhary, Girish, 2018. Research and development in agricultural robotics: a perspective of digital farming. Int. J. Agric. Biol. Eng. 11, 1–14.

Smith, P., Bustamante, M., Ahammad, H., Clark, H., Dong, H., Elsiddig, E.A., Haberl, H., et al., 2014. Agriculture, forestry and other land use (AFOLU). In: Edenhofer, O., Pichs-Madruga, R., Sokona, T., Farahani, E., Kadner, S., Seyboth, K., Adler, A. et al (Eds.), Climate change 2014: Mitigation of Climate Change. Contribution of Working Group III to the Fifth Assessment Report of the Intergovernmental Panel on Climate Change. Cambridge, UK. Cambridge University Press.

Spink, J., Chen, W., Zhang, G., Speier-Pero, C., 2019. Introducing the food fraud prevention cycle (FFPC): a dynamic information management and strategic roadmap. Food Control 105, 233–241.

Standards Australia, 2019. Developing Standards for Artificial Intelligence: Haring Australia's Voice, Discussion Paper https://www.standards.org.au/getmedia/aeaa5d9e-8911-4536-8c36-76733a3950d1/Artificial-Intelligence-Discussion-Paper-(004).pdf.aspx.

Stegmann, P., Londo, M., Junginger, M., 2020. The circular bioeconomy: its elements and role in European bioeconomy clusters. Resour. Conserv. Recycl. 6, 1–17.

Sundmaeker, H., Guillemin, P., Friess, P., Woelffle, S., 2010. Vision and challenges for realising the internet of things. Cluster of European research projects on the internet of things, European Commission.

Swinda, F., Pfau, J., Hagens, E., B., Dankbaar, Smits, A.J.M., 2014. Visions of sustainability in bioeconomy research. Sustainability 6, 1222–1249.

The Gate, 2019. Global Ag-Tech Ecosystem 2010. In: Sundmaeker, H., Guillemin, P., Friess, P., Woelfflé, S. (Eds.), Vision and Challenges for Realising the Internet of Things, EC/CERP-IoT. NSW Department of primary Industries https://www.thegate.org.au/data/assets/pdf_file/0008/1192328/DAF19001-DPI-Prospectus_mk6-3.pdf 2010.

The Guardian, 2019. https://www.theguardian.com/food/2019/nov/29/fake-penfolds-wine-seized-in-raid-on-bootlegging-operation-in-cambodia (accessed June 2020).

Tilman, D., Balzer, C., Hill, J., Befort, B.L., 2011. Global Food Demand and the Sustainable Intensification of Agriculture108. PNAS, pp. 20260–20264.

United Nations, Department of Economic and Social Affairs, Population Division, 2019. https://population.un.org/wpp/Download/Standard/Population/.

Verdouw, C., Sundmaeker, H., Tekinerdogan, B., Conzon, D., Montanaro, T., 2019. Architecture framework of IoT-based food and farm systems: a multiple case study. Comput. Electron. Agric. 165 (2019), 104939.

Verdouw, C.N., Wolfert, J., Beulens, A.J.M., Rialland, A., 2015. Virtualization of food supply chains with the internet of things. J. Food Eng. 17, 128–136.

Walsh, T., Levy, N., Bell, G., Elliott, A., Maclaurin, J., Mareels, I.M.Y., Wood, F.M., 2019. The Effective and Ethical Development of Artificial Intelligence: An Opportunity to Improve Our Wellbeing Report for the Australian Council of Learned Academies.

Webster, P., Van Oene, F., Kurosawa., M., Guillodo, L., 2014. The future of agri-food: harnessing innovation from adjacent industries to meet global challenges. Prism 2, 1–6.

White, M., 2018. Emerging Agricultural Technologies: Consumer Perceptions Around Emerging Agtech. AgriFutures, Australia Publication No. 18/048 Project No. PRJ-011141.

World Population Prospects, 2021. The 2012 Revision, Volume II, Demographic Profiles.

Yang, L., 2017. Industry 4.0: a survey on technologies, applications and open research issues. J. Industrial Inf. Integr. 6, 1–10.

Xi, M., Adcock, M., McCulloch, J., 2018. Future agriculture farm management using augmented reality. IEEE Virtual Reality 2018. Reutlingen, Germany.

Food Industry 4.0: Opportunities for a digital future

21

Serafim Bakalis[a], Dimitrios Gerogiorgis[b], Dimitrios Argyropoulos[c], Christos Emmanoulidis[d]

[a]*Department of Food Science, University of Copenhagen, Rolighedsvej Frederiksberg, Denmark*
[b]*School of Engineering, University of Edinburgh, Edinburgh, United Kingdom*
[c]*School of Biosystems and Food Engineering, University College Dublin, Dublin, Ireland*
[d]*School of Aerospace, Transport and Manufacturing, Cranfield University, Cranfield, United Kingdom*

21.1 Introduction

We have undergone three industrial revolutions and the food industry is currently undergoing a fourth one. The first, at the end of the 18th century, concerned the use of mechanization and industrialization of production. The second followed by use of new sources of energy, such as electricity, gas, and oil. The third took place in the second half of the 20th century, with the rise of electronics and robots giving rise to a high level of automation. Currently, the rise of digital technologies can blur the boundaries between the physical and the digital offering technologies such as artificial intelligence (AI), internet of things (IoT) with significant opportunities in the way the food chains operate to offer new value propositions to consumers.

The food sector has unique features that distinguish it from the rest of the manufacturing sectors, for example, automotive. It is very large in scale, being the largest manufacturing sector in terms of employment; for example, in the United Kingdom only there are 3.7 m people employed in the sector, and contribution to the economy, for example, about USD26b to the UK economy. Despite the large number of companies, the sector is dominated by a relatively small number of multinationals employing ~40% of the people and accounting for about 50% of the total turnover. These companies tend to manufacture foods globally in large factories, with a relatively small profit margin. As such, investment in capital is scrutinized and financial benefit is critical. As an example, the varied cost of labor across the globe often results in varied use for automation. Industry 4.0 can provide scalable solutions that can integrate the value chain and increase efficiencies. The food sector, similar to many fast moving consumer goods, involves processing of natural materials to deliver products to a wide range of consumers. The introduction of new products to the market is often a key to business success, leading to an increasing pressure to accelerate innovation and scale up. Furthermore, the relatively small margin gains per unit product impose significant limitations on investment to new infrastructure.

Food value chains consist of a wide range of versatile processes with a very diverse range of stakeholders situated all the way from primary and manufacturing producers to the end customer. At the low value end lie commodity products, characterized by high production and supply volumes but with small profit margins. It is increasingly common to identify prime products and services associated with targeting niche markets or even individual customers (Dreyer et al., 2016). A wide range of products

Food Engineering Innovations Across the Food Supply Chain. DOI: https://doi.org/10.1016/B978-0-12-821292-9.00011-X

and services populate the space between the two ends. Supply chain agility and resilience (Stone and Rahimifard, 2018; Ali et al., 2018; Rotz and Fraser, 2015), customer orientation (Todorovic et al., 2018; Giannikas et al., 2019), as well as traceability (Alfian et al., 2019; Badia-Melis et al., 2015) and transparency (Astill et al., 2019) are among the some of the important considerations when managing production. In terms of manufacturing optimizing production in terms of minimizing product rejections through advanced quality control, as well as minimizing downtime through reducing time required for product change overs. Extensive research and several surveys dealing with the digital transformation of supply chains have been published (Astill et al., 2019; Miranda et al., 2019; Büyüközkan and Göçer, 2018; Lezoche et al., 2020; Vanderroost et al., 2017). Such transformation is associated with dynamic and evolving business models (Spendrup and Fernqvist, 2019; Nosratabadi et al., 2020) but the extent to which different technology enablers of digital transformation contribute to innovative and, in particular, to consumer-driven food value chains is less well explored. This work will aim to contribute by (a) outlining the key challenges of consumer-driven food value chains; (b) discussing the main characteristics of demand-driven food chains; (c) mapping the extent to which different enabling technologies for digitalized food production chains contribute to addressing the challenges and to the desired characteristics of demand-driven food value chains.

21.2 Visual analytics on relevant literature

A comprehensive literature review for journal articles has been performed in both the Web of Science (WoS) and Scopus on "food production" and "supply chain," aimed at identifying the key concepts explored in the last 3 years. The intention was not to perform a detailed literature analysis but to inform a higher-level meta-analysis on the interrelationship between key challenges, requirements, and potential solution enablers. The results were analyzed through the graph network bibliometric visualization tool VOSviewer (VOSviewer.com, Leiden University, The Netherlands). Network graphs of terms used in article titles and abstracts was built. The key visualization features are determined by scores and weights and are visualized with bubbles and connectors. High scores correspond to more frequently used terms and are depicted with larger bubbles. Thicker connectors correspond to more frequent co-occurrence of terms. Trimming out weak connections and infrequent terms while not allowing significant cluster fragmentation, WoS produced two main clusters, one "red cluster" contained terms relevant to the overall "food ecosystem" (Fig. 21.1).

The green cluster was mostly relevant to resources and sustainability. There was very little evidence of Industry 4.0 technology enablers. The Scopus picture has similarities with WoS and displays three main clusters (Fig. 21.2). The green one is largely about resources and sustainability. Instead of a single food ecosystem cluster, two clusters are formed: one relevant to the stakeholders of the production ecosystem and associated processes and characteristics; and one relevant to biological aspects, including health, risk, and ingredients.

This reflects the somewhat different disciplines covered by the two databases, with Scopus covering more engineering and technology, compared to WoS. However, this is less significant compared to the observation that in none of the two cases can one identify any strong presence of digital enabling technologies. Both cases avoided explicitly including digitalization terms in the search so as to explore whether such terms are already prevalent in the literature, rather than directing the search toward them. This indicates that relevant research still has to catch up with employing such technologies.

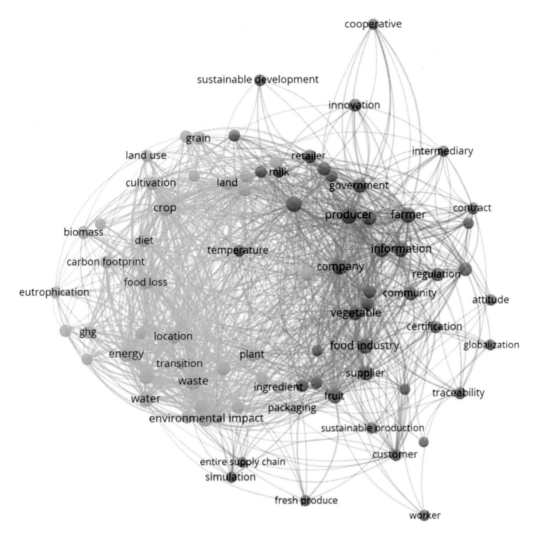

FIG. 21.1

WoS visual analysis of literature. (For interpretation of the references to color in this figure legend, the reader is referred to the web version of this article.)

21.3 Characteristics of resilient customer-driven food chains

Effective and sustainable food value chains with strong customer orientation are expected to exhibit some key characteristics, which are largely considered to fall under the broad categories of (i) food ecosystem; and (ii) performance, encompassing:

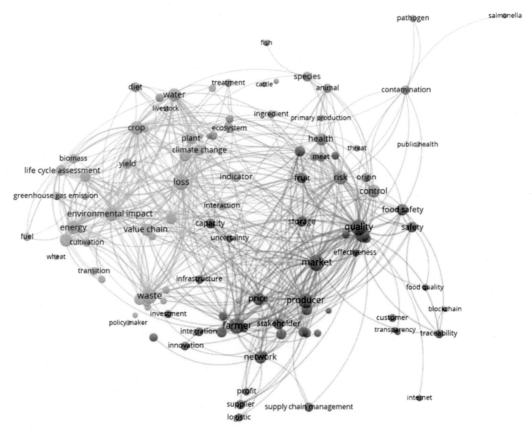

FIG. 21.2

Scopus visual analysis of food production and supply chain literature. (For interpretation of the references to color in this figure legend, the reader is referred to the web version of this article.)

Food ecosystem and market agility and resilience

- Food ecosystem stakeholder integration (Stancová and Cavicchi, 2018; Zondag et al., 2017).
- Business models and innovation (Todorovic et al., 2018; Spendrup and Fernqvist, 2019; Nosratabadi et al., 2020; Stancová and Cavicchi, 2018; Utami et al., 2019; Nakandala and Lau, 2019).
- Sustainability (Todorovic et al., 2018; Herrmann et al., 2014).
- Resilience/agility (logistics routing, suppliers, returns, last mile, local chain; Stone and Rahimifard, 2018; Ali et al., 2018; Rotz and Fraser, 2015; Utami et al., 2019).
- Personalized customer experience, including consumers in the design of products, and behavioral changes (Kumar et al., 2020).

Performance, quality, and safety assurance

- Performance and risk management (suppliers, speed, costs, inventories; Tsang et al., 2018).
- Transparency (Hsiao and Huang, 2016) and traceability (Alfian et al., 2019; Liu, 2015).
- Product assurance and compliance (Alfian et al., 2019; Aung and Chang, 2014; Shih and Wang, 2016; Fang et al., 2017; Bouzembrak et al., 2019).

While the above is not a complete list of key characteristics, it contains some of the most recognizable ingredients that can lead to high-performing food value chains. The birds-eye view of related literature indicates a clear need to link such characteristics with current and emerging technology enablers for digitalized food value chains. The present chapter focuses on several such technology enablers:

- Human interaction technologies (Giannikas et al., 2019; Miranda et al., 2019; Vanderroost et al., 2017). For example, developing data visualization methods to provide information to consumers and plant operators.
- Machine learning, and data analytics (Alfian et al., 2019; Elijah et al., 2018): For example, analyzing production data to optimize manufacturing conditions.
- Internet of things (with one or more of sensing, identification, communication, computing; Alfian et al., 2019; Miranda et al., 2019; Lezoche et al., 2020; Tsang et al., 2018; Liu, 2015; Bouzembrak et al., 2019; Gialelis et al., 2019; O'Sullivan, 2017; Verdouw et al., 2018). For example: develop affordable ways for devices to share data from remote locations and optimize a production network.
- Security, safety, and trust technologies and architectures (including hyperledger architectures; Bouzembrak et al., 2019; Verdouw et al., 2018; Mc Carthy et al., 2018; Doinea et al., 2015; Mondal et al., 2019; Leng et al., 2018; Caro et al., 2018). For example: ensure data, for example, from production facilities, will be used in a secure way.
- System integration (Shih and Wang, 2016; Verdouw et al., 2018). For example: integrate process data from different unit operations to optimize the entire production systems.
- Semantics, context, information management, and interoperability (Geerts and O'Leary, 2014). For example: use data obtained at different parts of the supply chain.
- Connectivity (physical layer, networking, application layer; Villa-Henriksen et al., 2020). For example: develop physical and digital infrastructure to share data.
- Augmented/virtual reality (AR/VR; Vanderroost et al., 2017). For example: consumers could use of augmented reality to obtain information about a product, for example, list of ingredients, directions of use.
- Simulation (Raba et al., 2019; Verboven et al., 2020). For example: use of computational fluid dynamics to understand the effect of different process parameters.
- Cloud-based platforms (Verdouw et al., 2016). For example: use of cloud computing to perform calculations in real time.
- Automation and robotics (Miranda et al., 2019): For example: developing flexible robotics for example for flexible packaging.
- Advanced/smart materials and packaging (includes embedded/printed electronics; Vanderroost et al., 2017; Fang et al., 2017; Ghaani et al., 2016; Majid et al., 2018). For example: developing of responsive packaging materials that provide contextual information.
- Supply chain optimization (Govindan et al., 2014; Mehdizadeh et al., 2012; Rong et al., 2011; Dabbene et al., 2008a, 2008b; Savastano et al., 2018; Bai et al., 2020; Akyazi et al., 2020; Tundys and Wiśniewski, 2020; Xu et al., 2020). For example: dynamic optimization of logistics to adjust for local conditions, for example, traffic.

All the above are typically associated with Industry 4.0. Some can be considered as part of auto-mation and robotics, but are listed separately to indicate the critical role they have within internal and external logistics, and retailing. Table 21.1 maps the extent to which the above technologies contribute to delivering the key characteristics. Food ecosystem depends heavily on the overall system integration, interoperability, and cross-layer connectivity. All of the above require cloud-based platforms. Security and trust technologies are of paramount importance. Technologies such as sensors and block chain can increase the trust of consumer on the products by addressing issues related to food fraud. Food fraud and adulteration in general are issues of significant importance both for consumers as well as manufac-turers as they can have a significant impact both in terms of hazards to the public as well as potentially bearing significant costs to the food manufacturers in terms of reputation and damaging high value brands. In this context trust is also relevant to the way data are stored and analyzed for decision-making processes. For example, data from manufacturing facilities are very sensitive in terms of confidentiality and reputational risks, and storage and analysis, especially from third parties, will bring different chal-lenges. Automation and robotics are key operational technologies that enable production and logistics workflows. Ensuring customer engagement is complex and relies to a significant extent on involving nearly all listed technology enablers, starting from human interaction ones. Performance and risk man-agement, as well as resilience and agility require the inclusion of higher added value enablers, such as AI, IoT, simulation, optimization, and advanced packaging, on top of others.

Supply chain optimization

Achieving higher profits by continuous network adaptation and aggressive cost reduction is a global trend. The focus of relevant publications has addressed lean manufacturing principles, fostering corpo-rate profitability, and providing direct (lower cost, shorter delivery time) benefits to wholesalers, retail-ers, and customers, at variable degrees. Supply chain optimization research and software development and implementation (especially in the past two decades) has accordingly focused on capturing, portray-ing, modeling, and modifying network structures, IT/transport/storage resources, and most importantly purchase order and product flows (e.g., via binary and/or continuous variables for network connections and resource requirements, respectively). Globalization of markets, transport networks, storage facili-ties, outsourcing capabilities, and venture/foreign capital solicitation emphasized this relentless quest, which however has invariably relied on the implicit assumption of adequate robustness and manage-able, distributed disruptions.

Enter Covid-19 pandemic, to serve only as an example of potential catastrophic effects on food provision security and sustenance, let alone sustainability. Global (in space) or prolonged (in time) disruptions expose systemic supply chain deficiencies, particularly when the latter are inevitable or deliberate. While in the past food chains have been optimized for efficiency, resilience will need to come into account. This could include developing an ecosystem of manufacturing facilities that will include local production of foods combined with global supply chains. In terms of manufacturing, we will need to further develop flexible equipment that could be adapted-repurposed to produce a range of products. Packaging equipment presents a significant challenge as currently has limited flexible. Food manufacturers should be prepared by identifying analyzing risks imposed from similar pressures, for example, diversifying in the supplier of ingredients and customers; formulate products with alternative locally sourced ingredients. It is important to keep in mind that existing food supply chains are heavily dependent on fossil fuels making them vulnerable. Existing and new technologies, including digital and physical, are likely to play key roles in the operation and maintenance of these enriched agrifood chains, as well as in ensuring food safety and hygienic practices. Domestic production (e.g., latitude/

Table 21.1 Mapping between technology enablers and food chain characteristics.

	Ecosystem and stakeholders integration	Personalized customer experience and engagement	Performance and risk management (suppliers, speed, costs, inventories)	Business models	Sustainability	Resilience and agility: logistics, routing, suppliers, returns, last mile	Transparency and traceability	Product assurance and compliance
Human interaction technologies	X	XXX	X	X	X	X	XX	
Machine learning-AI analytics		XXX	XXX	X	X	XX	XX	X
Internet of things (with one or more of sensing, identification, communication, computing)	XX	XXX	XXX	XX	XX	XXX	XXX	XXX
Security and trust technologies and architectures	XXX	XXX	XX	XXX	X	XX	XXX	XXX
System integration	XXX	XX	XXX	XXX	XXX	XXX	XX	X
Connectivity	XX	XX	XX	X	XXX	XXX	XXX	XXX
Physical layer Networking	XXX	XXX	XXX	XX	XXX	XXX	XXX	XXX
Application layer	XXX	XXX	XX	XXX	XX	XXX	XX	XX
Interoperability, including semantics and context information management	XXX	XXX	X	X	XX	X	X	XX
AR/VR	X	XXX	X	X	XX	XXX	XX	X
Simulation	XXX	XX	XXX	XX	XXX	XXX	XXX	XX
Cloud-based platforms	XXX	XXX	XXX	XXX	X	XXX	XXX	XX
Automation and robotics	XXX	XXX	XXX	XXX	x	XXX	XXX	XX
Advanced/smart materials and packaging (includes embedded/printed electronics)	X	XX	XXX	XXX	XXX	XXX	XXX	XXX

climate) limitations, consequent high or exclusive reliance on high-volume imports (especially of seasonal, sensitive, and/or perishable goods), minimized storage capacity for cost-optimal (e.g., Just-In-Time, JIT) distribution networks, and labor market (e.g., harvest season seasonality, Brexit) challenges all weaken regional and national food supply chains (Bakalis et al., 2020; Garnett et al., 2020). Therein emerges a clear need to understand the interplay (and strike a compromise) between short-term, cost-optimized profitability and long-term, risk-optimized resilience in food supply chain management. This multilayer problem has several levels of complexity, and it is all the more perplexed by the spatio-temporal variation of supply and demand, but even more so by consumer behavior, financial interests, national legislation (POST, 2021), and international regulatory landscapes (OECD, 2021).

Business models are enabled by application layer connectivity, integration, cloud platforms for stakeholder integration, as well as by automation technologies and smart packaging. Digital twins is key for sustainability regarding the design perspective, while operational delivery requires integration, connectivity, and advanced packaging to inform consumer behavior. IoT is key for transparency, traceability, and product assurance, together with security and trust, connectivity, and cloud platforms, integrating also automation technologies and smart packaging for transparent and efficient workflows. The mapping, currently based on literature, and the judgment of the authors, requires a systematic analysis that also include consumer behavioral studies.

21.4 Conclusions

Food value chains are increasingly driven by changing individual customer demands and are required to respond to highly disruptive market changes. Facing such challenges, food production ecosystems are undergoing digital transformation. Although relevant literature is rich with respect to the emerging characteristics of future resilient food value chains, as well as regarding the use of digitalization enablers, the contribution of such enablers toward consumer-oriented or -driven food value chains and their agile and resilient characteristics is less well-established. Customer-driven products specifically depend on more complex food value chains, compared to conventional supply models, and their dependence on efficient end-to-end digitization is evident. This chapter analyzed current literature in this area and proposed a mapping between relevant technology enablers that support such digital transformation. While the mapping takes into account current literature together with own judgment to produce a rated mapping of technology enablers contribution, more systematic research, which includes empirical findings is needed to establish its validity and indeed apply appropriate adjustments and extensions. Such a mapping is a valuable tool toward devising a model for prioritizing digitization interventions for food value supply chains.

References

Akyazi, T., Goti, A., Oyarbide, A., Alberdi, E., Bayon, F., 2020. A guide for the food industry to meet the future skills requirements emerging with industry 4.0. Foods 9, 492–507.

Alfian, G., et al., 2019. Improving efficiency of RFID-based traceability system for perishable food by utilizing IoT sensors and machine learning model. Food Control 110, 107016.

Ali, I., Nagalingam, S., Gurd, B., 2018. A resilience model for cold chain logistics of perishable products. Int. J. Logist. Manag. 29, 922–941.

Astill, J., et al., 2019. Transparency in food supply chains: a review of enabling technology solutions. Trends Food Sci. Technol. 91, 240–247.

Aung, M.M., Chang, Y.S., 2014. Traceability in a food supply chain: safety and quality perspectives. Food Control 39, 172–184.

Badia-Melis, R., Mishra, P., Ruiz-García, L., 2015. Food traceability: new trends and recent advances. Rev. Food Control. 57, 393–401.

Bai, C., Dallasega, P., Orzes, G., Sarkis, J., 2020. Industry 4.0 technologies assessment: a sustainability perspective. *Int. J. Prod. Econ.* 229, 107776.

Bakalis, S., et al., 2020. Perspectives from CO+RE: how COVID-19 changed our food systems and food security paradigms. *Curr. Res. Food Sci.* 3, 166–172.

Bouzembrak, Y., Klüche, M., Gavai, A., Marvin, H.J.P., 2019. Internet of things in food safety: literature review and a bibliometric analysis. Trends Food Sci. Technol. 94, 54–64.

Büyüközkan, G., Göçer, F., 2018. Digital supply chain: literature review and a proposed framework for future research. Comput. Ind. 97, 157–177.

Caro, M.P., Ali, M.S., Vecchio, M., Giaffreda, R., 2018. Blockchain-based traceability in agri-food supply chain management: a practical implementation2018 IoT Vert. Top. Summit Agric. — Tuscany, IOT Tuscany 2018, 1–4.

Dabbene, F., Gay, P., Sacco, N., 2008a. Optimisation of fresh-food supply chains in uncertain environments, Part I: background and methodology. *Biosyst. Eng.* 99, 348–359.

Dabbene, F., Gay, P., Sacco, N., 2008b. Optimisation of fresh-food supply chains in uncertain environments, Part II: a case study. *Biosyst. Eng.* 99, 360–371.

Doinea, M., Boja, C., Batagan, L., Toma, C., Popa, M., 2015. Internet of things based systems for food safety management. Inform. Econ. 19 (1), 87–97.

Dreyer, H.C., Strandhagen, J.O., Hvolby, H.H., Romsdal, A., Alfnes, E., 2016. Supply chain strategies for speciality foods: a Norwegian case study. Prod. Plan. Control. 27, 878–893.

Elijah, O., Rahman, T.A., Orikumhi, I., Leow, C.Y., Hindia, M.N., 2018. An overview of internet of things (IoT) and data analytics in agriculture: benefits and challenges. IEEE Internet Things J. 5, 3758–3773.

Fang, Z., Zhao, Y., Warner, R.D., Johnson, S.K., 2017. Active and intelligent packaging in meat industry. Trends Food Sci. Technol. 61, 60–71.

Garnett, P., Doherty, B., Heron, T., 2020. Vulnerability of the United Kingdom's food supply chains exposed by COVID-19. *Nat. Food* **1**, 315–318.

Geerts, G.L., O'Leary, D.E., 2014. A supply chain of things: the EAGLET ontology for highly visible supply chains. Decis. Support Syst. 63, 3–22.

Ghaani, M., Cozzolino, C.A., Castelli, G., Farris, S., 2016. An overview of the intelligent packaging technologies in the food sector. Trends Food Sci. Technol. 51, 1–11.

Gialelis, J., Theodorou, G., Paparizos, C., 2019. A low-cost internet of things (IoT) node to support traceability: logistics use case, ACM International Conference Proceeding Series, 72–77.

Giannikas, V., McFarlane, D., Strachan, J., 2019. Towards the deployment of customer orientation: a case study in third-party logistics. Comput. Ind. 104, 75–87.

Govindan, K., Jafarian, A., Khodaverdi, R., Devika, K., 2014. Two-echelon multiple-vehicle location-routing problem with time windows for optimization of sustainable supply chain network of perishable food. *Int. J. Prod. Econ.* 152, 9–28.

Herrmann, C., Schmidt, C., Kurle, D., Blume, S., Thiede, S., 2014. Sustainability in manufacturing and factories of the future. Int. J. Precis. Eng. Manufact.-Green Technol. 1 (4), 283–292.

Hsiao, H.I., Huang, K.L., 2016. Time-temperature transparency in the cold chain. Food Control 64, 181–188.

Kumar, A., Singh, R.K., Modgil, S., 2020. Exploring the relationship between ICT, SCM practices and organizational performance in agri-food supply chain. Benchmarking 27, 1003–1041.

Leng, K., Bi, Y., Jing, L., Fu, H.C., Van Nieuwenhuyse, I., 2018. Research on agricultural supply chain system with double chain architecture based on blockchain technology. Fut. Gener. Comput. Syst. 86, 641–649.

Lezoche, M., et al., 2020. Agri-food 4.0: a survey of the supply chains and technologies for the future agriculture. Comput. Ind. 117, 103187.

Liu, K., 2015. Research on the food safety supply chain traceability management system based on the internet of things. Int. J. Hybrid Inf. Technol. 8 (6), 25–34.

Majid, I., Ahmad Nayik, G., Mohammad Dar, S., Nanda, V., 2018. Novel food packaging technologies: innovations and future prospective. J. Saudi Soc. Agric. Sci. 17, 454–462.

Mc Carthy, U., Uysal, I., Badia-Melis, R., Mercier, S., O'Donnell, C., Ktenioudaki, A., 2018. Global food security – issues, challenges and technological solutions. Trends Food Sci. Technol. 77, 11–20.

Mehdizadeh, A., Shah, N., Raikar, N., Bongers, P.M.M., 2012. Food supply chain planning and quality optimization approach. In: Bogle, I.D.L., Fairweather, M. (Eds.), Computer Aided Chemical Engineering, 30. Elsevier, Amsterdam, pp. 1172–1176.

Miranda, J., Ponce, P., Molina, A., Wright, P., 2019. Sensing, smart and sustainable technologies for agri-food 4.0. Comput. Ind. 108, 21–36.

Mondal, S., Wijewardena, K.P., Karuppuswami, S., Kriti, N., Kumar, D., Chahal, P., 2019. Blockchain inspired RFID-based information architecture for food supply chain. IEEE Internet Things J. 6, 5803–5813.

Nakandala, D., Lau, H.C.W., 2019. Innovative adoption of hybrid supply chain strategies in urban local fresh food supply chain. Supply Chain Manage 24, 241–255.

Nosratabadi, S., Mosavi, A., Lakner, Z., 2020. Food supply chain and business model innovation. Foods 9 (2), 132.

OECD, 2021. Food Supply Chains and COVID-19: Impacts and Policy Lessons. http://www.oecd.org/coronavirus/policy-responses/food-supply-chains-and-covid-19-impacts-and-policy-lessons-71b57aea/.

O'Sullivan, M.G., 2017. A Handbook for Sensory and Consumer-Driven New Product Development: Innovative Technologies for the Food and Beverage Industry. Woodhead Publishing, Cambridge, UK.

POST, 2021. Effects of COVID-19 on the Food Supply System - POST. https://post.parliament.uk/effects-of-covid-19-on-the-food-supply-system/.

Raba, D., Juan, A.A., Panadero, J., Bayliss, C., Estrada-Moreno, A., 2019. Combining the internet of things with simulation-based optimization to enhance logistics in an agri-food supply chain, Proceedings of the Winter Simulation Conference 2019-December, 1894–1905.

Rong, A., Akkerman, R., Grunow, M., 2011. An optimization approach for managing fresh food quality throughout the supply chain. Int. J. Prod. Econ. 131, 421–429.

Rotz, S., Fraser, E.D.G., 2015. Resilience and the industrial food system: analyzing the impacts of agricultural industrialization on food system vulnerability. J. Environ. Stud. Sci. 5 (3), 459–473.

Savastano, M., Amendola, C., D'Ascenzo, F, 2018. How digital transformation is reshaping the manufacturing industry value chain: the new digital manufacturing ecosystem applied to a case study from the food industryLecture Notes in Information Systems and Organisation24. Springer, Heidelberg, pp. 127–142.

Shih, C.W., Wang, C.H., 2016. Integrating wireless sensor networks with statistical quality control to develop a cold chain system in food industries. Comput. Stand. Interfaces 45, 62–78.

Spendrup, S., Fernqvist, F., 2019. Innovation in agri-food systems—a systematic mapping of the literature. Int. J. Food Syst. Dyn. 10, 402–427.

Stancová, K.C., Cavicchi, A., 2018. Smart Specialisation and the Agri-food System: A European Perspective. MacMillan, New York, USA.

Stone, J., Rahimifard, S., 2018. Resilience in agri-food supply chains: a critical analysis of the literature and synthesis of a novel framework. Supply Chain Manage 23, 207–238.

Todorovic, V., Maslaric, M., Bojic, S., Jokic, M., Mircetic, D., Nikolicic, S., 2018. Solutions for more sustainable distribution in the short food supply chains. Sustainability 10 (10), 3481.

Tsang, Y.P., Choy, K.L., Wu, C.H., Ho, G.T.S., Lam, C.H.Y., Koo, P.S., 2018. An internet of things (IoT)-based risk monitoring system for managing cold supply chain risks. Ind. Manage. Data Syst. 118, 1432–1462.

Tundys, B., Wiśniewski, T., 2020. Benefit optimization of short food supply chains for organic products: a simulation-based approach. Appl. Sci. 10, 2783.

Utami, H.N., Alamanos, E., Kuznesof, S.: Co-creation benefits by re-configuring the value network in creative agri-food transformation through the SMEs e-commerce channel: a business market perspective, vol. 98, pp. 63–68 (2019).

Vanderroost, M., Ragaert, P., Verwaeren, J., De Meulenaer, B., De Baets, B., Devlieghere, F., 2017. The digitization of a food package's life cycle: existing and emerging computer systems in the logistics and post-logistics phase. Comput. Ind. 87, 15–30.

Verboven, P., Defraeye, T., Datta, A.K., Nicolai, B., 2020. Digital twins of food process operations: the next step for food process models? Curr. Opin. Food Sci. 35, 79–87.

Verdouw, C.N., Robbemond, R.M., Verwaart, T., Wolfert, J., Beulens, A.J.M., 2018. A reference architecture for IoT-based logistic information systems in agri-food supply chains. Enterp. Inf. Syst. 12, 755–779.

Verdouw, C.N., Wolfert, J., Beulens, A.J.M., Rialland, A., 2016. Virtualization of food supply chains with the internet of things. J. Food Eng. 176, 128–136.

Villa-Henriksen, A., Edwards, G.T.C., Pesonen, L.A., Green, O., Sørensen, C.A.G., 2020. Internet of things in arable farming: implementation, applications, challenges and potential. Biosyst. Eng. 191, 60–84.

Xu, W., Zhang, Z., Wang, H., Yi, Y., Zhang, Y., 2020. Optimization of monitoring network system for Eco safety on Internet of Things platform and environmental food supply chain. Comput. Commun. 151, 320–330.

Zondag, M.M., Mueller, E.F., Ferrin, B.G., 2017. The application of value nets in food supply chains: a multiple case study. Scand. J. Manag. 33, 199–212.

Potential applications of nanosensors in the food supply chain

S. Shanthamma, M. Maria Leena, J.A. Moses, C. Anandharamakrishnan

Computational Modeling and Nanoscale Processing Unit, Indian Institute of Food Processing Technology (IIFPT), Ministry of Food Processing Industries, Govt. of India, Thanjavur, India

22.1 Introduction

The food sector has seen revolutionary technological changes over the past few decades. Cognitive sciences, information technology, biotechnology are the most notable evolved sciences with implications of nanotechnology. More recently nanotechnology has entered the food sector and in a short duration widely applied in many sectors of the food supply chain from raw materials to distribution and tracking of food (Chaudhry et al., 2008). Nanomaterials are materials with at least one dimension smaller than 100 nm. These are further classified as (i) nanofilms and coatings (<100 nm in one dimension), (ii) nanotubes and wire (<100 nm in two dimensions), and (iii) nanoparticles (<100 nm in three dimensions). Because of the incredibly small size, nanomaterials exhibit unique physicochemical properties like increased surface area, improved electrical, thermal, optical, and mechanical properties that differ from their bulk counterparts and also from the same nanomaterials of different dimensions. This uniqueness offers more opportunity to customize sensors to discriminate one analyte from a complex mixture and also allows targeting of analytes that were previously inaccessible (Singh et al., 2007). Nanomaterials are broadly classified into inorganic nanomaterials, organic nanomaterials, and nanocomposites (Fig. 22.1; Baranwal et al., 2018). Inorganic nanomaterials include silver, gold, platinum, titanium, zinc, thallium nanoparticles, etc. Organic nanoparticle includes the nanomaterials formed using polysaccharides, proteins, natural, and synthetic polymers. The integration of organic and inorganic nanomaterials into bulk structures forms nanocomposites. Various nanomaterials like carbon nanotubes (CNTs), gold nanoparticles, quantum dots silicon nanowires, imprinted polymeric particles have been widely explored in sensing of toxic metal ions, toxic gases, pesticides, and hazardous chemicals with high selectivity, sensitivity, and fast response time.

The nanosensors can be classified into optical sensors, electrochemical nanosensors, chemical sensors, and biosensors based on their mechanism of detection. Nanosensors are extremely small compared to normal sensors. The mechanism of working of this nanosensor is that they will detect the physicochemical and biological signal and transfer them into an output or signal which can be pursued by humans (Omanović and Maksimović, 2016). Even though the implications of nanosensors in the food supply chain are relatively new, it has created a significant impact on every part of the food supply chain starting from farm to fork. The different applications of nanosensors in the food supply chain are shown in Fig. 22.2.

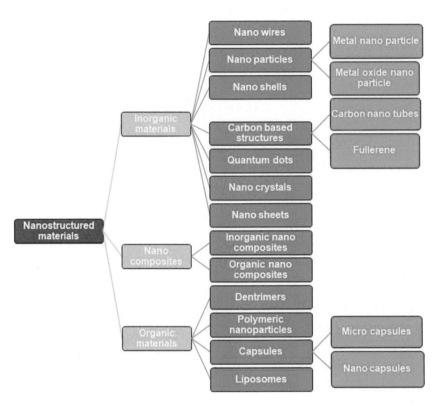

FIG. 22.1

Classification of nanomaterials (with permission from Baranwal et al., 2018).

Nanosensors have revolutionized the food packaging sector. Some examples of commercial sensors and time and temperature indicators used in the food supply chain are listed in the next chapter. The main function of food packaging is to maintain quality and extend the shelf life of food products. From the consumer point of view, the quality of packed food is known in the sense of its freshness, sensory, and safety. The food inside the package may be deteriorated but cannot be known by consumers until opened. The consumers are also uncertain about the processing condition and ingredients added to food. All these disadvantages of food packaging are overcome by applying knowledge of nanosensors in food packaging called smart packing. Nanosensors detect the presence of chemical contaminants, pathogens, aroma, and gases in food and ensure the consumers to buy safe and high-quality food (Omanović and Maksimović, 2016). The research carried out in the field of smart packing is discussed in the upcoming subunits of this chapter.

Quality assurance plays an important role in the food supply chain as the deterioration in the quality of food will cause harmful health effects. Hence it is the responsibility of the food industry to maintain quality assurance in their food products (Neethirajan and Jayas, 2011). Nanosensors reduce the detection time for microorganisms. The conventional method for the detection of microorganisms needs hours to days but nanosensors can detect the pathogens in seconds. Burris and Stewart (2012) worked

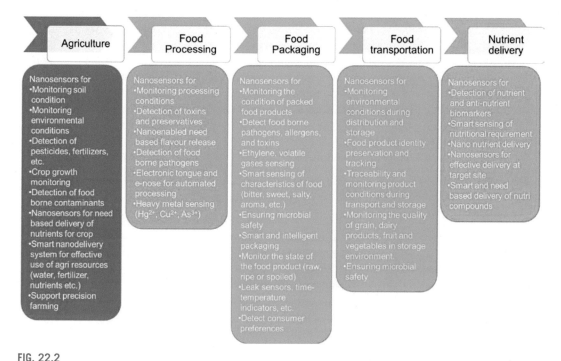

FIG. 22.2

Potential application of nanosensors in the food supply chain.

on the development of nanosensors to monitor the quality of food in the food supply chain. Nanosensors based on semiconductor quantum dots coupled with immunomagnetic nanoparticles have been successful in the simultaneous detection of three food-borne pathogenic bacteria, *Shigella flexneri*, *Salmonella typhimurium*, and *Escherichia coli* O157 in different liquid food samples (Zhao et al., 2009). Other examples of electrochemical nanobiosensors for food quality and safety monitoring are listed in the next chapter Table 1. Due to its greater sensitivity to gases released by spoiled food, these nanosensors have been used to monitor the freshness of the food products (Bowles and Lu, 2014). Nanosensors have also been used in monitoring the quality of packed food by monitoring oxygen (Mills, 2006) and CO_2 (Pal et al., 2007) in packed foods. Nanosensors have also proven their potential for the detection of allergens, nutrients, and temperature changes in foods (Danny, 2011). Considering the significance of nanosensor, this chapter essentially discusses different nanomaterials for nanosensors, types of nanosensors, and their application in the food supply chain.

22.2 Nanosensors

Today, the food industries is becoming more and more customer-oriented, hence the demand for efficient control and monitoring systems for maintaining food safety and getting faster response times to avoid food safety scandals is increasing. The production and distribution of unsafe or poor-quality food can be minimized by using good traceability. There is no guarantee that the traditional labeling system will provide authenticity, quality, and safety regarding food. On the other hand, foodborne disease is also the biggest problem to monitor and control in the food supply chain. The microorganisms responsible for foodborne illness are *Salmonella typhimurium*, *Escherichia coli*, *Campylobacter jejuni*, *Streptococci Legionella pneumophila*, *Staphylococcus aureus*, etc. The conventional detection of microorganisms is time consuming multistep procedures and laborious process too. Hence all these factors acting as the driving force for the invention and usage of sensors in the food supply chain (Ivnitski et al., 2000). Sensors can guarantee food safety by collecting, transferring, and providing authentic data about food (Aung and Chang, 2014).

The sensors should pass one of the below criteria in order to claim as nanosensors (Wiesendanger, 1995):

1. The display of detected parameters will be on the nanoscale. Example: nano-Newton, nanowatts.
2. Detection of objects with sensors will be on the nanometer scale.
3. The sensor's size will be on the nanoscale.

To apply nanosensors application in food supply it is important to know the structure and composition of nanomaterials. Nanosensors are lightweight and extremely small with large surface area, improved response time, and greater selectivity. The decrease in size will increases the surface area of nanosensors. Approximately, the surface area of a football field ($7140 \ m^2$) is equal to 4.5 g of CNTs ($1600 \ m^2/g$; Dahman, 2017). The advantage of having larger surface areas will enable an increase in signal intensity so allowing to sense trace amounts. The other properties that are responsible for making nanosensors a good sensor is that nanoscale materials that have many unique and desirable physical properties such as increased reactivity, optical absorption, catalytic efficiency, electrical conductivity, wear resistance, strength, and magnetic properties in comparison to the bulk matter of the same composition. Hence nanosensors play a vital role in the control and monitor the traceability of good quality food and provide nutrient food to consumers.

22.2.1 Classification of nanosensors

22.2.1.1 Optical nanosensors

An optical sensor is a device that converts light rays into electronic signals. Similar to a photoresistor, it measures the physical quantity of light and translates it into a form read by the instrument. One of the features of an optical sensor is its ability to measure the changes from one or more light beams. This change is most often based around alterations to the intensity of the light (Ahuja and Parande, 2012). Incorporation of nanomaterial into optical sensor has several advantages like high spectral specificity, improved contrast, photostability (Kneipp et al., 2010). Major progress in the fluorescence, quantum mechanics, luminescence was recorded in nanomaterial combined optical sensors (Rong et al., 2017). These sensors were used for the detection of foodborne pathogens, glucose detection, and pH sensing in the food industry (Jafarizadeh-Malmiri et al., 2019; Krishna et al., 2018; Caon et al., 2017).

22.2.1.2 Chemical nanosensors

Chemical sensors are devices used in the detection of various chemical processes and transform different chemical signals like concentration, composition, presence of a particular element or ion, particle pressure, etc., into a useful analytical signal. For example, Bustillo and Li (2017) reported a chemical sensor based on the field-effect transistor, consisting of the chemically sensitive region as a channel between the source and drain region. The occurrence of chemical reaction in the environment causes modulation of charges in the channel region and thus conductance and helps inaccurate sensing of analyte (Bustillo and Li, 2017). Sensitivity, specificity, and rapidness of sensors were improved by the integration of nanomaterial in chemical sensors (Jafarizadeh-Malmiri et al., 2019). Chemical nanosensors were used for the sensing of chemical hazards in the food such as heavy metals, dioxins, melamine, acrylamide, and agrochemicals like pesticides, veterinary drugs, and food additives like coloring agents and sweeteners (Kamalii et al., 2019).

22.2.1.3 Electrochemical nanosensors

Electrochemical sensors are based on potentiometric, amperometric, or conductivity measurements. In an electrochemical reaction, transfer of charge from an electrode to another take place. During this process, chemical changes occur at the electrodes and the charge is conducted through the bulk of the sample phase. This is served as the basis for the sensing process. Incorporation of nanomaterial in sensors improves sensitivity, selectivity, large surface area, electrochemical activity (Cash and Clark, 2010; Lin et al., 2016; Benvidi et al., 2017). These sensors were successfully employed for sensing of preservatives, additives, and antibiotics in the food products (Srivastava et al., 2018).

22.2.1.4 Bionanosensors

A biosensor is an analytical device which converts a biological response into an electrical signal. The desired biological material (example, enzymes, DNA, oligonucleotide, antibody, receptors, etc.) can be immobilized on sensing element by conventional methods like physical or membrane entrapment, noncovalent or covalent binding, or self-assembly. This immobilized biological material is in intimate contact with the transducer. The analyte binds to the biological material to form a bound analyte which in turn produces the electronic response that can be measured (Malhotra et al., 2017). Biosensors are categorized into enzyme-based, tissue-based, immunosensors, DNA biosensors, etc. (Mehrotra, 2016). The binding of nanomaterials with biosensors with making it advantages over simple biosensors in terms of better sensitivity, selectivity, and rapidity (Zhao et al., 2017; Khater et al., 2017). Bionanosensors were used for the detection of food pathogens, toxins, and adulterants in the food products (Yamada et al., 2014).

22.3 Potential applications of nanosensors in food supply chain

22.3.1 Detection of pesticides

Pesticides are the substances used for pest control (Kuswandi and Mascini, 2005). The common subclasses of pesticides consist of insecticides, herbicides, molluscicides, growth regulators, rodenticides, animal, or bird repellents (Kuswandi et al., 2017). The accumulation of pesticide residue in humans causes many undesirable harmful health effects like cancer, neurodegenerative reproductive and developmental toxicity, respiratory problems (Hanke and Jurewicz, 2004; Ntzani et al., 2017). In the food supply chain, the exposure to the pesticide can occur through drinking water and food. It is important to regulate pesticide residues in food to prevent harmful health disorders (Caetano and Machado, 2008; Kuswandi et al., 2017). Traditionally the pesticide residue is determined by accelerated solvent extraction, solid-phase microextraction, matrix solid-phase dispersion, supercritical fluid extraction, membrane extraction techniques (Wilkowska and Biziuk, 2011). These traditional methods are time-consuming, expensive, and complicated multistep procedures (Beyer and Biziuk, 2008). Hence, nano-sensors with superior properties have been designed for rapid, highly sensitive, simple, and low cost on-field detection of pesticides (Liu et al., 2008). Potential nanosensors for the detection of different pesticides are discussed in the following subsections.

22.3.1.1 *Detection of organophosphates*

Paraoxon, parathion, and coumaphos are the example for organophosphate (OP) pesticides. They are mainly used in agriculture to eradicate insects. It is an effective insecticide but has low environmental persistence. However, to feed the large word population tons of OP are used which result in serious food safety problem (Liu et al., 2008). Tuteja et al. (2014) reported organophosphate detection using silver nanoparticles based on the sensing mechanism of inhibition of cholinesterases, organophosphorus hydrolase (OPH)-based immunoassays. In enzyme inhibition mechanism pesticide acts as an inhibitor of cholinesterases. The presence of pesticides is detected by measuring the kinetic performance of the initial velocity of the reaction catalyzed by cholinesterases enzyme. The presence of pesticides shows the variation in the kinetic performance before and after an incubation step with the pesticide. To maintain enzymatic stability, activity, and selectivity of biosensors, various stabilization, and immobilization approaches have been examined using a different type of acetylcholinesterase mutants and immobilization matrices (Stein and Schwedt, 1993; Tran-Minh et al., 1990). Pesticide detection based on acetylcholinesterase is carried out by using different transduction schemes and different combinations of nanosensors such as optical biosensors, colorimetric indicators, fluorescent, and different transduction schemes such as amperometric, potentiometric, field effective transistors (Sotiropoulou and Chaniotakis, 2005; Velasco-Garcia and Mottram, 2003). Organophosphorus hydrolase is a homodimeric enzyme that catalyzes the hydrolysis of organophosphorus pesticides. The substrate specificity of an enzyme is very broad and hence it can hydrolyze many OPs like parathion, paraoxon diazinon, Dursban, methyl parathion, coumaphos, etc. A simple nano biosensor can be constructed using this enzyme-based hydrolysis mechanism (Mulchandani et al., 2001). This device works based on hydrolysis of OP which produces alcohol and two protons due to the cleavage of P–O, P–S, P–CN, P–F bonds, which are electroactive and chromophoric (Lai et al., 1994). However, the use of enzyme-based nanosensors are not widely used because it does not meet the standards of sensitivity, cannot be measured selectively (Liu et al., 2008).

Immunosensors are biosensors that use an antibody or antigen as a specific sensing agent. Immunosensors can be divided into indirect and direct assays. The indirect assays associate with detectable labels such as enzymes (ELISAs; Dankwardt, 2006; Hunter and Lenz, 1982), chemiluminescent compounds (chemiluminescent immunoassay), fluorescent chemicals (fluoroimmunoassays), or radioactive isotopes (radioimmunoassay) for sensing of an analyte. Among this, the ELISA method is commonly used for pesticide detection because low detection limits can be achieved and can avoid the use of radioactive materials (Heldman et al., 1985). Compare to enzyme-based nanosensors, immunosensors-based nanosensors give accurate and fast results in the detection of pesticides. In direct immunosensor assays, methods surface plasma resonance (SPR), quartz crystal microbalance (QCM), and impedimetric devices are found to detect direct binding of the analyte with the antibody. The disadvantage of this method is, only one analyte at a time is measured and the regeneration of sensors is not rapid due to strong bond formation with OP (Liu et al., 2008). These factors constraints its practical application in the food industry.

22.3.1.2 Detection of carbofuran

Carbofuran is a type of pesticide widely used to control insects and nematodes. It has wide application because of its low persistence and wide-ranging biological activity (Pogačnik and Franko, 2003). It is extensively used for the application in agriculture and nonagricultural field which results in the addition to pesticide residue in the food supply chain (Kuswandi et al., 2017). Carbofuran is responsible for health issues such as vomiting, sweating, diarrhea, weakness, breathing difficulty. At high dosages death incidents are reported due to breathing problems (Kuswandi et al., 2017). Sun et al. (2012) determined carbofuran concentration using an electrochemical immune nanosensor using gold nanoparticles and PB–multiwalled carbon nanotubes (MWCNTs)–CTS nanocomposite film coated with protein A (SPA, cell wall component of *Staphylococcus aureus*) through self-assembly process. This self-assembled SPA layer increases the binding capacity of the antibody which can detect carbofuran pesticide. Gold nanoparticles and MWCNTs increases electroactivity of the sensor and carbofuran specific antibody support immunosensing and the porous three-dimensional PB–MWCNTs–CTS nanocomposite film provided many amino groups and carboxyl groups to cross-link SPA and offered a large specific surface area to immobilize SPA which supports a wide linear range between 0.1 and 1 µg/mL with a low detection limit of 0.021 ng/mL of pesticide sensing. The proposed immune nanosensors offer high sensitivity, fast response, and good stability which indicates that it may be applied not only in the laboratory but also in the field for detecting the pesticide residues in food and environment.

22.3.1.3 Detection of dichlorodiphenyltrichloroethane

Dichlorodiphenyltrichloroethane (DDT) (1,1,1-trichloro-2-2-bis (4 chlorophenyl) ethane) is one of the most commonly used organochlorine pesticides. Lisa et al. (2009) developed the gold nanoparticle assisted immunoassay-based dipstick for DDT sensing (Fig. 22.3). In this technique, gold nanoparticles were conjugated with anti-DDT antibodies, which could specifically detect DDT pesticides. Different concentration of free DDT was treated with antibody-conjugated gold nanoparticles. The resulting immune complex was detected using nitrocellulose membrane preimmobilized with DDA-BSA antigen specific to the conjugated antibodies. Depending on the concentration of free DDT the biding of gold nanoparticles to strip will vary and provided visible color change which could sense DDT as low as 27 ng/mL.

This method provides a preliminary, qualitative result. This technique possesses several advantages in terms of rapidity, specificity, and cost-effectiveness and is suited for field applications in on-site testing of pesticides. The technique can also be applied for on-site testing of other classes of organochlorine pesticides and toxins depending on the antibody availability.

22.3.2 Detection of foodborne pathogenic bacteria

Foodborne pathogens are a severe threat to human life. Long detection time, low sensitivity, and low selectivity are the disadvantages of the traditional methods in the detection of foodborne pathogens. The emergence of nanomaterial enables the construction of nanosensors which possess the advantages of selectivity, sensitivity, specificity, rapidity, accuracy, and simplicity.

22.3.2.1 Salmonella sps

Salmonella is known for food poisoning with symptoms like diarrhea, vomiting, fever, nausea, stomach cramps. Increased rates of bioterrorism and other biotreats are demanding rapid and real-time detection methods. Joo et al. (2012) reported that the highly sensitive detection and separation

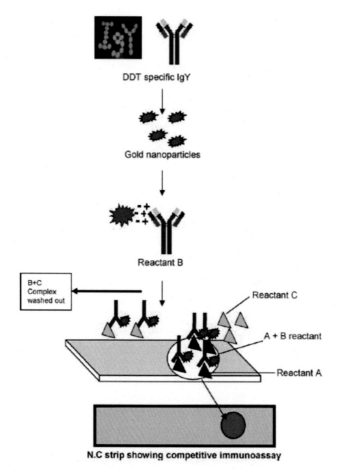

FIG. 22.3 Gold nanoparticles for detection of DDT in competitive immunoassay.

Mobilized DDA-BSA (Reactant A), IgY-conjugated gold nanoparticles (reactant B), and free DDT (reactant C; with permission from Lisa et al., 2009).

of *Salmonella* in milk using antibody-conjugated magnetic nanoparticles (MNPs). These antibody-conjugated nanoparticles could bind to *Salmonella* present in milk and form a complex which were then separated by applying an external magnetic field. These complexes were then reacted with antibody-immobilized TiO_2 nanocrystals (TNs) which act as an optical probe for quantification of the *Salmonella*. With this nanosensing approach rapid sensing of *Salmonella* in milk with a detection limit of 100 CFU/mL was achieved. In another study, amine-functionalized magnetic nanoparticles (AF-MNPs) were used for the separation of bacteria with higher efficacy up to 99.1%. This was achieved through strong electrostatic interaction between positive charges on the surface of AF-MNPs and negative charge on the surface of pathogenic bacteria (Huang et al., 2010). Recently, in a biomimetic approach *Salmonella typhimurium* specific DNA aptamers and antibodies were bound to aerosolized graphene which was then thermally annealed on to a polyimide tape for electrochemical sensing of the *Salmonella* in chicken broth. Based on resistance to charge transfer upon specific binding of *Salmonella* on to these bionanosensors, rapid sensing of the pathogenic organism in 17 min with a detection limit of ~5 CFU/mL was achieved. These graphene-based biosensors show high selectivity for the target pathogen of ~95% in chicken broth (Oliveira et al., 2019). Rapid advancements in nanosensing platform show great potential to replace current standard methods used by the food industry for rapid sensing and separation of foodborne pathogenic bacteria. MaxSignal, RapidChek, Gen-Probe, IQuum, Watersafe are the commercially available nanosensors for the detection of foodborne pathogens (Bülbül et al., 2015).

22.3.2.2 *Escherichia coli*
Maurer et al. (2012) developed a biosensor for the detection of *Escherichia coli* using CNT, GNPs, and ribonucleic acid (RNA). The thiolated RNA was bound to GNPs which was synthesized on growing CNT, having a graphite substrate. The CNT retained its electrical property and the incorporation of GNPs increased the surface area and its sensitivity. Selective RNA incorporated GNPs show higher *E. coli* capture efficiency, up to 185% when compared to uncoated particles. Chen et al. (2008) reported piezoelectric biosensor for real-time detection of *Escherichia coli* O157: H7 in a food sample, based on the mechanism of GNPs amplification and verification method. The piezoelectric biosensor surface was immobilized with a synthesized thiolated probe (Probe 1; 30-mer) specific to *E. coli* O157: H7 *eae A* gene. Exposing of the immobilized probe to *E. coli* O157: H7 *eae A* gene fragment (104-bp) amplified by PCR, will induce hybridization (Fig. 22.4). This results in a consequent frequency shift of the piezoelectric biosensor due to mass change. Gold (Au) nanoparticle is conjugated with second thiolated probe, complementary to the target sequence. This conjugated complex is used as a "sequence verifier" and "mass enhancer" to amplify the frequency change of the piezoelectric biosensor (Chen et al., 2008).

22.3.2.3 *Vibrio cholera*
Cholera is an acute infection of the intestine caused by vibrio cholera characterized by watery diarrhea may lead to severe dehydration and finally death if not treated promptly (Sack, 2011). Schofield et al. (2007) developed a calorimetric bioassay using GNPs of size ~16 nm. Specifically, the synthesized lactose derivative will self-assemble on gold nanoparticle to give a red color solution with max absorption at 524 nm. When cholera toxin (added as the B-subunit) (CTB) binds to the lactose derivative it induces aggregation of the nanoparticles and thus the color change in solution from red to purple color. This colorimetric nanosensor gives a direct measurement of CTB with a limit of detection of 3 μg/mL within 10 min. In another study, monoclonal anticholera toxin antibody bound gold nanoparticles were used for simple and fast sensing of cholera toxin based on color change. Antibody-conjugated gold nanoparticles aggregate in the presence of cholera toxin with high specificity over other diarrhetic toxins and cause a visible color change. Moreover, the degree of aggregation was quantified with the dynamic light scattering method. Visual detection of cholera toxin as low as 10 nM concentration in local lake water was achieved with this nanosensor (Khan et al., 2015).

22.3.3 Detection of food additives
Food additives are generally used for coloring, preservation, and increasing nutrient content in the food (Kuswandi et al., 2017). The use of preservatives in the food processing industry is increased due to

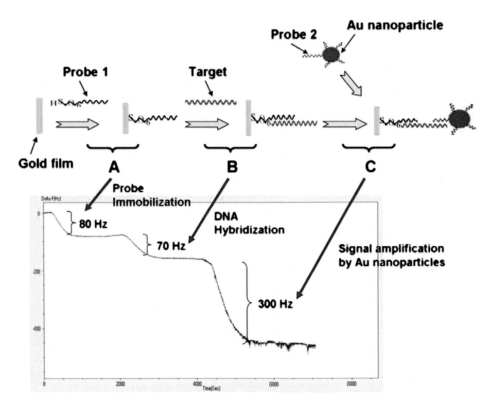

FIG. 22.4 **Time-dependent frequency changes in the circulating flow QCM sensor.**

(A) Addition of Probe 1 to self-assembly immobilization on the surface of the QCM sensor. (B) The complementary target oligonucleotides were subsequently introduced for DNA hybridization. (C) Additional treatment of the DNA hybridized QCM with Probe 2 capped with gold nanoparticles. These sequences of Probe 1 and Probe 2 are complementary to the two ends of the analyte DNA (i.e., target sequences; with permission from Chen et al., 2008).

increased demand for processed food. The main purpose of using preservatives is to increase shelf life. The commonly used additives to extend shelf life are benzoates of Na and K and benzoic acid and salts (Kuswandi et al., 2017). The benzoic acid and its salts have wide applications in the food industry. To maintain food quality, it is important to detect and quantify its concentration in food. The concentration above the permitted level will cause adverse health effects like chronic rhinitis, hyperactivity in children, genotoxicity, clastogenic, asthmatic reactions, mutagenicity (Wilson and Bahna, 2005). It was reported that 5 mg/kg may cause immunologic reactions in sensitive people. Hence it is important to measure the concentration of benzoic acid and benzoates in food (Sonawane, 2014). Shan et al. (2008) reported a sensitive and highly reversible amperometric nano biosensor for the determination of benzoic acid based on immobilization of tyrosinase (Tyro) by calcium carbonate nanomaterials (nano-$CaCO_3$). The high sensitivity and fast response of benzoic acid by inhibitor biosensor were found to be 1061.4 ± 13 mA/M/cm^2 and (<5 s) with a wide linear range of 5.6×10^{-7} to 9.2×10^{-5} M in the food samples like Sprite, Pepsi-Cola, Coca-Cola, and Yoghurt. The results were compared and in agreement with high-performance liquid chromatography analysis.

Selective detection of particular food additives in the mixture of multiple food additives is also made possible with nanosensors. For example, selective detection of vanillin in the presence of folic acid was achieved through electrochemical oxidation of vanillin using a nanomaterial electrode made of cadmium oxide nanoparticle decorated with single-wall carbon nanotubes (CdO/SWCNTs) and 1,3-dipropylimidazolium bromide (DPIB) as a binder (CPE/CdO/SWCNTs/DPIB) on carbon paste electrode. Real-time and simultaneous sensing of vanillin and folic acid in different food samples like chocolate, biscuit, and coffee milk was achieved with a detection limit of 9 nM and 0.06 µM, respectively (Cheraghi et al., 2017). Low cost, fast response time, high sensitivity, excellent reversibility, and ease of fabrication were made these nanosensors commercially valuable.

Naturally occurring antinutritional factors such as gluten, oxalate, glycoalkaloids normally hinder the absorption of nutrients in the small intestine. This will result in health disorders like Celiac disease and malnutrition and there is a need for fast detection methods like nanosensors. Nanohybrid film of gold colloidal nanoparticles and multiwalled carbon nanotubes (MWCNTs) were used to fabricate the amperometric oxalate biosensor. For this Cl^- and NO_3^-, insensitive oxalate oxidase was isolated from sorghum grain and immobilized on the electrode. The effect of oxalate concentration, temperature, and pH on electrode activity was studied. The electrode exhibited an optimum response within 7 s with a detection limit as low as 1 µM and the K_m value for the oxalic acid sensor was found to be 444.44 µM when stored at 4 °C. This nanoelectrode was employed for the measurement of oxalic acid in serum, urine, and foodstuffs (Pundir et al., 2011).

22.3.3.1 Dyes

Food dyes are added to improve color, selection criteria, and characteristics of food. The common dyes used in the food industry are brilliant blue, Erythrosine, Allura red, Tetrazine and sunset yellow, Indigotin, etc. In most cases, synthetic dyes are organic chemicals derived from coal tar, petroleum by-products. In the food industry, sometimes these synthetic dyes are mixed with natural dyes because of their intense color compared to natural color, stability, and low cost. However, these synthetic colors may cause cancer and other serious health issues. Hence it is important to maintain strict regulations for the permitted concentration of dyes. Ye et al. (2013) reported the synthesis of β-cyclodextrin-coated poly(diallyldimethylammonium chloride) functionalized graphene composite film (β-CD-PDDA-Gr) incorporated with l-ascorbic acid (l-AA) as the reducing agent at room temperature. This nanocomposite film is used for the highly sensitive and simultaneous detection of sunset yellow and tetrazine with a detection limit of 1.25×10^{-8} mol/L and 1.43×10^{-8} mol/L, respectively.

Similarly, Wang et al. (2014) developed an electrochemical sensor based on nano-Ag for the detection of sunset yellow and Tartrazine in drinks. This nanosensor had excellent reproducibility and repeatability with a linear response in the range of 4.0×10^{-9} to 1.0×10^{-6} for sunset yellow and 7.0×10^{-9} to 1.5×10^{-6} M for Tartrazine and detection limits of 5.2×10^{-10} and 8.3×10^{-10} M, respectively. Even till date, nanosensor developments are focusing on high sensitive detection of toxic food dyes in real food samples. Recently, Shah (2020) developed a nanosensor for the simultaneous detection of two toxic food dyes like fast green and metanil yellow in juice and water samples. Gold nanoparticles and calixarene were attached to the carbon electrode to enhance the electrochemical signals and detected potent dyes with a minimum limit of 9.8 and 19.7 nM for metanil yellow and fast green, respectively. All these nanoenabled sensors show reproducibility, high recovery, stability suggesting potent applicability for real food sample analysis.

22.3.3.2 Sweeteners

Sweeteners are substances that are used as alternatives to sucrose in food products. Isomalt, maltitol, lactitol, mannitol, saccharin, erythritol are some examples of the artificial sweeteners used in the food industry (Nikolelis et al., 2001). There are two types of sweeteners, intense and bulk sweeteners based on their sweetness level compared to sucrose. The intense sweeteners are generally synthesized through chemical processes and widely used in food products like carbonated beverages, candies, and baked foods, Jellies. The intense sweeteners increase the appeal and mouth feel of the product but do not add the energy to the food. However, alternative sugars are found to cause health disorders like cancer, oral decay, and cardiovascular diseases. They are also known for increasing the blood glucose level (Turkoglu and Fındıklı, 2014). Sonawane (2014) developed a voltammetric sensor that incorporates SWCNTs for the detection of neohesperidin dihydrochalcone (NHDC). The low detection limit was found to be 2×10^{-8} mol/L· with a linear range of 5×10^{-8} to 8×10^{-6} mol/L. The fabricated sensor was successful in getting satisfactory and high accuracy results in the analysis of beverages for NHDC determination. A saccharin conductometric sensor study was done using the nanohybrid membrane. The linear range of response was found to be 20–400 µM and the lower detection of the sensor is 6 µM. A CNT–quantum dot nanocomposite was used in the detection of Polycyclic aromatic compounds (PAHs) in river water. The sensor fabricated using single-walled CNT–quantum dot nanocomposites. The fluorescence of QD nanoparticles is compared with the SWCNT-QDs the results were found to be 3.6–5.5 times higher (Stradiotto et al., 2003). These sensors cannot be used in the daily routine analysis due to their low reproducibility and high cost. Voltammetric nanosensors have proven potential for sensing in the solid, semisolid, and liquid food samples (Jilani et al., 2020; Sinha et al., 2018; Bijad et al., 2013).

Nanoenabled sensors were also developed for heavy metal and melamine sensing in food samples which are discussed in the next chapter.

22.3.4 Nanosensors in food packaging

Food packaging acts as a barrier for oxygen, moisture, and other contaminants and prevents its interaction with packed food to ensure quality, wholesomeness, safety, and integrity of food products (Sabatini, 2019). The new food packaging technologies are constantly evolving fields in the food industry in response to demands from consumers, improved shelf life, traceability, flavor stability, the freshness of food products. Active and intelligent packaging materials are developed for the smart packing of food products. Active packaging technology is one of the most innovative and the main engine of nanomaterial applications in the food packaging industry (Basavegowda et al., 2020). Predominant factors that decide the packed food quality are oxygen, concentration, carbon dioxide concentration, relative humidity, and moisture content. Governing these factors are important to attain high-quality food and sensors incorporated in food packaging helps to monitor these variations and provide intelligent packaging solution throughout the food supply chain (Kalpana et al., 2019). Classification of nanoenabled packaging solution for smart food packaging are shown in Fig. 22.5. Nanosensors can be incorporated as contact-based sensing, contact less sensing and color changing smart packaging materials. Some applications of nanosensors in monitoring the quality of packed food are discussed below. Also some commercialized sensors for smart and intelligent packaging are discussed in the next chapter.

Gutiérrez-Tauste et al. (2007) developed a light-activated colorimetric oxygen indicator through the liquid deposition method. In the liquid phase deposition (LPD) technique Methylene blue (MB)/TiO_2 hybrid nanocomposite material was deposited on bare glass and indium tin oxide (ITO) covered glass.

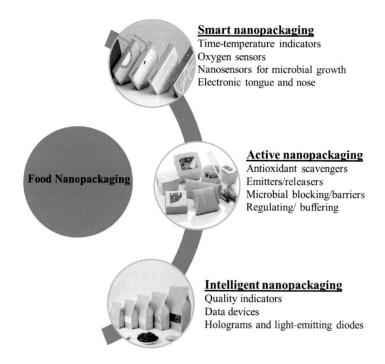

FIG. 22.5

Classification of nanoenabled food packaging (with permission from Shafiq et al., 2020).

They suggested an increased interest in the development of oxygen sensors in fields such as food packaging. The cyclometallated iridium(III) complexed with coumarins as nanobeads showed excellent luminescence intensity and sensing capability for very rapid changes in oxygen partial pressure. Signals are obtained by determination of luminescence in the frequency domain and in the time domain, and by ratiometric measurement of luminescence intensity. This nanosensors have been applied to sensing and imaging of dissolved oxygen, to monitor the consumption of oxygen during enzymatic oxidation of glucose, and to monitoring dissolved oxygen in a growing condition of *E. coli* (McEvoy et al., 2003).

Comini et al. (2005) developed a conductometric gas sensor using tin oxide nanobelts. Nano beads were found to be sensitive to oxygen and other polluting species like CO and NO_2 through their electrical characterization. The relative variation in conductance or resistance due to the introduction of the gas will define the sensitivity of sensors. The sensitivity of nano bead sensors was found to be 200% for 30 ppm for CO at 350 °C, 900% for 200 ppb NO_2 at 300 °C. The variation in the results was found to depend on the density of the nanobelts. The results demonstrate the potential of fabricating nanosize sensors using the integrity of a single nanobelt with sensitivity at the level of a few ppb and the necessity to control nanobelts density to optimize the sensing performances (Comini et al., 2005). Ruiz et al. (2005) prepared a nanosized La-TiO_2 sol–gel gas sensor that can be used for CO and ethanol detection during food spoilage. The gas test showed that the incorporation of 10 at.% of La resulted in an especially positive approach to a more selective ethanol sensor. Barreca et al. (2007) synthesized ZnO–TiO_2 nanocomposites with an average thickness of 140 nm. The composition and morphology directly affect the gas sensing performance of nanocomposites. The developed gas sensors are found to be more suitable for the detection of volatile compounds such as CH_3COCH_3, CH_3CH_2OH, and CO and hence can detect the spoilage of food (Barreca et al., 2007).

Ramgir et al. (2013) developed an ethanol sensing sensor fabricated from pure and Au modified ZnO nanowires. The incorporation of Au into ZnO nanowires was found to improve the reaction kinetics toward ethanol thus offers improved sensor response. The recovery time and response time for Au modified sensor were found to be 20 and 5 s at 325 °C toward 50 ppm ethanol. Gopalakrishna et al. (2013) optimized the ethanol concentration sensor fabricated from nanostructured CuO. The sensing was examined for 100 and 200 ppm at 350 °C with a fast response time within the range of 15–20 s and 15–18 s, respectively. This nanosensor also responded well at room temperature for varying concentrations of ethanol.

The capacitance of SWCNTs is highly sensitive to chemical vapors. The mechanism of sensing is in the presence of dilute chemical vapor, molecular adsorbates are polarized by the fringing electric fields radiating from the surface of the SWCNT electrode, which causes an increase in its capacitance. So,Cao and Rogers (2009) report that this capacitance can be used for constructing high-performance chemical sensors. These capacitors are fast, highly sensitive, and completely reversible and can be used as an indicator to know the composition of the vapor inside the packed food. nanofibers fabricated from an n-type organic semiconductor molecule, N-(1-hexylheptyl)perylene-3,4,9,10-tetracarboxyl-3,4-anhydride-9,10-imide were used as fluorescence sensory material with high sensitivity, selectivity, and photostability has been developed for vapor probing of organic amines (Che et al., 2008). Pimtong-Ngam et al. (2007) developed a thick film sensor using tungsten oxide and tin oxide nanocomposites for ethylene sensing. The author found that the incorporation of a small quantity of WO_3 (0.3 wt.%) will enhance the gas sensitivity of SnO. This composition of the sensor was very sensitive to 2–8 ppm ethylene at an optimum operating temperature of 300 °C. The results show that as the amount of WO_3 increases the sensitivity of the sensor is decreased.

Taccola et al. (2013) developed a humidity sensor containing iron oxide nanoparticle composite thin film. The sensors consist of nanocomposite ultrathin films fabricated from conductive polymer poly(3,4-ethylenedioxythiophene) polystyrene, sulfonate (PEDOT: PSS) embedded in iron oxide nanoparticles. PEDOT: PSS and PEDOT: PSS/iron oxide NP nanofilms were tested as resistive humidity sensors. The nanoparticle concentration is directly proportional to the sensitivity. Based on the results, the developed sensor was found to be a new simple, fast, and inexpensive technique for relative humidity detection.

22.4 Conclusion

In the food supply chain, maintenance of quality and safety of food is the most vital issue. The conventional methods and procedures involved in the quality determination of raw material and processed food are time consuming and expensive processes, which also lack traceability in the food supply chain. Hence, consumers are concerned about the origin, ingredients, process condition, and transportation of food. Nanosensors are a good solution to bring transparency to the food supply chain. Nanosensors are highly selective, sensitive, fast responsive, smaller size, and have low detection time. In the food supply chain, nanosensors are used in every sector for the detection of pathogens, pesticides, toxins, allergens, heavy metals, antibiotics, and also in the traceability of food products during distribution. Applications of nanosensors in food packaging and food quality assurance have been commercially established, while other sectors still require a significant amount of research for real-time applications.

References

Ahuja, D., and D. Parande. 2012. "Optical sensors and their applications." 1 (5): 60–68.

Aung, M.M., Chang, Y.S., 2014. Traceability in a food supply chain: safety and quality perspectives. Food Control 39, 172–184. https://doi.org/10.1016/j.foodcont.2013.11.007.

Baranwal, A., Srivastava, A., Kumar, P., Bajpai, V.K., Maurya, P.K., Chandra, P., 2018. Prospects of nanostructure materials and their composites as antimicrobial agents. Front. Microbiol. 9 (March), 422. https://doi.org/10.3389/fmicb.2018.00422.

Barreca, D., Comini, E., Ferrucci, A.P., Gasparotto, A., Maccato, C., Maragno, C., Sberveglieri, G., Tondello, E., 2007. First example of $ZnO-TiO_2$ nanocomposites by chemical vapor deposition: structure, morphology, composition, and gas sensing performances. Chem. Mater. 19 (23), 5642–5649. https://doi.org/10.1021/cm701990f.

Basavegowda, N., Mandal, T.K., Baek, K.H., 2020. Bimetallic and trimetallic nanoparticles for active food packaging applications: a review. Food Bioprocess Technol. 13 (1), 30–44. https://doi.org/10.1007/s11947-019-02370-3.

Benvidi, A., Nafar, M.T., Jahanbani, S., Tezerjani, M.D., Rezaeinasab, M., Dalirnasab, S., 2017. Developing an electrochemical sensor based on a carbon paste electrode modified with nano-composite of reduced graphene oxide and $CuFe_2O_4$ nanoparticles for determination of hydrogen peroxide. Mater. Sci. Eng.: C 75 (June), 1435–1447. https://doi.org/10.1016/j.msec.2017.03.062.

Beyer, A., Biziuk, M., 2008. Methods for determining pesticides and polychlorinated biphenyls in food samples – problems and challenges. Crit. Rev. Food Sci. Nutr. 48 (10), 888–904. https://doi.org/10.1080/10408390701761878.

Bijad, M., Karimi-Maleh, H., Khalilzadeh, M.A., 2013. Application of ZnO/CNTs nanocomposite ionic liquid paste electrode as a sensitive voltammetric sensor for determination of ascorbic acid in food samples. Food Anal. Methods 6 (6), 1639–1647. https://doi.org/10.1007/s12161-013-9585-9.

Bowles, M., Lu, J., 2014. Removing the blinders: a literature review on the potential of nanoscale technologies for the management of supply chains. Technol. Forecast. Social Change 82 (February), 190–198. https://doi.org/10.1016/j.techfore.2013.10.017.

Bülbül, G., Hayat, A., Andreescu, S., 2015. Portable nanoparticle-based sensors for food safety assessment. Sensors 15 (12), 30736–30758. https://doi.org/10.3390/s151229826.

Burris, K.P., Neal Stewart, C., 2012. Fluorescent nanoparticles: sensing pathogens and toxins in foods and crops. Trends Food Sci. Technol. 28 (2), 143–152. https://doi.org/10.1016/j.tifs.2012.06.013.

Bustillo, J., and S. Li. 2017. Methods for manufacturing well structures for low-noise chemical sensors, US patent US9841398B2, issued 2017.

Caetano, J., Machado, S.A.S., 2008. Determination of carbaryl in tomato 'in natura' using an amperometric biosensor based on the inhibition of acetylcholinesterase activity. Sens. Actuators B 129 (1), 40–46. https://doi.org/10.1016/j.snb.2007.07.098.

Cao, Q., Rogers, J.A., 2009. Ultrathin films of single-walled carbon nanotubes for electronics and sensors: a review of fundamental and applied aspects. Adv. Mater. 21 (1), 29–53. https://doi.org/10.1002/adma.200801995.

Caon, T., Martelli, S.M., Fakhouri, F.M., 2017. New Trends in the Food Industry: Application of Nanosensors in Food Packaging. Nanobiosensors. Elsevier. https://doi.org/10.1016/b978-0-12-804301-1.00018-7.

Cash, K.J., Clark, H.A., 2010. Nanosensors and nanomaterials for monitoring glucose in diabetes. Trends Mol. Med. 16 (12), 584–593. https://doi.org/10.1016/j.molmed.2010.08.002.

Chaudhry, Q., Scotter, M., Blackburn, J., Ross, B., Boxall, A., Castle, L., Aitken, R., Watkins, R., 2008. Applications and implications of nanotechnologies for the food sector. Food Additives Contam. - Part A Chem. Anal. Control Exposure Risk Assessment 25 (3), 241–258. https://doi.org/10.1080/02652030701744538.

Che, Y., Yang, X., Loser, S., Zang, L., 2008. Expedient vapor probing of organic amines using fluorescent nanofibers fabricated from an N-type organic semiconductor. Nano Lett. 8 (8), 2219–2223. https://doi.org/10.1021/nl080761g.

Chen, S.H., Wu, V.C.H., Chuang, Y.C., Lin, C.S., 2008. Using oligonucleotide-functionalized Au nanoparticles to rapidly detect foodborne pathogens on a piezoelectric biosensor. J. Microbiol. Methods 73 (1), 7–17. https://doi.org/10.1016/j.mimet.2008.01.004.

Cheraghi, S., Taher, M.A., Karimi-Maleh, H., 2017. Highly sensitive square wave voltammetric sensor employing CdO/SWCNTs and room temperature ionic liquid for analysis of vanillin and folic acid in food samples. J. Food Compos. Anal. 62 (September), 254–259. https://doi.org/10.1016/j.jfca.2017.06.006.

Comini, E., Faglia, G., Sberveglieri, G., Calestani, D., Zanotti, L., Zha, M., 2005. Tin oxide nanobelts electrical and sensing properties. Sens. Actuators B 111–112 (11), 2–6. https://doi.org/10.1016/j.snb.2005.06.031.

Dahman, Y., 2017. Nanosensors. Nanotechnology and Functional Materials for Engineers. Elsevier, Amsterdam, Netherlands. https://doi.org/10.1016/B978-0-323-51256-5.00004-6.

Dankwardt, A., 2006. Immunochemical assays in pesticide analysis. Encyclopedia of Analytical Chemistry. https://doi.org/10.1002/9780470027318.a1714.

Danny, M., 2011. Nanotechnology and the food sector: from the farm to the table. Emirates J. Food Agric. 23 (5), 387–403.

Gopalakrishna, D., Vijayalakshmi, K., Ravidhas, C., 2013. Effect of pyrolytic temperature on the properties of nano-structured Cuo optimized for ethanol sensing applications. J. Mater. Sci. Mater. Electron. 24 (3), 1004–1011. https://doi.org/10.1007/s10854-012-0866-7.

Gutiérrez-Tauste, D., Domènech, X., Casañ-Pastor, N., Ayllón, J.A., 2007. Characterization of Methylene Blue/TiO_2 hybrid thin films prepared by the liquid phase deposition (LPD) method: application for fabrication of light-activated colorimetric oxygen indicators. J. Photochem. Photobiol. A 187 (1), 45–52. https://doi.org/10.1016/j.jphotochem.2006.09.011.

Hanke, W., Jurewicz, J., 2004. The risk of adverse reproductive and developmental disorders due to occupational pesticide exposure: an overview of current epidemiological evidence. Int. J. Occup. Med. Environ. Health 17 (172), 223–243.

Heldman, E., Balan, A., Horowitz, O., Ben-Zion, S., Torten, M., 1985. A novel immunoassay with direct relevance to protection against organophosphate poisoning. FEBS Lett. 180, 243–248. https://doi.org/10.1016/0014-5793(85)81079-6.

Huang, Y.-F., Wang, Y.-F., Yan, X.-P., 2010. Amine-functionalized magnetic nanoparticles for rapid capture and removal of bacterial pathogens. Environ. Sci. Technol. 44, 7908–7913. https://doi.org/10.1021/es102285n.

Hunter, K.W., Lenz, D.E., 1982. Detection and quantification of the organophosphate insecticide paraoxon by competitive inhibition enzyme immunoassay. Life Sci. 30 (4), 355–361. https://doi.org/10.1016/0024-3205(82)90572-0.

Ivnitski, D., Abdel-Hamid, I., Atanassov, P., Wilkins, E., Stricker, S., 2000. Application of electrochemical biosensors for detection of food pathogenic bacteria. Electroanalysis 12, 317–325. https://doi.org/10.1002/(SICI)1521-4109(20000301)12:5<317::AID-ELAN317>3.0.CO;2-A.

Jafarizadeh-Malmiri, H., Sayyar, Z., Anarjan, N., Berenjian, A., 2019. Nano-sensors in food nanobiotechnology. Nanobiotechnology in Food: Concepts, Applications and Perspectives. Springer, Cham.

Jilani, B.S., Mounesh, P.M., Mruthyunjayachari, C.D., Venugopala Reddy, K.R., 2020. Cobalt (II) tetra methyl-quinoline oxy bridged phthalocyanine carbon nano particles modified glassy carbon electrode for sensing nitrite: a voltammetric study. Mater. Chem. Phys. 239 (1), 121920. https://doi.org/10.1016/j.matchemphys.2019.121920.

Joo, J., Yim, C., Kwon, D., Lee, J., Shin, H.H., Cha, H.J., Jeon, S., 2012. A facile and sensitive detection of pathogenic bacteria using magnetic nanoparticles and optical nanocrystal probes. Analyst 137 (16), 3609–3612. https://doi.org/10.1039/c2an35369e.

Kalpana, S., Priyadarshini, S.R., Maria Leena, M., Moses, J.A., Anandharamakrishnan, C., 2019. Trends in food science & technology intelligent packaging : trends and applications in food systems. Trends Food Sci. Technol. 93 (July), 145–157. https://doi.org/10.1016/j.tifs.2019.09.008.

Kamalii, A., Ahilan, B., Felix, N., Kannan, B., Prabu, E., 2019. Applications of nanotechnology in fisheries and aquaculture. J. Aquacult. Trop. 33 (3), 111–117. https://doi.org/10.32381/jat.2018.33.3-4.1.

Khan, S.A., DeGrasse, J.A., Yakes, B.J., Croley, T.R., 2015. Rapid and sensitive detection of cholera toxin using gold nanoparticle-based simple colorimetric and dynamic light scattering assay. Anal. Chim. Acta 892 (September), 167–174. https://doi.org/10.1016/j.aca.2015.08.029.

Khater, M., de la Escosura-Muñiz, A., Merkoçi, A., 2017. Biosensors for plant pathogen detection. Biosens. Bioelectron. 93 (15 July), 72–86. https://doi.org/10.1016/j.bios.2016.09.091.

Kneipp, J., Kneipp, H., Wittig, B., Kneipp, K., 2010. Novel optical nanosensors for probing and imaging live cells. Nanomed.: Nanotechnol. Biol. Med. 6 (2), 214–226. https://doi.org/10.1016/j.nano.2009.07.009.

Krishna, V.D., Wu, K., Su, D., Cheeran, M.C.J., Wang, J.P., Perez, A., 2018. Nanotechnology: review of concepts and potential application of sensing platforms in food safety. Food Microbiol. 75 (October), 47–54. https://doi.org/10.1016/j.fm.2018.01.025.

Kuswandi, B., Futra, D., Heng, L.Y., 2017. Nanosensors for the detection of food contaminants. Nanotechnology Applications in Food: Flavor, Stability, Nutrition and Safety, 307–333. https://doi.org/10.1016/B978-0-12-811942-6.00015-7.

Kuswandi, B., Mascini, M., 2005. Enzyme inhibition based biosensors for environmental monitoring. Curr. Enzyme Inhib. 1 (3), 207–221. https://doi.org/10.2174/157340805774580484.

Lai, H., Dave, K.I., Wild, J., 1994. Bimetallic binding motifs in organophophorus hydrolase are important for catalysis and structural organization. J. Biol. Chem. 269, 16579–16584.

Lin, X., Ni, Y., Kokot, S., 2016. Electrochemical and bio-sensing platform based on a novel 3D Cu nano-flowers/layered MoS$_2$ composite. Biosens. Bioelectron. 79 (15), 685–692. https://doi.org/10.1016/j.bios.2015.12.072.

Lisa, M., Chouhan, R.S., Vinayaka, A.C., Manonmani, H.K., Thakur, M.S., 2009. Gold nanoparticles based dipstick immunoassay for the rapid detection of dichlorodiphenyltrichloroethane: an organochlorine pesticide. Biosens. Bioelectron. 25 (1), 224–227. https://doi.org/10.1016/j.bios.2009.05.006.

Liu, S., Yuan, L., Yue, X., Zheng, Z., Tang, Z., 2008. Recent advances in nanosensors for organophosphate pesticide detection. Adv. Powder Technol. 19 (5), 419–441. https://doi.org/10.1163/156855208X336684.

Malhotra, S., Verma, A., Tyagi, N., Kumar, V., 2017. Biosensors: principle, types and applications. Int. J. Adv. Res. Innov. Ideas Educ. 3 (2), 3639–3644.

Maurer, E.I., Comfort, K.K., Hussain, S.M., Schlager, J.J., Mukhopadhyay, S.M., 2012. Novel platform development using an assembly of carbon nanotube, nanogold and immobilized RNA capture element towards rapid, selective sensing of bacteria. Sensors 12 (6), 8135–8144. https://doi.org/10.3390/s120608135.

McEvoy, K., Aisling, C.V.B., Mcdonagh, C., Maccraith, B., Klimant, I., Wolfbeis, O., 2003. Optical sensors for application in intelligent food-packaging technology. Proc. SPIE 4876, 806–815. https://doi.org/10.1117/12.464210.

Mehrotra, P., 2016. Biosensors and their applications – a review. J. Oral Biol. Craniofac. Res. 6, 153–159. https://doi.org/10.1016/j.jobcr.2015.12.002.

Mills, Andrew., 2006. Oxygen indicators and intelligent inks for packaging food. Chem. Soc. Rev. 34, 1003–1011. https://doi.org/10.1039/b503997p.

Mulchandani, A., Chen, W., Mulchandani, P., Wang, J., Rogers, K.R., 2001. Biosensors for direct determination of organophosphate pesticides. Biosens. Bioelectron. 16 (4), 225–230. https://doi.org/10.1016/S0956-5663(01)00126-9.

Neethirajan, S., Jayas, D.S., 2011. Nanotechnology for the food and bioprocessing industries. Food Bioprocess Technol. 4 (1), 39–47. https://doi.org/10.1007/s11947-010-0328-2.

Nikolelis, D.P., Pantoulias, S., Krull, U.J., Zeng, J., 2001. Electrochemical transduction of the interactions of the sweeteners acesulfame-K, saccharin and cyclamate with bilayer lipid membranes (BLMs). Electrochim. Acta 46 (7), 1025–1031. https://doi.org/10.1016/S0013-4686(00)00686-1.

Ntzani, E.E., Chondrogiorgi, M., Ntritsos, G., Evangelou, E., Tzoulaki, I., 2017. Literature review on epidemiological studies linking exposure to pesticides and health effects. EFSA Support. Publ. 10 (10), 1–159. https://doi.org/10.2903/sp.efsa.2013.en-497.

Oliveira, D.A., Stromberg, L.R., Pola, C.C., Parate, K., Cavallaro, N.D., Claussen, J.C., McLamore, E.S., Gomes, C.L., 2019. Biomimetic nanosensors for measuring pathogenic bacteria in complex food matrices (conference presentation). Smart Biomedical and Physiological Sensor Technology XV 11020. SPIE, p. 18. https://doi.org/10.1117/12.2519523.

Omanović, E., Maksimović, M., 2016. Nanosensors applications in agriculture and food industry. Bull. Chemists Technol Bosnia Herzegov. 43, 41–44.

Pal, S., Alocilja, E.C., Downes, F.P., 2007. Nanowire labeled direct-charge transfer biosensor for detecting *Bacillus* species. Biosens. Bioelectron. 22 (9), 2329–2336. https://doi.org/10.1016/j.bios.2007.01.013.

Pimtong-Ngam, Y., Jiemsirilers, S., Supothina, S., 2007. Preparation of tungsten oxide–tin oxide nanocomposites and their ethylene sensing characteristics. Sens. Actuators A 139 (1), 7–11. https://doi.org/10.1016/j.sna.2006.10.032.

Pogačnik, L., Franko, M., 2003. Detection of organophosphate and carbamate pesticides in vegetable samples by a photothermal biosensor. Biosens. Bioelectron. 18 (1), 1–9. https://doi.org/10.1016/S0956-5663(02)00056-8.

Pundir, C.S., Chauhan, N., Rajneesh, Verma, M., Ravi, 2011. A novel amperometric biosensor for oxalate determination using multi-walled carbon nanotube-gold nanoparticle composite. Sens. Actuators B 155 (2), 796–803. https://doi.org/10.1016/j.snb.2011.01.050.

Ramgir, N.S., Kaur, M., Sharma, P.K., Datta, N., Kailasaganapathi, S., Bhattacharya, S., Debnath, A.K., Aswal, D.K., Gupta, S.K., 2013. Ethanol sensing properties of pure and Au modified ZnO nanowires. Sens. Actuators B 187 (October), 313–318. https://doi.org/10.1016/j.snb.2012.11.079.

Rong, G., Kim, E.H., Poskanzer, K.E., Clark, H.A., 2017. A method for estimating intracellular ion concentration using optical nanosensors and ratiometric imaging. Sci. Rep. 7 (1), 1–10. June. https://doi.org/10.1038/s41598-017-11162-8.

Ruiz, A.M., Cornet, A., Morante, J.R., 2005. Performances of La–TiO_2 nanoparticles as gas sensing material. Sens. Actuators B 109 (1), 7–12. https://doi.org/10.1016/j.snb.2005.06.040.

Sabatini, K.C., 2019. Advanced food packaging. Nutr. Foodserv.

Sack, D.A., 2011. How many cholera deaths can be averted in Haiti? Lancet 377 (9773), 1214–1216. https://doi.org/10.1016/s0140-6736(11)60356-5.

Schofield, C.L., Field, R.A., Russell, D.A., 2007. Glyconanoparticles for the colorimetric detection of cholera toxin. Anal. Chem. 79 (4), 1356–1361. https://doi.org/10.1021/ac061462j.

Shafiq, M., Anjum, S., Hano, C., Anjum, I., Abbasi, B.H., 2020. An overview of the applications of nanomaterials and nanodevices in the food industry. Foods 9 (2), 1–27. https://doi.org/10.3390/foods9020148.

Shah, A., 2020. A novel electrochemical nanosensor for the simultaneous sensing of two toxic food dyes. ACS Omega 5 (11), 6187–6193. https://doi.org/10.1021/acsomega.0c00354.

Shan, D., Li, Q., Xue, H., Cosnier, S., 2008. A highly reversible and sensitive tyrosinase inhibition-based amperometric biosensor for benzoic acid monitoring. Sens. Actuators B 134 (2), 1016–1021. https://doi.org/10.1016/j.snb.2008.07.006.

Singh, S., Thiyagarajan, P., Mohan Kant, K., Anita, D., Thirupathiah, S., Rama, N., Tiwari, B., Kottaisamy, M., Ramachandra Rao, M.S., 2007. Structure, microstructure and physical properties of ZnO based materials in various forms: bulk, thin film and nano. J. Phys. D Appl. Phys. 40 (20), 6312.

Sinha, A., Dhanjai, Jain, R., Zhao, H., Karolia, P., Jadon, N., 2018. Voltammetric sensing based on the use of advanced carbonaceous nanomaterials: a review. Microchim. Acta 185 (2). https://doi.org/10.1007/s00604-017-2626-0.

Sonawane, S., 2014. Use of nanomaterials in the detection of food contaminants. Eur. J. Nutr. Food Safety 4 (4), 301–317. https://doi.org/10.9734/ejnfs/2014/6218.

Sotiropoulou, S., Chaniotakis, N.A., 2005. Lowering the detection limit of the acetylcholinesterase biosensor using a nanoporous carbon matrix. Anal. Chim. Acta 530 (2), 199–204. https://doi.org/10.1016/j.aca.2004.09.007.

Srivastava, A.K., Dev, A., Karmakar, S., 2018. Nanosensors and nanobiosensors in food and agriculture. Environ. Chem. Lett. 16 (1), 161–182. https://doi.org/10.1007/s10311-017-0674-7.

Stein, K., Schwedt, G., 1993. Comparison of immobilization methods for the development of an acetylcholinesterase biosensor. Anal. Chim. Acta 272 (1), 73–81. https://doi.org/10.1016/0003-2670(93)80377-W.

Stradiotto, N.R., Yamanaka, H., Valnice, M., Zanoni, B., 2003. Electrochemical sensors: a powerful tool in analytical chemistry. J. Braz. Chem. Soc. 14 (2), 159–173. https://doi.org/10.1590/S0103-50532003000200003.

Sun, X., Du, S., Wang, X., 2012. Amperometric immunosensor for carbofuran detection based on gold nanoparticles and PB-MWCNTs-CTS composite film. Eur. Food Res. Technol. 235 (3), 469–477. https://doi.org/10.1007/s00217-012-1774-z.

Taccola, S., Greco, F., Zucca, A., Innocenti, C., Fernández, C.D.J., Campo, G., Sangregorio, C., Mazzolai, B., Mattoli, V., 2013. Characterization of free-standing PEDOT:PSS/iron oxide nanoparticle composite thin films and application as conformable humidity sensors. ACS Appl. Mater. Interfaces 5 (3), 6324–6332. https://doi.org/10.1021/am4013775.

Tran-Minh, C., Pandey, P.C., Kumaran, S., 1990. Studies on acetylcholine sensor and its analytical application based on the inhibition of cholinesterase. Biosens. Bioelectron. 5 (6), 461–471. https://doi.org/10.1016/0956-5663(90)80035-C.

Turkoglu, S., Fındıklı, Z., 2014. Determination of the effects of some artificial sweeteners on human peripheral lymphocytes using the comet assay. J. Toxicol. Environ. Health Sci. 6 (8), 147–153. https://doi.org/10.5897/JTEHS2014.0313.

Tuteja, S., Kukkar, M., Kumar, P., Paul, A.K., Deep, A., 2014. Synthesis and characterization of silica-coated silver nanoprobe for paraoxon pesticide detection. BioNanoScience 4 (2), 149–156. https://doi.org/10.1007/s12668-014-0129-6.

Velasco-Garcia, M.N., Mottram, T., 2003. Biosensor technology addressing agricultural problems. Biosyst. Eng. 84 (1), 1–12. https://doi.org/10.1016/S1537-5110(02)00236-2.

Wang, X., Jiao, C., Yu, Z., 2014. Electrochemical biosensor for assessment of the total antioxidant capacity of orange juice beverage based on the immobilizing DNA on a poly L-glutamic acid doped silver hybridized membrane. Sens. Actuators B 192 (March), 628–633. https://doi.org/10.1016/j.snb.2013.11.025.

Wiesendanger, R., 1995. Recent advances in nanostructural investigations and modifications of solid surfaces by scanning probe methods. Jpn. J. Appl. Phys. 34 (60), 3388. https://doi.org/10.1143/JJAP.34.3388.

Wilkowska, A., Biziuk, M., 2011. Determination of pesticide residues in food matrices using the QuEChERS methodology. Food Chem. 125 (3), 803–812. https://doi.org/10.1016/j.foodchem.2010.09.094.

Wilson, B.G., Bahna, S.L., 2005. Adverse reactions to food additives. Ann. Allergy Asthma Immunol. 95 (6), 499–507. https://doi.org/10.1016/S1081-1206(10)61010-1.

Yamada, K., Kim, C.-T., Kim, J.-H., Chung, J.-H., Lee, H.G., Jun, S., 2014. Single walled carbon nanotube-based junction biosensor for detection of *Escherichia coli*. PLoS One 9 (9), 105767. https://doi.org/10.1371/journal.pone.0105767.

Ye, X., Du, Y., Lu, D., Wang, C., 2013. Fabrication of β-cyclodextrin-coated poly (diallyldimethylammonium chloride)-functionalized graphene composite film modified glassy carbon-rotating disk electrode and its application for simultaneous electrochemical determination colorants of sunset yello. Anal. Chim. Acta 779 (May), 22–34. https://doi.org/10.1016/j.aca.2013.03.061.

Zhao, W.-W., Xu, J.-J., Chen, H.-Y., 2017. Photoelectrochemical enzymatic biosensors. Biosens. Bioelectron. 92 (June), 294–304. https://doi.org/10.1016/j.bios.2016.11.009.

Zhao, Y., Ye, M., Chao, Q., Jia, N., Ge, Y., Shen, H., 2009. Simultaneous detection of multifood-borne pathogenic bacteria based on functionalized quantum dots coupled with immunomagnetic separation in food samples. J. Agric. Food Chem. 57 (2), 517–524. https://doi.org/10.1021/jf802817y.

Sensors for food quality and safety

Farshad Oveissi[a,b], Long H. Nguyen[a,b], Jacopo E. Giaretta[a,b], Zahra Shahrbabaki[a,b], Ronil J. Rath[a,b], Vitus A. Apalangya[c], Jimmy Yun[d,e], Fariba Dehghani[a,b], Sina Naficy[a,b]

[a]School of Chemical and Biomolecular Engineering, The University of Sydney, Sydney, NSW, Australia
[b]Centre for Advanced Food Enginomics, The University of Sydney, Sydney, NSW, Australia
[c]Department of Food Process Engineering, University of Ghana, Accra, Ghana
[d]School of Chemical Engineering, The University of New South Wales, Sydney, NSW, Australia
[e]Qingdao International Academician Park Research Institute, Qingdao, Shandong, PR China

23.1 Introduction

Accessing food quality and safety is a necessity in the increasingly complex food supply chain for reducing food waste and minimizing any health risk for consumers. According to the World Health Organization, nearly 600 million cases of foodborne diseases, among those, 420,000 deaths each year (W.H. Organization, 2015). To keep the contamination low in the food supply chain, food products must be shipped and distributed under highly controlled conditions. Hence, it is essential for food suppliers and consumers to identify the provenance and traceability of the food.

Sensors that indicate temperature, humidity, changes in gasses such as oxygen and carbon dioxide, and pH have been used for monitoring the food quality and safety. There are commonly four classes of analytes for designing advanced food sensors: (i) biological and chemical contaminants (Salvatore et al., 2017), (ii) allergens (Zhang et al., 2019), (iii) nutritional ingredients (Thong et al., 2017), and (iv) food additives which may be harmful at a higher dose (Manjunatha, 2018). The identification of some analytes can be used to determine the freshness of a desirable food product.

Food sensors for quality and safety could be wearable (Mishra et al., 2017; Ciui et al., 2018), constructed into miniaturized devices (Zhang et al., 2019), or incorporated into the food package (Wu et al., 2015). A few examples of food sensors that have been recently developed are depicted in Fig. 23.1. Sensors can also be implemented on a glove as a wearable device for detecting food taste (Mishra et al., 2017). Alternatively, sensors can be integrated into a portable device for the detection of food allergens. Nima, shown in Fig. 23.1B, is an example of such portable devices, that can grind, extract and detect gluten or peanut within the food sample (Zhang et al., 2019). Flexible sensors have also been implemented into food packaging materials that are based on electrochemical (Wu et al., 2015; Fig. 23.1C–D) or colorimetric (Fig. 23.1C–E) techniques for assessing the integrity, ripeness, and freshness of food products (Nguyen et al., 2019).

This chapter aims to discuss the current food sensors in the market and their method of detections. It reviews the recent advances in food sensors and the mechanism underlying these technologies.

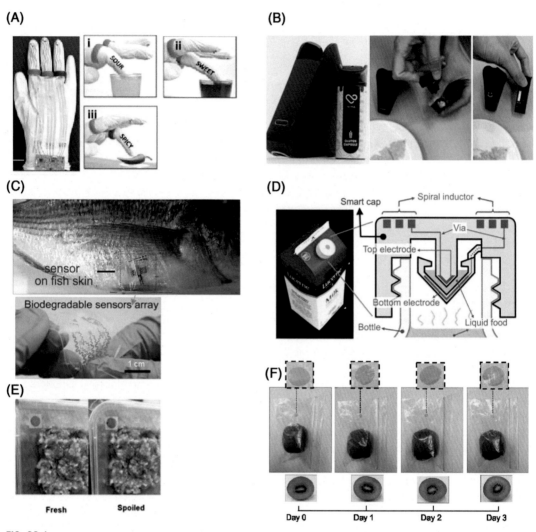

FIG. 23.1

Examples of sensors for food quality and safety: (A) Glove-based portable electrochemical sensor for detecting organophosphate compound detection (Ciui et al., 2018) Copyright 2018, American Chemical Society; (B) Commercial gluten detector device, Nima (Zhang et al., 2019) Copyright 2018, Elsevier; (C) Biodegradable, stretchable, foldable, and wireless temperature sensor (Salvatore et al., 2017) Copyright 2017, Wiley-VCH; (D) Wireless smart-cap sensor for detection of milk freshness (Wu et al., 2015) Copyright 2015, Springer Nature; (E) Ammonia sensor for meat spoilage (Nguyen et al., 2019) Copyright 2019, RSC; (F) Ethylene colorimetric sensor for detecting the ripeness of fruit (Nguyen et al., 2020) Copyright 2020, American Chemical Society.

23.2 Food sensors market

The current market of food sensors is dominated by colorimetric sensors as they are more suitable for end-user consumers and can be incorporated into packaging material. The main target analytes that have been the subject of research and development for food sensors include logging of temperature and, humidity, gases such as oxygen and carbon dioxide, pH, and certain chemicals and biochemicals generated from food spoilage microorganisms (Yousefi et al., 2019). Most food sensors that reach the market are colorimetric time–temperature indicators (Fig. 23.2). These colorimetric sensors aim to model the freshness or safety of the packaged food by showing their thermal histories such as potential temperature shocks and abuse, possible thawing cycles, or the extent of exposure to elevated temperatures (Zabala et al., 2015; Kim et al., 2016; Fu and Labuza, 1992). Thus, it is fair to say that the time–temperature sensors are passive indicators as they merely reflect the history of food packaging rather than measuring food's actual state.

Monitoring the humidity of packaged food such as baked products and cereals is important and can be carried out through humidity sensors. To this end, several humidity sensors have been developed to monitor the humidity of the packaged dry food in situ (Amin et al., 2013; Virtanen et al., 2010; Harrey et al., 2002; Feng et al., 2014). Humidity sensors for food packaging are often electrochemical sensors. One important manufacturing aspect of these sensors is using low-cost materials and wireless technologies such as radio frequency identification (RFID) that allow wireless communication to suppliers or consumers.

Change in the concentration of gases such as oxygen, ammonia, and carbon dioxide in food packaging can be taken as the indicators for food degradation and undesirable microbial activities. Redox food spoilage reactions such as fat and pigment oxidation alter the oxygen concentration in the packaging environment (Lee et al., 2015), while ammonia is released during the spoilage of protein-rich food such as meat. Similarly, variation in carbon dioxide concentration may represent the activity of undesired microorganisms or damage to the packaging (Sivertsvik et al., 2002). Hence, evaluating the concentration of oxygen, ammonia, or carbon dioxide in the packaging environment could indirectly indicate the quality of the packaged product and packaging integrity (Nguyen et al., 2019; Eaton, 2002; Barandun et al., 2019). For instance, Ageless Eye is a colorimetric sensor composed of glucose as the reducing agent in an alkaline medium and methylene blue as the redox dye (Fig. 23.3). In the presence of oxygen, colorless leucomethylene blue oxidizes to blue colored methylene blue, which could be reduced back to colorless leucomethylene blue by glucose (Eaton, 2002).

Sensors can also be used to identify the ripeness and freshness of packaged food by detecting the change in pH, or the presence of certain chemicals (e.g., ethylene) and metabolites that are produced in the process of ripening. A few commercial sensors that reached the market are shown in Fig. 23.4. Example of direct sensing includes RipeSense (Fig. 23.4A) that is a colorimetric sensor correlating the ripeness of fruit to the release of aromas in the packaging. RipeSense was developed in New Zealand by Jenkins Group in partnership with Plant and Food Research Institute. Another example is SensorQ (Fig. 23.4B), which is a pH-based sensor composed of a polymer matrix decorated with a green dye bromocresol sensitive to pH and is mostly used for identifying the freshness of meat (Chun et al., 2014). Unlike RipeSense, SensorQ was not commercially successful.

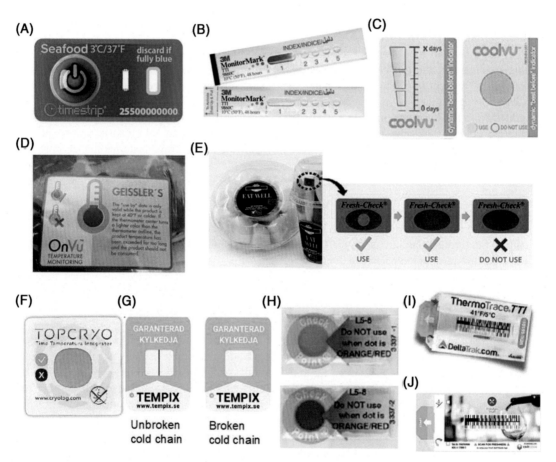

FIG. 23.2 Examples of commercial time and temperature indicating sensors.

(A) Timestrip; (B) MonitorMark by 3M; (C) CoolVu by FreshPoint; (D) OnVu by Ciba Specialty Chemical; (E) Fresh-check by Temptime Corp.; (F) TOPCRYO by Cryolog; (G) Tempix by Tempix AB; (H) CheckPoint by Vitsab; (I) ThermoTrace by DeltaTrak; (J) FreshCode by Varcode.

23.3 Colorimetric sensors for food quality and safety

The colorimetric sensors are preferable for end-user applications. The colorimetric mechanism is based on molecular interactions taking place between analytes and active substances, altering molecular conformations or colloidal rearrangement, which in turn result in a distinct color transitioning. Therefore, the design of colorimetric sensors revolves around stimuli-responsive materials which can change color upon physicochemical interactions with target analytes.

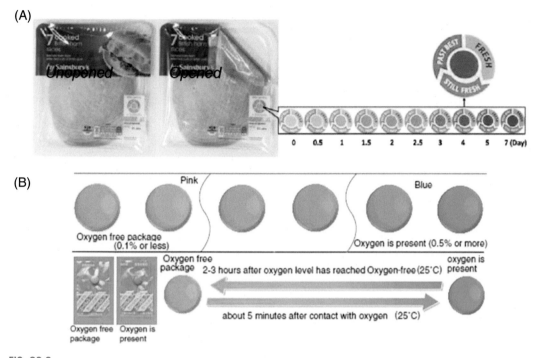

FIG. 23.3

Examples of commercial integrity indicating sensors: (A) Colorimetric sensor based on CO_2 exposure by Insignia Technologies Pty Ltd. (Yusufu et al., 2018), Copyright 2018, Elsevier and (B) Ageless Eye colorimetric oxygen sensor by Mitsubishi gas chemical company, Inc.

Colorimetric chemical sensors for food applications are usually used to detect contaminants or different compounds generated during food spoilage, including gases, volatile organic compounds (VOCs), toxic molecules, and heavy metals.

23.3.1 Detection of gases and volatile organic compounds

The level of VOCs and various gases such as ammonia, carbon dioxide, oxygen, and ethylene vary significantly during food spoilage, fruit ripening, or compromised packaging. Therefore, monitoring the concentration of such indicative gases and VOCs via colorimetric sensors provides the end-user with crucial information about the state of the food and packaging. For instance, Li et al. (2014) developed colorimetric sensors based on polyaniline (PANI) doped with poly(sodium 4-styrenesulfonate) (PSS) for detection of triethylamine (TEA)—a gas produced from spoilage of proteinaceous foods. PANI can change color via the acid–base interaction with the analytes and PSS can help stabilizing PANI in aqueous solutions. The PANI/PSS was able to differentiate between various concentrations of TEA,

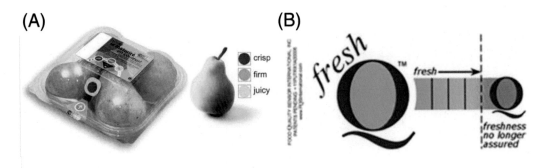

FIG. 23.4 Commercial sensor examples for identifying the food freshness.

(A) RipeSense by RipeSense and Ort Research and (B) SensorQ by DSM NV and Food Quality Sensor (Taoukis and Tsironi, 2016) Copyright 2016, Elsevier.

exhibiting a color transitioning from green to purple. Although the PANI/PSS sensors lacked specificity, they could detect TEA at low concentrations (approximately 200 ppm), which is adequate for the sensor to be used for monitoring food spoilage.

Trimethylamine (TMA), another food spoilage indicator, was used as a target VOC for food spoilage sensor arrays, made from eight pH indicator dyes and eight porphyrins or metalloporphyrins immobilized on TiO_2 nanoporous substrates (Xiao-wei et al., 2016). Compared to other substrates such as silica gel plates, the use of nanoporous TiO_2 resulted in more uniform color and dye distribution owing to the substrate's high surface area and microporous structure. These colorimetric sensor arrays were able to detect TMA down to very low concentrations (60 ppb), which is hardly identified by olfactory systems. The developed sensor arrays also successfully monitored the freshness of Yao-meat (salted pork in jelly, a traditional Chinese salted meat), indirectly, demonstrating their sensing potentials in complex food matrices.

Ammonia is another indicative gas released during the spoilage of protein-rich food. Nguyen et al. (2019) developed a flexible ammonia sensor based on polydiacetylene (PDA) liposomes immobilized in chitosan and cellulose films. The sensors were able to detect the presence of ammonia by changing color from blue to red, with a limit of detection of ~100 ppm.

Various colorimetric sensors have also been developed for detection of ethylene which is associated with the climacteric fruit ripening process. As an example, Kim and Shiratori (2006) developed an ethylene sensor based on a multistep reaction strategy in which ethylene was oxidized to CH_3CHO by reducing Pd^{2+} to Pd, which in turn is oxidized back by the reduction of Mo^{6+} to Mo^{5+}, resulting in a color change from white to blue. Similarly, Lang and Hübert (2011) developed a ripeness indicator based on the reducing effect of ethylene on molybdenum which resulted in a white/light yellow to blue color transition. The developed ethylene sensor, punched on Whatman filter paper, was used to identify the ripeness of apples by changing color from yellow to blue (Lang and Hübert, 2011).

For nonproteinaceous food sensing applications, Lin et al. used a combination of six different porphyrins and eight different metalloporphyrins to potentially monitor the ageing process of rice at different storage times (Lin et al., 2018). The dye array was able to detect a representative VOC of the ageing process in rice.

23.3.2 Detection of toxic molecules

Melamine is a triazine heterocyclic organic compound, which is very rich in nitrogen content (66% by mass). In the past, melamine has been added to food, especially milk, as an indicator for protein content—however, consumption of melamine results in the build-up of insoluble crystals in the kidney (Tyan et al., 2009). Thus, rapid and accurate detection of melamine in food is crucial. Cai et al. developed melamine detection sensors using 3-mercaptopriopionic acid-mediated gold nanoparticles (AuNPs) as sensing materials (Cai et al., 2014). The developed sensors were potentially able to detect melamine at concentrations as low as 0.4 µg/mL and 30 µg/mL in milk by UV–vis absorption spectroscopy and with the naked eye, respectively. In addition, the sensor exhibited excellent selectivity to melamine in the presence of interfering molecules and ions including vitamin C, Na^+, K^+, Mg^{2+}, Cu^{2+}, and Fe^{3+}. Built on a similar concept, a simple citrate mediated AuNP sensor was developed to detect melamine in milk (Kumar et al., 2014). Compared to the study of Cai et al., the citrate-mediated AuNP sensor exhibited an eightfold lower detection limit (0.05 µg/mL) of melamine, which is significantly lower than the permitted level of melamine by the Food and Drug Administration. The sensor was able to highly discriminate melamine from common adulterants and preservatives used in milk, including glucose, urea, formalin, hydrogen peroxide, sucrose, nitrate, and dextrose (Kumar et al., 2014).

In some agricultural and food processing practices, pesticides may remain in food products, posing a health risk to consumers. Tan et al. used a metal carbonyl-conjugated AuNP sensor for the detection of glyphosate pesticide (Tan et al., 2017). AuNPs were functionalized with organometallic osmium carbonyl clusters, which helped to reduce the signal-to-noise caused by interfering substances in complex biological samples. In the absence of glyphosate, acetylthiocholine was hydrolyzed to thiocholine by acetylcholinesterase (AChE), inducing aggregation of AuNP which resulted in color transitioning from red to purple. In the presence of glyphosate, the hydrolysis of acetylthiocholine to thiocholine by AChE was inhibited. Hence, AuNPs remained separated, and no color change was observed. The potential application of these sensors was demonstrated in the detection of organophosphorus pesticides in food such as beer.

23.3.3 Detection of heavy metals

Heavy metals such as mercury, cadmium, and lead can be present in certain foods. The removal of these metals is not always possible since they are ubiquitous and could be taken up by plants. The consumption of food containing excess amounts of heavy metals poses significant health risks such as neural disorder and cognitive deficits (Karri et al., 2016).

There are several examples of successful potential colorimetric sensors for the detection of neutral and ionic heavy metal species in food. For instance, poly(1,4-bis-(8-(4-phenylthiazole-2-thiol)-octyloxy)-benzene) has been used as the sensing material for the detection of Hg^{2+} cations in water and seafood (Pavase et al., 2015). The developed sensors were able to detect Hg^{2+} at concentrations as low as 1.86 nM in less than 2 min. In another example, AuNPs were functionalized with 3-mercaptopropionate acid and adenosine monophosphate to be used as Hg^{2+} sensors. Chen et al. designed a paper-based sensor to detect Hg^{2+} in which oligonucleotide functionalized AuNPs were immobilized in paper (Chen et al., 2014). Upon the formation of thymine-Hg^{2+} complex in the presence of Hg^{2+}, the functionalized AuNPs aggregated, resulting in a color change from red to blue. Following a similar concept, AuNPs have been functionalized with organic acids such as mercaptoundecanoic and meso-2,3-dimercaptosuccinic acid for the detection of Pb^{2+} (Kim et al., 2001) and Cr^{3+} (Chen et al., 2015), respectively. It is noteworthy that most sensors described in this section were developed in the liquid phase, which may limit their applications in the solid phase.

23.3.4 Detection of biomolecules

Colorimetric sensors can also be designed to detect specific biomolecules in food, including nucleic acids, proteins, carbohydrates, and lipids. Most biosensors, however, target nucleic acids and proteins which are considered as biomarkers for microbial or viral contamination, or food allergies. For instance, sensors for detection of *S. typhimurium* were developed based on AuNP and Fe_3O_4 magnetic nanoparticles (MNPs) each decorated with DNA molecules partially complementary to the target DNA sequence of *S. typhimurium* (Ma et al., 2017). In the presence of the target DNA, signal DNA-AuNPs were connected to capture DNA-MNPs via base pairing of the target DNA. This sandwich formation caused AuNP aggregation, resulting in the color transitioning from red to blue. The developed method was successfully used for microbial detection in milk samples, exhibiting its high potential for food safety applications (Ma et al., 2017).

Using noble metal nanoparticles as the sensing material, colorimetric enzyme-AuNP sensors have been developed for the detection of microbial contamination (Fig. 23.5; Miranda et al., 2011). In this method, amine functionalized AuNPs were bound to β-galactosidase (β-Gal) enzyme, which inhibits its enzymatic activity. However, in contact with *E. coli*, the β-Gal enzyme was released from the AuNPs. The unbound enzymes then reacted with chlorophenol red β-D-galactopyranoside (CPRG) in the solution, and the color of CPRG changed from pale yellow to red. The sensor was able to detect *E. coli* concentration as low as 10^2 CFU/mL (Miranda et al., 2011). Hence, by leveraging enzyme-nanoparticle assemblies, this system could determine bacteria concentrations as low as 100 CFU/mL in 10 min (Miranda et al., 2011). The authors claimed that their system could be used for other bacteria, albeit the detection sensitivity may vary within various species.

In another work, aptamer functionalized AuNP sensors were developed for detection of aflatoxin B1 (AFB1) (Hosseini et al., 2015). In the absence of AFB1, the aptamer was absorbed onto the surface of AuNPs, preventing their salt-induced aggregation. In the presence of AFB1, the AFB1 bonded to the aptamer causing its desorption from the AuNPs surface. The resulted nonprotected AuNPs were prone to salt-induced aggregation, resulting in the color change. The food application of this sensor was successfully demonstrated by exposing this colorimetric kit with peanut and rice extracts with results comparable to those of liquid chromatography methods.

23.4 Electrochemical sensors for food quality and safety

In electrochemical sensors, the signal induced by the receptor is transformed by a transducing element into an electrical signal. Based on the type of signal transduction, electrochemical sensors can be classified into voltammetric (or amperometric), potentiometric, impedimetric, and conductometric (or chemiresistive; Nikoleli et al., 2018). Generally, the analyte undergoes a redox reaction on the working electrode surface, which releases or subtracts electrons from the system—causing a variation of electron fluxes. The variation of electron movements generates the electrochemical signal that is measured by the detector (Viswanathan et al., 2009). In conductometric sensors, a chemical reaction, not necessarily

(A)

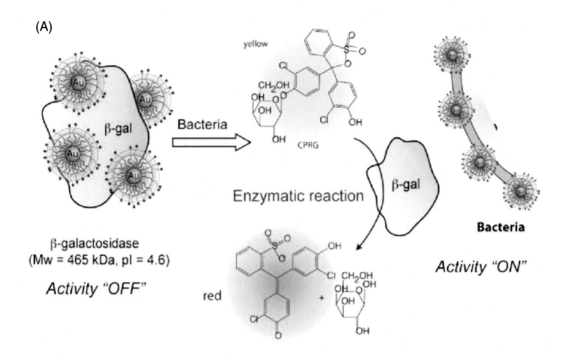

(B)

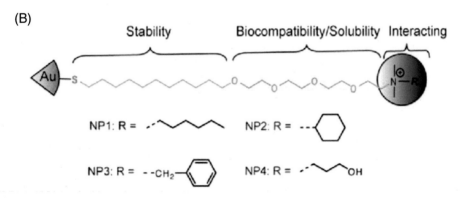

FIG. 23.5

(A) Enzyme-amplified sensing of bacteria, showing the relative sizes of the 2 nm core diameter NPs and β-Gal and (B) structures of ligands used for sensing studies (Miranda et al., 2011), Copyright 2011, American Chemical Society.

redox, modifies the conductivity of the sensing material. The electrodes used in electrochemical sensors are usually modified by adding nanostructured materials on their surfaces. Gold, silver, platinum, and carbon-based materials such as graphene and graphene oxide are the most common materials used for electrodes, with conductive polymers gaining importance in flexible sensor applications.

The main advantages of electrochemical sensors include fast response, high sensitivity (nanomolar), using relatively small amounts of an analyte, and quantitative response within the range of analyte concentration (Akyazi et al., 2018). Electrochemical sensors have been bulky and relied on using batteries and reading detectors, however, in the latest design, these shortcomings were addressed by implementing wireless communication (Dincer et al., 2019).

23.4.1 Voltammetric sensors

In voltammetric sensors, a varying potential is applied to the sensor, and the current is measured. Usually, this kind of sensors makes use of a three-electrode system: working electrode, reference, and auxiliary electrodes. The current is generated by the flow of electrons produced by the oxidation or reduction of the electroactive species. Thus, the current is proportional to the concentration of the analyte in the sample (Nikoleli et al., 2018). The potential profile can be varied in different ways, such as linear sweep, cyclic, and square wave. Amperometric sensors can be considered as a subcategory of voltammetric sensors, as the output signal is the same, but the potential is kept constant by an external energy source such as batteries. As an example, a voltammetric food sensor was developed from modified carbon paste electrodes to detect carmoisine and tartrazine, food dyes that can be harmful to children, in dried fruit and soft drink samples (Bijad et al., 2018). These voltammetric sensors were able to simultaneously determine carmoisine and tartrazine within the range of 70–650 μM and 0.1–750 μM, respectively, with low detection limits of 20 nM for carmoisine and 0.06 μM tartrazine.

23.4.2 Potentiometric sensors

Potentiometric sensors measure the potential of the sample between two electrodes. This is done at zero or nearly zero current, so the process has almost no effect on the sample solution, therefore, allowing quantitative measurements (Li et al., 2019). The change in potential is caused by the exchange of ions between the two electrodes and the sample. However, an ion-selective membrane is necessary if more than one type of ion is present in the solution, as multiple ions may alter the result and hence affect selectivity (Nikoleli et al., 2018). Potentiometric sensors are usually used for liquid, and they have been reported to be able to detect cadmium ions (Zhang et al., 2018; Rezvani Ivari et al., 2017), phenolic acid (Sun et al., 2019), biogenic amines (Minamiki and Kurita, 2019), and other ions (Mousavi et al., 2018) in food. The main drawback of this type of sensor is that ion-selective electrodes are usually delicate and bulky. New technologies such as the use of cotton thread (Mousavi et al., 2018) have been applied recently to overcome this hurdle.

23.4.3 Impedimetric sensors

The impedance of a system is usually calculated as the ratio between applied voltage, varying in time with a small amplitude, and the outcoming current. The impedance is a complex number, as the current not only changes the intensity but shows a phase shift (Lisdat and Schäfer, 2008). The impedance can be described by a modulus and a phase shift or by the real and the imaginary parts of the modulus.

When this method is applied for analytical purposes, the variation of an impedance element, resistance, or capacitance, is correlated to the change in the analyte concentration (Lisdat and Schäfer, 2008; da Silva et al., 2019; Zlatev et al., 2019; Brosel-Oliu et al., 2019). Impedimetric sensors are usually bulky because of the need for three electrodes and an alternating voltage source. They also require a complex analysis of the output data.

23.4.4 Conductometric sensors

The sensing mechanism of conductometric (or chemiresistive) sensors is the change in conductivity of the sensing material after the interaction with the analyte. Depending on the analyte and the sensing material, the interaction can be either chemical or physical. The most common material used in conductometric sensors is conductive polymers (Latif et al., 2018; Hartwig et al., 2018), carbon nanostructures such as nanotubes and graphene (Schroeder et al., 2019; Bag and Pal, 2019; He et al., 2019), and metal oxides (Srinivasan et al., 2019; Kanaparthi and Singh, 2019). Most of the conductometric sensors are developed for gas sensing applications, but they can also be used for liquid-phase samples. Conductometric sensors have high selectivity, sensitivity, and their structure is usually simpler and less bulky than other types of electrochemical sensors. Therefore, continuous monitoring and tag reading can be applied to these sensors. The latter is of particular interest for wireless sensors, where the voltage can be provided by the signal reader (Boada et al., 2019). A recent example is a paper-based electrical gas sensor which is connected to an RFID tag and can detect ammonia and carbon dioxide gases during fish or chicken spoilage (Barandun et al., 2019).

The configuration of a number of electrochemical sensors is shown in Fig. 23.6 that demonstrates the progress has been made in miniaturization and portability of these sensor systems. One of the innovative voltammetric sensors that have been manufactured for on-site detection of organophosphorus chemicals is depicted in Fig. 23.6A (Mishra et al., 2017). Since this sensor is voltammetric, there is a need for an external source of energy (e.g., battery) to use this device. Fig. 23.6B shows a wireless humidity sensor that has been proposed for in situ monitoring the taste of dry food (Tan et al., 2007). Although the sensor was conductometric and was built on a paper substrate, there is a need for an external network analyzer. When the target market of food sensors are consumers, it is more facile to use a network analyzer that is frequently used, such as smartphones. Fig. 23.6C shows a paper-based conductometric sensor for detecting ammonia and carbon dioxide (Barandun et al., 2019). This sensor can communicate with a smartphone via RFID communication (Barandun et al., 2019).

23.4.5 Electrochemical biosensors

In food sensors where selectivity is of great importance, bioreceptors play an important role as their interaction with analytes is extremely targeted (Rotariu et al., 2016). Biosensors are mainly aimed to detect unwanted contaminants in food, for example, biological toxins, allergens, antibiotic residues, and pathogenic microbes.

The first group of biosensors are enzymatic sensors. Enzymatic sensors utilize enzymes as their receptors and are used to determine food quality by directly detecting biogenic amines (Vanegas et al., 2018), tyrosine (Varmira et al., 2018), and sulfites (Jubete et al., 2017), or by indirectly detecting the metabolic by-products of bacteria (Shoja et al., 2017). The enzymes' activity could be compromised if their structures are altered by environmental factors such as temperature and contaminants.

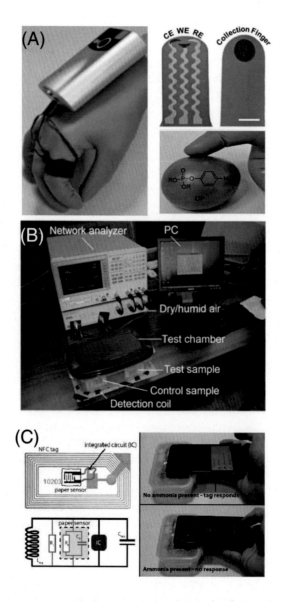

FIG. 23.6

Examples of electrochemical sensors for food quality and safety: (A) Portable and wearable sensors incorporated into gloves for detecting organophosphate (Mishra et al., 2017) Copyright 2017, American Chemical Society; (B) Wireless humidity sensor connected to a network analyzer (Tan et al., 2007); (C) A wireless, paper-based, and chemiresistive sensor connected to NFC tag for detecting ammonia in spoiled fish and chicken (Barandun et al., 2019) Copyright 2019, American Chemical Society.

Consequently, the enzymes must be immobilized and protected on a carrier scaffold to maintain their activity during sensing (Zhu et al., 2019). Immobilization with respect to the substrate can be either chemical or physical. In chemical methods, strong chemical bonds are formed between the enzyme and the substrate. In physical methods, the enzyme is connected to the substrate via hydrogen bonds and van der Waals forces (Bilal et al., 2018). While the chemical bonds are often more stable than the physical ones, the presence of chemical bonds may modify the enzyme's structure, decreasing its activity (Mohamad et al., 2015). The most common immobilization substrates are polymers, metal oxides, ceramics, nanoparticles, nanotubes, and electrospun fibers (Zdarta et al., 2018). Flexible substrates, such as hydrogels and paper, are favorable for food packaging applications (Nery and Kubota, 2016). One common strategy in enzymatic sensors is the use of multiple enzymes or different bioreceptors. Examples include the bi-enzyme systems, where an oxidase enzyme reacts with the analyte producing hydrogen peroxide. The hydrogen peroxide is then consumed by a peroxidase enzyme to produce a readable signal (Liu et al., 2019; Zhao et al., 2019; Kawai et al., 2019).

Sensors using an antibody as the receptor are called immunosensors. In immunosensors, the detection of bacteria or viruses takes place when the antibody–antigen complex is formed. This interaction is often difficult to detect, and therefore, many immunosensors make use of an enzyme labeled assay. In this case, two antibodies are present: the first is connected to the electrode, and the other is labeled with an enzyme. In the presence of an antigen, they both interact with it, forming a complex. In this way, the redox reaction occurring between the enzyme and the enzymatic substrate (present in the sensor structure) amplifies the signal (Angulo-Ibáñez et al., 2019). The labeling can also be achieved with inorganic materials, such as carbon nanotubes (Bhardwaj et al., 2017), gold nanoparticles, and quantum dots (Xu et al., 2017). Alternatively, it is possible to have unlabeled immunosensors if the output signal is the impedance. The formation of the antibody–antigen complex causes a change in the electrode surface impedance, which is detectable without using any assay (Malvano et al., 2019; Singh et al., 2018). It should be noted that emerging technologies such as microfluidics yielded in novel biosensors for detecting specific components in complex liquid samples. An example includes CYBERTONGUE that measures protein-degrading enzymes in food as well as the residual lactose in lactose-free milk.

Rather than enzymatic and immunosensors, there are other bioreceptors that have been studied in food sensors. DNA receptors are highly selective as their nucleotides sequence could be engineered to be complementary to a specific pathogen strand. However, the main drawback of DNA-based sensors is that DNA and RNA molecules are fragile, and therefore, these sensors lack stability and have a short shelf-life (Asal et al., 2018). Similarly, aptamers can be labeled without hindering the ligands and are more stable than antibodies (Mishra et al., 2018). Their strand is shorter than DNA or RNA strands, but they have a higher receptor density; hence they have higher binding efficiency (Kozitsina et al., 2018). They can also be added to other receptors or biotransducers to improve their selectivity. Phages, short for bacteriophages, are viral parasites that infect other cells to propagate and, therefore, can be used in bacteria detection. Most of them are strain-specific, meaning that they will infect only a specific subtype of bacteria; hence they are highly selective. Usually, they are used in colorimetric sensors, as the phages can be genetically modified to express a specific gene and produce chemiluminescent substances. A number of electrochemical phage sensors have been reported recently (Vanegas et al., 2017; Xu et al., 2019).

Table 23.1 Examples of electrochemical biosensors for food quality and safety.

Food sample	Analyte	Sensing material	Transduction	Limit of detection (CFU/mL)	Time (min)	Refs.
Blueberry	*Listeria monocytogenes*	Gold nanoparticles	Amperometric	2	N.A.	(Davis et al., 2013)
Apple juice	*Salmonella typhimurium*	PEDOT:PSS, antibody, paper	Potentiometric	6	<60	(Silva et al., 2019)
Vegetal broth	*Listeria monocytogenes*	Reduced graphene oxide, nanoplatinum, chitosan	Impedimetric	3	17	(Hills et al., 2018)
Skim and whole milk	*Salmonella typhimurium*	Gold electrode	Amperometric	10	N.A.	(Alexandre et al., 2016)
Milk	*Escherichia coli* K12	Gold nanoparticles	Voltammetric	10	120	(Zou et al., 2019)
Milk	*Salmonella typhimurium*	Gold screen printed electrode, antibody	Impedimetric	7×10^4	20	(Farka et al., 2016)
Milk	*Listeria monocytogenes*	Multiwalled carbon nanotubes fibers/ antibody-horseradish peroxidase	Voltammetric	1.7×10^2	30	(Lu et al., 2016)
Milk	*Shigella dysenteriae*	Aptamar, gold nanoparticle, glassy carbon electrode	Impedimetric	10	30	(Zarei et al., 2018)
Egg	*Salmonella typhimurium*	Glassy carbon electrode, nanoporous gold, aptamar	Impedimetric	650	40	(Riu and Giussani, 2020)
Chicken	*Salmonella pullorum*	Gold nanoparticles, antibody, reduced graphene oxide	Amperometric	89	50	(Fei et al., 2016)
Spiked fish	*Vibrio parahaemolyticus*	Gold nanoparticles, antibody	Amperometric	2	120	(Teng et al., 2017)
Ground beef	*Escherichia coli* O157:H7	Gold nanoparticles, modified graphene paper electrode	Impedimetric	104	90	(Wang et al., 2013)
Licorice extract	*Escherichia coli*	Gold, aptamar	Voltammetric	80	150	(Wang et al., 2019)

Bacterial pathogens such as *Escherichia coli*, *Listeria monocytogenes*, and *Salmonella* are often the main cause of foodborne diseases, albeit there are many more pathogens that can cause foodborne illness (Food and Drug Administration, 2012). Table 23.1 lists examples of electrochemical biosensors used for food applications along with their sensing mechanisms. As can be seen in Table 23.1, by using various sensing materials such as nanoparticles, carbon inks, and polymeric materials, a wide range of analytes such as *S. typhimurium*, *Listeria monocytogenes*, *Escherichia coli*, *Shigella dysenteriae*, *Vibrio parahaemolyticus* were detected. It is noteworthy that in the majority of the reported sensors, an external power source was needed, which is a limitation for their application in the food sector. Also, the main drawback in such sensors is the low stability and rapid degradation of the bioreceptor.

23.5 Recommendations and future direction

This chapter discussed various sensors and their mechanisms of detections that have been manufactured for assessing the safety and quality of food. In general, the current food sensors are either colorimetric or electrochemical. Most commercial food sensors in the market assess the food quality indirectly by monitoring environmental indicators such as time and temperature. The advances in sensing mechanism and materials provided the opportunity to design more selective sensors which can directly measure the chemicals, toxins, or unwanted pathogens in the food package. Majorities of manufactured electrochemical sensors are voltammetric and impedimetric, which demands an external energy source.

Although great advances have been reported in the field of food sensors, the current technology has several challenges including its relatively high cost of manufacturing, using noncompostable materials, being limited to a specific operating temperature (i.e., unable to sense accurately at all storage temperatures such as room, refrigeration, and subzero). To date, most sensors that have reached the market are limited to colorimetric sensors because of their ease of fabrication and compatibility with packaging materials.

The increasing global problem of waste generated from food and food packaging has fuelled the further development of sensors for food applications. Future food sensors should be fabricated from compostable, ingestible, bioresorbable, or even metabolizable materials preferably sourced from natural resources. Silk is one such example, offering natural abundance, biocompatibility, and suitable mechanical, electrical, and adhesive properties (Tao et al., 2012). Also, flexible electronics and new active materials can be employed in future electrochemical food sensors to enable their incorporation in flexible food packaging.

Cost is another defining factor for the success of food sensors in the market. Here, cost-effective fabrication technologies such as inkjet printing can be utilized for large-scale production of sensors directly applied to the food packaging. Economically viable materials such as cellulose and its derivatives can also be used to reduce production costs (Barandun et al., 2019; Nguyen et al., 2019). Another promising area in developing low-cost sensors for food is the use of nanomaterials in food supply change. This topic is comprehensively discussed in Chapter 24 (Tao et al., 2012).

The operational temperature window of food sensors should also be expanded to cover subzero temperatures. In this domain, specifically formulated compounds should be developed where the sensing mechanism is not impacted by subzero temperatures. One example is the recent work on polydiacetylene-based sensors that are operational at $-20\ °C$ (Nguyen et al., 2019).

Finally, by using electrochemical food sensors, near-real-time data for the quality and safety of the food could be achievable. There are ample opportunities to improve the measuring time of these sensors. While the topic of generating such data in the context of the internet of things in the food supply chain may have its challenges (e.g., cybersecurity), the main component must be reliable, accurate, selective, and wireless. Achieving real real-time or near-real-time data plays an important role in industry 4.0 as the food suppliers can monitor the quality and safety of the food during its supply change.

Acknowledgment

The authors acknowledge the financial support from the Australian Research Council (LP180100309) and the Centre for Advanced Food Enginomics at The University of Sydney. F.O. acknowledges the support of F.H. Loxton Postdoctoral Research Fellowship at the University of Sydney.

Abbreviations

AChE	acetylcholinesterase
AuNPs	gold nanoparticles
CFU	colony-forming unit
CPRG	chlorophenol red β-D-galactopyranoside
DNA	deoxyribonucleic acid
MNPs	magnetic nanoparticles
NPs	nanoparticles
PANI	polyaniline
PDA	polydiacetylene
PEDOT	poly(3,4-ethylenedioxythiophene)
PSS	poly(sodium 4-styrenesulfonate)
RFID	radio frequency identification
RNA	ribonucleic acid
TEA	triethylamine
TMA	trimethylamine
VOC	volatile organic compounds

References

Akyazi, T., Basabe-Desmonts, L., Benito-Lopez, F., 2018. Review on microfluidic paper-based analytical devices towards commercialisation. Anal. Chim. Acta 1001, 1–17.

Alexandre, D., Melo, A., Furtado, R., BORGES, M.d.F., Figueiredo, E., Biswas, A., Cheng, H., Alves, C., 2016. Amperometric biosensor for *Salmonella typhimurium* detection in milk, Embrapa Agroindústria Tropical-Artigo em anais de congresso (ALICE), Congresso Brasileiro de Ciência e Tecnologia de Alimentos, 25.

Amin, E.M., Bhuiyan, M.S., Karmakar, N.C., Winther-Jensen, B., 2013. Development of a low cost printable chipless RFID humidity sensor. IEEE Sens. J. 14 (1), 140–149.

Angulo-Ibáñez, A., Eletxigerra, U., Lasheras, X., Campuzano, S., Merino, S., 2019. Electrochemical tropomyosin allergen immunosensor for complex food matrix analysis. Anal. Chim. Acta 1079, 94–102.

Asal, M., Özen, Ö., Şahinler, M., Polatoğlu, İ., 2018. Recent developments in enzyme, DNA and immuno-based biosensors. Sensors 18 (6), 1924.

Bag, S., Pal, K., 2019. A PCB based chemiresistive carbon dioxide sensor operating at room temperature under different relative humidity. IEEE Trans. Nanotechnol. 18, 1119–1128.

Barandun, G., Soprani, M., Naficy, S., Grell, M., Kasimatis, M., Chiu, K.L., Ponzoni, A., Güder, F., 2019. Cellulose fibers enable near-zero-cost electrical sensing of water-soluble gases. ACS sensors 4 (6), 1662–1669.

Bhardwaj, J., Devarakonda, S., Kumar, S., Jang, J., 2017. Development of a paper-based electrochemical immunosensor using an antibody-single walled carbon nanotubes bio-conjugate modified electrode for label-free detection of foodborne pathogens. Sens. Actuators B 253, 115–123.

Bijad, M., Karimi-Maleh, H., Farsi, M., Shahidi, S.-A., 2018. An electrochemical-amplified-platform based on the nanostructure voltammetric sensor for the determination of carmoisine in the presence of tartrazine in dried fruit and soft drink samples. J. Food Measure. Charact. 12 (1), 634–640.

Bilal, M., Rasheed, T., Zhao, Y., Iqbal, H.M.N., Cui, J., 2018. "Smart" chemistry and its application in peroxidase immobilization using different support materials. Int. J. Biol. Macromol. 119, 278–290.

Boada, M., Lazaro, A., Villarino, R., Girbau, D., 2019. Battery-less NFC sensor for pH monitoring. IEEE Access 7, 33226–33239.

Brosel-Oliu, S., Abramova, N., Uria, N., Bratov, A., 2019. Impedimetric transducers based on interdigitated electrode arrays for bacterial detection – a review. Anal. Chim. Acta 1088, 1–19.

Cai, H.-H., Yu, X., Dong, H., Cai, J., Yang, P.-H., 2014. Visual and absorption spectroscopic detections of melamine with 3-mercaptopriopionic acid-functionalized gold nanoparticles: a synergistic strategy induced nanoparticle aggregates. J. Food Eng. 142, 163–169.

Food and Drug Administration, 2012. Bad Bug Book, Food borne Pathogenic Microorganisms and Natural Toxins. Second Edition.

Chen, G.-H., Chen, W.-Y., Yen, Y.-C., Wang, C.-W., Chang, H.-T., Chen, C.-F., 2014. Detection of mercury (II) ions using colorimetric gold nanoparticles on paper-based analytical devices. Anal. Chem. 86 (14), 6843–6849.

Chen, W., Cao, F., Zheng, W., Tian, Y., Xianyu, Y., Xu, P., Zhang, W., Wang, Z., Deng, K., Jiang, X., 2015. Detection of the nanomolar level of total Cr [(III) and (VI)] by functionalized gold nanoparticles and a smartphone with the assistance of theoretical calculation models. Nanoscale 7 (5), 2042–2049.

Chun, H.-N., Kim, B., Shin, H.-S., 2014. Evaluation of a freshness indicator for quality of fish products during storage. Food Sci. Biotechnol. 23 (5), 1719–1725.

Ciui, B., Martin, A., Mishra, R.K., Nakagawa, T., Dawkins, T.J., Lyu, M., Cristea, C., Sandulescu, R., Wang, J., 2018. Chemical sensing at the robot fingertips: toward automated taste discrimination in food samples. ACS Sensors 3 (11), 2375–2384.

da Silva, W., Ghica, M.E., Ajayi, R.F., Iwuoha, E.I., Brett, C.M.A., 2019. Impedimetric sensor for tyramine based on gold nanoparticle doped-poly(8-anilino-1-naphthalene sulphonic acid) modified gold electrodes. Talanta 195, 604–612.

Davis, D., Guo, X., Musavi, L., Lin, C.-S., Chen, S.-H., Wu, V.C., 2013. Gold nanoparticle-modified carbon electrode biosensor for the detection of *Listeria monocytogenes*. Ind. Biotechnol. 9 (1), 31–36.

Dincer, C., Bruch, R., Costa-Rama, E., Fernández-Abedul, M.T., Merkoçi, A., Manz, A., Urban, G.A., Güder, F., 2019. Disposable sensors in diagnostics, food, and environmental monitoring. Adv. Mater. 31 (30), 1806739.

Eaton, K., 2002. A novel colorimetric oxygen sensor: dye redox chemistry in a thin polymer film. Sens. Actuators B 85 (1-2), 42–51.

Farka, Z., Juřík, T., Pastucha, M., Kovář, D., Lacina, K., Skládal, P., 2016. Rapid immunosensing of *Salmonella typhimurium* using electrochemical impedance spectroscopy: the effect of sample treatment. Electroanalysis 28 (8), 1803–1809.

Fei, J., Dou, W., Zhao, G., 2016. Amperometric immunoassay for the detection of *Salmonella pullorum* using a screen-printed carbon electrode modified with gold nanoparticle-coated reduced graphene oxide and immunomagnetic beads. Microchim. Acta 183 (2), 757–764.

Feng, Y., Xie, L., Chen, Q., Zheng, L.-R., 2014. Low-cost printed chipless RFID humidity sensor tag for intelligent packaging. IEEE Sens. J. 15 (6), 3201–3208.

Fu, B., Labuza, T.P., 1992. Considerations for the application of time-temperature integrators in food distribution. J. Food Distrib. Res. 23 (856-2016-57005), 9–18.

Harrey, P., Ramsey, B., Evans, P., Harrison, D., 2002. Capacitive-type humidity sensors fabricated using the offset lithographic printing process. Sens. Actuators B 87 (2), 226–232.

Hartwig, M., Zichner, R., Joseph, Y., 2018. Inkjet-printed wireless chemiresistive sensors—a review. Chemosensors 6 (4), 66

He, M., Croy, R.G., Essigmann, J.M., Swager, T.M., 2019. Chemiresistive carbon nanotube sensors for N-nitrosodialkylamines. ACS Sensors 4 (10), 2819–2824.

Hills, K.D., Oliveira, D.A., Cavallaro, N.D., Gomes, C.L., McLamore, E.S., 2018. Actuation of chitosan-aptamer nanobrush borders for pathogen sensing. Analyst 143 (7), 1650–1661.

Hosseini, M., Khabbaz, H., Dadmehr, M., Ganjali, M.R., Mohamadnejad, J., 2015. Aptamer-based colorimetric and chemiluminescence detection of aflatoxin B1 in foods samples. Acta Chim. Slov. 62 (3), 721–728.

Jubete, E., Jaureguibeitia, A., Añorga, L., Lamas-Ardisana, P.J., Martínez, G., Serafín, V., Cabañero, G., Ramos, E., Salleres, S., Grande, H.J., Albizu, A., 2017. SO$_2$SAFE – enzymatic SO$_2$ biosensor for rapid food safety monitoring. Proc. Technol. 27, 51–52.

Kanaparthi, S., Singh, S.G., 2019. Chemiresistive sensor based on zinc oxide nanoflakes for CO_2 detection. ACS Appl. Nano Mater. 2 (2), 700–706.

Karri, V., Schuhmacher, M., Kumar, V., 2016. Heavy metals (Pb, Cd, As and MeHg) as risk factors for cognitive dysfunction: a general review of metal mixture mechanism in brain. Environ. Toxicol. Pharmacol. 48, 203–213.

Kawai, H., Kitazumi, Y., Shirai, O., Kano, K., 2019. Performance analysis of an oxidase/peroxidase-based mediatorless amperometric biosensor. J. Electroanal. Chem. 841, 73–78.

Kim, J.-H., Shiratori, S., 2006. Fabrication of color changeable film to detect ethylene gas. Jpn. J. Appl. Phys. 45 (5A), 4274–4278.

Kim, J.U., Ghafoor, K., Ahn, J., Shin, S., Lee, S.H., Shahbaz, H.M., Shin, H.-H., Kim, S., Park, J., 2016. Kinetic modeling and characterization of a diffusion-based time-temperature indicator (TTI) for monitoring microbial quality of non-pasteurized angelica juice. LWT-Food Sci. Technol. 67, 143–150.

Kim, Y., Johnson, R.C., Hupp, J.T., 2001. Gold nanoparticle-based sensing of "spectroscopically silent" heavy metal ions. Nano Lett. 1 (4), 165–167.

Kozitsina, A.N., Svalova, T.S., Malysheva, N.N., Okhokhonin, A.V., Vidrevich, M.B., Brainina, K.Z., 2018. Sensors based on bio and biomimetic receptors in medical diagnostic, environment, and food analysis. Biosensors 8 (2), 35.

Kumar, N., Seth, R., Kumar, H., 2014. Colorimetric detection of melamine in milk by citrate-stabilized gold nanoparticles. Anal. Biochem. 456, 43–49.

Lang, C., Hübert, T., 2011. A colour ripeness indicator for apples. Food Bioprocess Technol. 5 (8), 3244–3249.

Latif, U., Ping, L., Dickert, F.L., 2018. Conductometric sensor for PAH detection with molecularly imprinted polymer as recognition layer. Sensors 18 (3), 767.

Lee, S.Y., Lee, S.J., Choi, D.S., Hur, S.J., 2015. Current topics in active and intelligent food packaging for preservation of fresh foods. J. Sci. Food Agric. 95 (14), 2799–2810.

Li, F., Yu, Z., Han, X., Lai, R.Y., 2019. Electrochemical aptamer-based sensors for food and water analysis: a review. Anal. Chim. Acta 1051, 1–23.

Li, L., Ferng, L.-H., Wei, Y., Yang, C., Ji, H.-F., 2014. Highly stable polyaniline-poly (sodium 4-styrenesulfonate) nanoparticles for sensing of amines. J. Nanosci. Nanotechnol. 14 (9), 6593–6598.

Lin, H., Man, Z.-x., Kang, W.-c., Guan, B.-b., Chen, Q.-s., Xue, Z.-l., 2018. A novel colorimetric sensor array based on boron-dipyrromethene dyes for monitoring the storage time of rice. Food Chem. 268, 300–306.

Lisdat, F., Schäfer, D., 2008. The use of electrochemical impedance spectroscopy for biosensing. Anal. Bioanal. Chem. 391 (5), 1555.

Liu, L., Chen, C., Chen, C., Kang, X., Zhang, H., Tao, Y., Xie, Q., Yao, S., 2019. Poly(noradrenalin) based bi-enzyme biosensor for ultrasensitive multi-analyte determination. Talanta 194, 343–349.

Lu, Y., Liu, Y., Zhao, Y., Li, W., Qiu, L., Li, L., 2016. A novel and disposable enzyme-labeled amperometric immunosensor based on MWCNT fibers for *Listeria monocytogenes* detection. J. Nanomater., 2016.

Ma, X., Song, L., Xia, Y., Jiang, C., Wang, Z., 2017. A novel colorimetric detection of *S. typhimurium* based on Fe_3O_4 magnetic nanoparticles and gold nanoparticles. Food Anal. Methods 10 (8), 2735–2742.

Malvano, F., Pilloton, R., Albanese, D., 2019. Label-free impedimetric biosensors for the control of food safety – a review. Int. J. Environ. Anal. Chem. 100 (4), 468–491.

Manjunatha, J., 2018. A novel voltammetric method for the enhanced detection of the food additive tartrazine using an electrochemical sensor. Heliyon 4 (11), e00986.

Minamiki, T., Kurita, R., 2019. Potentiometric detection of biogenic amines utilizing affinity on a 4-mercaptobenzoic acid monolayer. Anal. Methods 11 (9), 1155–1158.

Miranda, O.R., Li, X., Garcia-Gonzalez, L., Zhu, Z.-J., Yan, B., Bunz, U.H., Rotello, V.M., 2011. Colorimetric bacteria sensing using a supramolecular enzyme–nanoparticle biosensor. J. Am. Chem. Soc. 133 (25), 9650–9653.

Mishra, G.K., Sharma, V., Mishra, R.K., 2018. Electrochemical aptasensors for food and environmental safeguarding: a review. Biosensors 8 (2), 28.

Mishra, R.K., Hubble, L.J., Martín, A., Kumar, R., Barfidokht, A., Kim, J., Musameh, M.M., Kyratzis, I.L., Wang, J., 2017. Wearable flexible and stretchable glove biosensor for on-site detection of organophosphorus chemical threats. ACS Sensors 2 (4), 553–561.

Mohamad, N.R., Marzuki, N.H.C., Buang, N.A., Huyop, F., Wahab, R.A., 2015. An overview of technologies for immobilization of enzymes and surface analysis techniques for immobilized enzymes. Biotechnol. Biotechnol. Equip. 29 (2), 205–220.

Mousavi, M.P.S., Ainla, A., Tan, E.K.W., Abd El-Rahman, M.K., Yoshida, Y., Yuan, L., Sigurslid, H.H., Arkan, N., Yip, M.C., Abrahamsson, C.K., Homer-Vanniasinkam, S., Whitesides, G.M., 2018. Ion sensing with thread-based potentiometric electrodes. Lab Chip 18 (15), 2279–2290.

Nery, E.W., Kubota, L.T., 2016. Evaluation of enzyme immobilization methods for paper-based devices—a glucose oxidase study. J. Pharm. Biomed. Anal. 117, 551–559.

Nguyen, L.H., Naficy, S., Chandrawati, R., Dehghani, F., 2019. Nanocellulose for sensing applications. Adv. Mater. Interf. 6 (18), 1900424.

Nguyen, L.H., Naficy, S., McConchie, R., Dehghani, F., Chandrawati, R., 2019. Polydiacetylene-based sensors to detect food spoilage at low temperatures. J. Mater. Chem. C 7 (7), 1919–1926.

Nguyen, L.H., Oveissi, F., Chandrawati, R., Dehghani, F., Naficy, S., 2020. Naked-eye detection of ethylene using thiol functionalized polydiacetylene-based flexible sensors. ACS Sensors 5 (7), 1921–1928.

Nikoleli, G.-P., Nikolelis, D.P., Siontorou, C.G., Karapetis, S., Varzakas, T., 2018. Chapter two -- Novel biosensors for the rapid detection of toxicants in foods. In: Toldrá, F. (Ed.), Advances in Food and Nutrition Research. Academic Press, Cambridge, MA, pp. 57–102.

Pavase, T.R., Lin, H., Li, Z., 2015. Rapid detection methodology for inorganic mercury (Hg2+) in seafood samples using conjugated polymer (1, 4-bis(8-(4-phenylthiazole-2-thiol)-octyloxy)-benzene)(PPT) by colorimetric and fluorescence spectroscopy. Sens. Actuators B 220, 406–413.

Rezvani Ivari, S.A., Darroudi, A., Arbab Zavar, M.H., Zohuri, G., Ashraf, N., 2017. Ion imprinted polymer based potentiometric sensor for the trace determination of cadmium (II) ions. Arab. J. Chem. 10, S864–S869.

Riu, J., Giussani, B., 2020. Electrochemical biosensors for the detection of pathogenic bacteria in food. TrAC Trends Anal. Chem. 126, 115863.

Rotariu, L., Lagarde, F., Jaffrezic-Renault, N., Bala, C., 2016. Electrochemical biosensors for fast detection of food contaminants – trends and perspective. TrAC Trends Anal. Chem. 79, 80–87.

Salvatore, G.A., Sülzle, J., Dalla Valle, F., Cantarella, G., Robotti, F., Jokic, P., Knobelspies, S., Daus, A., Büthe, L., Petti, L., 2017. Biodegradable and highly deformable temperature sensors for the internet of things. Adv. Funct. Mater. 27 (35), 1702390.

Schroeder, V., Evans, E.D., Wu, Y.-C.M., Voll, C.-C.A., McDonald, B.R., Savagatrup, S., Swager, T.M., 2019. Chemiresistive sensor array and machine learning classification of food. ACS Sensors 4 (8), 2101–2108.

Shoja, Y., Rafati, A.A., Ghodsi, J., 2017. Enzymatic biosensor based on entrapment of d-amino acid oxidase on gold nanofilm/MWCNTs nanocomposite modified glassy carbon electrode by sol-gel network: analytical applications for d-alanine in human serum. Enzyme Microb. Technol. 100, 20–27.

Silva, N.F., Almeida, C.M., Magalhães, J.M., Gonçalves, M.P., Freire, C., Delerue-Matos, C., 2019. Development of a disposable paper-based potentiometric immunosensor for real-time detection of a foodborne pathogen. Biosens. Bioelectron. 141, 111317.

Singh, C., Ali, M.A., Kumar, V., Ahmad, R., Sumana, G., 2018. Functionalized MoS_2 nanosheets assembled microfluidic immunosensor for highly sensitive detection of food pathogen. Sens. Actuators B 259, 1090–1098.

Sivertsvik, M., Rosnes, J.T., Bergslien, H., 2002. Modified atmosphere packaging. In: Ohlsson, T., Bengtsson, N., (Eds.), Minimal Processing Technologies in the Food Industry. Woodhead Publishing Limited, Cambridge, England, pp. 61–86.

Srinivasan, P., Ezhilan, M., Kulandaisamy, A.J., Babu, K.J., Rayappan, J.B.B., 2019. Room temperature chemiresistive gas sensors: challenges and strategies—a mini review. J. Mater. Sci. Mater. Electron. 30 (17), 15825–15847.

Sun, Z., Zhang, Y., Xu, X., Wang, M., Kou, L., 2019. Determination of the total phenolic content in wine samples using potentiometric method based on permanganate ion as an indicator. Molecules 24 (18), 3279.

Tan, E.L., Ng, W.N., Shao, R., Pereles, B.D., Ong, K.G., 2007. A wireless, passive sensor for quantifying packaged food quality. Sensors 7 (9), 1747–1756.

Tan, M.J., Hong, Z.-Y., Chang, M.-H., Liu, C.-C., Cheng, H.-F., Loh, X.J., Chen, C.-H., Liao, C.-D., Kong, K.V., 2017. Metal carbonyl-gold nanoparticle conjugates for highly sensitive SERS detection of organophosphorus pesticides. Biosens. Bioelectron. 96, 167–172.

Tao, H., Brenckle, M.A., Yang, M., Zhang, J., Liu, M., Siebert, S.M., Averitt, R.D., Mannoor, M.S., McAlpine, M.C., Rogers, J.A., 2012. Silk-based conformal, adhesive, edible food sensors. Adv. Mater. 24 (8), 1067–1072.

Taoukis, P., Tsironi, T., 2016. Smart packaging for monitoring and managing food and beverage shelf life. In: Kilcast, D., Subramaniam, P. (Eds.), The Stability and Shelf Life of Food. Woodhead Publishing Limited, Cambridge, England, pp. 141–168.

Teng, J., Ye, Y., Yao, L., Yan, C., Cheng, K., Xue, F., Pan, D., Li, B., Chen, W., 2017. Rolling circle amplification based amperometric aptamer/immuno hybrid biosensor for ultrasensitive detection of *Vibrio parahaemolyticus*. Microchim. Acta 184 (9), 3477–3485.

Thong, Y.J., Nguyen, T., Zhang, Q., Karunanithi, M., Yu, L., 2017. Predicting food nutrition facts using pocket-size near-infrared sensor, 2017 39th Annual International Conference of the IEEE Engineering in Medicine and Biology Society (EMBC). IEEE, pp. 742–745.

Tyan, Y.-C., Yang, M.-H., Jong, S.-B., Wang, C.-K., Shiea, J., 2009. Melamine contamination. Anal. Bioanal. Chem. 395 (3), 729–735.

Vanegas, D.C., Gomes, C.L., Cavallaro, N.D., Giraldo-Escobar, D., McLamore, E.S., 2017. Emerging biorecognition and transduction schemes for rapid detection of pathogenic bacteria in food. Comprehens. Rev. Food Sci. Food Safety 16 (6), 1188–1205.

Vanegas, D.C., Patiño, L., Mendez, C., Oliveira, D.A.d., Torres, A.M., Gomes, C.L., McLamore, E.S., 2018. Laser scribed graphene biosensor for detection of biogenic amines in food samples using locally sourced materials. Biosensors 8 (2), 42.

Varmira, K., Mohammadi, G., Mahmoudi, M., Khodarahmi, R., Rashidi, K., Hedayati, M., Goicoechea, H.C., Jalalvand, A.R., 2018. Fabrication of a novel enzymatic electrochemical biosensor for determination of tyrosine in some food samples. Talanta 183, 1–10.

Virtanen, J., Ukkonen, L., Björninen, T., Sydänheimo, L., 2010. Printed humidity sensor for UHF RFID systems, 2010 IEEE Sensors Applications Symposium (SAS). IEEE, pp. 269–272.

Viswanathan, S., Radecka, H., Radecki, J., 2009. Electrochemical biosensors for food analysis. Monats. Chem. – Chem. Monthly 140 (8), 891.

World Health Organization, 2015. WHO Estimates of the Global Burden of Foodborne Diseases: Foodborne Disease Burden Epidemiology Reference Group 2007-2015. WHO Press, Geneva.

Wang, H., Zhao, Y., Bie, S., Suo, T., Jia, G., Liu, B., Ye, R., Li, Z., 2019. Development of an electrochemical biosensor for rapid and effective detection of pathogenic *Escherichia coli* in licorice extract. Appl. Sci. 9 (2), 295.

Wang, Y., Ping, J., Ye, Z., Wu, J., Ying, Y., 2013. Impedimetric immunosensor based on gold nanoparticles modified graphene paper for label-free detection of *Escherichia coli* O157: H7. Biosens. Bioelectron. 49, 492–498.

Wu, S.-Y., Yang, C., Hsu, W., Lin, L., 2015. 3D-printed microelectronics for integrated circuitry and passive wireless sensors. Microsyst. Nanoeng. 1, 15013.

Xiao-wei, H., Zhi-hua, L., Xiao-bo, Z., Ji-yong, S., Han-ping, M., Jie-wen, Z., Li-min, H., Holmes, M., 2016. Detection of meat-borne trimethylamine based on nanoporous colorimetric sensor arrays. Food Chem. 197, 930–936.

Xu, J., Chau, Y., Lee, Y.-k., 2019. Phage-based electrochemical sensors: a review. Micromachines 10 (12), 855.

Xu, N., Wang, Y., Pan, L., Wei, X., Wang, Y., 2017. Dual-labelled immunoassay with goldmag nanoparticles and quantum dots for quantification of casein in milk. Food Agric. Immunol. 28 (6), 1105–1115.

Yousefi, H., Su, H.-M., Imani, S.M., Alkhaldi, K., Filipe, C.D.M., Didar, T.F., 2019. Intelligent food packaging: a review of smart sensing technologies for monitoring food quality. ACS Sensors 4 (4), 808–821.

Yusufu, D., Wang, C., Mills, A., 2018. Evaluation of an 'After Opening Freshness (AOF)' label for packaged ham. Food Packaging Shelf Life 17, 107–113.

Zabala, S., Castán, J., Martínez, C., 2015. Development of a time–temperature indicator (TTI) label by rotary printing technologies. Food Control 50, 57–64.

Zarei, S.S., Soleimanian-Zad, S., Ensafi, A.A., 2018. An impedimetric aptasensor for *Shigella dysenteriae* using a gold nanoparticle-modified glassy carbon electrode. Microchim. Acta 185 (12), 538.

Zdarta, J., Meyer, A.S., Jesionowski, T., Pinelo, M., 2018. A general overview of support materials for enzyme immobilization: characteristics, properties, practical utility. Catalysts 8 (2), 92.

Zhang, J., Portela, S.B., Horrell, J.B., Leung, A., Weitmann, D.R., Artiuch, J.B., Wilson, S.M., Cipriani, M., Slakey, L.K., Burt, A.M., 2019. An integrated, accurate, rapid, and economical handheld consumer gluten detector. Food Chem. 275, 446–456.

Zhang, W., Xu, Y., Zou, X., 2018. Rapid determination of cadmium in rice using an all-solid RGO-enhanced light addressable potentiometric sensor. Food Chem. 261, 1–7.

Zhao, M., Li, Y., Ma, X., Xia, M., Zhang, Y., 2019. Adsorption of cholesterol oxidase and entrapment of horseradish peroxidase in metal-organic frameworks for the colorimetric biosensing of cholesterol. Talanta 200, 293–299.

Zhu, Y.-C., Mei, L.-P., Ruan, Y.-F., Zhang, N., Zhao, W.-W., Xu, J.-J., Chen, H.-Y., 2019. Chapter 8 – Enzyme-based biosensors and their applications. In: Singh, R.S., Singhania, R.R., Pandey, A., Larroche, C. (Eds.), Advances in Enzyme Technology. Elsevier, Amsterdam, pp. 201–223.

Zlatev, R., Stoytcheva, M., Valdez, B., Montero, G., Toscano, L., 2019. Simple impedimetric sensor for rapid lipase activity quantification. Talanta 203, 161–167.

Zou, Y., Liang, J., She, Z., Kraatz, H.-B., 2019. Gold nanoparticles-based multifunctional nanoconjugates for highly sensitive and enzyme-free detection of *E. coli* K12. Talanta 193, 15–22.

Re-engineering bachelor's degree curriculum in food engineering: Hypothesis and proposal

24

Keshavan Niranjan

Department of Food and Nutritional Sciences, University of Reading, Reading, United Kingdom

24.1 Introduction

It would not be an exaggeration to state that engineering degrees have lost some of their shine this century. This is also true of the food engineering degree which seems to have got significantly devalued in this century, especially in comparison with the value and prestige that it commanded in the last century when it was seen to be a key driver for growth in agrifood business. Businesses made money through technology and engineering; and the leaders of major agrifood companies were engineers and technologists. As a result, engineering skills were very much valued; graduate engineers received much higher salaries in comparison to other disciplines and were very much respected. This also encouraged talented 16–18-year olds to embark on doing degrees in engineering. Unfortunately, the circumstances changed considerably as we entered the 21st century, when the dynamics of food business also changed in such a way that engineering was no longer considered to be the key driver for growth in food business.

With globalization, amalgamation of companies through mergers and acquisitions, and retailers holding the highest leverage in the food chain, there is considerable pressure to lower costs which has resulted in lesser investment in research and development as well as downsizing of engineering departments in many food companies. The way food manufacturing is conducted has also changed significantly with outsourcing of key engineering activities such as managing plant hygiene and maintenance of equipment. Indeed, the very way in which businesses view capital assets has changed where companies are not keen to buy plant equipment, instead preferring to buy the benefits of the equipment. Such changes, although innovative and in some cases cost reducing, have inadvertently lowered the pre-eminence of food engineers and engineering skills within the business hierarchy, so much so that food engineering is not valued in the business as much as it was earlier, and food engineers are not enjoying salary premiums that they once enjoyed. Companies also tend to see engineering as a cost to their business; not the reason for their profits. Thus, engineering is merely seen as a service; not as a strategic business driver.

There is yet another feature of food business that has hit food engineering very hard. Food business is characterized by the presence of few very large companies and a very large number of small and medium enterprises. The larger companies seem to have undermined food engineering by taking the view that any engineering graduate can fulfill the role of a food engineering graduate, implying that the core skills of a food engineer can be "picked up on the job"! The small and medium

Food Engineering Innovations Across the Food Supply Chain. DOI: https://doi.org/10.1016/B978-0-12-821292-9.00016-9

enterprises, on the other hand, find graduate food engineers to be overqualified and too expensive, and prefer to employ science/technology graduates to fulfill a food engineer's role. Both these approaches are severely flawed, because the people responsible for manufacturing foods may not have adequate knowledge of the very food products that they are responsible to produce. Thus, food engineering has been hit very hard by such unhelpful attitudes within business. It is worth noting that there is a crying need in the food industry for personnel who understand engineering as well as foods, and most importantly, the interactions between engineering and foods. The need for such expertise is hardly in doubt. By not recognizing it, we are putting at risk the very health of a nation. The time has therefore come for this discipline to introspect and re-engineer itself, so that it becomes fit-for-purpose in a contemporary world.

24.2 Hypothesis

It is worth asking, what is food engineering in this day and age? Just as any other branch of engineering, food engineering is about being able to *design* using mathematical and quantitative principles. The next logical question is, what does a food engineer design? Is it the *food product*? Or the process to *manufacture it*? This is not a binary choice: processes and products are two sides of the same coin and the design must be sensitive to both. The last century saw major developments in unifying the principles of various processes used in food and chemical processing—such as momentum, hear, and mass transfer—which strengthened process design but took the emphasis away from food products. For example, the principles of designing a distillation column, as taught in the last century, were the same regardless of whether it was crude oil or a brewery mixture that was being distilled. With human health and environment being the key drivers of new product development in the food industry, there is a clear recognition that process design must be sensitive to product attributes such as quality, safety and efficacy, as well as to the environmental impact of the processes and products.

In the context of product design, food engineers can take a product brief and source raw materials sustainably—both in terms of quality and quantity. They can design transformation processes which have the lowest environmental legacy, while considering in their product design: (1) material and energy balances, (2) product and process safety (systems approach or otherwise), (3) product quality (sensory—visual, mouth-feel, etc.), (4) health efficacy and well-being aspects of the products, and (5) packaging for product integrity during transportation and shelf-storage—embedding a low environmental legacy approach. Food engineering must therefore deal with process as well as product design, using as its basis, the principles of mathematics and quantitative methods. In a nutshell, *process design meant to realize a product in quantity, quality and efficacy is at the heart of food engineering design.* This must also be done while minimizing costs so that the products are affordable. As a hypothesis, *food product realization engineering* must be at the core of the discipline. *Food engineering is therefore the work of designing, formulating and manipulating food products which have desired sensory, satiety, health and well-being responses; and developing—across various operational scales—designs for the lowest environmental impact processing, packaging and storage systems which are capable of realizing the products and attributes* (Niranjan, 2016).

The acceptance of such a definition of the subject, which encompasses and embeds quality, safety, health, environment, and even security aspects will not only enable robust curriculum development, but also give the discipline a strong, distinct, and purposeful identity with considerable opportunities for employment in a wide range of food related sectors—which typically constitutes around 20% of national manufacturing GDP of most countries. In addition, the discipline will also be able to better respond to the needs of 18-year-old entrants, who find engineering subjects to be too dry and tedious, mainly because current programs overload too much science and maths, even before the students are able to see the bigger picture. *Food product realization engineering* will hopefully enable them to see the bigger picture and make them realize why they are doing all the quantitative subjects.

24.3 Designing a curriculum for degree programs

It is important to recognize that, very broadly speaking, there are two aspects to the design of an engineering degree curriculum which will impart: (1) core subject specific competencies and (2) core professional competencies. The latter includes character building and aspects such as: responsibility, management, leadership, communication, interpersonal skills, and professional commitment. The new generation of food engineers should also have an awareness of other skills such as social sciences to understand the impact of the food system on the consumers, community, and the environment. While such competencies are common to all branches of engineering, and indeed other disciplines, the hypothesis presented above forms the basis of imparting core subject specific knowledge, skills, and competencies relating to food engineering discipline. The core subjects that should be taught in order to gain subject specific expertise in food engineering are summarized in Fig. 24.1 in the form of a portico, which contains five themes covering all the core subject areas which a food engineer needs to know (Niranjan, 2016). It is however important to recognize that the way food engineering will be applied in, say, an industrial context, will involve an integration of up to all five themes. Therefore, the program must provide the opportunity to the students to synthesis the knowledge they gain under each theme; this forms the pediment over the five themed columns in Fig. 24.1. It is equally important to recognize that the five themes stated in Fig. 24.1 cover a broad range of subdisciplines, which will require the students to learn a range of enabling courses depending on their qualifications and backgrounds at the point of program entry. These enabling courses will lay the foundation to launch the five themes. Further, it is worth noting that the interdisciplinary approach taken to cover the subject competencies, inherently broadens the science base needed to study food engineering. In addition to conventional *engineering sciences* (including mathematics), the discipline will need *biophysical, biochemical, health*, and environmental sciences, making food engineering a very highly *science-based engineering*. The following sections will focus on Themes 3, 4, and 5 stated in Fig. 24.1.

24.3.1 Food product realization engineering (Theme 3)

This must be core to the discipline of food engineering. It defines the main purpose of engineering in food, that is, to realize a food product in terms of quantity, quality, and possibly efficacy—if it has a health functionality, while leaving behind the lowest possible environmental legacy (Fig. 24.1). Food product realization engineering must therefore set the agenda and the blueprint for a new product to be manufactured. Although the development of flowsheet and the design of each process and equipment will form a key part of engineering design, it cannot be the be all and end all. Product realization is only possible when, in addition to the process, operations management principles and hygiene are applied to plant and equipment; packaging materials and systems are designed and selected so that there is a healthy balance between functionality and environmental impact, right across from the supply chain through to food distribution; and the whole operation is legally compliant with food regulations as well

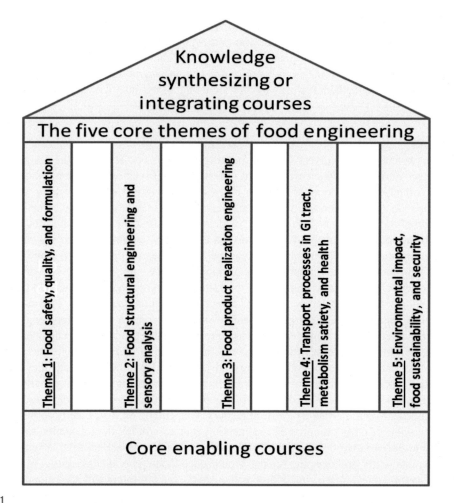

FIG. 24.1

The food engineering edifice consisting of five themes (Niranjan, 2016, reproduced with permission).

as with health and safety. Product realization engineering must also include assessment of economic viability of the various processes. Thus, product realization engineering covers virtually all the technical aspects manufacture—which requires a computer-based production management system which will not only seek to providing complete manufacturing transparency at a *shop-floor level*, but also serve to communicate with the business at an *enterprise level* (Chen and Voigt, 2020). The main learning outcomes of courses in product realization engineering have been listed as follows (Niranjan, 2016):

1. to be able to draw up a clear statement of requirements for products and processes;
2. to define/characterize safe and hygienic designs for recipe, process and packaging formats for a given product concept;

3. to be able to apply HACCP system for food safety management and use good manufacturing practice and prerequisite programs to control the product manufacturing environment;
4. to understand engineering factors influencing hygienic equipment design and operation, and embed these within the manufacturing environment;
5. to identify product processing stages for a given product, and design individual processes, as well as the overall manufacturing outfit;
6. to be able to measure, control, and assess individual equipment, and the overall plant operational performances, and implement performance improvement measures;
7. to be able to analyze and assess logistics and informatics relating to sustainable supply chain for ingredients as well as traceability, and do the same for downstream distribution;
8. to gain sufficient background knowledge of process economics in order to assess the economic viability of the project, and be able to chalk out project management pathways using appropriate software; and finally;
9. to assess the environmental impact of the production facility as well as the environmental lifecycle of products and packaging from cradle to death.

The theories needed to realize the above learning outcomes are, by and large, available in published literature, but it is quite possible that the examples to which such theories are applied may belong to industrial sectors other than foods. It is therefore our primary responsibility to develop the theory and examples covering food and drink sectors—which must be our main concern. Academics are generally not very good at creating practically relevant and real-life situation examples, and an appeal must be sent out to colleagues in industry who will help generate relevant examples for training the next generation of food engineers. Such examples derived from industrial practice motivate students much more into learning, than ivory tower examples created by academics. The benefits of student placements in industry and industry-based projects (as part of the course) will address this need, and if properly managed, will be beneficial to students as well as industry.

24.3.2 Transport processes in the gastrointestinal tract, metabolism, satiety, and health (Theme 4)

With health impact being a key driver for growth in food business, it is absolutely essential that food engineering must broaden its vision span to cover not merely the "farm and the fork" paradigm but also include the post consumption fate of food (Fig. 24.1). This will give an engineer valuable input for product design and formulation and help develop ingredients which can be delivered to targeted areas of the gastrointestinal tract. The engineering "tool kit" for analyzing the passage of food and its metabolism are unfortunately not yet fully developed but there have been some publications attempting to focus on modeling oral processing of foods (Gray-Stuart, 2016), gastric digestion (Bornhorst and Singh, 2014), flow and absorption through small intestine (Hari et al., 2012), and flow and biochemical reactions in the colon (Spratt et al., 2016). Most models for flow through the lower gastrointestinal tract are in silico and have yet to be validated in vivo. We still have a long way to go for this theme to be fully developed, but this must not deter curriculum developers from introducing the theme at an undergraduate level because it will get engineers interested in the subject at an early age.

24.3.3 Environmental impact, food sustainability, and security (Theme 5)

Historically, most courses around the world addressed environmental impact by dealing with wastewater treatment and considering options for recovering material and energy resources from food wastes (Fig. 24.1). Although these two topics are essential, the environmental agenda is much broader today and courses must include sustainability, not only in respect of products and processes used by food industry, but also in respect of such holistic issues as the food–water–energy nexus. Food engineers must be able to collaborate effectively with other disciplines to solve major environmental problems facing humanity.

The concepts of food sustainability and food recovery hierarchy developed by the United States Environmental Protection Agency (United States Environmental Protection Agency (EPA), 2020) to prevent and divert wasted food must be discussed. Further, the relationship between food hierarchy and food processing wastes must also be explored within the course. Packaging, which is an integral part of food manufacture, storage, and distribution, has a significant environmental impact, and a course addressing this theme must consider how its environmental impact, and indeed the environmental impacts of food products and processing options are quantified and compared using techniques such as life cycle analysis.

Although embedding sustainability right across the food chain will contribute significantly to food security, we often find food security in several parts of the world being threatened by natural disasters (such as earthquakes and those caused by extreme weather conditions) as well as man-made disasters (such as war, terrorism, and other accidents). Food security can also be threatened by emerging pandemics like COVID-19. When such disasters strike in a region—often with little or no prior notice—provision of safe, good quality, and culturally appropriate food to the affected regions must be made as an immediate response, and food engineering has a very key role to play in such efforts. *The approach, methodology, and processes involved in planning, designing, and implementing the provision of food and drinking water, in response to a disaster, is known as "humanitarian food engineering."* The role of food science and technology in disaster management has recently been addressed by Bounie (2020). This chapter recommends a transformative change in our approach to food aid. More often than not, regions become dependent on external aid for a long time after being struck by disaster. The way to avoid this dependency is to embrace innovation across the whole food system with an increased emphasis on science and technology that addresses *local* food security issues, thus generating employment and contributing to local economy and development. For instance, if a new nutritious food needs to be developed or the nutrition content of an existing product needs improving, can this be done by using locally grown ingredient, say, a local source of protein? Can equipment that is low cost, easy to operate and maintain, and resistant to wear be developed so that it can be easily be adapted to function efficiently in the region? Can food safety and management systems be developed that can run effectively in the region? The adaption of science and technology sensitively to local conditions, with a time-bound plan to be weaned off international aid and become sustainably self-reliant are critical in any humanitarian response. A key point—as Bounie (2020) note—is that policy makers, donors, governments, nongovernmental agencies, and other humanitarian stakeholders are not aware of the value that food scientists and engineers could potentially bring to their relief efforts. Such agencies must be made aware that food science and food engineering can bring an integrated approach across the food chain and lower the environmental legacy of relief and rehabilitation efforts. At the same time, universities and educational institutions must recognize *humanitarian food engineering* as a formal academic subdiscipline and develop suitable courses within Theme 5. This will make food engineering much more purposeful.

24.4 Course content vis a vis management of student learning experience

Historically, like most academic programs, food engineering programs in higher educational institutions (universities and institutions teaching at this level) were *curriculum driven*. In other words, the program primarily specified, and formally stated, the core academic content to be delivered or taught through lectures, tutorials, and laboratories. The student was essentially an *independent learner* who was supported by *teaching*, with resources offered by library and laboratory infrastructures. With the advent of *quality assurance in education*, the emphasis in program delivery changed from being *teaching led* to *student learning led*. In other words, student learning of the content takes precedence and becomes the ultimate goal, while teaching is just one of the many methods of student learning. While teachers always recognized, albeit informally, that students pick-up knowledge in many different ways, this change in emphasis has turned university academics, formally, into *designers and providers of student learning experience*; they are not merely teachers and professors! The increasing predominance of quality assurance in higher education has made the *management of student learning experience* a highly time-consuming activity—putting it into conflict with *content review and development*, instead of the two working in harmony. Furthermore, it is worth noting the quality assurance is essentially audited in most institutions; not assessed and evaluated, which has essentially made it a tick-box exercise. Academics are therefore preoccupied with such audits and are not finding time and opportunity to develop innovative course contents and learning tools. It is therefore important that a healthy balance is struck between contents, resources, and management of learning—which clearly recognizes that a student is essentially an *independent learner albeit supported*.

The recent pandemic lock-down imposed right across the globe has compelled educational institutions to move course delivery on-line, and universities have been caught off-guard, due to the lack of availability of appropriate content for on-line delivery as well as the questionable learning experience offered by the on-line methods currently available. Food engineering discipline must adopt rapidly to on-line course delivery and develop appropriate content. A good start has already been made. For example: Singh (2020) has developed online courses, video lectures and tutorials, and virtual experiments; and Datta (2020) has developed simulation-based learning on selected topics in food engineering. In general, learning resources for food engineering curricula are not sufficiently extensive and adequately challenging, in comparison with other branches of engineering.

If this discipline must develop a strong identity of its own, for its own survival and sustainability, academics must discard the 20th century mindset, and put considerable effort into developing innovative and intellectually challenging content and resources—which reflects how food engineering is practiced in today's digital world. One of the key focus areas in the food industry now is "*smart digitized manufacturing*"; that is, producing products that satisfy consumer needs in the most efficient manner using digital tools to increase efficiencies, reduce waste, minimize the use of resources, thereby lowering the environmental legacy and enhancing sustainability. Undergraduate courses must therefore cover aspects such as manufacturing execution systems (Chen and Voigt, 2020), and connected technologies—which embed the internet of things combined with Industry 4.0 and the cloud, that are capable of pulling in large amounts of data, analyzing it, and providing valuable business information (Food Engineering, 2020). Industry 4.0 is a term that refers to the fourth industrial revolution that has happened in manufacturing, where computers connect and communicate with one another to ultimately make decisions without human involvement.

Undergraduate courses must therefore introduce ideas of artificial intelligence, and machine and ensemble learning to develop in silico models—which will increasingly be used in food processing to overcome the limitations of conventional phenomenological modeling. The relationships between the four corner-stones defining the scope of food engineering modeling: (1) food product formulation, (2) operating parameters of food processes, (3) transportation and storage parameters, and (4) food safety, quality, and efficacy are so complex that phenomenological models based on the interactions between physicochemical and life science phenomena will be challenging to construct and apply. We must move toward in silico models that take a *systems-based approach* and employ machine and ensemble learning which can be used in digital manufacturing with relative ease. It is arguable whether *smart digital manufacture* must become a theme in its own merit in Fig. 24.1, but there is no doubt that it is a key part of Theme 3 in Fig. 24.1. Yet another culture change that will be required is to recognize that learning does not stop with the end of an undergraduate program. It may be the end of formal supported learning, but the skills gained must enable students to become life-long learners (mainly on the job) to solve problems in a rapidly changing world. Having an adaptable mindset and working effectively in multidisciplinary teams is critically essential for food engineers.

24.5 Status of food engineering programs around the world

Undergraduate food engineering programs commanding significant student enrolments currently operate in Brazil, Turkey, Mexico, Chile, and Thailand, where the discipline is still seen to be very attractive. In addition, food engineering programs also exist in Southern Europe (Spain, Portugal, Italy, and Greece), Japan, India (predominantly within agricultural engineering), China (where food science and engineering seem to be combined), and in certain other countries in Africa and Latin America. The relative inactivity of USA and Northern Europe in respect of full-blown undergraduate programs in food engineering is somewhat striking because these countries are the traditional torchbearers of most academic disciplines, especially when it comes to program and learning resource developments. In a number of countries, food engineering is piggybacking on agricultural/biosystems engineering, chemical engineering, and/or mechanical engineering. Food engineering is also taught—to different extents—in food science and technology, and in human nutrition programs. Although dissemination of food engineering knowledge is always welcome, it is important to note that such programs do not produce professional food engineers—who are badly needed in every country. In addition, there are very few national/regional institutions representing food engineering as a discipline/subject—which also needs to be redressed.

24.6 Concluding remarks

This chapter highlights the need to re-visit and re-conceptualize food engineering for the 21st century and beyond. Throughout the 20th century, this discipline has operated as an application sector for other branches of engineering such as mechanical, chemical, etc. But the time has come for food engineering to become an independent discipline with its own core competencies and skillsets. To enable this, a proposal that includes an hypothesis and a new definition of food engineering is presented in this chapter, and the basis of an undergraduate curriculum has been presented in terms of a food engineering edifice consisting of five themes at the heart of which is a theme called food product realization engineering—which essentially encompasses the design principles and methodology which enables an engineer to realize a food product sustainably, in quantity, quality, and efficacy, at the lowest cost. Food is critical for the health and well-being of any society. Without food engineering, it will be impossible to deliver nutrition and well-being on a scale that the society all over the world needs so badly. It is therefore necessary that food engineering—as an independent discipline—is cultivated, nourished, and allowed to flourish in every country.

References

Bornhorst, G.M., Singh, R.P., 2014. Gastric digestion in vivo and in vitro: how the structural aspects of food influence the digestion process. Annu. Rev. Food Sci. Technol. 5, 111–132. https://doi.org/10.1146/annurev-food-030713-092346.

Bounie, D., Arcot, J., Cole, M., Egal, F., Juliano, P., Mejia, C., Rosa, D., Sellahewa, J., 2020. The role of food science and technology in humanitarian response. Trends Food Sci. Technol. 103, 367–375. https://doi.org/10.1016/j.tifs.2020.06.006.

Chen, X., Voigt, T., 2020. Implementation of the manufacturing execution system in the food and beverage industry. J. Food Eng. 278, 109932. https://doi.org/10.1016/j.jfoodeng.2020.109932.

Datta, A.K., http://blogs.cornell.edu/edusim/description-and-goals-2e/ (accessed September 13, 2020).

Food Engineering. https://www.foodengineeringmag.com/articles/97170-how-the-industrial-internet-of-things-is-affecting-food-processing (accessed September 13, 2020).

Gray-Stuart, E.M., 2016. Modelling food breakdown and bolus formation during mastication. PhD thesis. Massey University, New Zealand.

Hari, B., Bakalis, S., Fryer, P.J., 2012. Computational modelling and simulation of the human duodenum, Excerpt from the Proceedings of the 2012 COMSOL Conference in Milan https://www.comsol.com/paper/download/151975/hari_paper.pdf.

Niranjan, K., 2016. A possible reconceptualization of food engineering discipline. Food Bioprod. Process. 99, 78–89.

Singh, R.P., http://www.rpaulsingh.com/learning.html (accessed September 13, 2020).

Spratt, P., Nicollela, C., Pyle, D.L., 2016. An engineering model of the human colon. Food Bioprod. Process. 83 (C2), 147–157. https://doi:10.1205/fbp.04396.

United States Environmental Protection Agency (EPA). https://www.epa.gov/sustainable-management-food (accessed September 13, 2020).

Experience-based learning: Food solution projects

Myriam Loeffler[a], Maarten van der Kamp[b]

[a]*KU Leuven, Department of Microbial and Molecular Systems, MTSP, Ghent Technology Campus, Ghent, Belgium*
[b]*European institution in Leuven, Leuven, Belgium*

25.1 Introduction

Experiential learning or experience-based learning (EL) is understood as learning through reflection on doing and allows students to move from the traditionally more passive to a more active role (Wankat, 2015). Within the idealized EL cycle, the learner goes through several phases consisting of concrete experiences, reflective observations, abstract conceptualizations, and active "experiments" (Kolb, 1984, 2005). Experimental learning as a part of higher education has gained prominence in recent years, as reflected in the number and scope of new EL-based projects and programs. For instance, courses are offered that provide "real-world" experiential learning by developing food products as part of a food industry case study (Hollis and Eren, 2016). Studies have shown that this type of teaching and learning format has a positive impact on higher order thinking and increases retention of information (Leveritt et al., 2013; Markant et al., 2016).

EIT Food is a Knowledge and Innovation Community (KIC) established by the European Institute for Innovation & Technology (EIT) to drive innovation and entrepreneurship across Europe. It is a network formed by universities, industry partners, and research centers as well as agri-food start-ups that are working together to make the food system more sustainable, healthy, and trusted. Universities that are part of this community have the possibility to conduct "Food Solution" projects that are focused on industrial challenges such as the utilization of side streams, holistic use of raw materials, and development of more sustainable packaging concepts. This extracurricular activity promotes the idea of experience-based learning in the setting of multidisciplinary student teams with strong academic and industrial mentorship and allows students to convert knowledge into valuable skills enabling them to tackle key challenges across the food system. The industry-provided challenges, the international setup and the competitive nature of the projects are unique in the educational landscape.

25.2 EIT Food

EIT Food's mission is to catalyze the transformation of the food system for it to become more sustainable, healthy, and trusted (EIT Food, 2018).

With its objective to change how food is produced, consumed and valued by society, EIT Food is responding to a number of global challenges that put pressure on current practices and systems.

i. First, the food system needs to be capable of feeding a growing world population, with projections to reach 9.8 billion in 2050, and 11.2 billion in 2100.

ii. Second, with more than 2 billion people overweight or obese, and 800 million undernourished (Webb et al., 2018) and an expectation that half of the global population will be overweight by 2030 (Dobbs et al., 2014), there is an urgent need to create healthier diets that address global nutritional needs, while respecting local and regional food practices.

iii. Third, the rise of food-related noncommunicable diseases—such as type-2 diabetes almost doubling over the past 30 years (World Health Organization, 2016)—puts pressures on health systems around the globe.

iv. Fourth, food production is the largest cause of global environmental change: agriculture occupies approximately 40% of global land (Foley et al., 2005); food production is responsible for up to 30% of global greenhouse gas (GHG) emissions (Vermeulen et al., 2012) and 70% of freshwater withdrawals come from the food industry; food system industrial activities requires approx. 26% of EU's energy consumption (EC, 2016); the expected 76% rise in the global appetite for meat and animal products by 2050 could increase greenhouse gases by 80%. Moreover, approximately one third of all food is lost or wasted (FAO, 2019). Finally, highly publicized food contamination and authenticity scares have led to consumer concerns over the complexity of the global food system, undermining confidence in the transparency, safety, and integrity of the food value chain.

Against this backdrop, EIT Food is driven by the understanding that the food system challenges can only be tackled if competence networks and knowledge across all sectors of the food supply chain, consumers' inputs and expectations, and the emerging technologies from outside agri-food sector's traditional toolbox are fully leveraged. Thus, EIT Food approaches every actor of the food system as equally responsible for the future of food and as co-owners of the transformation process. Specifically, this nonprofit organization is driving impact by creating and scaling-up agri-food start-ups to deliver new food innovations and businesses, developing talent and leaders to transform the food system, launching new innovative products and ingredients to deliver healthier and more sustainable food, and engaging the public so they can become the agents of change in the food system. Specifically, the community aims to achieve six strategic objectives (EIT Food, 2018):

1. Overcoming low consumer trust: EIT Food informs, motivates, and supports citizens to become change agents. This is supported by the measurement of consumers' trust to identify how to best reconnect people to their food in the transition toward a food system that is healthy, inclusive, and trusted.

2. Creating citizen valued food for healthier nutrition: EIT Food supports innovation for healthier and more nutritious food, and co-creates technologies with citizens to help them make personalized nutrition choices for improved health, environment, and wellbeing.

3. Building a consumer-centric connected food system: EIT Food develops innovative tools that improve nutrition, increasing transparency, safety, and quality while ensuring authenticity and preventing fraud, reducing food waste, and saving resources.

4. Enhancing sustainability: Through knowledge exchange and acceleration, EIT Food's tailored programs are introducing next-generation technologies to reduce food waste, energy- and water-use

and prevent the depletion of valuable finite resources to the benefit of farmers, producers, consumers, and the environment.

5. Educating future generations to engage, innovate, and advance: EIT Food develops open-access as well as tailored educational opportunities to benefit different segments of the society, such as young parents, school children, students, career professionals, young farmers, and life-long learners. EIT Food equips them with the latest knowledge on food production and consumption in order to close the skills gap and create a new generation of active citizens, food pioneers, and business entrepreneurs.

6. Catalyzing food entrepreneurship and innovation: EIT Food supports start-ups and pools resources to invest in the most promising businesses and transformative ideas to deliver solutions on the market.

To address the particular challenge of trust in the food system, EIT Food has added a fourth element to the knowledge triangle (characterizing the interaction between research, education, and innovation/business creation in a knowledge-based society as set out in the Lisbon declaration, European Council, 2000) in the form of Public Engagement activities to strengthen its consumer-centric approach.

25.2.1 Education at EIT Food

Mainly addressing Strategic Objective 5, EIT Food's education activities aim to develop a skilled, innovative workforce that can play a significant role in how the global challenges outlined above are addressed. This requires shifts in the skill sets that are taught to ensure that individuals can play an effective role as innovators in a rapidly changing world (e.g., Council of the European Union, 2019; Bughin et al., 2018). This means shifting from more traditional, disciplinary approaches toward multi- and interdisciplinary approaches, and a strong focus on innovation and entrepreneurship as underpinning capabilities to support domain-specific technical capabilities.

To support this shift, EIT Food has developed a robust framework for the food system to offer clear pathways for employees, entrepreneurs and job seekers to acquire the relevant skills and knowledge to aid their careers as innovators and entrepreneurs in the food system. Designed to promote lifelong learning, the EIT Food competency framework gives employers the tools to recruit top talent, enhance the available skill set of the workforce, and aid staff retention. The framework sets out eight core competences that innovators and entrepreneurs need to possess to drive effective change in the food system (Fig. 25.1).

To demonstrate how new skill sets can be taught, EIT Food has developed a range of advanced training programs targeting both students in postgraduate programs (Masters and PhD levels), and professionals and entrepreneurs already working in, or wishing to transition their career into, a part of the food system. All program types are based on innovative, experiential learning approaches in which creating solutions for sustainability challenges in the food system is a central feature. Furthermore, all programs are designed to be robust in how they help learners create impact in the sector. This means that learning outcomes are clearly articulated and assessed to be able to demonstrate mastery. Also, the design of learning opportunities needs to be based on the latest insights and innovations in approaches to teaching and learning.

The learning opportunities offer the latest insights about the state of the food system, including system dynamics, sustainability trends, the organization of value chains, deep technical skills required for innovation, business models, broad competences for innovation and entrepreneurship, and food quality, safety, and integrity.

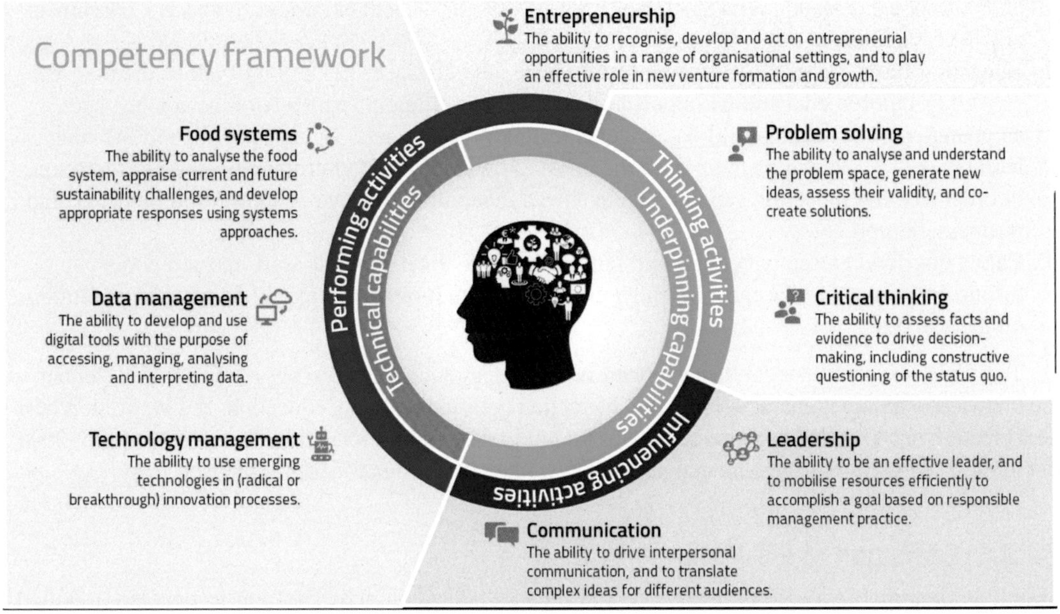

FIG. 25.1 EIT Food competency framework

(image supplied by EIT Food).

Finally, the programs need to be recognizable in the sector as hallmarks of excellent education. To aid employability, it is essential that evidence of attained competence levels, in the form of certificates, is recognized as valuable by employers, employees and job seekers alike.

25.3 Food solution programs

Within the portfolio of programs, EIT Food has developed the concept of Food Solution programs to fully integrate all three aspects of the knowledge triangle. The programs provide a challenge-based learning opportunity for postgraduate students that is either offered as an extracurricular activity, or in some cases for a set number of credits in the European Credit Transfer and Accumulation System as codified by the Bologna process. The challenges that the students work on must fit EIT Food's Strategic Agenda priorities and are set by one or more of EIT Food's industrial partners to ensure industrial relevance. This also enables students to access some of the industrial partner's resources, such as background intellectual property (e.g., research and insights), and to facilities and materials.

Supported by academic staff and industrial mentors, teams of four to five students engage in an innovation process that typically lasts about 9–12 months in which to research the challenge, devise possible solutions, build and test prototypes, and scope out the commercialization strategy for the proven solution; in most cases this leads to new venture formation by some or all of the team members involved in creating the solution. Depending on the organizing universities, the intellectual property generated during the Food Solutions programs either belongs to the students, or to the university where the students are registered. In the latter case, the relevant technology transfer offices are involved to grant the students access rights to commercialize their work if they want to. The programs are set up with a minor competitive element, so that the solution deemed to be most innovative and with the most commercial potential is awarded a prize. Follow-up support for viable ventures is then provided by EIT Food's Business Creation services, such as preseed funding and access to further support.

25.3.1 Food solutions: program design at universities

Food Solution projects are jointly planned and organized by the participating universities and challenge-giving industry partner/s with one of the participants being the leading partner who takes care of the deliverables that must be achieved and provided to EIT Food during the 1-year project. This program is supported by several universities and industry partners from among EIT Food's partnership of 100+ partners (for an up to date list of partners please visit https://www.eitfood.eu/partners). After confirmation of the projects and the inclusion of the project proposals in the respective EIT Business Plan (project budget covers personal costs, material/other costs as well as costs linked to travel and subsistence of the students, mentors and consortium partners), the universities start recruiting the student teams. This phase (1) is followed by a mentored solution development (2) and a competition event (3), where the students pitch and present their products and business cases in front of the other teams and a jury formed by judges from EIT Food/Business Creation, industry, and/or retail.

25.3.1.1 Phase 1: Recruitment and selection of student teams

Depending on the Food Solution projects, 4–8 teams from at least two but usually three or even four different universities across Europe are tackling the same industry challenge in a competitive setup. Student recruitment and team building is done at the respective universities and is usually based on a portfolio including a motivation letter, CV, and/or transcripts of records. Students are selected on the basis of several criteria:

Discipline:
Food Solution projects require multidisciplinary approaches to be solved effectively, and therefore students are sought from different backgrounds (e.g., Food Science and Engineering, Bio-economy, Food Chemistry, Agricultural Sciences…).

Entrepreneurial mindset:
Students need to demonstrate that they can solve problems creatively, possess resilience in seeing projects through, and can overcome challenges and learn from mistakes. This is a key feature of the required motivation letter.

Academic excellence:
As Food Solutions programs are extracurricular, students need to demonstrate rigor in their studies so that their academic results are not negatively affected by the additional workload.

25.3.1.2 Phase 2: Solution development

At the beginning of the solution development phase, the students are taking part in a 2-days Kickoff event hosted at one of the partner locations. Here, they are introduced to the actual industry-challenge, receive insights into successful product development and get to know the partner network and participating teams. Given tutorials during the event usually cover information on consumer trends, market analysis, and concept ideation, building the base for a guided idea development phase. At the end of the Kickoff or shortly after, the students pitch their initial ideas to the consortium in order to get feedback on the innovative potential, technical feasibility and other relevant factors that need to be considered for a successful product- and business case development. Afterward students start to work on their proposed solutions that they present for feedback at an intermediate (virtual or physical) event, before finalizing the products and the business case. Depending on the challenge, this phase also includes the development of suitable packaging, marketing, and communication concepts. Additionally, food safety challenges as well as economic and environmental impacts are taken into account. Throughout the solution development phase, students are supported by their supervisor/s at the respective university and by the industry mentors that can be directly approached for expert guidance.

Moreover, from 2021 onward, a specialized consortium formed by educators from academia and industry provides students the basis to drive innovation and commercialization in an entrepreneurial context. These accompanying, virtual education sessions aim to generate a better understanding of commercialization pathways, different markets, key components of business modeling and planning, IP, basics of entrepreneurial finance, and effective communication.

25.3.1.3 Phase 3: Final competition

The final competition is an event to showcase and judge the prototypes and business cases and to award the best idea, which is done by a jury of 5–7 members from EIT Business Creation & industry.

25.3.2 Food solutions: examples from 2018 to 2020

25.3.2.1 Circular Food Generator Track

The Circular Food Generator Track was a competition, which challenged Master and PhD students to develop solutions for and from food losses of production facilities and retail activities. The proposed (former) waste streams were bread, bananas and potatoes, which still have significant potential for use. Each of the participating multidisciplinary student teams was challenged to create innovative products with high commercial potential as well as a valid business case. Experts from the food industry and retail as well as mentors from the universities acted as advisors to ensure valid and realistic "solutions" were developed.

25.3.2.2 Foodio and FoodMio food solutions master class

"Foodio" and "FoodMio" served as co-creation platforms for tangible and sustainable solutions to real-life product development challenges. More specifically, the focus of "Foodio" was on the use of plant-based side streams/by-products (e.g., citrus or apple fibers), which are generated during juice and olive oil production or pectin extraction. About 40 students from five academic organizations formed multidisciplinary teams and met this challenge by developing innovative dairy products while addressing the technological challenges associated with the use of fiber in complex food matrices. In contrast, "FoodMio" focused on attempts to complete a circular food sector that maximizes the exploitation of side-streams from vegetable and fruit manufacturing or novel raw materials (algae- or insect-based) in meat products. About the same number of students tackled this challenge and developed innovative MeatHybrid prototypes and realistic business cases. Throughout both projects, the students were offered online lectures on dairy or meat technology, sensory evaluation, marketing concepts and business plan development, and all teams were supported by mentors from academia and from several business partner organizations.

25.3.2.3 Building student skills in microalgae processing

To tackle the globally rising food shortage, this Food Solution project aimed at providing food science, bioeconomy, and engineering students from three universities with the knowledge and skills necessary to process, fractionate, and characterize functional properties of algae components, with a focus on algae proteins. The teams also developed innovative food products containing significant amounts of the fractionated protein. The students were jointly supervised by academic and industrial mentors, assuring industrial, and commercial relevance.

25.3.2.4 Tasty macronutrients

A new class of plant-based ingredients based on combinations of proteins and polysaccharides (physically, chemically, or enzymatically formed) had been developed by the participating partners that could be used to support various product development platforms. Prior the Food Solution project "Tasty Macronutrients," it was unknown which combinations are best suited to deliver desired taste and texture functionalities in a specific product category. This was a challenge ideally suited for multidisciplinary student teams to develop concepts on how to optimize texture and flavor simultaneously to improve consumer choice and acceptability. The proposed team exercise provided students with an opportunity to work collaboratively with industry and academic mentors on both an innovative product and a realistic business case development.

25.3.2.5 EcoPack

During EcoPack, Masters and PhD students of three universities were challenged to develop innovative solutions to reduce the amount of plastic bags used by consumers in supermarkets to pack on-the-go items (e.g., fruits, vegetables, breads) and/or to replace shrink-wrappings that are usually applied to pack two or more stock units, by using more sustainable packaging materials (e.g., leaf-based, fiber-based or grass-based) to develop carriers that are both convenient for consumers and feasible for retailers. Multidisciplinary teams tackled these challenges by developing prototypes and a business case for each proposed solution. Experts from the packaging industry, retail, and academia ensured that valid and realistic "solutions" were developed.

25.3.2.6 From leaf to root—holistic use of vegetables

Many vegetables including sweet corn, carrots, artichokes, and cabbage are not holistically used in food production, which is due to several reasons including harvesting procedures having been designed for conventional products, consumer unawareness, and a lack of holistic product concepts. In the "from leaf to root" project, student teams tackled these challenges by developing product prototypes based on plant parts that are to date not or only very infrequently utilized, thereby promoting both a more sustainable use of resources and healthier nutrition. As for the other Food Solutions, the students pitched their ideas to a jury and provided a detailed business plan.

25.3.2.7 Product concepts for less refined ingredients

The manufacture of "less-refined" ingredients represents a true paradigm shift in manufacture and use of sustainable food ingredients due to them not only requiring no solvent, and less water and energy but them also exhibiting new and highly interesting techno functionalities (e.g., emulsification gelation, etc.). For example, a combination of an air milling, classification, and electroseparation process is used to fractionate protein-rich and fiber-rich fractions from peas or lentils. As such, the industrial challenge to be worked on by student teams was to develop new prototypes and food concepts that holistically use both of these "less-refined" fractions to develop new healthy food or beverage concepts that give consumers more choice while improving the sustainability of the food system.

As previously mentioned, the presented experience-based learning program also aims to create the next generation of innovators and entrepreneurs, which is also reflected in the number of start-ups emerging directly from the Food Solution projects.

25.4 Intended learning outcomes

The intended learning outcomes and competencies are agreed upon the different consortium partners. After successfully completing the EIT Food Solution program, the students are able to:

- Define challenges being of key importance in product design and development. Depending on the Food Solution challenge this can refer to a food product and/or packaging or a digital solution
- Think creatively and "out of the box" by incorporating ideas and viewpoints from different disciplines (multidisciplinary teams/mentors from industry and research organizations)
- Collect, analyze, interpret, and report information to develop sustainable solutions to current and future challenges
- Describe the essential steps in developing products/solutions including feasibility and/or sustainability aspects
- Turn ideas into action
- Competently use appropriate technologies to contribute to food system innovations
- Effectively manage projects
- Develop and pitch a (valid) business case
- Work in multidisciplinary and intercultural teams and access new networks, thereby contributing to co-creation processes to develop solutions
- Communicate with industry partners (engage in two-way communication with stakeholders about their concerns) and target consumers/customers
- Recognize how innovations can contribute to achieving societal impact

After the final event, the students are asked to participate in an online survey that focuses on the learning outcomes, the actual experience, and on general feedback. The results are evaluated and used to further improve prospective Food Solution projects.

For instance, in the surveys, students are asked what skills they were able to improve as a result of the projects. Fig. 25.2 summarizes the most frequently mentioned answers from various EIT Food Solution questionnaires.

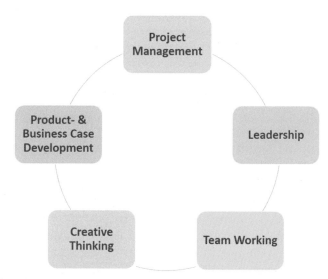

FIG. 25.2 The top five skills that participants were able to improve through the EIT Food Solution projects.

25.5 Conclusion

The integration of the Food Solution program into the curriculum varies from university to university, which is also reflected in the number of credits (European Credit Transfer and Accumulation System; ECTS) that can be obtained after completing such a program. Food Solution programs can for instance be implemented as elective modules or project modules. Although the ECTS may vary for the participating students, the general feedback clearly shows that the other benefits associated with the projects, such as the opportunity to work on a real challenge that allows participants to both generate the idea and turn the idea into a product/solution, the networking with colleagues from academia and people from industry, and last but not least the opportunity to improve relevant skills for their professional career, play a much more important role than the number of ECTS associated with the projects as part of the curriculum.

The students generally consider the experience they gain during the EIT Food Solution projects to be extremely positive for themselves and for their professional careers. Some students have already started their own business (e.g., ZBS-Food UG) or are in the process of starting their own business, others are in the process of patenting their inventions. In addition, new collaborations for research projects and/or master's theses have already emerged from this program.

References

Bughin, J., Hazan, E., Lund, S., Dahlström, P., Wiesinger, A., Subramaniam, A., 2018. Skill Shift: Automation and the Workforce. McKinsey Global Institute, Atlanta, USA.

Council of the European Union, 2019. Resolution on Further Developing the European Education Area to Support Future-Oriented Education and Training Systems Council publication 13298/19, 24 October.

Dobbs, R., Sawers, C., Thompson, F., Manyika, J., Woetzel, J., Child, P., McKenna, S., Spatharou, A., 2014. Overcoming Obesity: An Initial Economic Analysis. McKinsey Global Institute, Atlanta, USA.

EIT Food Strategic Agenda, 2018. https://www.eitfood.eu/media/documents/EIT-StrategicAgenda-Booklet-A4-Final_disclaimer.pdf.

European Commission, 2016, European Research & Innovation for Food & Nutrition Security: FOOD 2030 High-Level Conference Background Document. European Commission, Brussels.

European Council, 2000. Presidency Conclusions, Lisbon European Council, 23 and 24 March 2000. Available at: https://www.consilium.europa.eu/uedocs/cms_data/docs/pressdata/en/ec/00100-r1.en0.htm.

Foley, J.A., DeFries, R., Asner, G.P., Barford, C., Bonan, G., Carpenter, S.R., Chapin, F.S., Coe, M.T., Daily, G.C., Gibbs, H.K., Helkowski, J.H., Holloway, T., Howard, E.A., Kucharik, C.J., Monfreda, C., Patz, J.A., Prentice, I.C., Ramankutty, N., Snyder, P.K., 2005. Global consequences of land use. Science 309, 570–574.

Food and Agriculture Organization, 2019. The State of Food and Agriculture 2019. Moving Forward on Food Loss and Waste Reduction. FAO, Rome.

Hollis, F.H., Eren, F., 2016. Implementation of real-world experiential learning in a Food Science course using a food industry-integrated approach. J. Food Sci. Educ. 15 (4), 109–119.

Kolb, D., 1984. Experiential Learning: Experience as the Source of Learning and Development. Prentice Hall, Englewood Cliffs, NJ.

Kolb, A.Y., Kolb, D., 2005. Learning styles and learning spaces: enhancing experiential learning in higher education. Acad. Manage. Learn. Educ. 4, 193–212.

Leveritt, M., Ball, L., Desbrow, J., 2013. Students' perceptions of an experiential learning activity designed to develop knowledge of food and food preparation methods. J. Food Sci. Educ. 12 (3), 56–60.

Markant, D.B., Ruggeri, A., Gureckis, T.M., Xu, F., 2016. Enhanced memory as a common effect of active learning. Mind, Brain Educ. 10 (3), 142–152.

Narain, S., Nishtar, S., Murray, C.J.L., 2019. Food in the anthropocene: the EAT-Lancet Commission on healthy diets from sustainable food systems. Lancet 393, 447–492.

Vermeulen, S.J., Campbell, B.M., Ingram, J.S.I., 2012. Climate change and food systems. Annu. Rev. Environ. Resour. 37, 195–222.

Wankat, P.C., Oreovicz, F.S., 2015. Teaching Engineering, second ed. Purdue University Press, West Lafayette, Indiana, USA.

Webb, P., Stordalen, A.S., Singh, S., Wijesinha-Bettoni, R., Shetty, P., Lartey, A., 2018. Hunger and malnutrition in the 21st century. Br. Med. J. 361, k2238.

Willett, W., Rockström, J., Loken, B., Springmann, M., Lang, T., Vermeulen, S., Garnett, T., Tilma, D., DeClerck, F., Wood, A., Jonell, M., Clark, M., Gordon, L.J., Fanzo, J., Hawkes, C., Zurayk, R., Rivera, J.A., De Vries, W., Majele Sibanda, L., Afshin, A., Chaudhary, A., Herrero, M., Agustina, R., Branca, F., Lartey, A., Fan, S., Crona, B., Fox, E., Bignet, V., Troell, M., Lindahl, T., Singh, S., Cornell, S.E., Reddy, K.S. https://www.who.int/publications/i/item/9789241565257.

World Health Organization, 2016. Global Report on Diabetes. WHO Press, Geneva.

Link: https://www.un.org/development/desa/en/news/population/world-population-prospects-2017.html, accessed 06/04/2020.

Link: http://www.fao.org/aquastat/en/overview/methodology/water-use, accessed 06/04/2020.

Link: https://ec.europa.eu/education/resources-and-tools/european-credit-transfer-and-accumulation-system-ects_en, accessed 06/04/2020.

Food engineering innovations across the food supply chain: debrief and learnings from the ICEF13 congress and the future of food engineering

26

Pablo Juliano[a], José I. Reyes-De-Corcuera[b]

[a]CSIRO Agriculture and Food, Australia
[b]Department of Food Science and Technology, CAES, Athens, GA, United States

26.1 Introduction

The 13th edition of the International Congress on Engineering and Food gathered the top world-class food engineers. In a world with increasingly fine disciplinary granularity, the congress theme of "Engineering Innovations for Food Supply Chains" integrated the most recent advances in the broad spectrum in food engineering. Indeed, the congress included keynote talks, oral presentations, and e-posters on food production and operations, postharvest processing, long-term storage, consumer drivers, and digestion. The expansion and increased reliability of the internet along with increasing computer power has tremendously increased the efficiency and productivity of food engineering research and the transfer of technology. As with any development, the use of "new" or "emerging" to qualify a technology is risky because purely creative output is rare. Food engineering innovation is generally a combination of existing technologies or their application to new products. Typically, leaps in food engineering occur when a technology previously unexplored in foods, or a significant advance in a technology is ingeniously combined with others to produce better products more efficiently and/or making them more accessible. These types of innovations have a tremendous impact in the food supply chain. The minds that contributed to ICEF13 are arguably among the best in the world of food engineering. In this concluding chapter, we reflect on the main chapters of this book, and attempt to summarize some of the most relevant contributions from abstracts and presentations submitted to ICEF13.

With attendees from all regions of the world, the engineering themes were judiciously entwined with the global and local realities and within the wide spectrum of diverse economies. Among many topics, the congress covered the role of food engineers in one of the most pressing human challenges: food security. In modern times, reduced food access resulting in hunger is an unfortunate but persistent anomaly of the global food supply chain that has become more evident in recent years with emerging humanitarian challenges associated with war, climate change, cultural bigotry, and most recently the COVID-19 pandemic. At the other end of the spectrum, automation and the virtual world of Industry 4.0 and space exploration shed light into the cutting-edge of technology. Significant emphasis was given to high end technologies that mostly suit the realities of developed countries, which make foods affordable only by a few. However, an increasing number of research efforts on processing for improved nutrition and digestion systems were also presented. Far from making a moralist statement, this emphasizes the breadth and scope of food engineering.

As mentioned in the Preface, ICEF13 was an outstanding success with more than 750 abstracts (ICEF13 International Congress on Engineering and Food, n.d.), 280 posters presented, and 550 delegates attending from 40 countries and five continents. It is impossible and it is not our goal to be exhaustive and reference everyone from such a large crowd of outstanding engineers. Instead, we highlight some of the topics that knit the fabric of ICEF13 and the regions of the world where the studies were made. In this chapter, after an overview of most of the topics covered in ICEF13, we summarize the trends and raise the questions that the food engineering community will be addressing and that hopefully will prepare us for ICEF14 in Nantes, France. Furthermore, we refer to the chapters included in this book and articles in special issues generated from the congress in *Food Engineering Reviews* (*FERE*) (Buckow et al., 2021) and *Journal of Food Engineering* (*JFE*) (Bhandari et al., 2021) from authors who contributed to the ICEF13 program.

Each section provides an initial overview of the most recent trends gathered under each theme from the abstracts, presentations, and mentioned manuscripts. Then, a list of distilled bullet points summarizes new industrial implementations, recently scaled up or game changing practical applications, and promising technology innovations. ICEF13 also hosted sessions devoted exclusively to the mathematical modeling of food processes; the challenges associated with these models and their complexity deserve special attention and are mostly captured in Chapter 16. This chapter incorporates reference to modeling presentations within sections describing the object or process being modeled.

26.2 Biosystems engineering for food security and sustainability

Global biological ecosystems are continuously being impacted by several human factors such as population growth, urbanization, geopolitical events, reckless depletion of natural resources (e.g., fresh water, sources of energy, forests, and land), extended life expectancy, changes in consumer attitudes, among several others. Consequently, there are almost one billion hungry people, 2.1 billion obese people, 2 billion people deficient in key micronutrients, 1 in 4 children stunted, and 240 million displaced people (refugees, or forced migrants because of violence, drought, famine, land seizures, environmental disasters, and internal conflict). In addition, climate, and environmental changes, will continue to impact biodiversity, and change our food systems, which are interconnected with the way that we produce, preserve, distribute, and consume food. Pests and diseases of agricultural products and livestock as well as human pandemics are likely to continue disrupting the food supply chain and pose a threat to food security (Bounie et al., 2020). There is global agreement on the need for transforming food systems to deliver improved nutrition and health of the entire human population; however, tangible efforts are also required to contribute to reducing the environmental pressures on our ecosystems. Food insecurity situations are either acute, when there is interruption of access to food arising from transitory situations (e.g., hikes in food prices, conflict, natural disasters), or chronic conditions, when there is a persistent lack of availability, access, utilization, and stability of food.

Five action tracks have been recently proposed to transform food systems: (1) access to safe and nutritious food, (2) sustainable consumption, (3) nature-positive production, (4) equitable livelihood, and

(5) resilience to shocks and stress (Herrero et al., 2021). Chapter 1 describes a framework to understand how the structure of food supply networks may enhance resilience when facing disruptions, particularly given the new challenges brought about by the ongoing COVID-19 pandemic. Chapter 2 describes sustainable food systems, considerations for their implementation, and how food engineering innovations may assist in solving issues within these systems. Chapter 3 exemplifies how the sustainability of the food supply chain and its dependence on conservation of resources can be achieved by re-engineering energy and water consumption, as well as waste. Other system aspects and articles addressing cyber systems and other Industry 4.0 solutions, including sensors, are covered in Chapters 20–23 and the FERE special issue (Buckow et al., 2021).

26.2.1 Engineering safe and efficient food access, sustainable nutrition, and health for all

Following the recent Food Systems Dialogues for Developing Nations (Food Loss and Food Waste, n.d.), toward the development of the United Nations 2030 Agenda for Sustainable Development (Food Systems Dialogue Summit, 2021), the question which appeared most often was on the need to develop technologies suitable for the realities of developing regions to tackle access, utilization, and stability challenges. A few engineering examples for developing regions are covered below in Section 26.6 on drying; however, a gap was clearly identified in this field during the congress and recommendations were given to raise this issue to the top of the global agenda. Within this context, the global demand for meat and dairy continues to increase and is not being sustainably matched with current production practices. To meet the need for sustainable nutrition, in particular protein, there is a renewed interest in plant, insect, algal, and cell-based sources (animal, yeast, fungi, or microalgal cells) of sustainable nutrients.

In addition to the above-mentioned chapters, ICEF13 presentations from Australia, France, Germany, India, Indonesia, Mexico, Thailand, Mozambique, Myanmar, New Zealand, Samoa, Switzerland, United Kingdom, USA covered a few biosystems engineering topics, in an area that promises to grow, including:

- Humanitarian food systems engineering, including processing as a fit-for-purpose solution to address specific needs in the value chain, with examples on modular containerized lines designed for fast installation in remote locations, locally made ready-to-use supplementary foods, and food (safety, quality, or scarcity) incident management strategies.
- The importance of mapping supply chains, and time fluctuations across the years, to identify processing interventions and market incentives for change, human-centered design including engineering (e.g., optimizing washing tables for vegetables in Cambodia, dragon fruit washing and packaging in Vietnam to open export market opportunities and create jobs, and intermediate level mechanization of small rice farmer households).
- Processes that enable systems decarbonization by using renewable energies for electricity generation (e.g., ohmic and microwave [MW] heating), positive interventions from local rather than distributed manufacture, personalization strategies including alternative protein sources, and optimizing retail energy consumption.
- The higher water scarcity footprinting of lower quality diets based on assessments using the ISO 14046:2014 water footprint standard, although consumers will need labeling guidance to identify low water scarcity footprint foods. Water footprint does not correlate with carbon footprint for the creation of sustainable foods.

- Personalized food systems including, resource friendly eating, food as a people-connecting element, food indulgence and anxiety, eating for self-optimization, and fast refill foods requiring digital twins for optimization of process parameters based on chemical and sensory fingerprints (covered further in Sections 26.7 and 26.10).
- Engineering of digestion models, which is further covered in Section 26.9.

Section 26.11 will cover educational opportunities in systems engineering, including examples presented on a masters' degree on sustainable food systems engineering, or e-learning opportunities for sustainability.

26.2.2 Alternative sustainable food sources

Plant-based nutrition has been presented as the next megatrend, where fruits, legumes, whole grains, nuts, and vegetables are the foods that deserve greater emphasis (Food Planet Health, 2019). In addition, production of plant- and cell-based meat to replace animal-based meat, either as stand-alone product or within recipes, present through-chain opportunities for raw material optimization via breeding or engineering, functionalization via processing, and food formulation. The exponential growth of companies across the globe in the meat analogue space with an up to $370B market by 2035 (Bushenell et al., 2020) has been highlighted and also well exemplified in the "new protein" companies map (Fox Cabane, n.d.). Chapter 6 addresses the engineering and other strategies to mitigate protein deficit in view of the food security limitations arising from the increasing demand for animal-based proteins such as ruminant and porcine meat and dairy. The chapter addresses alternative protein sources as a way of mitigating such demand. In addition to plant-based sources such as vegetables, pulses and seeds, other sources are included such as meat and fish by-products, microbes, insects, algae, and in vitro meat. Innovative engineering solutions for alternative protein extraction and development of novel protein-based ingredients, concentrates and isolates and the key determinants for consumer acceptance and health considerations are also covered. Other engineering approaches and findings related to the utilization and transformation of sustainable food sources were introduced in ICEF13 by authors from Australia, Belgium, Chile, France, Germany, Japan, Nigeria, South Africa, Switzerland, The Netherlands, and USA.

26.2.2.1 Life cycle assessment

- A quality-based lifecycle assessment method using protein quality indicators (protein efficiency ratio, protein digestibility corrected amino acid score, and biological value), attempted to challenge the *status quo* by showing that pork and egg production may have a lower environmental impact than soybean.

26.2.2.2 Algae

- Industrial integrated aquaculture cropping of green seaweed, protein extraction, and processing of seaweed glycans to separate alginate, sugars, and bioactives.
- A low-cost photobioreactor for microalgal production.
- Heterotrophic cultivation of microalgae as an economic and sustainable alternative. Microalgae can be employed as a whole-cell ingredient or protein can be extracted giving specific functionalities.
- Options for pulsed electric field growth stimulation of microalgal cells and microalgal "milking" through reversible pulsed electric field-assisted extraction of protein, oil or other bioactives, while maintaining cell viability.

- Growth stimulation and bioactive production enhancement by culturing microalgae in air nanobubble water (examples included increased production of astaxanthin in *Haematococcus lacustris* and lipids in *Botryococcus braunii*).
- Enzymatic digestion of blue–green spirulina cyanobacteria, with a 95% increase in human gastric and duodenal protein digestibility from human model systems, showing potential as a functional food ingredient.
- Homogenization and thermal processing can increase viscosity of *Porphyridium* and *Chlorella* microalgae, while no changes were observed in *Nannochloropsis,* providing options for structural ingredients.

26.2.2.3 Insects

- Consumer waste recycling for black soldier fly production and food product development at industrial scale.
- Food safety and processing considerations to make cricket powder.
- Mealworm and cricket freeze-dried protein extracts with comparable functionalities (solubility) to whey and pea protein at selected pH ranges.
- Attributional (economic) lifecycle assessment of black soldier fly production, separation of protein-rich side streams for feed applications, and protein concentrate production, showed lower environmental impacts than animal food.

26.2.2.4 Alternative crops

- Various processing operations were suggested for central African shea fruit processing including shea kernel recovery steps, shea kernel butter production, shea butter processing into various products.
- Western and South African Bambara groundnut rich in starch-soluble dietary fiber and polyphenols with various ethnomedicinal uses. Bambara contributed to softening corn tortillas. Another study showed that combination of soluble dietary fiber and native starch can create a stable emulsifier.

26.3 Sustainable food supply through-chain engineering for food waste reduction and transformation

One third of food produced for human consumption is lost or wasted globally, which amounts to 1.3 B tons per year (Transforming our world: the 2030 Agenda for Sustainable Development, n.d.), while more than 820 million people do not have enough to eat (Transforming our world: the 2030 Agenda for Sustainable Development, n.d.). Postharvest engineering was presented first from a waste reduction perspective and a few presentations addressed the topic within the context of food security in developing countries. Often in these countries, the large postharvest loss is due not only to the lack of an infrastructure that ensures the continuity of the supply chain but also to the disconnection between interventions by developed countries and the sociocultural realities of the smallholders. From the perspective of equipment and instrumentation infrastructure, likely solutions require a combination of low cost, high and low-tech, self-powered deployable systems. Solutions on this front are not in the classical understanding of engineering but rather at the interface of engineering, systems, operations research, social sciences, economics, law and policy, and politics, as further covered in examples in Chapters 1, 2, and 3 in this book and in a special issue manuscript on cyber-physical systems (CPSs; Smetana et al., 2021). Therefore, transitioning from subsistence to hybrid (commercial/subsistence) agriculture and the creation of circular economies will require integration of economics and engineering. A collaborative rather than a prescriptive approach is needed. A critical missing element of support is the almost nonexistence of academic programs that aim at such integration and context within the food supply chain and Chapter 24 provides further discussion on re-structuring food engineering programs within this context. The food waste reduction and transformation topic had contributions from research teams from across the globe working mostly in Australia, Belgium, Brazil, France, Canada, China, Colombia, Italy, Malaysia, Mexico, Mozambique, New Zealand, and South Africa, Switzerland, United States, and Uruguay.

26.3.1 Food waste reduction through product safety, quality, and preservation technologies

Presented advances in postharvest processing in developed countries included studies in safety, quality, and food preservation pathways (see Chapters 2 and 3). Several novel and combination technologies were proposed at ICEF13 to improve the quality retention and safety of produce:

- Plasma functionalized sanitation water for fresh lettuce (Schnabel et al., 2021).
- Dielectric barrier discharge generated cold plasma microbial inactivation (Arserim et al., 2021).
- High-pressure processing and other technologies to extend the shelf-life of fish fillets (Giannoglou et al., 2021).
- Photocatalytic removal of ethylene.
- Active packaging with ClO_2 (see details in Chapter 17).
- Magnetic field treatment coupled with cold storage of fruits and vegetables to extend shelf-life and of fresh vegetables by affecting respiration rate and enzyme activities.

Reduction of pathogens, spoilage microorganisms, and larvae infestation were also addressed through in abstracts presenting MWs or RF applications (Chapters 8 and 9). Metabolomics were applied to better understand the effects of modified atmosphere packaging. To establish food preservation and safety performance at consumer level, deterministic and probabilistic modeling were applied to create a food preservation indicator for home refrigerators (de la Cruz Quiroz et al., 2021). Advances in mathematical modeling and computational fluid dynamics (CFD) in conjunction with improved in situ measurements were shown to contribute to improved controlled atmospheres and to the prediction of produce freshness. Chapter 16 summarizes computer-aided engineering frameworks extended to quality and safety or packaging design by incorporating Multiphysics approaches in practical applications of computer-aided food processing.

26.3.2 Other food waste prevention strategies

Arguably, in the battle for a more sustainable food supply chain and to ensure food security, the elusive yet low-hanging fruit is food waste reduction. The reliance on staple grain foods was questioned in view of the micronutrient deficiencies in many populations and the increasing understanding of health and nutrition. It was acknowledged that the problem is not only technical but that there are several opportunities to minimize waste through adequate production and preservation systems throughout the food supply chain (Chapters 2, 3, and 4). All the elements of the multidimensional nature of the food supply chain must be analyzed to create sustainable systems (Smetana et al., 2021) (Chapters 20 and 21). A conceptual framework to decrease waste during transportation was discussed (Chapter 2).

A holistic international analysis with multiple indicators covering nutrition, environment, food affordability, food availability, sociocultural well-being, resilience, food safety, and waste was presented. This analysis confirmed anecdotal evidence that wealthy countries score poorly in food waste, the environment, and some health-related nutrition aspects. Indeed, the developed world wastes a substantial portion of processed food that has already made an impact in the environment (Food and Agriculture Organization of the United Nations, 2019). However, food processing companies benefit economically from the consumers' wasteful practices. The obesity epidemic is evidence of poor dietary choices among the wealthiest and, ironically, the most educated. A challenge to ensure that all the food production is optimally used, that is, waste is minimized, requires an accurate production inventory and complex mathematical models to assess the impact of different scenarios and the consequences of different decisions along the supply chain (Chapters 1 and 21). This type of assessments is critical to food policy makers, food companies, and for the consumer to make informed, sound decisions.

In developed countries, an important portion of losses still occurs along the supply chain and before the product is purchased (Food and Agriculture Organization of the United Nations, 2019). However, the cost associated with that waste continues to stimulate the development of technologies and efficient processes that make the food business more profitable and environmentally sustainable (Chapters 2 and 4). Other approaches to improve the sustainability of food production systems include the reduction of water consumption and the implementation of renewable energy in factories. There is also the opportunity to not only making existing processes more efficient and streamlining energy and water recovery and reuse, but also by reformulating foods (Chapter 3). A well-known example of water recovery includes increasing practices to recover and repurpose the water evaporated during concentration and drying back into the factory.

In contrast, another significant portion of the food is wasted after it has been purchased. On the surface, there seems to be little that food engineers can do to change certain wasteful behaviors of consumers. However, there is an important opportunity to refine the "expiration," "best before," or "use/consume by" dates of products. Indeed, real-time shelf-life indicators are being developed for many perishable and nonperishable foods (see Chapters 22 and 23 and Section 26.9). Standardizing date labels to best inform the consumer is also critical (Chapter 2). In the future, if not yet existing, engineers must have a role in ensuring that technologies exist for in-packaging reprocessing of perishable foods by retailers and consumers. There are opportunities to redesign difficult to empty packaging to reduce product thrown away with the package such as those containing highly viscous and sticky materials. At the consumer end, the shift in eating habits to foods that require less water and other resources paves the way to making sustainable foods.

26.3.3 Upcycling of by-products to higher value co-products

Finally, a common, yet always ingenious, approach to reduce waste is the conversion of on-farm losses and industrial by-product streams into co-products (Chapters 2 and 4). Traditionally, the conversion of waste streams to by-products was incentivized by the cost of waste-associated government fees and fines. However, novel co-products that have significant added value, such as nutraceuticals, are gaining momentum driven by a health-conscious market in developed countries, and the growing awareness of the value of circular economies. One can expect increasing numbers of opportunities to convert by-products into co-products from the plant-based industry. By-products are typically of lesser value than the main product. For example, the annual production of brewer's spent grain from increasingly expanding craft brewers is estimated to be approximately 30 million tons worldwide. Some examples employing engineering approaches for extraction, hydrolysis, drying, or structuring were presented at ICEF13 include:

- The use of brewer spent yeast to produce a protein hydrolysate that can be used as emulsifier.
- The use of artichoke waste to extract bioactive compounds.
- Defatting acai waste by supercritical CO_2 extraction.
- Extruded micronutrient-rich waste stream from the fruit and vegetable processing industry into snacks.
- Oil and flour produced from mango seeds using local varieties from Mozambique.
- Mango peel powder as a functional ingredient in yogurt.
- Natural colors from different carrot varieties and their pomaces.
- Xylo-oligosaccharides from sugar cane straw and coffee peel.
- Bakery leftovers enzymatically hydrolyzed to recover sugars.
- Bioactive peptides obtained by a combination of enzymatic hydrolysis, pulsed electric fields, and electro membrane technology.
- Fibers with different structural characteristics from date pits (Al-Mawali et al., 2021) or spent grains.
- Bioactive compounds extracted from olive pomace.
- Flour from orange peel.
- Application of pulsed electric fields to improve waste valorization in industrial tomato processing.
- Utilizing MW and rotating-pulsed fluidized bed drying for to produce okara powders.
- High-pressure processed fruit peels dietary fibers.
- Tangential membrane filtration of small cheese maker sweet whey to develop synbiotic dairy beverages or salad dressings.
- Powdered creamed apple pomace as a source of dietary fiber and as a thickening agent.

26.3.3.1 Green extraction methods

The use of green extraction methods was exemplified by applying high voltage electrical-discharge-plasma to extract bioactives from leaves in with water or aqueous ethanol (Nutrizio et al., 2021). Megasonic waves were also used for the reduction of oil losses and phospholipid (lecithin) extraction during seed oil degumming (Gaber et al., 2021). Furthermore, combination of MW and megasonic flow through reactors enabled developing an industrial process to improve the yield in extra virgin olive oil processing.

Others used MW-assisted extraction of bioactives from barley malt and seaweeds, which can be commonly repurposed to co-products. Analytical modeling and optimization were also applied in this context to scale-up supercritical CO_2 extraction of lipids from prepressed oil seeds (del Valle et al., 2021) and examples were presented on starch, and curcuminoids recovery from turmeric wastes and other specialty oil and colorant extraction applications.

Pulsed electric field (PEF) processing was mentioned broadly for extraction examples, including grape anthocyanin and sugar beet extraction, improved extraction of fruits and vegetables for juice manufacturing, tomato carotenoids, olive oil, and algal oil. One example presented was the improved extraction of phenolic compounds and flavor of jujube wine, by PEF treatment of jujube mash before fermentation.

The understanding of food loss data (types of foods, location, volumes, quality) at regional levels while considering the key enterprises (on-farm producers, transport and storage companies, and food processors) before retail can lead to decentralized enterprise associations where critical by-product biomass exists (Chapter 2). Regional investment of various players along the chain and other stakeholders will enable wider implementation of these innovative technologies to reduce food waste into smart food processing hubs using today's Industry 4.0 connectivity tools within hub and spokes by-product flow models (Chapters 1 and 2). Other opportunities may include mobile processing facilities on-farm, all of which will improve regional economies through the creation of new sustainable ingredients and food products.

26.4 Advances in refrigeration, freezing, and thawing

Heavily linked to postharvest engineering are refrigeration and freezing, with contributions from research teams from across the globe, namely, Belgium, China, Japan, France, Germany, Mexico, New Zealand, United Kingdom, and USA.

26.4.1 Refrigeration

Refrigeration is among the most energy intensive operations in the food industry. Therefore, it is a major challenge in the development of a sustainable supply chain of perishables. A detailed analysis presented in the congress revealed that the costs of product loss, refrigerant, source of energy, etc. determined the sustainability of such operations from a life-cycle approach. However, the incredibly challenging aim of zero net emissions may be possible through the recovery of waste heat from refrigeration systems, the incorporation of renewable sources of energy and the elimination of failures in the cold supply chain. However, in addition to cutting edge engineering, political will, social commitment, and economic incentives are needed to properly develop or support cold chains.

The purchasing power of wealthy countries continues to increase and the desire for "fresh" foods conflicts with ecologically friendly approaches, precisely because of increased use of refrigeration. For example, in the USA, the market has rejected UHT milk, which results in increased refrigeration costs. Similarly, the market for not-from-concentrate orange juice has increased since the early 1990s which, compared to frozen orange juice concentrate, also represents considerably greater refrigeration costs. In contrast, the lack of refrigerated infrastructure has limited the preservation of food in many developing countries, which has led to humanitarian crises. There are several challenges and opportunities in such countries for the development of low-cost refrigeration technologies. It is unclear what share of the total energy costs of food processing and transportation refrigeration will have in the next few years.

Chapter 10 describes the state-of-the-art of refrigeration technologies, and includes a section on chilling, super chilling, and supercooling alternatives. Presentations on refrigeration covered the whole supply chain including:

- An on-farm ice thermal rapid-cooling storage system that reduced energy cost and improved milk grade was presented and is described in Chapter 11, together with other rapid cooling options.

- Refrigerated sea-freight, including the description of refrigerated container transport (containerization) and bulk shipment in refrigerated holds in specialized reefers, and further methods for refrigeration performance improvement.
- Probabilistic modeling applied to assess an index for food preservation performance of residential refrigerators, with examples on shelf-life of milk and meat, which has potential to contributing to develop food preservation performance standards (de la Cruz Quiroz et al., 2021).

Chapter 10 provides an update on food freezing, in particular technologies to reduce freeze damage mostly by controlling the size of ice crystals. Controlling crystal size is important to reduce cell wall damage, particularly in fruit, vegetable, and meat freezing. The chapter includes descriptions of noninvasive innovative freezing methods beyond traditional blast air and cryogenic freezing, including pressure-shift freezing and pressure-assisted freezing, opportunities to apply electric and magnetic fields, MW, and radiofrequency (RF)-assisted freezing and ultrasound-assisted freezing. A special issue review on methods to control ice nucleation for subzero preservation presents a complementary update on emerging freezing techniques (You et al., 2021). Presentations connected to this topic covered:

- Progressive freeze concentration and its application to new food products.
- The Cells Alive System (CAS), which applies small amounts of energy to the food using a magnetic field to maximize the formation of small crystals (also discussed in Chapter 10).
- Crystallization into small ice crystals using MW, RF, static electric fields, and magnetic fields (Chapter 10).
- Freezing of palletized food and time-to-freeze prediction.
- Measuring freezer burn of foods by X-ray computed tomography.
- Freezer shape design using constrained optimization techniques for ice cream production (Sarghini and De Vivo, 2021).

Substances regulating the freezing process and final product quality introduced or brought in contact with the sample to manipulate the freezing process and reduce the freeze damage are also described in Chapter 10 and in the special issue review (You et al., 2021). An example is impingement freezing to significantly reduce time in blast freezers.

26.4.2 Tempering and thawing

Frozen food, such as meat, fish, and seafood, is often stored at −20°C to −30°C and requires rapid tempering (to final temperatures of −5°C to −2°C) or thawing (to 0°C), before final consumption of these products, or their use in a processing line. However, slow tempering or thawing can lead to quality degradation. A transient heat transfer model was developed to optimize the time to temper frozen raw meat based on simulations (Shan and Heldman, 2021).

RF heating can complement air or water-based processes to accelerate tempering and decrease food quality degradation. Chapter 9 provides an overview of RF tempering and thawing with several examples.

MW thawing was presented by a group from Japan, who showed that the rate of heating of frozen products highly depended on whether water was bound or frozen. This has important implications in food formulation and thawing time. Also, to address heating nonuniformities, a system with staggered through-field electrodes was studied to thaw frozen tuna. A mathematical model for thawing, which accounts for the cavity geometry and rotation, was also presented as a tool for scale-up.

Another innovative example was the ultrasound-assisted thawing of mango pulp.

26.5 Thermal and nonthermal processing for food safety and preservation

Unsafe food causes 600 million cases of foodborne diseases and 420,000 deaths worldwide every year (Aalaei et al., 2021). At the heart of food safety and stabilization are the advances in thermal and nonthermal processing. Both heating and cooling are energy intensive processes with associated economic and environmental impacts. Australia, Belgium, Brazil, Czech Republic, China, France, Greece, Germany, India, Italy, Iran, Ireland, Japan, Malaysia, Mexico, New Zealand, Oman, Philippines, Portugal, Spain, Switzerland, Turkey, United Kingdom, and the United States are among the countries that presented recent progress in thermal and nonthermal processes.

Cross-comparisons on energy requirements were presented among thermal and nonthermal treatments including high-pressure processing (HPP), ultrasound, and PEF for applications in beer, apple juice, and strawberry puree pasteurizations. While results pointed at the advantages of lower energy consumption of nonthermal processes such as HPP and PEF, comparisons need to carefully consider the relevance of industrial throughputs and the advantages that thermal processing provides in enabling high throughput continuous processing through a life cycle assessment and whole cost structure versus benefit evaluation including other requirements such as maintenance inputs and packaging.

26.5.1 Traditional thermal processing

Presentations provided some advances in canning through heat transfer modeling, microstructural understanding, or kinetics. CFD simulations that optimize thermal sterilization of liquid canned foods containing headspace were demonstrated. The influence of food microstructure on inactivation kinetics was presented in reciprocal agitated retorts. Furthermore, the accuracy of D- and z-value calculations was revisited, by challenging the errors created in the current approach due to the dynamic nature of temperature measurements. Instead, a systematic kinetic approach was developed, by using several dynamic temperature profiles of different types, to generate accurate kinetic parameters with small confidence intervals. Further research considered how oil-rich foods provide further protection to *Salmonella* and other pathogens as impacted on their D-values, providing further challenges in pathogen control in low moisture foods. Steam applications were shown to achieve microbial inactivation in cricket flour production, although further research is needed to meet microbial standards including improving thermal uniformity and sublethal damage of microorganisms.

26.5.2 High-pressure thermal processing

After decades of investment to develop an affordable high-pressure vessel with heating elements, able to maintain compression heat, with several related high-pressure thermal processing (HPTP) trials, an alternative solution was found. A polymeric canister has been developed which can be packed and loaded prior to undergoing a preheating stage. The canister can then be inserted into a cold commercial high-pressure vessel without significantly losing compression heat during pressure holding time, as demonstrated by both temperature distribution trials and CFD modeling (Chapter 19). The technology opens the opportunity to innovate in the chilled ready-to-eat meal market, which is rapidly growing in the developed world, while also offering opportunities to develop shelf-stable foods richer in nutrients and flavors, and with alternative or more desirable textures.

26.5.3 Dielectric heating

Innovations in MW and RF heating continue to appear with increased levels of granularity and continue to find new niches when combined with other technologies. MW and RF research presented in ICEF13 demonstrated the increased understanding and the practical applications of this technology, as shown across this chapter. Both MW and RF heat volumetrically within food components based on combined dipole rotation and conduction. The typical RF range is about 20–360 times longer than MW frequencies, thus enabling heat to penetrate deeper into food product with lesser level of heat uniformity, depending on the product application.

26.5.3.1 Microwave pasteurization, sterilization, and enzyme inactivation

Two excellent examples of bridging research to industrial applications were presented at ICEF13. The first one was the continuous flow MW processing of viscous and particulate products described in Chapter 8. The second, the in-package atmospheric MW-assisted pasteurization system (MAPS) of chilled ready-to-eat meals that was presented within the context of the US Food Safety Modernization Act. The MAPS offers another alternative, aligned to HPTP, for the growing ready-to-eat meal market. Adequate heating of nonuniform multicomponent foods was followed by fiber optic temperature sensors that are "transparent" to dielectric heating, capable of recording nonuniform heating. The same systems are used for MW-assisted thermal sterilization (MATS) to produce shelf stable low-acid foods, operate under hydrostatic pressure, and therefore require a pressure pump and extra temperature gauges. Metal shielding and the effect of tray location in a MW-assisted sterilization system, and the effect of salt and lipid composition were presented as ways to improve heating uniformity. Some papers studied the dielectric properties of different foods such as mango products and a mathematical model for the simulation of continuous flow heating and thermal kill in mango puree.

ICEF13 included a presentation on the combination of four MW frequencies to blanch cabbage that better retained ascorbic acid and chlorophyll compared to hot water. A continuous MW-assisted system was used to inactivate enzymes in apple juice. Conformational studies of enzymes inactivated by MW or conventional heating did not reveal differences in the denatured states of the enzymes but suggested a greater extent of inactivation with MW, which can probably be attributed to more efficient and uniform heating. MW heating was also used to increase the temperature and the yield of the enzymatic hydrolysis of fish protein. A detailed kinetic analysis of enzyme inactivation and heat transfer allowed for the prediction of enzyme inactivation at other processing conditions for potential optimization.

26.5.3.2 Radiofrequency applications

Chapter 9 is a collection of contributions from presenters in a session specially dedicated to RF applications for innovative thermal processing. In addition to thawing and tempering applications, RF can be applied for disinfestation of agriculture products and pasteurization of food products. RF is applicable for bulk size product processing at industrial scale by making use of the larger penetration depth of the RF energy, compared to MW energy. The chapter includes a review on the industrial systems with various electrode configurations and computational mathematical modeling approaches for process innovation, design, and optimization. In addition, to the contributions covered in Chapter 9, other ICEF13 presentations included RF applications under development for:

- Pasteurization of agricultural products (e.g., nuts, fruits, vegetables, wheat).
- RF pasteurization of seeds and food powders (black and white pepper, corn flour, paprika, cumin, wheat)—although nonuniformity in temperature distribution remains a challenge for industrial application.
- RF heating for sterilization of foods packaged in pouches (e.g., fish).
- Synergistic combinations of low and high RF for *E. coli* inactivation.
- Treatments with RF electric fields on orange juice processing.

Advances in instrumentation, mathematical modeling, dielectric properties characterization, and applications of MW and RF are accelerating. Industrial implementation of MW and RF is expected to grow. However, uneven heating uniformity of RF remains a challenge and limits wider practical use of RF energy. It is critical to explore methods that improve heating uniformity through developing RF-related combinations with other traditional heating or drying methods. The heating nonuniformity has been studied in many food and agricultural products, such as dry nuts, fresh fruits and vegetables, grains, and meats. Drying applications of MWs were also presented and are covered further below.

26.5.4 Nonthermal processing

Three key technologies utilizing low-temperature processing were highlighted across several presentations describing generic aspects of technology advances or specific findings: cold plasma, HPP, and PEF. A special session discussed the advances in nonthermal food processing in China.

26.5.4.1 Cold plasma

Cold plasma is known for its potential applications for microbial decontamination, pest control, toxin elimination, food and package functionalization, and many others while improving product quality and functionality. Plasma can be formed with a wide variation of discharges in the form of corona, spark, or arc in gas or liquid media using pulsed power with various kinds of reactor configurations. A presentation discussed scaling approaches for the most widely tested plasma designs including plasma jets and surface dielectric barrier discharge units (DBDs) to realize uniform plasmas of higher density in larger volumes. Other ICEF13 contributions included applications with gas or functionalized water:

- The use of atmospheric air as the inducer gas with potential of employing current modified atmosphere processing (MAP) food gases.
- Dried product decontamination with a two-stage pilot scale MW excited plasma torch and air as process gas showing effectiveness in peppermint tea leaves.
- Shelf-life extension of seam bream fillets using DBD generated cold atmospheric plasma (CAP) without altering color and texture, unlike HPP, pulsed electromagnetic fields (PEMF), and ozonation.
- Computational fluid dynamics methods predicted inactivation of *E. aerogenes* using DBD generated CAP.

- Inactivation of *Listeria monocytogenes* and *Salmonella typhimurium* biofilms by combining CAP and hydrogen peroxide, resulting log-reductions of up to 6 log (cfu/cm^2) and partial removal of biofilms.
- A pilot scale prototype producing plasma functionalized water with compressed MW plasma air applied to sanitize lettuce and other fresh cut produce.
- The functionalized liquid platform (PlaSmarter) is based on open-air spark or glow discharge to control the principal reactive species composition, applied to apple juice, fresh produce, and grains.
- Microbial inactivation resistance and tailing of *Listeria monocytogenes* and *Salmonella typhimurium* when applying various forms of plasma-activated liquid.

26.5.4.2 High-pressure processing

HPP is an industrial, globally established cold pasteurization, inactivation, and structuring technology for in-pack food applications. Prepackaged pouches, tubs, trays, or bottles are introduced to a vessel and subjected to a high level of hydrostatic pressure (300–700 MPa). Innovative uses of HPP were highlighted in ICEF13 presentations:

- HPP enhanced the activity of superoxide dismutase, a compound used for treating pain and inflammation and other health applications, by up to 38%.
- Low-salt applications for sausage manufacturing, shelf-life extension, with synergistic effects between HPP and two spice extracts.
- A lifecycle assessment study showed the superior benefits of HPP combined with biopreservation, versus utilizing nitrites in ham production, toward impacting climate change, resource depletion, and ecotoxicity.
- Successful nonthermal pasteurization of wine without sulfur dioxide addition to prevent *B. bruxellensis* yeast and other microbial spoilage.

26.5.4.3 Pulsed electric fields

PEF is an industrially established technology, originally developed for cold pasteurization and extraction using membrane electroporation caused by electric field pulse charge polarization effects, with applications in biotechnology, medicine, food, and effluents. A broad overview was presented showing more than 120 systems installed worldwide mostly among four key companies (Diversified Technologies, Elea, Energy Pulse Systems, and Scandinova). Innovative industrial examples presented included:

- Cold pasteurization of low pH juices.
- Structure modification of vegetables for French fry and chips manufacturing.
- Fruit and vegetable porosification for drying applications.
- Increased yield in algal extraction.

PEF applications presented at ICEF13 beyond food preservation are described in Sections 26.1, 26.6, and 26.7 in this chapter. Some innovative uses of PEF for bacterial inactivation and food preservation highlighted in ICEF13 presentations included:

- Nonthermal pasteurization of wine without sulfur dioxide addition to prevent *B. bruxellensis* yeast and other microbial spoilage such as acetic acid bacteria, while avoiding release of metallic particles and shortening fermentation time.

- Application to heat-sensitive protein solutions shown by the inactivation of *Listeria innocua* and *Escherichia coli* with PEF and reduced heat on liquid whey protein without impacting heat sensitive immunoglobulins and vitamin A and C, or *E. coli* inactivation in soy protein isolates.
- Molecular dynamic simulations to understand the effects of electric fields on enzymes.
- Production of extended shelf-life milk products by combining high-temperature short time pasteurization and PEF.
- PEF pretreatments of tough meat for reducing 60°C sous vide tenderization time and improving meat quality.
- Improved extraction of phenolic compounds and flavor of jujube wine, by PEF treatment of jujube mash before fermentation.
- Ultrasound radiation in combination with natural antimicrobials and heat was used to inactivate vegetative cells, spores, and enzymes.

During the last decades we have witnessed the successful development of product applications, large-scale equipment development, and industrial uptake of technologies such as HPP and PEF around the world, beyond belief of many. This provides confidence in seeing higher capacity equipment ready for applications in cool plasma and other emerging technologies beyond proof of concept.

26.6 Drying, predrying, and separation, technologies for preservation, and the incorporation of bioactives for health

It is impossible to conceive a congress in food engineering without presentations on drying. Among, if not the oldest, method of preservation, drying continues to play a critical role in the food industry and research. Despite the hundreds of different foods that have been dried, the creative efforts of drying experts are driven by the need to:

- Improve the affordability and quality of dried foods.
- Reduce high energy consumption and carbon footprint, within the context of the product life cycle, by
  - new technology development,
  - developing new pretreatment interventions or technologies,
  - drying time optimization, or
  - using renewable energies.
- Develop new dried products with improved health properties and functionalities.

Two types of innovations were discussed (a) food material concentration or pretreatment to improve drying efficiencies and (b) simultaneous application of traditional heat transfer-based drying combined with ultrasound or electromagnetic energy. Attention was also given to advancing research on concentration and separation of actives, particularly at developments on membrane technologies, the incorporation of these actives through encapsulation or other technologies, and improvement in powder properties. ICEF13 gathered experts from Australia, Belgium, Brazil, Canada, Chile, China, Denmark, France, India, Ireland, Japan, New Zealand, Nigeria, Poland, Portugal, South Africa, Sri Lanka, Taiwan, Thailand, The Netherlands, United States, Uruguay, and other countries to bring different advances and solutions in the areas above.

26.6.1 Drying for developing regions

Drying appears to bring the most attractive and sustainable proposition for developing regions requiring stabilization of food at regional levels in farming regions. However, only a few presentations from authors based in developing regions addressed the development of low-scale low-cost technologies. Examples presented included applications of sun drying, drum drying, heat pump, or hot air drying, including:

- An example of village level food processing included a low-cost "Royal solar dryer" for solar-assisted drying of fruits, meats, and seafood to improve quality, shelf-life and safety, and bringing training opportunities for women.
- Controlled direct sun drying of nectarines designed by predicting drying rates based on real-time ambient weather factors; the method is implementable with a weather station with a solar pyranometer, but some level of predictions can be achieved with a hygrometer.
- A solar power hot air food drier design, to be implemented near fields in Ethiopia, improved by simulating the air flow resistance characteristics of sliced sweet potatoes.
- Heat pump drying, as an alternative to tray and sun drying, to shorten drying time and deliver higher quality on quality of Arabica green coffee in Thailand, although another presentation showed that this process is economical only at temperatures below 100°C unless renewable resources are used to provide additional heat.
- An experimental drum dryer and optimization of drying in whole fish or fish parts in Nigeria.
- A pumpkin thin slice drum dryer system was also presented as a shorter time alternative to make pumpkin powder.
- A small-scale steam parboiling and hot air-drying unit designed, built, and integrated into rice processing areas in rural communities in Nigeria, as an alternative to conventional sun drying, bringing greater technical simplicity, ease of operation and maintenance, and technoeconomic benefits to smallholder rice processors.
- A hybrid-solar-vacuum dryer to produce crisp fruits and vegetables with zero greenhouse gas emissions.
- A low-temperature vacuum drying technique to dry papaya in Chile using temperatures below 40°C.

Other technologies (including osmotic dehydration, specialized belt dryers such as refractance window) described below, may suit, or be adapted to, the context of developing regions.

26.6.2 Advanced spray drying

Spray drying has been long established in the food industry as a high throughput continuous process for several soluble food materials, and generally uses hot temperatures to dry a concentrated liquid. However, drying of heat sensitive or high sugar materials remains a challenge. ICEF13 examples addressed the challenges of drying heat sensitive probiotics or materials like honey, which are high in monosaccharides and highly viscous:

- Direct spray drying of probiotics, as a viable and more economical alternative to conventional freeze drying, by modulating the stress tolerance of the target probiotic before drying. *Lactobacilli* strains tolerance was increased by increasing culturing temperatures by 3–5°C, beyond the standard temperature for each strain, and by adding calcium chloride to the feed, showing higher survival after spray drying.

- Honey could be spray-dried by using modified starch as a carrier and nitrogen flushing, giving a powder with acceptable wettability.
- A monodisperse droplet spray dryer developed at laboratory scale was demonstrated as a preliminary option for improving spray drying formulations.

Electrostatic spray drying (PolarDry) was introduced as a low-temperature spray drying technology commercially available. It combines gas–liquid atomization and electrostatic charge in a streamlined one-step process for the conversion of liquid feed to powder. Its development and commercialization were presented, and the technology is also covered in detail in Chapter 13. In short, the technology works with a spray dryer adjusted with an electrostatic two-fluid nozzle operating at low operating temperatures (inlet: <100°C, and outlet: <60°C). Electrostatic charges enable water diffusion to the core of droplets at low temperatures. The technology promotes the agglomeration of dried particles and eliminates the need for a postdrying agglomeration step. Examples are provided for drying probiotic microorganisms, microorganisms associated with the human microbiota, heat sensitive proteins, microalgae, and other pharmaceutical products, and combinations with other components to achieve high oil load emulsions.

Another innovative technology presented, also designed for low-temperature operations, was the extrusion porosity technology. The technology produces powders with high aroma retention and improved rehydration properties, without the need to agglomerate. A high-solid's feed goes into a twin-screw extruder where the concentrate is modified using an intellectually protected method and then fed into a spray dryer to produce highly porous dried particles.

26.6.3 Batch low-temperature drying for high value-added products

Three low-temperature drying strategies presented included some improvements in freeze drying, ultrasound atmospheric freeze drying, and MW vacuum drying to improve drying time and energy consumption and/or improving quality.

26.6.3.1 Advances in freeze drying

Freeze-drying processing has been recognized to be the most expensive process for manufacturing dehydrated products. The process consists basically of freezing food pieces or slurries at −20°C, following sublimation of the ice and finally desorption of the remaining water content. The application of CO_2-laser microperforations was presented as an innovative way of reducing drying time, with up to 14 h reduction of drying time in strawberries, with negligible energy inputs from laser use. However, the technology did not deliver drying time reduction in blueberries, although it greatly reduced the number of defects in blueberries caused by shrinkage by 40%.

Another advance presented proved that blanching fruits pieces such as pomegranate arils, lowered drying time, improved redness retention, and showed better retention of antioxidant capacity, in an intermediate moisture product. PEF pretreatment of fruits was also shown to enhance porosification and better preservation of product shape after freeze drying. It was achieved in fruits like mango with up to 96% reduction of drying time, compared to hot air drying, while retaining natural flavor and color.

26.6.3.2 Ultrasound atmospheric freeze drying

This technology has been introduced as an alternative to freeze drying without the use of energy intensive vacuum for ice sublimation. The process can operate at −5°C and has shown to achieve similar drying times to freeze drying while only using one third of the energy cost. Even though the technology was presented as batch, it displays the potential to develop a continuous process that delivers similar quality to freeze drying at reduced energy requirements. A pilot air drying chamber was designed with ultrasonic plate transducers to provide indirect transmission of ultrasonic energy through to the material. A computer-based ultrasonic drying setup was built to allow continuous recording of the process variables, in real time, which also enabled simulation of drying to be accomplished under controlled conditions over a range of drying parameters. The drying unit can also work with slightly higher temperatures and has shown to be effective for products including cannabis, fruits, and vegetables and meat.

26.6.3.3 Microwave-assisted drying

MW-assisted drying received significant attention during ICEF13 as a technology that enables near room to low-temperature drying (<60–70°C product core temperature) due to rapid volumetric heating of MWs. A few companies are offering commercial prototypes for MW vacuum drying (e.g., Enwave). Many presentations highlighted its benefits and methods to understand and optimize the process. These include, new product applications, combination with other techniques such as fluidized bed or osmotic dehydration, product shrinkage reduction with novel visualization methods, in situ pasteurization, process simulation, and modeling for process understanding and optimization; more specifically:

- MW-assisted convective drying retained color, vitamins, and antioxidants in bitter gourd, which rehydrated faster than convective drying.
- A pulsed fluidized bed dryer combined with MW showing advantages for processing okara at 70°C, a residue from soybean processing, increasing health compounds such as aglycones, over malonyl-glycosides; the process did not sufficiently inactivate trypsin inhibitors.
- Combination of MW radiation with osmotic dehydration of sliced mango to achieve improved color, size, shape, and cell structure than each technology applied alone; the study was supported by computer vision and confocal microscopy.
- Other visualization of product structural changes after drying with computed tomography on a micron scale.
- Further understanding of shrinkage reduction in roots like radish, carrot, and potato during room temperature MW-vacuum drying using MRI distribution of moisture contents, and glass transition temperature comparisons.
- Inactivation of vegetative cells and spores during freeze drying achieved in situ by simultaneously treating a model food with high frequency dielectric heating, leading to plasma generation between the electrodes, while keeping the samples at temperatures below freezing.
- Intermittent MW drying, simulated, and validated through a mathematical model that combined computational fluid dynamics, heat, and mass transfer to further understand process dynamic transients.

Chapter 12 includes recent experimental and numerical studies combining electric or electromagnetic fields with conventional air to reduce drying time and thereby increase the quality of dried products. Examples are provided on MW-assisted drying, RF drying, and electrohydrodynamic drying and the concept of Multiphysics models is discussed for the design and optimization of such innovative drying processes.

26.6.4 Membrane separation for nonthermal concentration and bioactive separation

Besides the drying pretreatments mentioned earlier, separation of bioactives and concentration are fundamental steps to facilitate the economics and efficiencies of drying. Membranes are commonly used as a nonthermal concentration technique before drying and can also reduce the costs of storage and shipping of liquids in concentrated form.

The application of forward osmosis for nonthermal concentration, as a cheaper alternative to high energy demanding reverse osmosis was presented, including:

- Nonthermal whey concentration using aquaporin-based hollow fiber membranes by using cheese salt brine as a draw solution.
- Coconut water concentration using aquaporin-based hollow fiber and tubular forward osmosis membrane, reaching beyond concentrations achievable by reverse osmosis, freeze concentration, and similar concentrations to vacuum/thermal evaporation.

Progressive freeze concentration was also presented as a continuous method for concentrating heat sensitive melon and watermelon juices, for further fermentation applications (e.g., fruit wines such as Japanese sake).

The dairy industry is the most advanced industry that achieved large-scale fractionation of milk into dairy ingredients with health attributes by using durable ceramic tubular membranes or spiral-wound membranes. Technologies such as electrodialysis, electrodialysis with bipolar membranes, electrodialysis with ultrafiltration, capacitive de-ionization, eutectic freeze crystallization, and membrane distillation continue to be explored to tackle the challenges from the concentration and disposal of saline wastewater. Examples of a few of these and other technologies presented in the conference include:

- Isolation of immunoglobulin from dairy whey by electrodialysis with filtration membrane process. A filtration membrane across which an electrical current was applied enabled high bovine serum albumin flux and high rejection for immunoglobulin.
- Efficient sweet whey demineralization achieved by PEF pretreatment, leading to reduced fouling, and scaling during electrodialysis, with decreased energy consumption.
- PEF pretreatment of protein molecules in by-product streams enhancing hydrolysis and bioactive peptide release, followed by electrodialysis with filtration membrane to recover these compounds.
- Spiral-wound membrane prototype with increased protein filtration performance, by changing the architecture and spacer nets to decrease flux, while increasing protein transmission.
- Hollow fiber membranes, rarely used in dairy so far, as a cheaper alternative to costly membranes used at present, which have shown to be more efficient with whey protein at lower transmembrane pressures through five laboratory membrane modules connected in series.
- Reverse osmosis membrane performance with olive oil-in-water emulsions, showing that oil concentration in the emulsion did not decrease the permeate flux with cellulose acetate membranes but did affect the performance of polyamide membranes.
- Rotary disk column as a low-cost alternative to recover coffee aromas.

26.6.5 Incorporation of bioactives, probiotics, and synbiotics

The area of encapsulation as a method to incorporate bioactives and probiotics was widely presented during ICEF13. The industrial interest in this area is increasing toward the development of sustainable therapeutic foods and nutraceuticals. Encapsulation methods tackled a number if challenges to ensure incorporation of health compounds in food products:

- Increasing protection and stability of sensitive bioactives or probiotics in powders and food formulations.
- Masking off flavors and taste in food formulations.
- Gastric protection from acidic compounds.
- Delivery and bioavailability in the human body through, for example, through controlled intestinal release.

Topics presented can be organized into three key areas: dairy-based or plant-based encapsulation of bioactives, and encapsulation of probiotics. Chapter 14 takes the application of spray drying further into the microencapsulation applications for dairy. It presents an overview of atomization-based micro-encapsulation technology by using spray drying and further expands on other techniques such as spray chilling, and fluidized bed coating applicable beyond dairy. Chapter 19 mentions the Australian commercialization of Micromax as a dairy-based encapsulation technology for stabilization and protection of omega-3 oils, probiotics, and phospholipidic bioactives.

ProGel is another Australian patented technology for encapsulation, which consists of impinging aerosol process to continuously produce alginate microgel particles prepared by mixing sodium alginate, $CaCl_2$, the target bioactive and then atomizing in a chamber for rapid gelation and trapping of the bioactive in the gel matrix, and coating. Applications also include encapsulation of fish oil, probiotics, lactoferrin and microalgae, or pharmaceuticals such as ibuprofen.

26.6.5.1 Dairy-based encapsulation

Examples of dairy ingredients used as encapsulants or wall materials (dairy proteins, lactose, milk fat, mixtures) and the entrapped bioactive compounds (lactoferrin, peptides) are also included in Chapter 14. Some ICEF13 presentations highlighted that:

- Milk fat globules can effectively encapsulate micronutrients such as vitamins A and D and phytochemicals including curcumin and quercetin, while providing barrier properties. Even though process conditions influencing encapsulation efficiency and yield of bioactives need to be optimized, it was argued that 1 g of milk fat globules can deliver 100% of the recommended daily intake of a vitamin such as vitamin D.
- Milk fat blending fish oil with milk fat could be a viable practice for improving omega-3 oxidative stability.
- Omega-3 fatty acid from chia seed oil and fish oil can be encapsulated with whey protein after low-temperature spray drying, and incorporated in finger millet flour biscuits without changing its flavor.
- Novel protein powders were obtained from an aqueous combination of sodium caseinate and blackcurrant concentrate. Powders were used to fortify cookies, with increased antioxidant activities that provided hypoglycemic effects (Wu et al., 2021).

- Water-soluble bioactives such as anthocyanins were protected in hardened whey and fruit extract beads. The technology consists of forming stable particles by mixing concentrated whey proteins and sugars, forming droplets in a silicon tubing, hardening wall materials in oil at 100°, vacuum oven drying, and storage of beads under 33% relative humidity, maintaining 0.1–0.2 water activity.

26.6.5.2 Plant-based encapsulation

Other ICEF13 presentations included the following innovations using plant-based materials:

- Microencapsulation of healthy Eurasian ethanol polyphenol extracts (e.g., turmeric) into spherical microparticles using core materials such as gums, alginate and modified chitosan, and spray drying. In vivo bioperformance was followed by kinetic models for controlled release.
- Encapsulation of curcumin using rice starch modified by 4-α-glucanotransferase as a host material for improving the encapsulation efficiency of curcumin by complexation, greatly improving curcumin oxidative, and pH stability and solubility, compared to conventional encapsulants (maltodextrin, β-cyclodextrin).
- Production and stabilization of bioactive peptide-loaded double emulsions with hydrophilic surfactant mixtures with Tween80 and chitosan using novel homogenization techniques.
- Encapsulation of therapeutic and antiaging α-linolenic acid and α-lipoic acid together in nanoliposomes, using soy phosphatidylcholine and Tween 80, solvent evaporation, followed by probe sonication, enabling good antioxidant potency and sustained release.

26.6.5.3 Encapsulation for probiotic protection

There is a present need for encapsulation of lactic acid bacteria or probiotics to (a) protect them during processing and in product applications or preparation; (b) preserve them during storage, particularly in dry mixtures stored at room temperature; and (c) enable the delivery of live cultures to the large intestine. Market trends show the need to develop synbiotic powders with probiotics and prebiotics (i.e., food for bacteria, not digested by the human host in the large intestine) such as resistant starch to maximize bowel performance. The following innovations included in ICEF13 presentations can be highlighted:

- Spray freeze drying of synbiotic mixtures of *Lactobacillus plantarum* and fructo-oligosaccharide, with whey protein and maltodextrin encapsulants, and liquid nitrogen, provided higher encapsulation efficiency, cell viability retention, and survivability under simulated gastrointestinal conditions, than spray drying, and refractance windows drying; although refractance windows drying appears as a promising relatively effective low-cost technique for encapsulation (Yoha et al., 2020).
- Spray drying encapsulation of *Lactobacillus rhamnosus* in milk hydrolyzed with rennet (chymosin enzyme) enzyme before atomization led to good bacterial survival and water insoluble microparticles in 40°C water and increase protection or water soluble microparticles at 8°C for bacterial release.
- *Lactobacillus casei* protection was increased by increasing skim milk concentration to 20–30% solids, demonstrating significant improvement in cell survival and heat stability at 65°C for 10 min, suggesting an interesting pretreatment method (Suo et al., 2021).

26.6.6 Powder properties and functionality

The study of powder bulk properties continues to be of interest, particularly due to the challenges that concentrates, infant formula, and amorphous powders present during storage (e.g., caking, ageing) or at the time of rehydration. Some of these topics were covered during ICEF13:

- Milk protein concentrate (MPC) powder rehydration properties were improved by injecting high-pressure nitrogen into the liquid MPC prior to spray drying at pilot scale, although regular and agglomerated powders both had reduced bulk density and flowability.
- An in-line hydrodynamic cavitation system tackled the challenge of instant and simultaneous rehydration of MPC powders dispersed in water at semi-industrial pilot scale.
- Better ageing control of spray-dried infant milk formula powders was achieved by addition of maltodextrin, increasing the glass transition temperature (Tg) and decreasing the crystallinity of powders, at specific temperature and relative humidity during storage, to maintain target solubility and avoid caking.
- Camel milk powder is becoming relevant for its compositional profile closer to human milk, compared to milks from other animals, and has shown higher wettability than bovine milk powder.
- A predictive model for caking of amorphous powders, using model freeze-dried blends of maltodextrin and increasing the amount of plasticizers (glucose maltose and sorbitol) to increase caking behavior, was proposed by establishing critical water activities from Tg curves and water isotherms of the powders, and studying the viscosity dependency on temperature (Alvino Granados et al., 2021).
- Insights on the hardness of thermally compressed maltodextrin, in combination with crystalline additives (NaCl, monosodium glutamate monohydrate, sucrose, lactose monohydrate, lauric acid, and stearic acid) as a function of Tg were discussed and published (Mochizuki et al., 2021).

Drying, assisted with predrying and separation technologies, will continue to play a critical role in food processing and preservation. In addition, new food formulations and the combination of drying technologies for encapsulation of added-value nutraceuticals are expected to continue to grow.

26.7 Innovative technologies for food structuring and product enhancement

Food structuring technologies are providing opportunities for personalized nutrition, to match individual biological needs and preferences, catering to an increasing diversity of new and specific consumer population segments (e.g., children, elderly, undernourished, specialty healthy or prepared foods, or personalized catering). For example, the young age markets in developed countries are finding new ways of restructuring vegetables into fun or more appealing formats to increase their consumption, or the incorporation of health compounds within the food matrix in a way that they will be delivered to the blood stream and provide therapeutic effects. New ways to apply traditional, or still emerging technologies, able to modify the food and ingredient physicochemical properties, and therefore their physical and nutritional functionality, where presented during ICEF13 and highlighted below. Contributions were provided by presenters from many countries including Australia, Brazil, China, Italy, Indonesia, France, Germany, New Zealand, Philippines, South Africa, Russia, and USA.

26.7.1 Extrusion for novel ingredient incorporation and texturization

Twin-screw extrusion transitioned from the plastic industry to the food industry in the early 1970s to precook flours and process flat crispy bread. Since then, this technology has been leveraged in the food and feed areas with hundreds of products being produced using single and twin-screw extrusion. The technology has been key to the global development of mainstream food products including breakfast cereals and snacks, as well as pet food and fish feed. Extrusion is also a key tool in the ingredient industry, for instance, to make encapsulated flavors.

Extrusion has evolved from a simple forming/shaping process to a highly flexible and advanced process with many unit operations. Chapter 7 includes applications of extrusion to design sustainable food systems such as textured vegetable protein from legumes (mainly for soybean, pea, faba bean, and chickpea) or protein rich oilseeds for meat analogue production, to cater to the rapidly growing market for plant-based meal items (e.g., mince, hamburgers, sausages). The chapter brings a discussion on engineering tools to control this process and the corresponding products toward a less empirical and more mechanistic approach. Specific innovations in ICEF13 included uses of extrusion for meat analogues and for food loss value addition into functional ingredients:

- Improving the control of extrusion processing of soy-based meat analogues by characterizing the rheological properties of extruded samples of various compositions, at extrusion-like conditions, with a closed cavity rheometer, to target specific meat analogue textures.
- Protein aggregation in extrudates of soybean protein isolate and canola meals was characterized in terms of color, hardness, tensile strength, protein extraction, in vitro digestibility as a function of specific mechanical energy input and barrel moisture.
- Modification of brewer's spent grains dietary fiber structure and proteins via extrusion to increase sensory acceptability and bioaccessibility of amino acids.
- Value addition to broken rice residues from dehulling and polishing through extrusion processing to produce a staple rice product with lower digestibility.
- Value addition to apple pomace residue from apple juice processing by optimizing extrusion parameters to improve water binding, water absorption, and thickening properties.

26.7.2 New structures through 3D printing

Linked to extrusion is the area of 3D printing of food, also called food layered manufacturing, which allows for the creation of complex and novel shapes and designs, personalized nutrition, and production of customized dishes. It provides a sustainable way of reducing storage and distribution costs for consumer-based processing with applications in chocolate, cake frosting, meat purees, dough, hydrocolloids, among others. Significant research has been published from China and Australia, and challenges were mentioned in terms of the research carried out in China (a) identifying suitable gel material systems, (b) accuracy for high precision printing, and (c) long printing time for the technology and need for better software development and scalable equipment. Chapter 15 provides an overview of 3D printing and describes four 3D printing types of hardware (printers), with one of them being extrusion, and the future of 4D printing for where food changes after processing with time. 3D printing advances presented in ICEF13 included:

- Commercial applications in China using plant material sources for fish analogues and multigrain snacks, or by using animal sources for surimi or dairy gels.

- Edible and printable formulations of dairy materials using extrusion-based 3D printing by changing preprocess parameters and casein-whey protein mixtures for protein denaturation to come up with multilayered printed gels.
- Structural enhancement of 3D printed surimi using sweet potato starch.
- Using dual-nozzle extrusion to structure meat products with composite layers of lean beef paste and lard.
- Other meat structuring improvements through: (a) potential viscosity enhancers (hydrocolloids and fat) were suggested to improve the flow during extrusion; (b) cold-set binders (transglutaminase enzyme, calcium/alginate system, and plasma protein system) and heat set binders (blood plasma proteins and hydrocolloids) were suggested to improve mechanical stability during depositions.
- Low-cost 3D printed algal extracellular polysaccharides, with algae cultivated and harvested using a cost efficient photoautotrophic polymeric-based reactor.
- Use of 3D food printing to deliver synbiotics through food systems.
- Innovative cereal-based biscuits designed using wheat and rice flour and milk butter and vegetable oil to achieve customized textured properties for new applications (e.g., fragile biscuits for the elderly).
- Ohmic heating applied in a rectangular 3D printing channel head to control heating during prebaking of cake batter.
- Stabilized micron-size CO_2 nanobubbles, which may provide further porosification opportunities for structuring 3D printed products (Phan et al., 2021).

26.7.3 Pulsed electric field structuring

PEF's novel opportunities for structure modification were presented. For example, a highlight of a PEF equipment manufacturer was the PEF-assisted softening of potatoes, sweet potatoes, and other vegetables for smooth cutting in French fry and chip production, which allows even colored products and better adhesion of seasoning. The technology brings opportunities for new shapes of sweet potatoes and brings significant reduction in energy and water consumption during potato processing with less starch in the process water. Tubers, roots, vegetables, and fruits can be treated with PEF belt systems (e.g., Smoothcut) up to 70 tons per h capacity, while continuous liquid treatment systems have capacities of up to 10,000 L/h and clean-in-place (CIP) ready. Other research advances or innovative uses of PEF were highlighted in ICEF13 presentations included:

- A first-order kinetics model of color development during frying of PEF pretreated potato.
- PEF strengthened cell walls of blanched carrots added with $CaCl_2$, and in vivo bioaccessibility of β-carotene.
- Reduced 60°C sous vide tenderization time and improved meat quality after PEF pretreatments of tough meat.
- Oat flour fractions can improve their gelatinization properties after PEF treating flour suspensions.
- Edible films with improved tensile strength, thermal stability, and morphology were developed after PEF pretreatment of zein, chitosan, and polyvinyl alcohol composite dispersions.

26.7.4 High-pressure processing structuring

The application of HPP for structure modification has been studies for several decades. However, the area continues being of interest with the following presentations highlighting:

- Meat tenderization using low-pressure and moderate to high-temperature processing.
- HPP enhanced calcium infusion in baby carrot, celery, and mango using calcium lactate gluconate was explored at a microstructural level, even though the produce cutting force was decreased (Gosavi et al., 2021).

26.7.5 Cold plasma: from legume plant germination to powder functionalization

Cold atmospheric pressure plasma was also found to:

- Increase germination rate and elongate shoots and roots of legumes like mung bean and soybean while decontaminating mung beans from microbes.
- Alter the swelling ability of testa flour, protein flour, and protein isolate from grain pea.
- Increased the water binding capacity of protein flour opening opportunities for functionalization.

26.7.6 Precision fermentation and other methods for probiotic incorporation

There is renewed interest in the application of fermentation to a wider range of plant- or animal-based materials to enhance the availability or incorporate target health compounds. Another area of interest is the application of synthetic biology to produce food components such as bioactives, proteins, peptides, fats, color compounds, and carbohydrates, without the need for complex separations to recover them from existing raw materials. New companies are using these concepts presented as sustainable alternatives to conventional agriculture. Chapter 5 includes a description of novel plant-based fermentation and bioreactor design. Developments in synthetic biology for the manufacture of ingredients analogues to animal derived and other food ingredients is introduced as a technology with untapped potential for a sustainable and food secure future. Some concepts covered in ICEF13 included:

- Rapid lactic acid bacteria fermentation of broccoli puree and red cabbage, preprocessed with mild heating or ultrasound, greatly increased the production of sulforaphane, an isothiocyanate that has been shown to have antidiabetic and anticancer effects, polyphenols and antioxidant activity, protein levels and soluble fiber, while increasing the bioavailability of these compounds.
- Lactic acid bacteria fermentation of second grade apples increased the abundance of polyphenols, fatty acid, and aminoacids as indicated by metabolomic methods.
- White grain sorghum's nutritional value was improved via solid state fermentation with co-culture of *Aspergillus oryzae* and *Bacillus subtilis*, with an increment in crude protein and reduction of antinutritional phytic acid by 95%.
- Value addition of finger millet mallet flours by lactic acid bacteria fermentation showed increased total phenolics and antioxidant activity, and the process was suggested for a nonalcoholic finger millet malt beverage.

- A probiotic apple snack was developed by pretreating the diced apple pretreated with ultrasound, low-temperature fluidized bed drying of apple, and by coating apple particles with the probiotic *Bacillus coagulans* at the final stages of the drying process.
- Fermented soybeans have better protein digestibility and antioxidant potential compared to digested fractions of raw, cooked, and fermented soybeans, with higher essential amino acid bioactivity through digestion.

26.8 Sustainable packaging innovations for increased food safety, stability, and quality monitoring

Plastics have the bad reputation of contributing to environmental pollution during both their production, with a considerable carbon footprint, and when not disposed properly after use, because in most cases they are not biodegradable. Today, developed countries are taking action on plastics to tackle plastic pollution and marine litter, and to accelerate the transition to a circular plastics economy (Plastics, n.d.). Cultural and technological shifts are required to address this issue.

Presentations from Australia, Brazil, China, Germany, Japan, New Zealand, Portugal, Spain, The Netherlands, and United States covered the areas of packaging for microbial contamination, physical and barrier properties of new packaging materials, including biodegradable materials, and in-packaging sensors. Relevant research on creating new functionality for active/intelligent packaging for food safety (Chapter 17) or developing respiring films that respond to respiring foods (Turan, 2021) and adjusting the physical properties of packaging for in package processing (Chapter 18) reflects the continuous progress in this multidisciplinary field.

26.8.1 Biodegradable packaging materials

Presentations on biodegradable packaging materials included soil applications in agricultural systems, nonhazardous packaging manufacturing, and compostable food packaging, including:

- Biodegradable plastic mulch (BDM) used in specialty crop production systems to reduce soil water evaporation, as alternatives to the current application of conventional polyethylene. Sprayable BDMs have shown up to 28% decrease in soil irrigation requirements and weed control and can be tilled into soil after use where resident microorganisms degrade the plastic. Common bio-based polymers used in BDMs include polylactic acid, starch, cellulose, and polyhydroxyalkanoates. Fossil-sourced polyesters used in BDMs include poly(butylene succinate), poly(butylene succinate-co-adipate), and poly(butylene-adipate-co-terephthalate) (Bandopadhyay et al., 2018).
- Organic active film biopolymers developed from gelatin, chitosan, and sodium caseinate, and an antimicrobial extract of boldo-of-chile (used as moisture barriers for nuts, cheese, and meet patties) required only a day for compost disintegration in selected biopolymer combinations.
- A new renewable nonisocyanate starch-polyurethane films, as an alternative to the conventional polyurethane counterparts that require toxic isocyanate as a starting material for their industrial synthesis, including some biodegradable polyurethane-based formulations.
- Hybrid biodegradable materials made of plasticized starch combined with polyurethane as a low-cost compostable material with improved flexibility and water barrier properties.

26.8.2 Packaging for in-pack microbial decontamination

Active antimicrobial packaging is of increasing interest due its advantage to release gases or additives to the food surface through the contact time during storage. Chapter 17 presents the use of active food packaging applications and other microbial decontamination applications using chlorine dioxide (ClO_2). The chapter summarizes multiple chemistries that can produce ClO_2 gas or liquid and describes their combination with polymers for ensuring microbial safety, extending shelf-life, and reducing waste of perishable foods such as fresh fruits, vegetables, and berries. The chapter describes a system that produces ClO_2 in-package. A second application includes a superabsorbent hydrogel polymer that enables humidity-activated, time-released controlled and sustained production of low concentrations of gaseous ClO_2 in-container. ICEF13 also included studies in:

- Lignin-polyurethane films with antimicrobial properties and suitable for different processing as potential cost-effective alternative.
- Novel materials such as fatty acid modified cellulose nanofiber, biodegradable chitosan films with controlled release antimicrobial essential oils.
- Natural antimicrobial carvacrol, mainly present in oregano essential oil, was incorporated into cellulose acetate films, and used to form active packaging and its diffusion into model food systems was demonstrated through kinetic modeling.
- Other antimicrobial agents were tested including rosemary extract and green tea extract, showing that oregano essential oils produced the highest concentration of volatile compounds (mainly carvacrol), and the latter had the highest antimicrobial effectiveness when incorporating it in poly(3-hydroxybutyrate-co-3-hydroxyvalerate) films.

26.8.3 High gas barrier properties of packaging for advanced food processing

Chapter 18 describes the advances in polymer packaging able to withstand in-pack thermal pasteurization technologies. The chapter describes flexible packaging material options that maintain visual integrity post-thermal processing, and low and medium gas barrier properties to provide pasteurized products with a shelf-life of 10 days to 12 weeks. Gas barrier and migration properties resulting from both conventional and MW-assisted pasteurization are discussed, including efforts in developing EVOH and metal oxide coated PET-based films subjected to sterilization processes. Storage studies and the impact of oxygen migration on food components such as pigments, vitamins, and lipids are also discussed. Other aspects of the physical and barrier properties of packaging presented at ICEF13 include:

- Accounting for the gas permeability of freshly produced polymers but considering the effects of in-package processing such as MW heating or high-pressure processing.
- Water condensation that results from temperature fluctuations in packaging of sensitive produce such as strawberries shortens their shelf-life. Packaging materials with high water permeability and absorbent pads were compared. A new method and simpler method to measure water vapor permeability of polar polymer was demonstrated.
- Like plastics, edible films with functional properties have been developed. For example, films based on modified chitosan and anthocyanin nanocomplexes capable of reducing the rate of oxidation of olive oil was developed.

26.8.4 In-packaging sensors for real-time response

Key to the development of packaging materials that extend the shelf-life of products is the development of instruments or bioindicators to monitor shelf-life. Section 26.10 further describes sensing technologies covered in this book and during ICEF13 under a supply chain digitalization Industry 4.0 framework. In packaging sensing technologies presented at ICEF13 included:

- A respirometer to detect the senescence of packaged produce consisting of a wireless technology that enabled in-package, real-time monitoring of produce respiration rates.
- A low-cost polydiacetylene film-based, colorimetric, ammonia sensor that can be incorporated to food packaging to indicate meat (and other food) spoilage at low temperatures between frozen and room temperature.
- Highly flexible pH-sensitive film sensor, with electrical conductivity linearly related to pH, to be embedded in flexible packaging materials to monitor food spoilage.

26.8.5 Other aspects of packaging

Like in other fields of engineering, modeling is a powerful tool for packaging development. A genetic algorithm was used to optimize pallet load, hand-hole location, and shape of boxes. Also, packaging affects the rate of cooling in storage. A model that uses computational fluid dynamics was presented and compared to experimental measurements. In addition, a compression test including a depth camera was developed to evaluate compressive strength of corrugated fiberboard packaging.

26.9 In vitro and in vivo digestive systems

Food engineers have extended the scope of their research to bridge the fields of food formulation and manufacture, preservation, storage, and sensory into model digestion systems that mimic all stages from oral processing, to delivery of macromolecular nutrients. This extension aligns with the increasing demand for personalized foods with target oral processing features, digestibility, bioaccessibility, and bioavailability to improve or sustain health. Therefore, food engineers must play a critical role in the public interest of our global society. A clear understanding of transport phenomena, thermodynamics, and chemical and biochemical kinetics are needed to design processes and food matrices that deliver the desired nutrition, body function, and health. The nutraceutical boom that started about two decades ago has produced countless studies on bioavailability, bioaccessibility, and interactions of a myriad of food compounds establishing a platform for the study of digestion models.

ICEF13 presentations covered engineering aspects of processing for improved body function and delivery, oral processing, digestion and absorption of food structures and compounds, and highlighted the progress achieved in developing in vitro models that adequately represent in vivo conditions. The latter aims at addressing the technical, ethical, and financial challenges that mice, rats, pigs, monkeys, and human clinical trials present, by finding quick, repeatable, and low-cost models that can represent the human body machinery. Given the extensive number of abstract submissions in this topic, the *Journal of Food Engineering* dedicated an ICEF13 special issue following selected presentations (Bhandari et al., 2021). Presentations from Australia, Belgium, Canada, Chile, China, Denmark, France, Greece, India, Israel, Japan, France, New Zealand, South Korea, South Africa, Spain, Switzerland, The Netherlands, United Kingdom, and United States covered these topics.

26.9.1 Biological and human-driven processing for improved in vivo processing

The biological processes occurring when growing food on farm, postharvest, and industrial operations, or home cooking, create specific multiscale structures that need to be further processed in vivo to deliver targeted molecular functions in the human body. In addition to topics covered in Sections 26.6.5 and 26.7 dealing with structuring, preservation, and incorporation of bioactives, probiotics, and synbiotics, ICEF13 brought other presentations covering processing aspects to improve body function:

- β-Galactosidase-loaded whey protein isolates microparticles were designed to deliver β-galactosidase in the duodenum, an enzyme lacking in lactose intolerant people, and demonstrated gastric resistance and intestinal release, to improve digestion of dairy products.
- Soaking of beans beyond 8 h and cooking with potassium carbonate avoided losses of key micronutrients including calcium, iron, zinc, and total phenolics, by reducing cooking time by 1.2 h. This technique partially addresses the 30–60% bean micronutrient losses seen during bean production, stabilization, and storage.

26.9.2 Oral processing impacts on bolus formation and digestion

Oral processing is governed by the individual's eating methods, in-mouth chewing, mixing, and grinding, saliva dosage, and their interactions with in-mouth processing variables, including portion size per mouthful, and water present. These biomechanical and other fluid dynamic factors at the mouth stage, influence gastric and intestinal digestion outcomes such as the glycemic index response. The structural and compositional changes during mastication can use three-dimensional mathematical models of the mouth, in combination with bolus dynamics, to design food formulations, shapes, and structures. Several oral processing researches were presented at ICEF13:

- A hierarchical framework for mastication model development was proposed to predict particle size change, bolus saturation, and taste and flavor release during chewing in real food systems. Aspects of mixing and selection of food for occlusion, softening the structure through mechanical damage, particle size reduction and/or pasting, saliva uptake and hydrolysis, caused by changing areas and volume ratios between food and saliva and air, were discussed. Such models can predict the influence of these variables on aroma release and perception and sensory acceptability and palatability, and may combine the interactions between nasal cavity, oral cavity, and pharynx.
- Mechanistic selection and breakage function models used to model industrial particle size reduction were applied to predict particle size distribution during human mastication. A fast method relating the total occlusal area of the teeth to the projected area of food particles of various sizes, as measured with a fast-optical method, was validated with human subjects using chewing gum.
- A chewing simulator controlling the flow rate of artificial salivary fluid was used to test the oral processability of extruded pea flour snacks of various densities and protein solubility. The study related bolus viscosity, saliva uptake, and particle size reduction as a function of chewing process time and protein solubility (Kristiawan et al., 2021).
- Chewing of rice was modeled by predicting particle size, pasted starch, and saliva content of the bolus as a function of chew number. The model allowed evaluating the influence of portion size per mouthful, initial water content, and salivary flowrate.
- To tackle the dysphagia in aging populations, an in vivo study characterized the chewing of meat cooked to different textures by varying the number of chews and characterizing boluses for saliva

impregnation and particle size distribution, showing that increasing chewing cycles increased particle size breakdown with more saliva being required, with a group of the human subjects showing greater swallowing difficulty.

- Another study followed textural changes during chewing of beef bolus samples, which were characterized with a compression test and a slip extrusion test.
- In vivo texture mastication demonstrated that PEF treated blanched carrots did not experience texture softening and that β-carotene bioaccessibility of treated carrots was not compromised by PEF treatment, based on assessments of the masticated carrot bolus in an in vitro gastrointestinal digestor.
- Microstructural changes of sponge-cake and brioche upon compression, mimicking chewing, were modeled using 3D finite element modeling. The model was validated with 3D ultra-fast X-ray microtomography that monitored the changes of the porous cellular structure under compression.
- Bread and starch disintegration during oral processing were higher with higher bread porosity. The bread structure (baked bread, steamed bread, and baguette) influenced starch hydrolysis in an in vitro intestine model. However, physical cutting, cut-and-pestle, blending, and grinding of bread did not affect the extent of starch hydrolysis (Gao et al., 2021).

26.9.3 Artificial stomachs and in vitro gastric digestion

Presentations on the design and in vivo validation of "near human" artificial stomachs and in vitro gastric digestion models were applied to determine food materials buffering capacity, gastric breakdown, mixing requirements, and to establish gastric digestion systems for particular population segments:

- A J-shaped 3D printed dynamic in vitro human stomach developed with a similar stomach morphology, dimension, and wrinkled inner structure to the in vivo stomach. The stomach is composed of a series of motors, rollers, eccentric wheels to produce peristaltic contractions, and mimicked well in vivo stomach emptying with model beef stew mixed with orange juice and cooked rice.
- Buffering capacity of heat-induced egg and whey protein gels in the stomach was modeled by changing protein concentration and surface area format (liquid dispersion, pureed gel, and gel cubes) and adding HCl to the samples to reach a pH of 1.5. Eggs have higher buffering capacity than whey, and high protein and larger surface area have higher buffering capacity, therefore requiring higher amounts of HCl to reach the target pH.
- The link between breakdown, gastric pH, and mixing was evaluated in a human gastric simulator, a dynamic gastric model that provided the pH profile at 13 intragastric locations, and particle size distribution was measured by image analysis after set digestions times. Mixing requirements on blanched and fried sweet potatoes demonstrated mixing requirements to achieve more homogenous pH distribution and particle breakdown.
- In vitro gastric models of adult and the elderly have shown that gastric digestion of caseins in the elderly model occurred at a lower rate than in adults. The gastric models showed that in general UHT treated beta-lactoglobulin was better digested than pasteurized beta-lactoglobulin (Aalaei et al., 2021).
- In vitro gastric proteolysis (pepsinolysis) and pH-induced protein aggregation of dairy proteins demonstrated that beta-lactoglobulin and alpha-lactalbumin generate more bioactive peptides

under adult conditions contrary to lactoferrin, which is a better source for bioactive peptides for infants and the elderly.
- In vitro digestion in a toddler model and proteomic analysis showed that carrageenan addition in a whey protein isolate drink affected levels of essential amino acids in the digestome as carrageenan modulated digestive proteolysis.
- Similar work on silk moth flour using digestomic analyses showed that digestion released antimicrobial peptides in adults but to a lesser extent in elderly adults.

26.9.4 In vitro gastric and intestinal digestions

Dynamic in vitro gastric and intestinal digestion models were showcased as integrated tools to study antibiotic degradation and release in the body, reducing glucose concentration and bioaccessibility, selecting formulations to decrease gastric digestion time and emptying, assessing digestibility of new protein and fiber sources, measuring effects on polyphenol bioaccessibility, and bacterial survivability in the digestive track, and assess the effect of processing in including HPP, thermal processing, and fermentation.

26.9.4.1 Release of antibiotic compounds

- In vitro gastric and intestinal digestion study suggested that antibiotic compounds in foods may be released into the body during digestion. Plant phytometabolites derived from antibiotic applications were shown to survive gastric and intestinal digestion, with 70% more antibiotic degradation products in the intestinal digestion liquid.

26.9.4.2 In vitro starch digestion and glycemic index reduction

- Digestion of saliva treated potato starch with agar or gellan gum hydrogel particles in a continuous gastric digestion simulator demonstrated that agar hydrogel was disintegrated and emptied more largely. The subsequent digesta into small-intestine digestion model demonstrated 90% intestinal digestion of starch.
- Bread made with whole wheat flour of coarse and medium particle sizes showed higher in vitro glucose release rate and lower final glucose concentration compared to bread made from fine whole wheat flour and white flour, although a larger particle size provides a more compact bread structure and harder texture.
- Starch digestion in bread and white rice formulated with fiber hydrocolloids in in vitro gastric systems showed slower bolus disintegration, compared to those without fiber, and hydrocolloid structuring of the gastric digesta resulted in reduced glucose bioaccessibility.

26.9.4.3 Processing evaluation

- In vitro digestion of common beans sized reduced and thermally processed to produced intact cells encapsulating starch had slower digestion rates with higher amount of remaining protein.
- In vitro gastrointestinal digestion of raw, cooked, and fermented soybean showed that fermented soybeans had better digestibility, availability of essential amino acids, and antioxidant activity than raw and cooked soybean and may benefit human health.

- In vitro gastric and intestinal digestion studies of unprocessed, high-pressure processed, and pasteurized strawberry extract, demonstrated that while thermal decreased polyphenol bioaccessibility especially in intestinal conditions, high-pressure processing mostly preserved the original polyphenols bioaccessibility (Eran Nagar et al., 2021).
- In vitro salivary, gastric, and small-intestine digestion of spray-dried and spray–freeze-dried encapsulated synbiotic *L. plantarum* powder with fructo-oligosaccharide, whey protein and maltodextrin, showed that spray-freeze drying provided better probiotic survivability during digestion (Yoha et al., 2020).

26.9.4.4 Evaluation of new material sources

- Gastrointestinal models predict the behavior of promising novel ingredients such as nanocellulose as a fiber source or delivery vehicles for drugs or nutraceuticals (Liu and Kong, 2021). Ex vivo approaches using animal intestinal sections and in vitro assays were used to study the flow behavior and adhesion of nanocellulose capsules on mucosal surfaces and showed that three types of tested nanocellulose exerted hypoglycemic potential, delayed lipid digestion, and free amino nitrogen absorption at high nanocellulose concentration.
- A blue–green microalga (cyanobacteria, *Arthrospira maxima*) presented as a potential food ingredient with 53% protein, 4.5% leucine content, and omega-6, exhibited more than 95% digestibility after subjecting it through an in vitro dynamic gastric and duodenal model.

26.9.4.5 Oil emulsion digestion

- In vitro lipolysis of 5% oil in water emulsions stabilized with an emulsifier or other agents showed that lipolysis was faster in small droplets. Larger droplets caused unstable emulsions in the small intestine resulting in slower or incomplete lipolysis; linseed oil, rich in polyunsaturated fatty acids digested slower than olive oil, rich in monounsaturated fatty acids; pectin emulsifier provided slower or incomplete lipolysis than tween or sucrose esters; a validated multiresponse kinetic model of in vitro digestion of emulsified lipids follows triglyceride hydrolysis and free fatty acid release, and can be used to predict lipolysis kinetics.

26.9.5 Intestinal digestion, prototype, and modeling

Other work solely focused on intestinal digestion through in vivo assessment, development of an engineered intestine model, and a Multiphysics predictive model:

- An in vivo assessment showed that porridge prepared from oat flakes gave a lower glucose response to porridge prepared with more processed oat flower.
- An engineered small intestine was developed and predicted the human glycemic concentration curve against in vivo data and is suggested as an effective tool to study glycemic responses (Priyadarshini et al., 2021).
- A Multiphysics model numerically predicted glucose absorption rate during start or maltodextrin digestion in a human small intestine simulated by a two-dimensional axisymmetric fluid flow induced by peristaltic waves (Karthikeyan et al., 2021).

Even though these in vitro systems are quite different from in vivo conditions, they provide very valuable insights on food product development and the potential impact of food ingredients in human digestion. Digestion models remain a huge challenge for food engineers. On the one hand, it is very difficult to make fair comparisons among research groups because standard methods only exist for simple models that do not reflect the biological complexities of, for example, the gut microbiota and active transport. On the other hand, we have an incomplete understanding of the impact of the myriad of compounds up taken during digestion and their interactions. Not only age, but cultural and psychological factors affect digestion. This is not a critique but a statement of admiration of the courageous and immensely creative minds that have undertaken the task of assembling this enormous puzzle and from whom we hope to learn more at ICEF14. Digestion models are among the emerging topics that are become now part of some university curricula as discussed in Section 26.11.

26.10 Industry 4.0 and sensor technologies to develop integrated food chain cyber-physical systems

Food production and consumption undergo enormous biological- and human-driven variations that challenge the efficient integration and integration of food supply chains. Lack of suitable and reliable real-time automated control measures, and the intrinsic digital sensing systems enabling consistent data acquisition at various levels of the chain, challenge its integration and the efficient use of resources during food production and related operations (Smetana et al., 2021). The following subsections provide a snapshot of aspects of Industry 4.0 integration across the supply chain together with advances in real-time sensing technologies, covered in this book, in ICEF13 special issues, and during ICEF13.

26.10.1 Industry 4.0 for digital integration and real-time supply chain response

Chapter 20 introduces the concept of digitization of the agri-food ecosystem to support the bioeconomy and the application of Industry 4.0 including digital technologies to food manufacturing. Specific applications of internet of things (IoT) for sorting and inspection, food safety systems, machine learning, robotics, data visualization and analysis, and decision making with integrated IoT systems are included. The chapter also identifies the barriers to adoption by the agri-food industry and some of the potential benefits and impacts. Chapter 21 delves further into the Industry 4.0 topic by mapping the key features of high performing food production chains and enabling Industry 4.0 technologies. The Industry 4.0 technology enablers listed in the chapter are predicted to increase the performance of customer-oriented food chains and their specific food chain characteristics. Implementation of machine learning algorithms for the collection, integration, and analysis of data associated with biomass production and processing on different levels from molecular to planetary, leads to the precise analysis of food systems and estimation of upscaling benefits, as well as possible negative rebound effects associated with societal attitude. Moreover, such data-integrated assessment systems allow transparency of chains, integration of nutritional and environmental properties, and construction of personalized nutrition technologies.

Another article emanating from the congress (Smetana et al., 2021), introduces the concept of CPSs, well established in automotive and the pharmaceutical industries, but needing definition and implementation in the food industry. CPSs are generally defined as intelligent autonomous systems going through learning human behaviors with the help of artificial intelligence, together with the development of extensive and advanced physical architecture-based information systems (physical world, transducers, control components, robotics, data analytics elements, computation elements, and communication components). The article recognizes the need to bridge CPS developed for precision agriculture into agri-food supply chain networks, including traceability, and "virtualization" of food supply chains by including food processing through a number of "system levels." A few examples that fit underneath each level include, wireless networks for bread manufacturing optimization, drying control and optimization systems, digital twins for remote factory operations, robotics for packaging, picking, placing pelletizing, inspection testing, catering with intelligent appliances like kitchens and stoves, collaborative robots, and intelligent packaging.

During ICEF13 others covered on different topics with presentations from Australia, Germany, Japan, New Zealand, and the United Kingdom:

- An overview of disruptive digital technologies including drones, sensors, robots, artificial intelligence, virtual reality, augmented reality, and block chain underpinning the IoT. In-factory examples were brought on digitally automated pumps, temperature control, conductivity, and pH sensors with data logging and cloud computing, predictive maintenance platforms and machine learning engines for adaptive cleaning and fouling prevention. Other health analytic data platforms using human genotype and phenotype data were mentioned.
- One of the world's largest dairy exporters showcased their broad scope in modernizing dairies from traditional automation systems to rapidly evolving areas such as IoT, operation technology cyber security, predictive analytics, and edge computing systems.
- An Industry 4.0 solutions supplier provided examples on digitalization in food and beverage, as a way of improving productivity, reducing cost pressures, dealing with changing consumer demand, and the use of a digital twin in brewing.
- A major beverage manufacturer brought examples of (a) a new line with robots run by "process technicians" (rather than operators) for automation and paperless production with 85% improved efficiencies, (b) predictive monitoring of equipment performance with vibration sensors, and (c) real-time analysis and downtime monitoring. They are moving from prevention activities to prescriptive maintenance by incorporating big data and machine learning into their data systems networks.
- Introduction to blockchain to add value by trust, process optimization through value addition by design digital tools, CIP algorithms with sensors at cleaning point, data-driven consumer understanding and use of virtual reality to understand consumers.
- The meat industry is now optimizing livestock genetics from processing feedback, systems implemented for meat yield prediction per carcass, automated robotic cutting lines for precise clean cutting, advanced grading with 3D X-ray CT tomography. They also brought new technologies such as exo-suits to follow operator's movements, augmented vision, collaborative robots (or cobots) working with humans, and artificial intelligence algorithms.
- Companies like Not Co. use artificial intelligence to formulate ingredient combinations of dairy analogues or Just Inc. use high tech robotics for high throughput assays and chemical data, and machine learning to formulate egg analogues.

26.10.2 Sensors for supply chain digitalization

A feature of digitalization and data generation across the supply chain is the ability to implement adequate sensing technologies. Chapter 22 introduces micro- and nanosensors (optical, chemical, electrochemical, biological) applicable for quality, safety, and traceability of food from farm to fork. It presents the nanomaterials required and potential applications across the supply chain including crop cultivation, food processing, packaging, and traceability of food, including applications in detection of pesticides, pathogenic bacteria, food additives, dyes, sweeteners, and food packaging. Chapter 23 complements Chapter 22, by describing the various sensors currently manufactured for assessing food quality and safety and their mechanisms of detection (i.e., colorimetric or electrochemical), beyond time–temperature indicators. Applications described include detection of gases and volatile organic compounds, toxic, and antinutritional molecules, or unwanted microbial pathogens in-pack, many of which are commercialized, and others are at proof-of-concept stage. Besides the novel in-packaging flexible sensors described in Section 26.8, other examples presented in ICEF13 include presentations from Canada, Ireland, Sweden, and Turkey on process analytical technologies:

- Development of an industry scale prototype fluorescence optical sensor for real-time monitoring of rennet-induced milk coagulation kinetics and prediction of cheese cutting time.
- Rapid in-line detection of tryptophan in dairy products with fluorescence-based process analytical technology to discriminate between dairy ingredients with of various compositions, measure soluble protein and predicting heat treatment temperatures and soluble protein in dairy systems.
- A review of FTIR spectroscopy coupled with chemometrics applied to measure meat adulteration, spoilage, and shelf-life by measuring changes in proteins and lipids (Candoğan et al., 2021).
- The use of hyperspectral imaging to rapidly measure intramuscular fat in meat (Kucha et al., 2021).
- Strawberry freshness quality parameters (water content, firmness, sweetness, acidity, and sensory quality) and shelf-life using the reflective visible spectrum can be predicted with near-infrared spectroscopy chemometric models.

26.11 Re-engineering food engineering education to accommodate technological advances and societal challenges

Food engineering education like all other engineering and scientific disciplines is evolving at a very fast pace. Within the time constraints of Bachelors' degrees (4–5 years depending on the country), today academics face the challenge of incorporating additional topics and balancing course content with the conventional topics, or bud-off more specialized degrees or new areas of emphasis at the undergraduate and graduate levels. Approaching food engineering education from the optics of the food supply chain appears overwhelming. Yet ICEF13 gathered passionate educators and visionaries from Belgium, France, Germany, Greece, Ireland, Israel, Italy, Malta, New Zealand, Spain, United Kingdom, and United States, who clearly expressed their passion and optimism for an in-depth review of food engineering curricula.

Most efforts have focused on curriculum development to fill the demand for new skills required by the food industry. Also, big societal challenges such as ending hunger and addressing the environmental impact of the food industry were discussed. Furthermore, the COVID19 pandemic forced academics to focus on the methods of delivery with online content increasingly available. Some have taken advantage of this challenge as an opportunity to reflect on the pedagogical dimension of teaching.

The Society of Food Engineering carried out a survey with responses from 100 industry professionals, recognizing that the disciplinary broadening of the already multidisciplinary field is very much needed. New hires should have knowledge of food laws, supply chain logistics, and raw materials sources, and six sigma quality management. Other topics included hygienic design of equipment, allergen management and traceability, and food microbiology hazards. Exposing students to artificial intelligence, the IoT, Industry 4.0, and other technological tools that will contribute to a sustainable food supply chain is critical. Food engineering education can no longer ignore climate change, nutrition, and health.

Chapter 24 provides an overview of where food engineering education is now and suggests how food engineering degrees should be re-engineered for graduates to be able to effectively tackle strategic innovation challenges required in the manufacture of food in the 21st century. The chapter provides a new definition of food engineering aimed at bettering students' understanding of their future fit within a larger food systems scheme and across a wide range of food related sectors. A food engineering edifice was proposed with core enabling courses as a foundation and five key themes: (1) food safety quality and formulation, (2) food structural engineering and sensory analysis, (3) food product realization engineering, (4) transport processes in GI tract, metabolism, satiety, and health, and (5) environmental impact, food sustainability and security. The program suggests providing students with international exchange opportunities and exposure to experiential leadership and entrepreneurship training through industry placement and social science training.

The Erasmus program was described during ICEF13 and leads efforts in international exchange education, bridging many of the disciplines and innovations critical to the global food supply chain. For example, Erasmus also covered an international collaboration of Ireland, Belgium, France, Spain, Greece, and Malta that developed the Q-Safe graduate-level educational program that combined predictive modeling with life cycle and risk assessment. The creation of a repository (library) of educational materials in biosystems and food engineering (Biological Engineering Digital Library) has been created to facilitate the internationalization of curricula. This along with other increasing number of resources that are being made available online (some free of charge) will leverage new ideas and contribute to some level of internationalization. However, harmonizing curricula from different countries is quite challenging as local standards and priorities often dictate the curricula.

Another multicountry education initiative is the European Institute for Innovation & Technology Food Solutions programs for *experience-based learning* that promotes extracurricular project activities for students to receive industry exposure where they co-create innovative circular bioeconomy solutions, while receiving mentoring from professionals in food production and retail (Chapter 25). These programs aim at reducing food waste, increasing sustainability, and training students in the transfer of technology. One-year long projects through industry-academy partnerships allowed students to compete to solve "real-world" problems integrating technical, managerial, and business skills. The program also provides entrepreneurial and innovation training and promotes the commercialization of ideas. Chapter 25 brings successful competitive program examples delivered in recent years.

Chapter 24 also includes a critical assessment of the available resources, including online courses and content. Another ICEF13 presentation describes a senior course proposed at UC Davis to remedy the disconnect between food engineering and digestive systems, which takes advantage of new knowledge in the field gained by several groups working in digestion models. With an "open source" philosophy, materials for this course are available for free.

With the current technology, learner-centered, active learning approaches are more accessible and are easier to implement. It is easier to flip the class, to create virtual laboratories, and to make simulations that we could only dream of 30 years ago. However, academics also need to relearn how the youth learns today. The z-generation is very different from previous generations, not only in the way they learn but also in the source of their motivation. Academics can no longer organized to be dispensers of knowledge.

Indeed, students in the developed world have access to more information than they can properly interpret to make sound technical decisions. Therefore, it has become more important than ever, to devote teaching efforts to train students to solve industrial problems through active learning. An excellent example of the new tools available for enhanced learning was the simulation-based modules on food safety for food engineers that incorporate among other aspects, risk assessment, predictive microbiology, and food processing. Another great effort that will serve not only to students but the industry in general is the creation of an interactive database with food data properties and predictive equations for such properties that is being built as an international crowdsourcing effort. Furthermore, simulations were tested in summer experiences with students from 24 different countries. There is an excellent opportunity to offer these simulation tools through the internet.

However, education innovation is needed not only in the fast-paced developed world. Education is at the core and arguably the most powerful tool that developing countries need. Small entrepreneurs have been trained in West Africa using containerized training processing plants where they learn the technical and business aspects of a small food processing plant, which can have a huge impact in reducing food waste. Ironically, a gap in our ability to support the education in developing countries is that there are very few programs that train food engineers to address humanitarian issues. To address this, a Franco-Belgian collaboration initiated the development of an education platform in humanitarian food engineering, involving approaches, methods, and processes to plan, design, and implement the provision of food and drinking water in response to a disaster. That platform is based on case studies and offers an excellent opportunity to educators to become aware of this challenge and become involved. Beyond the traditional university framework, an excellent presentation on the development of courses for industry personnel was presented which proposed virtual tools and many online opportunities for self-paced learning.

Finally, the food engineering educator yet faces challenges in selecting content and in dedicating time to delivering high quality teaching while sacrificing publication output.

1. Selecting food engineering content: More new knowledge is being developed than obsolete knowledge discarded. If one compares the 1944 Hougen and Watson's Chemical Process Principles (Hougen and Watson, 1947) to any of the current food engineering textbooks, one can see the vast overlap in the "principles" and how little time is left to teach any "advances." As mentioned in the description of food engineering programs around the world in Chapter 24, standalone undergraduate degrees are practically nonexisting in the United States and Northern Europe. The undergraduate food engineering curriculum varies from country to country and often runs parallel to chemical engineering. Anecdotally, at least in the United States, it appears that industry prefers to hire chemical engineers that know something about food than food engineers. Therefore, there is still room in the United States for the creation of undergraduate food engineering programs but there appears not to be the critical mass in Colleges of Engineering to pursue that, precisely because chemical engineers appear to fill industrial needs. In contrast, graduate programs in food engineering continue to grow because they lead the engineering developments in the area.

2. High quality teaching: The growing "publish or perish" paradigm in academia has led even engineers to spend less time learning how to teach effectively, preparing to teach, and becoming aware of practical industrial needs. Also, technology and attitudes toward learning are changing the teaching paradigm. Good teachers are more respected by their students than by their peers and university administrators. This paradigm has existed for decades and appears to be exacerbated as time passes. There is little to no incentive to change that trend in many countries. This might as well explain why there is a lack of critical mass to create undergraduate food engineering programs in many countries. This is not always true in some institutions where teaching remains highly valued at all levels. To leverage teaching efforts, various individuals and organizations have created open access teaching resources that include content and teaching methods. These resources are leveraged by institutional teaching and learning centers and will continue to help both, beginning and well-seasoned food engineering instructors.

26.12 Concluding remarks

This chapter demonstrates the value, display of creativity, and diversity of thought that the ICEF13 congress provided within the expanding field of food engineering to address future food chain challenges. It became evident across the program that a wider and more integrated biosystems engineering vision for problem-solving needs to be further refined and developed as various systems, tools, and schemes are being developed and deployed throughout the food supply chain. Food engineering needs to partake in solutions that suit the global food needs and challenges of the planet as we navigate the anthropocene through the next centuries. Food production, manufacturing, and consumption is posing a significant risk to break our planetary boundaries impacting on climate change, freshwater use, ozone depletion, ocean acidification, loss of biodiversity, and introduction of novel entities such as chemicals and plastics leading to ecotoxicity (Lade et al., 2020). The systematic solutions found in one region of the planet can be extrapolated to other regions facing similar challenges. Engineers need constantly updated education, training, and leadership empowerment to have a strong voice in the development of government policies that promote regional technological benchmarking strategies to promote extension between regions and nations at global scale. Speakers from key food multinationals have shown that their companies play an important role in supporting technological deployment into the regions through socially and environmentally responsible strategies.

Food chain data are currently being mapped geographically and dynamically, to different extents, in both developed and developing regions, to bring to light the food availability, access, utilization, safety, and food waste challenges being faced across regional and cross-regional chains. Such geographical and logistic maps require pools of dynamic data, including the time-dependent variability of food quality and through-chain volume flow of production, utilization, consumption, and by-product streams being generated across the chain. Data, managed with Industry 4.0 sensing and analytic frameworks, will enable accurate stochastic resiliency modeling and blockchain technology to directly identify bottlenecks requiring engineering solutions to ensure access to safe, sustainably produced food for everyone. The climate change-driven fluctuations in the regions including drought, floods, cyclones, pandemics such as COVID-19, or the positive impact of human interventions to mitigate climate change, will better allow addressing the food availability, access, utilization, and stability of food challenges. For example, improvement in the decarbonization of the food chain by developing local manufacture for decentralized food processing was mentioned. Coupling these solutions with strategies for decarbonization of thermal processing technologies operating with renewable energy will provide synergistic solutions to carbon emissions. New standards and methodologies for lifecycle assessment of foods require the measurement of wider planetary parameters beyond carbon emissions, such as eutrophication, water scarcity, and impact on biodiversity.

The importance of humanitarian food systems and engineering to define fit for purpose solutions to address the food challenges that 240 million people being displaced through crisis was highlighted. Creative solutions including the installation of modular containerized food processing facilities for long-term capacity building, the personalized development of ready-to-use supplementary foods, and the food safety engineering interventions leading to avoid food incidents were highlighted. Such concepts will also serve and be translatable to the context of long-duration space missions and ensure the food system meets crew health requirements on spaceflight vehicles and upon landing in new exoplanets.

As consumers shift to more sustainable food with lesser impact on the environment, engineers will have an important role in developing and implementing the technologies that transform new food sources to meet the protein gap before 2050. Ingredients from pulses and seeds, nuts and roots, fruits, and vegetables, micro and macroalgae, larvae and insects will become increasingly transformed, value added, and used as environmentally friendly food sources for health impact. For example, companies showcased their macroalgal and insect developments, and others the opportunities for low-cost photobioreactors and heterotrophic cultivation of algae, which could be enhanced with pulsed electric field treatments.

Several strategies were suggested to address the global 1.3B tons food waste problem through reduction and transformation of food by-products. Some of these included circular economy solutions such as feeding retail waste to insects or into microalgal reactors. Preservation technologies play an important role in reducing wastage across the supply chain and novel and combined technologies such as cold plasma, HPP, active packaging, and magnetic field treatment were mentioned. In packaging thermal or nonthermal processing will play a role in reducing food waste generation at retail and consumer levels when food spoilage is the limiting factor for shelf-life. Another driver for circular economy presented was the upcycling of by-products to higher value products and a long list of examples from various sources using engineering approaches for traditional and green extraction, hydrolysis, drying, or structuring were presented. Decentralized food processing following regional hub and spokes models, as well as on-farm mobile processing, will play a role in tackling the food loss issue before retail. Postharvest engineering also includes the advances in refrigeration and freezing, with systems that are more energy efficient and can operate with renewable sources, by decreasing shares in energy costs during transportation. Energy optimized refrigeration systems were discussed on-farm, during seafreight and in households. Novel technologies ensuring minimum ice crystal formation during freezing, or rapid tempering, were also discussed by using electromagnetic energy and by using advanced X-ray for online characterization.

Food stabilization through thermal and nonthermal processing will continue to be at the heart of food engineering. Thermal processing parameters will include online dynamic temperature profiles, as instrumentation and data integration into Industry 4.0 platforms will enable increased accuracies, and CFD modeling will enable improved designs for industrial uptake. High-pressure thermal processing will become realized via a polymeric canister design to produce high quality ready to eat products with extended shelf-life. Industrial applications of MWs for continuous pasteurization (MAPS), sterilization

(MATS) and enzymatic inactivation will continue to drive the chilled and shelf stable ready-to-eat meals production. RF energy will continue to address industrial bulk heating challenges in thawing and tempering applications, and for disinfestation of agriculture products and food pasteurization (e.g., seeds, flours, powders, nuts, fruit, vegetables, in-pouch applications). The accelerated industrial deployment seen in HPP and PEF systems for nonthermal pasteurization provides greater confidence for cold plasma gas or functionalized water to accelerate its scale-up for industrial uptake for microbial decontamination, pest control, toxin elimination, food and package functionalization, among other applications.

Drying technologies will continue to evolve to improve the affordability, quality, and value of dried foods and reduce high energy consumption. Dryers will be adapted to operate with renewable energies to enable cost reductions and minimize environmental impact. Technologies will reduce drying time by (a) developing new pretreatment interventions or technologies, (b) in situ drying optimization through on-line sensing and advanced process control or technology improvements, or (c) introducing new batch drying technologies alternative to freeze drying, potentially with mobile or modular options. Dried product value will also be increased by incorporating or improving bioactives with health properties and functionalities. Food engineers will need to provide greater focus on developing low-cost high-tech solar, drum, vacuum, or heat pump, perhaps osmotic dehydration or specialized belt drying solutions, for decentralized drying in villages in developing countries. Spray drying operations with electrostatic nozzles will be greatly utilized to dry heat sensitive probiotics or emulsions, other similar turbine driven, or combustion technologies will be applied to dry highly viscous materials, and extrusion aided spray drying will enhance porosification and solubility of high feed solids.

While freeze-dried foods continue to grow, costs of drying will be decreased via porosification using PEF or CO_2 laser microperforations, by indirect ultrasound transmission, or by adopting MW-assisted systems with low-temperature drying. Membranes will increasingly play an important role for concentration and bioactive separation applications, and low-cost techniques such as hollow fiber membranes and forward osmosis will be adopted. Pretreatments with PEF will be applied to reduce fouling or combinations with membranes to increase peptides application. Increasingly, used proteins from algal, insect, pulses, or synthetic biology sources will require the removal of off flavors and aromas with low-cost rotary disk columns technologies to manufacture ingredients for alternative meat, dairy, or egg analogue applications.

The continued growth of the total wellness and medicinal food sector will drive the creation of more specialty stable bioactive probiotic ingredients, with masked flavors for a wider application of food formulations, providing gastric protection from acidic compounds, and controlled delivery and bioavailability. Traditional dairy-based encapsulation methods will continue to enable developing new encapsulants to protect water soluble bioactives or probiotics, while plant-based microencapsulation for synbiotic powder production will become more popular by using spray drying or other novel drying and microstructuring techniques for enhanced bulk properties. Development of fortified medicinal foods will help address nutritional deficiencies in the world population, where currently 1.62 billion people, particularly young children and pregnant women are suffering from anemia (World Health Organization, 2008).

The engineering behind personalized food systems will follow the pace of growing and ageing populations and urbanization, to meet demands for resource friendly eating, targeted indulgence, eating for self-optimization and anxiety management, and fast refill foods. Food structuring methods will allow recombining bioactives or synbiotics with plant- and cell-based ingredients through extrusion and 3D printing at various scales. Extrusion will continue to play a role in creating textured vegetable proteins or protein fibration to develop high quality meat analogues as the process becomes more controlled through a more mechanistic understanding of process parameters and rheological properties. 3D printing will move from commercial uses for meat, fish, analogues, or chocolate applications into other egg or dairy analogue applications as more suitable material systems, accurate high precision printing and better software allows for scalability and controlled sensory delivery. Commercial PEF equipment will continue finding opportunities for food structuring in vegetable-based foods and meat products. Precision fermentation or bioprocessing reactors will receive greater applications for either nutritionally enriching or improving bioactivity or biodelivery of plant-based materials. Precision fermentation through synthetic biology will emerge for industrially tailored animal biomolecule or cell-based meat production. Personalized food systems will contribute to reduce a concomitant population of 450 million people suffering from diabetes, including 1.5 million yearly deaths (Diabetes, n.d.), due to the high sugar-driven dietary imbalances. It is expected that personalized nutrition will decrease risks of cardiovascular diseases, which currently add up to 18 million deaths annually (Cardiovascular diseases, n.d.).

Packaging innovations will need to react to future legislation that is likely to phase out of LDPE and oxo-biodegradable plastics due to their nondegradable nature and associated toxicity. Biodegradable, bio-based, and compostable plastics in both soil and food packaging applications will start to appear as alternatives for compliance. We are witnessing the growth startup companies coming with creative biodegradable alternatives, including wrap films derived from potato waste (Great wrap, n.d.; Arıkan and Bilgen, 2019) for triple bottom line solutions including ecofriendly packaging, food waste transformation and reduced emissions. The role of food engineering in the development of food packaging has been the subject of multiple discussions. The further disciplinary extension of an already very broad field appears unwise to some and vital to others. Regardless of how the profession continues to evolve on this front, there is still room to further tailor the barrier properties and the ability of polymers to withstand different processing and environmental conditions. Active and smart packaging for the controlled release of existing and novel antimicrobials will continue to grow. As mentioned on Chapter 17, judicious chemists are needed to develop effective antimicrobial generation chemistries.

More advanced mouth, neck, and below the neck in vitro apparatuses that reproduce human anatomy hardware and software operations will continue to better mimic and correlate with in vivo food processing. Advanced in vitro food processing equipment will provide a faster, more ethical, and low-cost assessments of the fluid dynamics of chewing, swallowing, digestion, and bioactive delivery. In vitro processing provides a huge potential to accelerate studies that improve body function and address personalized nutrition needs. Mouth, gastric, and intestinal models will increasingly incorporate inbuilt sensors to develop system automation and will be correlated with biosensors developed for in vivo tracking in the human body and external assessment. This area is expanding beyond biomimicking human systems to different animals, due to their relevance in our food chain and potential to improve the understanding on their systems to improve our systems. Nevertheless, efforts need to be made to utilize these tools in the context of the big humanity challenges brought by malnutrition and obesity. While the mouth, the gut, and organ functions like the heart, and blood health have been considered in the equation, little has been done on engineering for brain health. The mental health issue has become an increasing pandemic, with 792 million people in 2018 living with a mental health disorder (Ritchie et al., 2018), and compounds that deliver better mental health (e.g., cannabinoids and gamma-amino butyric acid compounds) are becoming increasingly relevant.

Supply chain system integration efficiencies, cost pressures, food authenticity and traceability, and environmental outcomes will be improved by enabling Industry 4.0 technologies across food chains. Equipment and system integration across the chain will require real-time automated control measures and intrinsic digital systems for data acquisition. Sorting and inspection, food safety, predictive analytics, machine learning, data visualization and analysis, and decision-making systems will become an integral part of food production, manufacturing, distribution, and consumption. Sensing drones and other sensors, robots and cobots, artificial intelligence, virtual reality, augmented reality with digital twin factories, and block chain platforms will become part of engineers' everyday toolkits as they collect, transport, sort, sanitize, convey, pump, thermally, nonthermally, or bioprocess, separate, fractionate, dry, or package foods or for factory maintenance and CIP. Digital twins for optimization of process parameters based on chemical and sensory fingerprints will pivot opportunities for personalized manufacturing by connecting with health analytic data platforms using human genotype. Perhaps one day humans will be bioconnected to the supply chain with internal sensing devices.

Digital micro- and nanosensors for rapid detection and analysis of compounds in crop cultivation, food processing, packaging, and traceability of food, including applications in detection of pesticides, pathogenic or spoilage microorganisms, enzymes, food additives, dyes, sweeteners, toxins, and antinutritional factors, and food packaging will become essential. Other visualization methods such as X-ray technology will provide efficiencies in various industries. Multidisciplinary collaborations will be needed for the integration of wireless sensors capable of assessing the quality and safety of packaging materials throughout the entire supply chain into the IoT. Even before that is achieved, there is a great need for the development of time–temperature integrators that accurately and in real-time inform the consumer of the shelf-life of perishable foods. These sensors need to be compatible with both the food and the packaging material. On the waste reduction front, there are many opportunities for in-package re-processing of foods to reduce waste. While in-package processing of foods is commonplace for shelf-stable products, there is an opportunity to develop packaging materials and economically viable processes that can be used for re-processing of foods near their expiration date.

Food engineering education is being re-engineered by academics with feedback from industry participants to update the curricula by adding new coursework and experiences that prepare the future engineering graduate to meet the requirements of the fast-changing food sector. Disciplinary broadening will be required by integrating product and process engineering with the broader industrial context of food law, environmental management, and supply chain logistics and the provenance requirements for more diverse and sustainable raw material sources. Future engineers will be savvy on digitalization and integration of Industry 4.0 systems across the supply chain. They will not only be environmental experts and advocates, but also have deeper understanding on the transport processes occurring during oral processing, digestion, and delivery of food, as online based tools and exchange programs continue to create more content across countries. New experience-based learning programs including industry placements will not only provide the expected industrial preparedness but will also develop socially and environmentally conscious food entrepreneurs that bring innovative solutions to consumers in both developed and developing countries and across humanitarian contexts.

Great progress has been made in the integration and sophistication of the food supply chain. While some of its components have reached a mature stage where there is little room for further improvement, there are several emerging dimensions in which food engineers will play a pivotal role. The broader applications of artificial intelligence will eventually influence most, if not all the links of the food supply chain. It will also become a foundational component of food engineering education. Food engineers will continue to produce incremental developments that refine existing and emerging technologies and bring them to the market. Heightened social and ethical awareness will direct the efforts of some food engineers to address the unacceptable issues of hunger, inequality between developed and developing countries and related food cultural bigotry. Arguably this highest moral priority will be leveraged by the new generations of food engineers that are ecology and ethically conscious. This level of awareness and social commitment will permeate into educational programs as well. ICEF13 participants feasted on all the topics listed here. The authors and editors of this book are looking forward to seeing the gastronomic delicacies and the new technologies that will be unveiled at ICEF14 in Nantes, France.

References

Aalaei, K., Khakimov, B., De Gobba, C., Ahrné, L., 2021. Digestion patterns of proteins in pasteurized and ultra-high temperature milk using in vitro gastric models of adult and elderly. J. Food Eng. 292, 110305.

Al-Mawali, M., Al-Habsi, N., Rahman, M.S., 2021. Thermal characteristics and proton mobility of date-pits and their alkaline treated fibers. Food Eng. Rev. 13 (1), 236–246.

Alvino Granados, A.E., Mochizuki, T., Kawai, K., 2021. Effect of glass transition temperature range on the caking behavior of freeze-dried carbohydrate blend powders. Food Eng. Rev. 13 (1), 204–214.

Arıkan, E.B., Bilgen, H.D., 2019. Production of bioplastic from potato peel waste and investigation of its biodegradability. Int. Adv. Res. Eng. J. 3 (2), 93–97.

Arserim, E.H., Salvi, D., Fridman, G., Schaffner, D.W., Karwe, M.V., 2021. Microbial inactivation by non-equilibrium short-pulsed atmospheric pressure dielectric barrier discharge (cold plasma): numerical and experimental studies. Food Eng. Rev. 13 (1), 136–147.

Bandopadhyay, S., Martin-Closas, L., Pelacho, A.M., DeBruyn, J.M., 2018. Biodegradable plastic mulch films: impacts on soil microbial communities and ecosystem functions. Front. Microbiol. 9, 819.

Bhandari, B., Juliano, P., Knoerzer, K., Nguyen, M., Buckow, R., 2021. ICEF13 special issue on food engineering for nutrition and digestion. J. Food Eng. 308, 110668.

Bounie, D., Arcot, J., Cole, M., Egal, F., Juliano, P., Mejia, C., Rosa, D., Sellahewa, J., 2020. The role of food science and technology in humanitarian response. Trends Food Sci. Technol. 103, 367–375.

Buckow, R., Barbosa-Cánovas, G.V., Candoğan, K., Welti-Chanes, J., Roos, Y., 2021. Food engineering reviews special issue based on the 13th International Congress on Engineering and Food – (ICEF 13). Food Eng. Rev. 13 (1), 1–2.

Bushenell, C., Ignaszewski, E., Brody, P., Gaan, K., 2020. Plant-Based Strategies for Retail: An Overview of Leading Plant-Based Assortment, Merchandising, and Marketing Tactics at Top U.S. Retailers. The Good Food Institute. https://gfi.org/images/uploads/2020/09/Webinar_Plant-based-strategies-for-retail-1.pdf (accessed 11/12/2021).

Candoğan, K., Altuntas, E.G., İğci, N., 2021. Authentication and quality assessment of meat products by Fourier-transform infrared (FTIR) spectroscopy. Food Eng. Rev. 13 (1), 66–91.

Cardiovascular diseases. https://www.who.int/health-topics/cardiovascular-diseases/#tab=tab_1 (accessed 07/18/21).

de la Cruz Quiroz, R., Fagotti, F., Welti-Chanes, J., Torres, J.A., 2021. Food preservation performance of residential refrigerators: pasteurized milk and ground beef as animal food models. Food Eng. Rev. 13 (1), 104–114.

del Valle, J.M., Núñez, G.A., Díaz, J.F., Gelmi, C.A., 2021. Radial variations in axial velocity affect supercritical CO_2 extraction of lipids from pre-pressed oilseeds. Food Eng. Rev. 13 (1), 185–203.

Diabetes. https://www.who.int/news-room/fact-sheets/detail/diabetes (accessed 7/18/21).

Eran Nagar, E., Berenshtein, L., Hanuka Katz, I., Lesmes, U., Okun, Z., Shpigelman, A., 2021. The impact of chemical structure on polyphenol bioaccessibility, as a function of processing, cell wall material and pH: a model system. J. Food Eng. 289, 110304.

Food and Agriculture Organization of the United Nations, *Moving Forward on Food Loss and Waste Reduction*, Food and Agriculture Organization of the United Nations: Rome, 2019.

Food Loss and Food Waste. http://www.fao.org/food-loss-and-food-waste/flw-data (accessed 11/15/21).

EAT Healthy Diets from Sustainable Food Systems. Food Planet Health. Summary Report of the EAT-Lancet Commission 2019.

Food Systems Dialogue Summit 2021. https://foodsystemsdialogues.org/ (accessed 11/15/21).

Gaber, M.A.F.M., Juliano, P., Mansour, M.P., Tujillo, F.J., 2021. Entrained oil loss reduction and gum yield enhancement by megasonic-assisted degumming. Food Eng. Rev. 13 (1), 148–160.

Gao, J., Tan, E.Y.N., Low, S.H.L., Wang, Y., Ying, J., Dong, Z., Zhou, W., 2021. From bolus to digesta: how structural disintegration affects starch hydrolysis during oral-gastro-intestinal digestion of bread. J. Food Eng. 289, 110161.

Giannoglou, M., Dimitrakellis, P., Efthimiadou, A., Gogolides, E., Katsaros, G., 2021. Comparative study on the effect of cold atmospheric plasma, ozonation, pulsed electromagnetic fields and high-pressure technologies on sea bream fillet quality indices and shelf life. Food Eng. Rev. 13 (1), 175–184.

Gosavi, N.S., Polunas, M., Martin, D., Karwe, M.V., 2021. Effect of food microstructure on calcium infusion under high pressure. Food Eng. Rev. 13 (1), 36–53.

Great wrap. https://www.greatwrap.co/ (accessed 11/15/21).

Herrero, M., Hugas, M., Lele, U., Wira, A., Torero, M., Shift to healthy and sustainable consumption patterns. In: *Food Systems Summit 2021*, Nations, T.U. (Ed.), 2021, pp. 1-25.

Hougen, O.A., Watson, K.M., 1947. Chemical Process Principles. Wiley, New York, pp. 1107.

ICEF13 International Congress on Engineering and Food, Congress Handbook. https://www.aifst.asn.au/resources/Documents/AFEA/ICEF13%20program%20and%20book%20of%20abstracts.pdf (accessed 11/15/21).

Karthikeyan, J.S., Salvi, D., Karwe, M.V., 2021. Modeling of fluid flow, carbohydrate digestion, and glucose absorption in human small intestine. J. Food Eng. 292, 110339.

Kristiawan, M., Della Valle, G., Réguerre, A.L., Micard, V., Salles, C., 2021. Artificial oral processing of extruded pea flour snacks. Food Eng. Rev. 13 (1), 247–261.

Kucha, C.T., Liu, L., Ngadi, M., Gariépy, C., 2021. Assessment of intramuscular fat quality in pork using hyperspectral imaging. Food Eng. Rev. 13 (1), 274–289.

Lade, S.J., Steffen, W., de Vries, W., Carpenter, S.R., Donges, J.F., Gerten, D., Hoff, H., Newbold, T., Richardson, K., Rockström, J., 2020. Human impacts on planetary boundaries amplified by Earth system interactions. Nat. Sustain. 3 (2), 119–128.

Liu, L., Kong, F., 2021. The behavior of nanocellulose in gastrointestinal tract and its influence on food digestion. J. Food Eng. 292, 110346.

Mochizuki, T., Alvino Granados, A.E., Sogabe, T., Kawai, K., 2021. Effects of glass transition, operating process, and crystalline additives on the hardness of thermally compressed maltodextrin. Food Eng. Rev. 13 (1), 215–224.

Nutrizio, M., Maltar-Strmečki, N., Chemat, F., Duić, B., Jambrak, A.R., 2021. High-voltage electrical discharges in green extractions of bioactives from oregano leaves (*Origanum vulgare* L.) using water and ethanol as green solvents assessed by theoretical and experimental procedures. Food Eng. Rev. 13 (1), 161–174.

Fox Cabane, O. Creating a New Protein. http://protein.ketmaps.com/ (accessed 7/19/21).

Phan, K.K.T., Truong, T., Wang, Y., Bhandari, B., 2021. Formation and stability of carbon dioxide nanobubbles for potential applications in food processing. Food Eng. Rev. 13 (1), 3–14.

Plastics. https://ec.europa.eu/environment/topics/plastics_en (accessed 08/31/2021).

Priyadarshini, S.R., Arunkumar, E., Moses, J.A., Anandharamakrishnan, C., 2021. Predicting human glucose response curve using an engineered small intestine system in combination with mathematical modeling. J. Food Eng. 293, 110395.

Ritchie, H., Roser, M., 2018. Mental health. Our World Data. https://ourworldindata.org/drug-use.

Sarghini, F., De Vivo, A., 2021. Application of constrained optimization techniques in optimal shape design of a freezer to dosing line splitter for ice cream production. Food Eng. Rev. 13 (1), 262–273.

Schnabel, U., Handorf, O., Stachowiak, J., Boehm, D., Weit, C., Weihe, T., Schäfer, J., Below, H., Bourke, P., Ehlbeck, J., 2021. Plasma-functionalized water: from bench to prototype for fresh-cut lettuce. Food Eng. Rev. 13 (1), 115–135.

Shan, S., Heldman, D.R., 2021. The influence of operation parameters and product properties on time-to-temper for frozen raw meat based on simulation. Food Eng. Rev. 13 (1), 225–235.

Smetana, S., Aganovic, K., Heinz, V., 2021. Food supply chains as cyber-physical systems: a path for more sustainable personalized nutrition. Food Eng. Rev. 13 (1), 92–103.

Suo, X., Huang, S., Wang, J., Fu, N., Jeantet, R., Chen, X.D., 2021. Effect of culturing lactic acid bacteria with varying skim milk concentration on bacteria survival during heat treatment. J. Food Eng. 294, 110396.

Transforming our world: the 2030 Agenda for Sustainable Development. https://sdgs.un.org/2030agenda (accessed 11/15/21).

Turan, D., 2021. Water vapor transport properties of polyurethane films for packaging of respiring foods. Food Eng. Rev. 13 (1), 54–65.

World Health Organization, *Worldwide Prevalence of Anaemia 1993-2005*, World Health Organization: 2008.

Wu, G., Hui, X., Wang, R., Dilrukshi, H.N.N., Zhang, Y., Brennan, M.A., Brennan, C.S., 2021. Sodium caseinate-blackcurrant concentrate powder obtained by spray-drying or freeze-drying for delivering structural and health benefits of cookies. J. Food Eng. 299, 110466.

Yoha, K.S., Moses, J.A., Anandharamakrishnan, C., 2020. Effect of encapsulation methods on the physicochemical properties and the stability of *Lactobacillus plantarum* (NCIM 2083) in synbiotic powders and in-vitro digestion conditions. J. Food Eng. 283, 110033.

You, Y., Kang, T., Jun, S., 2021. Control of ice nucleation for subzero food preservation. Food Eng. Rev. 13 (1), 15–35.

Index

Page numbers followed by "*f*" and "*t*" indicate, figures and tables respectively.

Printed in the United States
by Baker & Taylor Publisher Services